8701-2967
TD
196
073
L37
1994
NORTHWEST COMMUNITY
COLLEGE

REACTION MECHANISMS IN ENVIRONMENTAL ORGANIC CHEMISTRY

Richard A. Larson
Eric J. Weber

Library of Congress Cataloging-in-Publication Data

Larson, Richard A.
Reaction mechanisms in environmental organic chemistry / Richard A. Larson and Eric J. Weber
p. cm.
Includes bibliographical references and index.
1. Organic compounds—Environmental aspects. 2. Environmental chemistry. 3. Chemical reactions. I. Weber, Eric J. II. Title.
TD196.073L37 1994
628.5—dc20 93-1622
ISBN 0-87371-258-7

Lewis Publishers is an imprint of CRC Press

International Standard Book Number 0-87371-258-7

Library of Congress Card Number 93-1622

Printed in the United States of America
1 2 3 4 5 6 7 8 9 0

Printed on acid-free paper

RAL:

To the memory of James Wright
1928–1980

Poet, educator, sage

Morir com'esso, ma morir segundo te.

EJW:

To my wife, Jodi, and children, Joel and Sarah, for their love, support and patience during the writing of this book, and to my chemistry mentors, Dr. Christopher J. Dalton at Bowling Green State University and Dr. Scott E. Denmark at the University of Illinois, who provided me with a fundamental education in organic chemistry.

Richard A. Larson (BA, Chemistry, University of Minnesota, 1963: PhD, Organic Chemistry, University of Illinois, 1968) has had extensive research experience over the past 20+ years in the area of environmental chemistry. He has been author or coauthor of well over 100 papers, presentations, and reports in this period, including over 70 peer-reviewed manuscripts. In addition, he is the author, coauthor, or editor of three books.

After postdoctoral appointments at Cambridge University and the University of Texas, Dr. Larson worked for several years at the Academy of Natural Sciences of Philadelphia. Since 1979, when he joined the faculty of the Institute for Environmental Studies at the University of Illinois, Dr. Larson has held a joint appointment in the University's Department of Civil Engineering. During the academic year 1985–1986, he studied free radical reactions in water as a National Research Council senior fellow in collaboration with Dr. Richard Zepp at the U.S. Environmental Protection Agency research laboratory in Athens, Georgia.

Dr. Larson has worked principally in the specific research areas of environmental photochemistry (kinetics, mechanisms, and products of light-induced reactions of environmental significance), disinfectant chemistry (ozone, chlorine, and chlorine dioxide and their reactions with organic compounds), and natural product chemistry. He is especially interested in the reactions of polar organic compounds of potential environmental health significance.

Eric J. Weber (BS, Chemistry, Bowling Green State University, 1980; PhD, Organic Chemistry, University of Illinois, 1985) received his initial training in synthetic and physical organic chemistry. During his PhD program he developed an interest in environmental chemistry after taking a course from his current co-author, Dr. Richard Larson, focusing on the fate of organic chemicals in aquatic ecosystems. Upon completion of his PhD, Dr. Weber furthered his training in environmental chemistry as a Research Associate with the National Research Council at the U.S. Environmental Protection Agency research laboratory in Athens, Georgia. In 1986, he joined the staff at the Athens laboratory as a Research Chemist. Dr. Weber's research has focused on transformation pathways of organic chemicals at the sediment-water interface with a primary emphasis on the identification of reaction products. He has also developed an interest in elucidating the reaction mechanisms by which organic chemicals form covalent bonds with natural organic matter.

PREFACE

Environmental organic chemistry is a rapidly expanding subject and one that allows many perspectives. Environmental chemistry historically grew out of analytical chemistry and the ability of analytical chemists to detect very low concentrations of pollutants, especially chlorinated organic compounds, in complex matrices such as soils, atmospheric particles, and animal tissues. The discovery that such pollutants are transported throughout the world, and that some are highly persistent in the environment, led to increasing interest in the fates of such compounds in nature.

The physical and chemical factors that govern the transport of organic compounds in the environment have been intensely studied. Thanks to the work of Sam Karickhoff, Donald Mackay, Cary Chiou, Louis Thibodeaux, and many others, we now have a group of sophisticated modeling tools with which to investigate the movement of organic materials within and between various environmental compartments—air, water, soils and sediments, and biota. Organic reactions that transform particular chemicals into by-products, however, have received less attention. There are several reasons for this. First of all, most investigations of organic chemical reactions have been performed in the absence of water. Rigorous procedures for the exclusion of moisture, and often, oxygen from reaction mixtures are commonplace in the organic laboratory. Secondly, organic reactions can be extremely complex. Even in purified solvents using carefully controlled conditions, many products can be formed whose identification may tax the ingenuity of the investigator. Finally, in many environmental situations, readily identified organic compounds are present only in extremely small concentrations in the presence of a complex matrix. In order to study the fate of pollutants under these conditions, early practitioners of environmental organic chemistry found it difficult enough merely to determine the rates of disappearance of their substrates, let alone to determine the mechanisms and products of the reasons that they were undergoing.

Recent years have seen an expansion of interest in studying organic reactions under environmental conditions. Many studies have shown that the environmental alteration products of some organic molecules are much more hazardous than their precursors; for example, treatment of natural waters with chlorine causes potentially toxic or mutagenic organochlorine compounds to be formed. Moreover, a general curiosity about how the global environment functions has led to a desire for intellectual re-examinations of fundamental scientific issues, such as the carbon cycle and the effects of human activities on it. To acquire this fundamental knowledge, it is necessary that we understand the forces that drive these global processes. As a consequence, many scientists throughout the world are turning their attention to investigating some well-known chemical reactions in detail, with an eye to being able to use the knowledge gained to predict the fates of unknown synthetic chemicals that may be released in significant concentrations in the future.

It is the purpose of this book to assist this process by giving an overview of the environment, of the principal organic chemical species in it, and of the processes and reactions that tend to transform these species. The organization of the book features, first, an introductory chapter that lays out the three principal environmental compartments—air, water, and solid phases—and surveys the conditions found in each of them that tend to promote chemical reactions. The remainder of the book is a survey of the principal types of organic reactions that may occur under environmental conditions, with discussions of the particular structural features of organic molecules that may make them more or less susceptible to each type of reaction. Chapter 2 deals with hydrolyses and nucleophilic reactions, with many examples chosen from the literatures of pesticide chemistry, industrial chemistry, and physical organic chemistry. Chapter 3 covers reduction, a process that until recently has been neglected from an environmental perspective, but one that is being shown to be an increasingly important route for converting many compounds once thought to be "persistent" to products. Oxidation, the subject of Chapter 4, takes place in a range of environments from the upper atmosphere to the surfaces of sediments, and encompasses a plethora of oxidizing agents, from transient free radicals with lifetimes of microseconds to mundane minerals such as iron oxide. In Chapter 5, disinfection is addressed; these reactions and their projects are the subjects of public debate in virtually every community where water treatment is practiced. Sunlight-induced reactions are covered in Chapter 6, on photochemistry. These reactions are also sure to come under increasing scrutiny, as the world tries to adjust to life under a different regime of solar energy, featuring higher levels of short, energetic UV-B wavelengths. Finally, Chapter 7 introduces a few other reactions that do not fit under the previous categories, but nevertheless could be significant for the fates of many classes of compounds.

The production of this book has been the outcome of many hours of discussions over the years. The two coauthors have learned a great deal from each other as well as from our many colleagues, students, and friends. An incomplete list of the most important people to whom we owe debts of gratitude would include Mike Barcelona, Michael Elovitz, Bruce Faust, Chad Jafvert, Karen Marley, Gary Peyton, Frank Scully, Alan Stone, Paul Tratnyek, Lee Wolfe, Ollie Zafiriou, and Richard Zepp. Invaluable help with the manuscript was provided by Jean Clarke, Tori Corkery, Jennifer Nevius, and Heather Walsh. Finally, special thanks are due to the students of Environmental Studies 351 at the University of Illinois, who have provided indispensable suggestions about the subject matter of this book over the years.

CONTENTS

Chapter

If the Lord Almighty had consulted me before embarking upon the Creation, I should have recommended something simpler.

—ALPHONSO X OF CASTILE ("THE WISE")

Strange events permit themselves the luxury of occurring.

—"CHARLIE CHAN" (CREATED BY EARL DERR BIGGERS)

The map appears to us more real than the land.

—D. H. LAWRENCE

Organic chemistry just now is enough to drive one mad. It gives the impression of a primeval, tropical forest full of the most remarkable things, a monstrous and boundless thicket, with no way of escape, into which one may well dread to enter.

—FRIEDRICH WÖHLER (1845)

There is something fascinating about science. One gets such wholesale return of conjecture out of such a trifling investment of fact.

—MARK TWAIN

Reality may avoid the obligation to be interesting, but hypotheses may not.

—JORGE LUIS BORGES

God loves the noise as much as the signal.

—L. M. BRANSCOMB

This world, after all our science and sciences, is still a miracle.

—THOMAS CARLYLE

CHAPTER 1

ORGANIC CHEMICALS IN THE ENVIRONMENT

A. ENVIRONMENTAL FATES OF ORGANIC CHEMICALS

This book will mainly be about **environmental fate processes**, and in particular about a certain subset of these fate processes; namely, organic chemical reactions. Specifically, if a particular organic chemical is introduced into the environment, what will happen to it? How much can we tell from physical measurements of the chemical's properties, how much can we learn from lab experimentation, and how much do we need to learn directly from measurements on the chemical in the actual environment? The sort of questions that have been asked are:

1. Where does it go?
2. How long will it remain?
3. What are the products of its reactions?

We need this information for two reasons: the first is **intellectual**; that is, the knowledge we gain from such studies helps us to explain the functioning of the natural world and the cycling of naturally occurring materials; secondly, from a **practical** standpoint, we need the information for large-volume synthetic organic chemicals in order to predict their effects on human health and on ecosystem functioning. In principle, it should be possible to use chemical concepts derived from studies of the natural environment to forecast the fates of chemicals in the human, or engineered, environment; or, possibly, the flow of information could proceed in the opposite direction.

Philosophically, environmental organic chemists make use of traditional reductionist assumptions and arguments. An organic compound, when discharged into a milieu that manifests a given array of chemical and physical conditions, should, it is believed, respond in a predictable manner to the constraints of those conditions. Although these responses may depend on an apparently bewildering assortment of chemical, physical, and biological qualifications, given sufficient information the fate of the compound should be predictable.

The subject matter of this book is an attempt to classify and organize what is known about the reactions of environmentally important organic compounds, using concepts and data largely drawn from traditional mechanistic and physical organic chemistry. We hope this approach will help the reader understand these reactions and their importance for the environmental fates of organic compounds of many types. The book has a molecular and mechanistic emphasis. We will take particular organic molecules and look at their fates in an aquatic ecosystem context. We will discuss their reactions in terms that an organic chemist would use. However, we will need to bring in concepts from biology, ecology, geochemistry, and environmental engineering. The purpose of this introductory section is to give background data to assist the reader's understanding of organic chemicals and their fates under environmental conditions.

1. The Carbon Cycle

In order to begin a consideration of the fate of organic compounds in nature, it is worthwhile to take a look at the carbon cycle. The discussion of the carbon cycle which follows is largely drawn from Woodwell and Pecan (1973), Bolin (1979) and Bolin and Cook (1983). A diagram of the carbon cycle (Figure 1.1) is intended to show the interconversions and movements of carbonaceous species, both organic and inorganic, throughout the earth's gaseous, liquid, and solid phases, as well as processes mediated by living organisms. Inorganic compounds are located principally on the left and top sides of the diagram, and organic matter is localized in the lower right portions. The boundary between inorganic and organic carbon species is rather arbitrary; metal carbides and cyanides intuitively seem to be inorganic compounds, but salts of organic acids do not. Carbon disulfide, $S=C=S$, is normally considered an "organic solvent," yet carbonyl sulfide, $O=C=S$, has an inorganic quality. Regardless of these borderline cases, in geochemical terms inorganic carbon is overwhelmingly dominated by carbon oxides and carbonates. Similarly, compounds of carbon containing covalent bonds to C, H, O, N, S, P, and halogens constitute the vast majority of organic compounds in the carbon cycle.

Fundamentally, the carbon cycle is a series of linked chemical reactions, both biological and abiotic. Many are redox reactions. Although the principal source of energy that drives the global redox system is sunlight, humans have not only diverted naturally occurring sources of energy and carbon to their own use, but are also

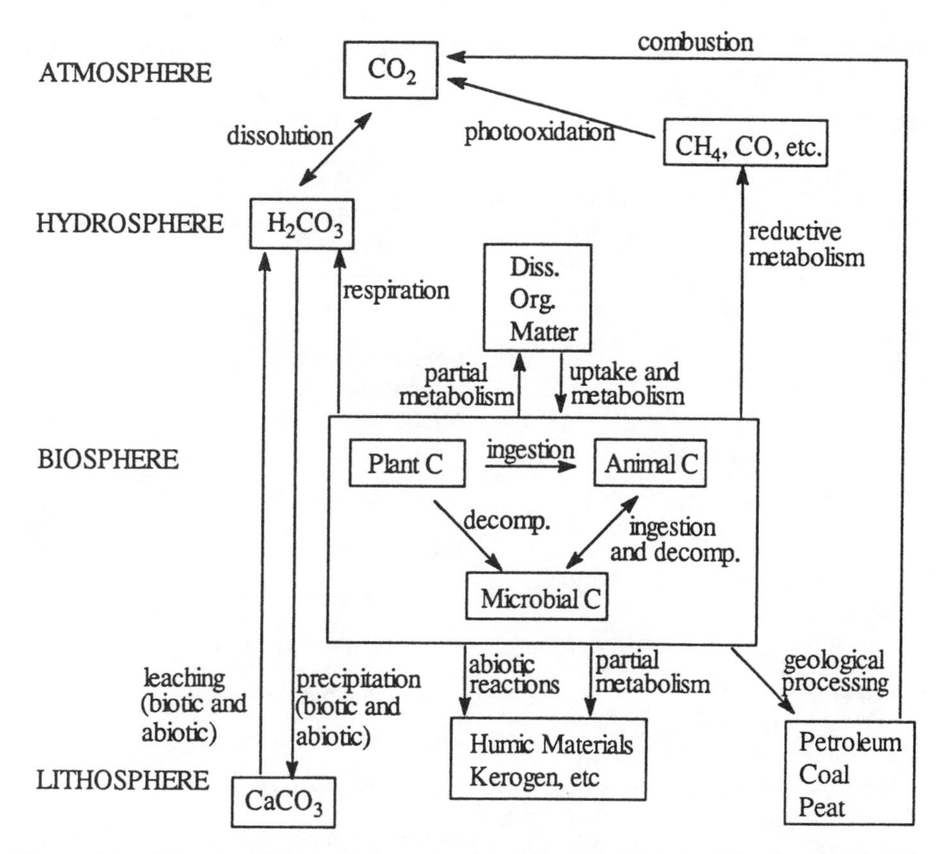

Figure 1.1. Diagram of the carbon cycle, showing movement of oxidized and reduced carbon species between the atmosphere, hydrosphere, biosphere, and geosphere.

contributing ever-increasing amounts of energy (and volatile carbon) to the system by virtue of fuel-burning and managed agriculture. In the Northern Hemisphere, anthropogenically generated energy now exceeds biotic energy flux (photosynthesis). This phenomenon has been called the "civilization engine" (Stumm and Morgan, 1981).

The carbon cycle is not complete—there are some sinks or areas where compounds accumulate, or are at least very slowly turned over. The approximate masses of carbon in the various atmospheric and terrestrial carbon pools and some of their approximate annual rates of conversion are given in Table 1.1.

The most oxidized species, CO_2, exists in the atmosphere as a gas whose concentration far exceeds that of other carbon-containing substances. In water, it takes part in a series of equilibrium reactions involving hydration, ionization, and precipitation:

Table 1.1. Some Components of the Carbon Cycle (Estimated magnitudes in grams of C)

Reservoirs		Annual Transport Rates	
Atmospheric		Atmosphere to oceans	1×10^{17}
CO_2	7.3×10^{17}	Land to oceans	
CO	2.27×10^{14}	Inorganic C	4×10^{14}
CH_4	3×10^{15}	Dissolved org. C	1×10^{14}
Nonmethane organic C	5×10^{13}	Particulate org. C	6×10^{13}
Freons	1×10^{11}	Net land primary production	6×10^{16}
Aquatic		Net oceanic primary production	6×10^{16}
Inorganic C	1×10^{20}	Animal respiration	8×10^{16}
Dissolved organic C		Microbial respiration	4×10^{16}
0–200 m	$9 x 10^{16}$	Plant litter production	5×10^{16}
>200 m	9×10^{17}	Algal excretion	4×10^{16}
Particulate organic C	3×10^{16}	Human harvest (cereals)	6×10^{14}
Plankton	3×10^{15}	Human harvest (wood products)	5×10^{14}
Bacteria	2×10^{14}	Fossil fuel combustion	5×10^{15}
Terrestrial		Synthetic organic chemical production	1×10^{14}
Rocks and sediments	2×10^{22}		
Coal, oil and peat	7×10^{18}		
Soil humic material	2×10^{18}		
Organisms (total)	7×10^{17}		
Living phytomass	5.6×10^{17}		
Dead phytomass, litter	9×10^{16}		
"Wild animals"	3×10^{15}		
Livestock	1.2×10^{14}		
Humans	2.4×10^{13}		
Bacteria, fungi	5×10^{15}		

Sources: Woodwell and Pecan (1973), Bolin (1979), and Bolin and Cook (1983).

$$CO_2\,(g) \rightarrow H_2CO_3 \rightarrow HCO_3^- \rightarrow CO_3^{2-} \rightarrow CaCO_3 \tag{1.1}$$

that transport it throughout the aqueous and solid phases. Carbon dioxide gas dissolves in water to form carbonic acid, a weak acid. Therefore, pure water containing CO_2 becomes slightly acidic; the pH of rain water at equilibrium with CO_2 is 5.5. (Obviously, atmospheric water containing dissolved nitrogen or sulfur oxides will be much more acidic.) The first proton of carbonic acid ionizes to the monovalent anion, bicarbonate, with a pKa of 6.4. Thus at equilibrium at pH 6.4, the concentrations of carbonic acid and bicarbonate will be equal. The second ionization to the divalent anion, carbonate, occurs with a pKa of 10.4. However, in water containing calcium ions or other ions that form insoluble carbonates, carbonate anion will rapidly be removed. Calcium carbonate exists in several forms, including limestone.

The photosynthetic activities of plants and, especially, algae that live in water remove some CO_2 from the water directly, and also increase the pH to such an extent that more carbonate occurs and precipitates out. The "shorthand" equation for photosynthesis explains the direct loss of CO_2:

$$n\,CO_2 + n\,H_2O \rightarrow (HCOH)_n + n\,O_2 \qquad (1.2)$$

This formation of reduced carbon species and the simultaneous release of oxygen from carbon dioxide by plants is called "primary production" in ecological jargon.

A more accurate equation for photosynthesis (Stumm and Morgan, 1981) also explains the pH increase in natural waters containing photosynthesizing organisms:

$$CO_2 + 0.2\,NO_3^- + 0.01\,HPO_4^{2-} + 1.2\,H_2O + 0.2\,H^+ \rightarrow CH_{2.6}ON_{0.2}P_{0.01} + 1.4\,O_2 \qquad (1.3)$$

The complex expression on the right-hand side of the equation is the average composition of algae. It can be seen that the process of photosynthesis results in a net consumption of hydrogen ions, and thus an increase in pH.

There are mechanisms for reconverting $CaCO_3$ to soluble forms. One is simply to redissolve it using acid, such as acidic precipitation. A second way is to convert it back to bicarbonate using CO_2:

$$CaCO_3 + CO_2 + H_2O \rightarrow Ca^{2+} + 2\,HCO_3^- \qquad (1.4)$$

The reverse of this reaction also occurs, for example when bicarbonate-containing water evaporates.

Looking at atmospheric CO_2 and its cycling, there is a total of 2.3×10^{18} grams of CO_2 in the atmosphere. Since the atmosphere weighs about 6.7×10^{21} g, this works out to 0.034% or 340 ppm. This has been increasing at about 1–2 ppm per year since at least 1957 (direct measurement), and probably well before. Samples of trapped air from the preindustrial period show concentrations between 260 and 295 ppm.

Concentrations of CO_2 are highest in the Northern Hemisphere in winter and drop sharply during the spring and summer. This is consistent with the increase being due to fossil fuel combustion. Estimates of the magnitude of present combustion-produced CO_2 are about 5×10^{15} g of C, (about 3 ppm of atmospheric CO_2) over and above natural CO_2 production by respiration, which is about 1.2×10^{17} g C/yr. Estimates of the cycling of naturally produced CO_2 are that 14% (1×10^{17} g C/yr) enter the oceans and 16% (1.2×10^{17} g) is taken up by photosynthetic organisms. Therefore, almost all of the atmospheric CO_2 is cycled in a three-year period. The extra carbon from combustion clearly does not all stay in the atmosphere; otherwise the observed increase would be 3 ppm/yr rather than 1–2 ppm/yr. It is still not certain whether the majority of the "missing" CO_2 dissolves in the ocean or is taken up by plants. The increased CO_2 concentration probably, however, will not translate directly into increased biomass of plants around the world, because CO_2 is not usually the limiting substrate for plant growth; most ecologists believe that water and trace mineral nutrients are usually more important.

The respiration (metabolism) of the reduced carbon produced by plants returns it to the atmosphere as CO_2. It is assumed that photosynthetic fixation of CO_2 from the atmosphere by plants and its return by respiration are in exact balance, but there is really no good way of telling. All that we know is that atmospheric carbon is increasing. An increase in CO_2 may affect the earth's surface temperature because the sun emits not only light (visible radiation energy), but also ultraviolet (UV) and infrared (IR or heat) radiation (see Section 6.A). When, for example, visible energy strikes the earth's surface, it loses energy and is partly converted to the lower-energy heat (IR) radiation, some of which is reflected back into space. Carbon dioxide is transparent to visible energy, but it strongly absorbs IR, so some of the heat generated near the earth's surface doesn't escape into the atmosphere. The net result is that the lower atmosphere becomes warmer.

A consideration of terrestrial carbon shows that inorganic carbon (carbonate rocks, mostly) predominates to a tremendous degree over organic carbon. There are very large reserves of "dead" organic carbon (coal, oil, peat, and soil humus), all ultimately derived from animals and plants which have died. Dead carbon (about 10^{19} g) exceeds living carbon by about 14:1. These materials, taken as an aggregate, are not rapidly recycled at the present rates of human utilization, although readily useful fossil fuels are exploited on a large scale.

Living carbon is largely (more than 80%) in higher plants. Most of this "phytomass" is in trees. About 30% of the land area of the earth is forested, but this proportion is decreasing. Tropical forests, which still constitute about 1/3 of the total forest area, are rapidly being cut down as the increasing population in developing countries exerts its requirement for living space and fuel.

The mass of the five billion or so living humans, about 240 billion kilograms, is only a few hundredths of a per cent of the total living biomass. Mankind's domestic livestock herds outweigh us by a factor of about 5, and all the "wild creatures," including all birds, mammals, lizards, fish, etc., by only about 20:1. The biomass of microorganisms, although difficult to estimate precisely, probably amounts to only a few per cent of all the living carbon.

Annually, we harvest about 10^{15} g, or about 50 times our own weight, in plant products including wood, fiber, and food. About 10% of the earth's land area is now being used for agriculture (including forestry). Other, urban, human institutions (housing, roads, industrial plants) consume between 1% and 2% of the surface of the globe (it has been estimated that 1% of the United States is paved). Densely populated countries have much more of their land area in urban use; for example, the Netherlands has about 9%. Synthetic organic chemical production has increased dramatically over the last 50 years, and now about 4×10^{14} g of such chemicals are produced annually (an amount close to the annual use of wood products).

To summarize, although the absolute effect of humans on the global quantity and flux of carbon has perhaps been modest, our contribution to changing the pattern of the cycle has increased significantly in the last century or so. Given the serious lack of knowledge of the feedback mechanisms that tie various elements of the carbon system together, it would appear to be an urgent priority that we increase our understanding of the effects of our activities on these important planetary operations.

2. Translocation of Organic Chemicals

The fates of organic molecules (whether naturally or anthropogenically produced) include, first, **translocation**, in which the molecular structure of the chemical is not changed; a molecule will be carried between air, surface water, groundwater, organisms, aquatic sediments, and soils by various processes. Rates and equilibria can ideally be obtained to describe these transport processes, often by chemical engineering concepts like mass transfer equations, and to predict their extent. A good summary of environmental transport processes is given by Thibodeaux (1979).

Volatilization

Transport of organic compounds from the solid or aquatic phases to the gas phase (and back again) is now known to be a highly important process for the dispersion of chemical compounds around the globe. Dissolution into and volatilization from the aqueous phase is an elaborate process that depends on solubility, vapor pressure, turbulence within the two phases, and other physical and chemical factors. Volatilization of materials from the earth's surface into the troposphere can result in their long-range transport and redeposition, with the outcome being that measurable quantities of such substances can be detected far from their point of release.

Many chemicals escape quite rapidly from the aqueous phase, with half-lives on the order of minutes to hours, whereas others may remain for such long periods that other chemical and physical mechanisms govern their ultimate fates. The factors that affect the rate of volatilization of a chemical from aqueous solution (or its uptake from the gas phase by water) are complex, including the concentration of the compound and its profile with depth, Henry's law constant and diffusion coefficient for the compound, mass transport coefficients for the chemical both in air and water, wind speed, turbulence of the water body, the presence of modifying substrates such as adsorbents in the solution, and the temperature of the water. Many of these data can be estimated by laboratory measurements (Thomas, 1990), but extrapolation to a natural situation is often less than fully successful. Equations for computing rate constants for volatilization have been developed by Liss and Slater (1974) and Mackay and Leinonen (1975), whereas the effects of natural and forced aeration on the volatilization of chemicals from ponds, lakes, and streams have been discussed by Thibodeaux (1979).

Once a chemical becomes airborne, atmospheric mixing processes on regional, elevational, and global scales come into play. East-west mixing of air masses is much more efficient than north-south mixing. Because of the intra-hemispheric constraints on the prevailing winds, air masses seldom mix efficiently across the equator. The atmosphere becomes completely mixed only over very long time scales; for organic compounds with lifetimes of even several years, Northern and Southern Hemisphere variations are measurable if (as is usually the case) one hemispheric source predominates. Compounds of industrial origin are usually localized in the Northern Hemisphere, whereas substances derived from marine processes are usually more abundant in the Southern Hemisphere.

We know from studies of gases in solution that the solubility of a gas which does not react with its solvent depends to a considerable degree on its vapor pressure at a given temperature. We can extend these studies to other solutes if we can measure their vapor pressures at higher temperatures and extrapolate them to lower, environmentally realistic temperatures. For the case of air-water partitioning, a simple equation describes the behavior of many substances:

$$\mathbf{H} = \mathbf{P}/\mathbf{C} \tag{1.5}$$

where **H** is the Henry's law coefficient for the chemical, **P** its vapor pressure, and **C** its water solubility. If we know or can estimate the quantities on the right-hand side of the equation, we can obtain **H**, and this will allow us to estimate the magnitude of the air-water partition.

Henry's law constants for chemicals of environmental interest have been tabulated by many authors, including Mackay and Shiu (1981), Burkhard et al. (1985), Gossett (1987), Murphy et al. (1987), Hawker (1989), and Brunner et al. (1990). If **H** has a relatively large value for a particular compound, it means that it has a large tendency to escape from the water phase and enter the atmosphere. To get a large value for **H**, obviously either a high **P** or a low **C** (or both) is required. Thus, for example, *sec*-butyl alcohol and decane have vapor pressures that differ by a factor of 10, with the alcohol being the higher, but because the hydrocarbon's water solubility is negligible, it is much more likely to enter the gas phase than is the alcohol. Similarly, although the pesticide DDT is essentially nonvolatile, its water solubility is far less even than decane's. As a result, a small quantity will be volatilized; this accounts for the widespread detection of DDT in environments far from the sites where it was applied. Another heavily applied chemical, the herbicide atrazine, is a little more volatile than DDT, but it is far more soluble, so its tendency to enter the atmosphere is negligible.

The movement of a chemical substance within the vapor phase occurs by the combined driving forces of flow and diffusion. An illustration of these effects can be visualized by considering a smokestack plume; in the absence of wind, the plume will rise vertically in a more or less uniform column until it reaches an elevation where density considerations result in its spreading out into a relatively broad and flat mantle. When wind is factored into the equation, the plume may move in a more nearly horizontal direction, more or less parallel to the surface of the ground, and at certain wind speeds the plume structure can break up into loops or bends due to turbulent aerodynamic effects such as eddy formation. In addition, small eddies can result in the breakdown of the coherent plume structure, with the formation of vertical or horizontal regions of increasingly large cross-section and lower concentrations of plume constituents.

Transport Within the Aqueous Phase

The three-dimensional dispersion of a completely soluble organic solute within a volume of pure water will be governed by its rates of diffusion within the water column and by the flow characteristics of the water itself (also called *convection* or *advection*). In actual water bodies, complicating factors include the presence of particles of various sizes within the aqueous phase and the effects of boundary layers such as those associated with the air-water and sediment-water interfaces. Further complications occur in soil-water and groundwater systems in which the aqueous phase is a minor component in the presence of an excess of solid material (Thibodeaux, 1979).

Movement of a soluble chemical throughout a water body such as a lake or river is governed by thermal, gravitational, or wind-induced convection currents that set up laminar, or nearly frictionless, flows, and also by turbulent effects caused by inhomogeneities at the boundaries of the aqueous phase. In a river, for example, convective flows transport solutes in a nearly uniform, constant-velocity manner near the center of the stream due to the mass motion of the current, but the friction between the water and the bottom also sets up eddies that move parcels of water about in more randomized and less precisely describable patterns where the instantaneous velocity of the fluid fluctuates rapidly over a relatively short spatial distance. The dissolved constituents of the water parcel move with them in a process called eddy diffusion, or eddy dispersion. Horizontal eddy diffusion is often many times faster than vertical diffusion, so that chemicals spread sideways from a point of discharge much faster than perpendicular to it (Thomas, 1990). In a temperature- and density-stratified water body such as a lake or the ocean, movement of water parcels and their associated solutes will be restricted by currents confined to the stratified layers, and rates of exchange of materials between the layers will be slow.

The other method of diffusion of a chemical through a liquid phase, molecular diffusion, is driven by concentration gradients. It is normally orders of magnitude slower in natural waters than eddy-driven processes, unless the water body is abnormally still and uniform in temperature (Lerman, 1971). Such situations are found only in isolated settings such as groundwaters and sediment interstitial waters. Even here, however, empirical measurements often indicate that actual dispersion exceeds that calculated from molecular diffusion alone.

The transport of a substance through a water body to an interface may involve eddy or molecular diffusion through aqueous sectors of differing temperature, such as those characteristic of stratified lakes (cf. Section 1.B.2c), and through interfacial films such as air-water surface layers (Thomas, 1990). Conditions near a phase boundary are very difficult to model accurately. The resistance to diffusion through various regions may vary by large amounts, and the overall transfer rate is governed by the slowest step, which usually occurs in a thin film or boundary layer near the interface where concentration gradients are large and molecular diffusion becomes influential.

Transport of a dissolved substance through a porous medium like a sandy soil, in which interaction between the solute and the solid phase is negligible, is governed by laws of mass transport that are similar to those that apply in solutions. When interactions with a solid phase such as a soil become significant, a situation similar to solid-liquid chromatography develops; solutes with less interaction with the "support," or soil, are moved along with the "solvent front" of water leaching through the medium, whereas others are held back in proportion to their degree of binding. Studies of this phenomenon in artificial microcosms such as soil columns or thin-layer chromatography plates are useful in helping to predict which compounds are likely to contaminate groundwater (see Section 1.B.2e). The predictions can be tested in field studies using wells or lysimeters.

Partition into Solid Phases

The transfer of molecules from solution into an environmental solid phase such as a soil or sediment is referred to as sorption, with the reverse process usually called desorption (Karickhoff, 1984; Weber et al., 1991). A variety of solid phases are available in the aquatic environment: small suspended particles, both living and nonliving, the anatomical surfaces of larger biota such as fish, and bulk soils and bottom sediments. Even colloidal organic "solutes" such as humic macromolecules might be thought of as separate phases to which a dissolved molecule could be sorbed. Each of these surfaces may be thought of as a source or a sink for compounds in solution.

The passage of a compound from solution into a solid environment can be promoted or inhibited by a variety of factors. Sorption and desorption equilibria are, for example, strongly temperature-dependent. In addition, the surface area of the solid, as well as its physicochemical characteristics (charge distribution and density, hydrophobicity, particle size and void volume, water content) are major factors that determine the importance and extent of sorption for a particular solute. In thermodynamic terms, for sorption to occur, the energy barrier associated with bringing the interacting species into proximity must be overcome by a greater decrease in free energy in the sorbed system. By measuring the heat of adsorption, some insight can be gained as to whether the sorption process is primarily due to physical (van der Waals-type) uptake or to chemical reaction, with physical uptake usually involving much lower (< 50 kJ/mol) energy differentials than chemical reactions, which have heats of adsorption in the range of 150 to 400 kJ/mol.

Although distinctions are sometimes made between adsorption (uptake of compounds by the surface of a solid phase) and absorption (diffusion of molecules into the interior of a solid), it is usually not possible to distinguish between these cases in environmental situations. A complicating factor in sorption studies is that natural solid phases are not only not chemically and physically homogeneous, but are normally coated with extraneous materials such as transition metal oxides, microorganisms and their excretion products, and humic substances that often almost completely disguise the sorption properties of the underlying mineral.

Sorption is important from the viewpoint of chemical reactivity, as well. A compound that is sorbed usually goes from a situation in which it is entirely surrounded by water molecules to one in which it is in a mineral environment rich in organic matter. In fact, a chemical substance in a suspension of natural particulate matter will exist in a complex equilibrium in which a fraction of the material is dispersed into several disparate phases that may contribute differently to the reactions the substance may undergo.

Studies of the uptake of organic compounds by many types of natural solid phases (soils and sediments) in the presence of water have clearly shown that only two types of interactions are important: first, a coulombic interaction, in which organic compounds of opposite (positive) charge are sometimes taken up by the (usually) negatively charged solid material; and, generally more important, a hydrophobic interaction in which nonpolar organic compounds are attracted into the solid phase.

Among the most important constituents of most natural soils and sediments are the clay minerals (see Section 1.B.3a). These minerals usually exist as very fine (<1 μM) particles with high surface area and (usually) negative charge. This makes them potent adsorbents for cations, either inorganic or organic, and leads to the possibility of cation-exchange displacement reactions. There may also be important pH effects at clay surfaces, especially in soft waters where cations other than H^+ are not abundant. It has been found that the pH near the surface of certain types of clay may be as much as 2 units lower ($[H^+]$ 100-fold higher) than in the associated solution (McLaren, 1957; see Section 1.B.3a). Obviously, there may be significant effects on the rate of reactions requiring protonation or acid catalysis in such environments.

In general, for both naturally occurring compounds and pollutants,

1. Hydrophobically bound adsorbates are most strongly bound;
2. Cationic adsorbates are next most strongly bound;
3. Anionic species are most weakly bound.

Uptake by clays of charged organic materials is termed **hydrophilic sorption**. As an example, an organic cation like the herbicide paraquat (**1**) is very readily

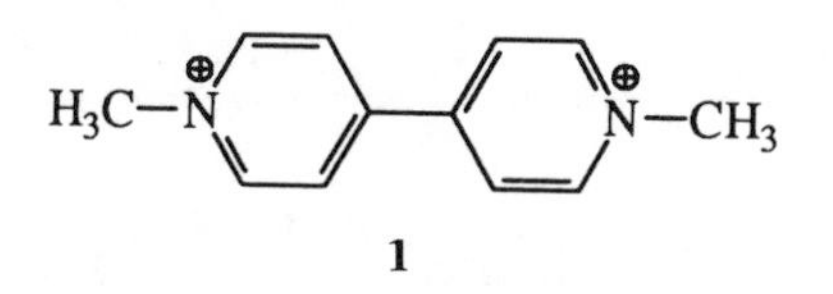

1

taken up by clays, despite its high solubility in water, because of these strong electrostatic interactions. There is also a possibility of weak adsorption of anions by "bridging cations." By this mechanism, anionic compounds like some proteins, carboxylic acids, and humic materials may be associated near the water-solid interface. Generally, however, the direct uptake of cations by clays is much more important than the indirect (bridging) uptake of anions.

Mention should also be made of another weak mechanism of indirect uptake of strongly hydrogen-bonding materials. Water molecules are quite strongly oriented in the vicinity of certain clays because of attraction between the lone pairs on the oxygen atom of water and the positive charges on cations at the clay surface. This means that an excess of hydrogen atoms will be facing out into solution, and in the presence of molecules with lone pairs (hydroxyl groups, ether oxygen, carbonyl groups, etc.), hydrogen bonding will occur (Figure 1.2), and through these "water bridges," these molecules may, in favorable cases, build up to quite a high degree of adsorption.

Another general type of uptake of organic molecules by solid surfaces is called **hydrophobic sorption**. This interaction is quite general for natural sediments and soils, and leads to a high degree of concentration of hydrophobic material near the interface. Hydrophobic adsorption is strongly correlated with the organic carbon content of the sediment or soil. It has elements of partitioning; many investigators have, in fact, shown a very clear correlation between the extent of uptake of a chemical by a natural solid phase and the **partition coefficient**, K_p, of the chemical; that is, its ratio of concentration or activity between an organic solvent, often octanol, and water:

$$K_p = [\text{solvent}]/[\text{water}] \tag{1.6}$$

The partition coefficient is not the same as the ratio of the solubilities of a chemical in the two pure solvents, because at equilibrium the solvent phase contains some water and the water phase contains some solvent. Values of K_p sometimes vary with solute concentration, but are seldom much affected by temperature.

Many data are specifically available on the octanol/water partition coefficients (K_{ow}) of organic molecules (see, for example, Tewari et al., 1982; Miller et al., 1984, and Lyman, 1990a), and it has been repeatedly demonstrated that chemicals with high K_{ow}'s are very readily sorbed by natural sediments. For example, the extremely nonpolar compound DDT has a K_{ow} of about 10^6 (it is a million times more soluble in octanol than in water), and it is almost completely associated with the solid phase in a two-phase water-sediment system. For atrazine, a chemical of intermediate polarity, the K_{ow} is still high (3×10^3), but far less than that for DDT, so the extent of

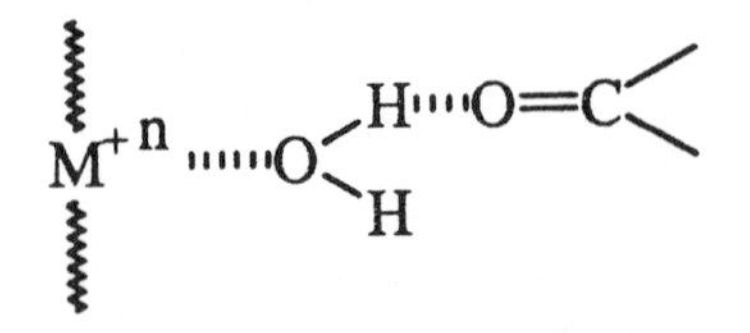

Figure 1.2. A water bridge, showing how hydrogen bonding may assist in bringing certain organic molecules into the vicinity of clay surfaces.

association with sediments would be expected to be far less pronounced. For quite polar organic compounds, such as acetic acid, the octanol-water partition coefficient is far lower (0.5), and the distribution in the two-phase system would favor water. Because soil organic matter is so highly oxidized, current thinking is that it must have extensive nonpolar regions, perhaps alkyl chains, that are responsible for the partitioning (see Section 1.B.3c). (The quantity K_{ow} has also been shown to correlate well with other environmental parameters that depend on distribution between hydrophobic and aqueous phases, such as bioconcentration in aquatic organisms, uptake by mammalian skin, water solubility, and toxicity within a given series of compounds; Verschueren, 1983).

For soils and sediments of differing organic matter content, the useful concept of K_{oc} has been introduced; this form of the partition coefficient makes the simplifying assumption that only the organic carbon is active in the sorption process, and the partition coefficient expression can be rewritten

$$K_{oc} = K_p/[\text{fraction organic carbon}] \tag{1.7}$$

Values of K_{oc} for selected chemicals have been tabulated (Lyman, 1990b). Data for many herbicides and polycyclic organic compounds (Walker and Crawford, 1968; Hassett et al., 1980) have confirmed the general applicability of this expression. However, at low organic carbon concentrations, such as are found in sandy soils and some clays, sorption still occurs for many chemicals, and the above equation does not fit the data particularly well.

Octanol-water partitioning and aqueous solubility are closely related, and one can be predicted fairly accurately from the other using a relationship devised by Mackay et al. (1980):

$$\ln K_{ow} = 7.494 - \ln C_w \tag{1.8}$$

where C_w is the water solubility in moles/L. Many other forms of this equation have been promulgated (cf. Lyman, 1990a) that appear to predict solubility more or less accurately for a given series of compounds.

At equilibrium, the uptake of a dissolved compound can often be expressed in the simple terms of the Freundlich isotherm equation,

$$[A]_s = K_p\,[A]_w^{1/n} \tag{1.9}$$

where $[A]_s$ and $[A]_w$ are, respectively, concentrations of the compound in the solid and water phase; K_p is an appropriate partition coefficient; and $1/n$ is an empirical exponential factor, often close to 1.0. Achievement of equilibrium, however, is often difficult to measure in studies with natural soils or sediments; kinetics of uptake may be complex, with a fraction of the solute rapidly taken up and a residual uptake period that may last for days or weeks. Pollutant uptake and release (desorption) kinetics are dependent on particle chemical characteristics, mass transfer properties for the solute in the sorbent phase, aggregation state of particles, and ability of the solid state to swell or shrink after incorporation of organic matter (Karickhoff, 1984; Karickhoff and Morris, 1985).

3. Transformation of Organic Compounds

a. Reaction Mechanisms

The transformation of organic chemicals most often occurs in several molecular events referred to as *elementary reactions*. An elementary reaction is defined as a process in which reacting chemical species pass through a single transition state without the intervention of an intermediate. A sequence of individual elementary reaction steps constitutes a *reaction mechanism*. For example, the overall reaction for the hydrolysis of a Schiff base is written:

$$R_1(H)C{=}N{-}R_2 + H_2O \rightleftharpoons R_1(H)C{=}O + R_2NH_2 \quad (1.10)$$

The reaction mechanism for this seemingly simple reaction is in fact quite complex, and is composed of four elementary reactions (Cordes and Jencks, 1963).

$$R_1(H)C{=}N{-}R_2 + H^+ \underset{k_{-1}}{\overset{k_1}{\rightleftharpoons}} R_1(H)C{=}N^{\oplus}(H){-}R_2 \quad (1.11)$$

$$R_1(H)C{=}N^{\oplus}(H){-}R_2 + H_2O \underset{k_{-2}}{\overset{k_2}{\rightleftharpoons}} R_1{-}C(OH)(H){-}N(H){-}R_2 + H^+ \quad (1.12)$$

$$R_1(H)C{=}N^{\oplus}(H){-}R_2 + OH^- \underset{k_{-3}}{\overset{k_3}{\rightleftharpoons}} R_1{-}C(OH)(H){-}N(H){-}R_2 \quad (1.13)$$

$$R_1{-}C(OH)(H){-}N(H){-}R_2 \xrightarrow{k_4} R_1(H)C{=}O + R_2NH_2 \quad (1.14)$$

A reaction mechanism is actually a hypothesis or model that has been constructed from experimental evidence. As new experimental evidence is obtained, changes in the proposed reaction mechanism may be required. Chemical kinetics is probably the most powerful tool for the investigation and development of reaction mechanisms. The consistency of a reaction mechanism can be verified in the laboratory by determining the dependence of reaction rates on concentration.

Once a reaction mechanism consisting of a sequence of individual elementary reactions has been proposed it is possible to develop rate equations, which predict the dependence of the observed reaction rate on concentration. The *principle of mass action*, which states the rate at which an elementary reaction takes place is proportional to the concentration of each chemical species participating in the molecular event, is used to write differential rate equations for each elementary reaction in the proposed reaction mechanism. The goal is then to obtain explicit functions of time, which are referred to as integrated rate laws, from these differential rate equations. For simple cases, analytical solutions are readily obtained. Complex sets of elementary reactions may require numerical solutions.

It is useful to classify elementary reactions according to their molecularity, which is defined as the sum of the exponents appearing in a rate equation for a single elementary reaction. The term *unimolecular reaction* is used to describe an elementary reaction involving one chemical species. A *bimolecular reaction* involves the interaction of two chemical species. The interaction of three chemical species, or a *termolecular reaction*, is quite rare and will not be considered for further discussion. Molecularity is often confused with the *order* of a reaction, which refers to the sum of the exponents appearing in an experimental rate equation.

b. Kinetics

Rate expressions. Initially, to determine how rate equations are developed from a proposed reaction mechanism. we will consider simple reaction mechanisms consisting of only one elementary reaction For example, the differential rate equation for the hypothetical reaction in Equation 1.15

$$A \rightarrow B \tag{1.15}$$

can be written as

$$-d[A]/dt = k_1[A] \tag{1.16}$$

For experimental determination of k, the integrated form of Equation 1.16 is conveniently written

$$[A] = [A]_o e^{-kt} \tag{1.17}$$

or

$$\ln A = \ln A_o - kt \tag{1.18}$$

Thus a plot of ln A versus t should be linear for a first order reaction with a slope of -k, which has units of time^{-1}, and a y-intercept of $[A]_o$. When experimental data obtained in the laboratory are in close agreement with this theoretical expectation, we can say that the reaction obeys first-order kinetics. This allows us to speak of an experimental rate law as opposed to a theoretical rate law based on mechanistic considerations.

When comparing the reactivity of chemicals, it is convenient to speak in terms of the half-life ($t_{1/2}$) of a reaction or the time for 50% of the chemical to disappear. For first-order reactions, if we set $[A] = {}^1/_2[A]_o$, and substitute into Equation 1.18, the expression for $t_{1/2}$ becomes:

$$t_{1/2} = 0.693/k \tag{1.19}$$

The half-life for a first-order reaction, therefore, is independent of the initial concentration of the chemical of interest.

For the more complex reaction containing two reacting chemical species, A and B (Equation 1.20)

$$A + B \rightarrow C \tag{1.20}$$

the rate of disappearance of A is given by Equation 1.21.

$$d[A]/dt = k_2[A][B] \tag{1.21}$$

The integrated form of Equation 1.21 is:

$$\ln[(B_o{*}A)/(A_o{*}B)] = k_2t(A_o{-}B_o) \tag{1.22}$$

Thus for bimolecular reactions, plotting $\ln[B_o(A)/A_o(B)]$ versus t gives a line with slope $k_2(A_o{-}A_b)$.

If one of the concentrations of a chemical species in a second-order reaction is present in excess, and as a result does not effectively change during the course of the reaction, the observed behavior of the system will be first order. Such a system is said to follow pseudo-first-order kinetics. If the concentration of the chemical species B in Equation 1.20 were in large excess of A, the disappearance of A could be written as:

$$d[A]/dt = k_{obs}[A] \tag{1.23}$$

where

$$k_{obs} = k_2[B] \tag{1.24}$$

Reaction mechanisms for the transformation of organic chemicals usually involve several elementary reactions. The kinetic expressions resulting from such mechanisms can become quite complex and difficult to handle. Often, simplifying assumptions can be used to analyze these systems. We shall examine one simplifying assumption referred to as the *steady-state approximation*. It can be illustrated with the hypothetical reaction scheme:

$$A \underset{k_{-1}}{\overset{k_1}{\leftrightarrows}} B \tag{1.25}$$

$$B \xrightarrow{k_2} C \tag{1.26}$$

If B is a reactive, unstable species its concentration will remain low. Applying the steady-state assumption, we assume that the rate of formation of the reactive species is equal to its rate of destruction:

$$k_1[A] = k_{-1}[B] + k_2[B] \tag{1.27}$$

Solving for the steady-state concentration of B, we find that:

$$[B] = k_1[A]/(k_{-1} + k_2) \tag{1.28}$$

If we assume that the conversion of B to C is *rate-limiting* or the *rate-determining step,* the overall rate of the reaction will be:

$$d[C]/dt = k_2[B] \tag{1.29}$$

Substituting the term for the steady-state concentration of B (Equation 1.28) into Equation 1.29 gives:

$$d[C]/dt = (k_2k_1[A])/(k_{-1} + k_2) \tag{1.30}$$

If the experimental rate law was determined to be $d[C]/dt = k_{obs}[A]$, then k_{obs} would be related to the elementary reaction rate constants by:

$$k_{obs} = k_2k_1/(k_{-1} + k_2) \tag{1.31}$$

Arrhenius equation. The measurement of rate constants in the laboratory for transformation processes that are exceedingly slow ($t_{1/2}$ on the order of months to years) can be quite difficult. To circumnavigate this problem, kinetic measurements are made at elevated temperatures to accelerate reaction kinetics. Of course, it is then necessary to extrapolate measured rate constants to environmentally significant temperatures. The temperature dependence of observed rate constants is given by the Arrhenius equation:

$$k_{obs} = A\ \exp(-E_a/RT) \tag{1.32}$$

where A is the *preexponential factor*, E_a is the activation energy, R is the gas constant, 1.987 cal K^{-1} mol^{-1}, and T is the temperature in degrees Kelvin. From the logarithmic form of the Arrhenius equation it is apparent that the preexponential term A and the energy of activation E_a can be calculated from logarithms of the observed rate constants versus the reciprocal of the absolute temperatures. In the absence of experimental data to calculate E_a, a useful approximation for extrapolation is that rate constants will vary by a factor of 10 for each 20°C change in temperature. This corresponds to an activation energy of 20 kcal/mol.

c. *Linear Free Energy Relationships*

Correlation Analysis. Numerous empirical models have been developed in organic chemistry that describe relationships between structure and reactivity. The most successful and intensively investigated are the linear free energy relationships (LFER), which correlate reaction rate constants and equilibrium constants for related sets of reactions. As Hammett stated: "From its beginning the science of organic chemistry has depended on the empirical and qualitative rule that like substances react similarly and that similar changes in structure produces similar changes in reactivity" (Hammett, 1970). "Linear free energy relationships constitute the quantitative specialization of this fundamental principle" (Chapman and Shorter, 1970). The origin of the LFER can be better understood by considering the relationship between the changes in free energy involved in the kinetic and equilibrium processes. If two reactions exhibit a LFER, we can write:

$$\log k - \log k_o = m(\log K - K_o) \tag{1.33}$$

Substituting for k and K with the appropriate terms for free energy of activation (from transition-state theory) and free energy of reaction gives:

$$-\Delta G^{\ddagger}/2.3RT + \Delta G_o^{\ddagger}/2.3RT = m(-\Delta G/2.3RT + \Delta G_o/2.3RT) \tag{1.34}$$

$$-\Delta G^{\ddagger} + \Delta G_o^{\ddagger} = m(-\Delta G + \Delta G_o) \tag{1.35}$$

$$\Delta\Delta G^{\ddagger} = m\Delta\Delta G \tag{1.36}$$

From this analysis, it is apparent that a relationship between k and K (at constant temperature) is essentially a relationship between free energies. The LFER therefore indicates that the change in free energy of activation ($\Delta G^{\ddagger}$) exerted by a series of substituents is directly proportional to the change in free energy of reaction (ΔG).

The use of LFERs constitutes one of the most powerful means for the elucidation of reaction mechanisms. LFERs also provide us a means to predict reaction rates or bioactivity from more easily measured equilibrium constants such as octanol-water partition coefficients (K_{ow}), ionization constants (K_a), or acidity constants (K_{HB}). Brezonik (1990) has summarized the major classes of LFERs applicable to reactions in aquatic ecosystems (Table 1.2). These empirical correlations pertain to a variety

Table 1.2. Major Classes of LFERs Applicable to Reactions in Aquatic Systems

Relationship	Types of Reaction or Reactants	Basis of LFER
Brønsted	Acid- or base-catalyzed reactions: hydrolysis, dissociation, association	Rate related to K_a or K_b of product or catalyst
Hammett (sigma)	Reactions of para- or meta-substituted aromatic compounds: hydrolysis, hydration of alkenes, substitution, oxidation, enzyme-catalyzed oxidations; some photooxidations	Electron withdrawal and/or donation from/to reaction site by substituents on aromatic rings via resonance effects
Taft	Hydrolysis and many other reactions of aliphatic organic compounds	Steric and polar effects of substituents
Marcus	Outer-sphere electron exchange reactions of metal ions, chelated metals, and metal ion oxidation by organic oxidants such as pyridines and quinones	The three components of energy needed to produce transition state; for related redox reactions, ln k proportional to E^o

of transformation processes, including hydrolysis, nucleophilic substitution, reduction, and oxidation.

Hammett equation. One of the first methods for relating structure and reactivity was developed by Hammett (1937). Hammett found that the reactivities of benzoic acid esters were directly related to the ionization constants, K_a, of the corresponding benzoic acids (Figure 1.3). Using substituted benzoic acids as his standard reference reaction, Hammett developed a LFER in the form:

$$\log(k_x/k_o) = \rho\log(K_x/K_o) \tag{1.37}$$

where O denotes unsubstituted benzoic acid and x denotes differently X-substituted benzoic acids in the m- and p-positions and ρ is the *reaction constant*. Hammett defined the log of the ratio of the ionization constants as a *substituent constant*, σ,

$$\sigma_x = \log(K_x/K_o) \tag{1.38}$$

Substitution of Equation 1.38 into Equation 1.37 gives Equation 1.39:

$$\log(k_x/k_o) = \rho\sigma_x \tag{1.39}$$

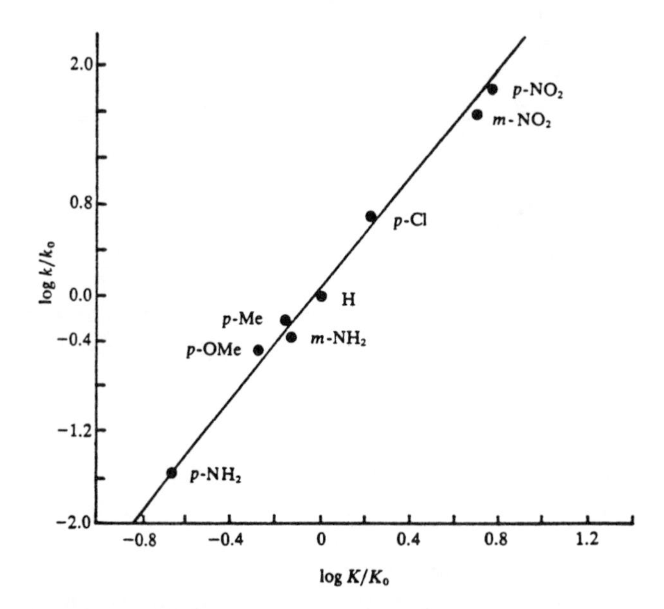

Figure 1.3. Correlation of acid dissociation constants of benzoic acids with rates of alkaline hydrolysis of ethyl benzoates. Reprinted from Hammett (1937) by permission of the American Chemical Society.

Values for the substituent constant, σ, have been determined for a large number of substituent groups by measurement of the dissociation constant of the substituted benzoic acids. Select values of σ_m and σ_p for substituents in the meta- and para-positions, respectively, are summarized in Table 1.3. These substituent groups perturb the electron density at the reactive center through resonance, inductive, or field effects. In principle, σ values are independent of the nature of the reaction that is being investigated. The Hammett equation does not apply to ortho substituents because these groups may affect the reaction center through steric interactions. The sign of σ reflects the effect that a particular substitutent has on developing charge at the reactive site. Electron-withdrawing substituents have positive values and electron-donating substituents have negative values.

The value of ρ for the ionization of benzoic acids, the standard reaction, is arbitrarily assigned a value of 1. The value of ρ for a given reaction is determined from the slope of the line of a plot of $\log(k_x/k_o)$ versus σ. A straight line indicates that the linear free energy of Equation 1.39 is valid. The magnitude of ρ reflects the susceptibility of a reaction to substituent effects. The sign of ρ is of diagnostic value because it indicates the type of charge development in the transition state for the rate limiting step. Reactions with negative ρ values (e.g., the hydrolysis of benzyl chlorides: Table 1.4) have positive charge development occurring in the transition state. Accordingly electron-donating substituents will stabilize the transition state, result-

Table 1.3. Substituent Constants[a]

Substituent Group		σ_m	σ_p	σ^+	σ_t	F	R
Acetamido	AcNH	0.21	−0.01	−0.25		0.47	−0.27
Acetoxy	AcO	0.39	0.31	0.18		0.68	−0.07
Acetyl	Ac	0.38	0.50	0.57	0.28	0.53	0.20
Amino	NH_2	−0.16	−0.66	−1.11	0.13	0.04	−0.68
Bromo	Br	0.39	0.23	0.02	0.45	0.72	−0.18
t-Butyl	$(Me)_3C$	−0.10	−0.20	−0.28		−0.10	−0.14
Chloro	Cl	0.37	0.23	0.04	0.47	0.69	−0.16
Cyano	CN	0.56	0.66	0.67	0.56	0.85	0.18
Ethoxy	EtO	0.10	−0.24	−0.58		0.36	−0.44
Ethyl	Et	−0.07	−0.15	−0.22		−0.07	−0.11
Fluoro	F	0.34	0.06	−0.25	0.52	0.71	−0.34
Hydrogen	H	0.00	0.00	0.00	0.00	0.00	0.00
Hydroxy	OH	0.12	−0.37	−0.85	0.30	0.49	−0.64
Methoxy	MeO	0.12	−0.27	−0.65	0.30	0.41	−0.50
Methyl	Me	−0.07	−0.17	−0.26	−0.05	−0.05	−0.14
Nitro	NO_2	0.71	0.78	0.74	0.63	1.11	0.16
Phenyl	Ph	0.06	−0.01	−0.08		0.14	−0.09
Trifluoromethyl	CF_3	0.43	0.54	0.58	0.42	0.63	0.19
Trimethylammonio	$(Me)_3N^+$	0.88	0.82	0.64	0.90	0.146	0.00

[a]*Sources:* C. G. Swain and E. C. Lupton, Jr., *J. Am. Chem. Soc.* 90, 4328 (1968), and P. R. Wells, *Linear Free Energy Relationships,* Academic Press, New York, 1968.

Table 1.4. Reaction Constants[a]

Reaction	ρ
$ArCO_2H \rightleftarrows ArCO_2^- + H^+$, water	1.00
$ArCO_2H \rightleftarrows ArCO_2^- + H^+$, EtOH	1.57
$ArCH_2CO_2H \rightleftarrows ArCH_2CO_2^- + H^+$, water	0.56
$ArCH_2CH_2CO_2H \rightleftarrows ArCH_2CH_2CO_2^- + H^+$, water	0.24
$ArOH \rightleftarrows ArO^- + H^+$, water	2.26
$ArNH_3^+ \rightleftarrows ArNH_2 + H^+$, water	3.19
$ArCH_2NH_3^+ \rightleftarrows ArCH_2NH_2 + H^+$, water	1.05
$ArCO_2Et + {}^-OH \rightarrow ArCO_2^- + EtOH$	2.61
$ArCH_2CO_2Et + {}^-OH \rightarrow ArCH_2CO_2^- + EtOH$	1.00
$ArCH_2Cl + H_2O \rightarrow ArCH_2OH + HCl$	−1.31
$ArC(Me)_2Cl + H_2O \rightarrow ArC(Me)_2OH + HCl$	−4.48
$ArNH_2 + PhCOCl \rightarrow ArNHCOPh + HCl$	−3.21

[a]*Source:* P. R. Wells, *Linear Free Energy Relationships*, Academic Press, New York, 1968, pp. 12, 13.

ing in an acceleration of the reaction rate. On the other hand, electron-withdrawing groups will inhibit the rate of reaction by destabilizing the transition state.

Taft equation. Taft (1956) has extended the Hammett-type correlation to aliphatic systems. Because steric effects of substituents in aliphatic systems cannot be ignored as they were for *m*- and *p*-substituted benzene compounds, Taft recognized the need to develop separate terms for the polar and steric effects for substituent constants. Based on the observation that the acid-catalyzed hydrolysis of meta- and para-substituted benzoic acid esters are only slightly affected by the electronic nature of the substituent group (ρ values are near 0), Taft concluded that the acid-catalyzed hydrolysis of aliphatic esters would also be insensitive to polar effects of substituent groups. Any effect on rate due to substituent groups could therefore be attributed to steric effects. Taft defined a *steric substituent constant*, E_s, by:

$$E_s = \log(k_x/k_o)_A \tag{1.40}$$

where k_x and k_o are the hydrolysis rate constants for XCOOR and CH_3COOR, respectively, and the subscript A denotes acid-catalyzed hydrolysis. The large ρ values measured for basic hydrolysis of benzoate esters demonstrates that the polar effects of substituents cannot be ignored. Assuming that the steric demands of the transition states for the acid- and base-catalyzed hydrolysis of esters are approximately equal, the polar parameter, σ^*, can be defined by:

$$\sigma^* = (\log(k_x/k_o)_B - \log(k_x/k_o)_A)/2.48 \tag{1.41}$$

where B denotes base-catalyzed hydrolysis and the factor 2.48 is used to normalize σ^* to Hammett's σ. The general Taft equation for LFERs in aliphatic systems (Pavelich and Taft, 1957) can then be written as:

$$\log(k_x/k_o) = \sigma^*\rho^* + \delta E_s \tag{1.42}$$

Several E_s and σ^* constants are listed in Table 1.5.

Not many environmentally oriented investigations have applied the Taft equation to the treatment of their results. One of the few such studies that have been reported is actually not on a purely aliphatic system. Based on the general Taft equation (1.42), Wolfe et al. (1980b) established a LFER for the alkaline hydrolysis of phthalate esters that is described by:

$$\log k_{OH} = 4.59\sigma^* + 1.52E_s - 1.02 \tag{1.43}$$

This LFER has been useful for the environmental assessment of phthalate esters for which hydrolysis rate constants have not been measured (Wolfe et al., 1980a).

Table 1.5. Steric and Polar Parameters for Aliphatic Systems[a]

X	E_S	σ^*
H	+1.24	+0.49
CH_3	0.00	0.00
CH_3CH_2	−0.07	−0.10
i-C_3H_7	−0.47	−0.19
t-C_4H_9	−1.54	−0.30
n-C_3H_7	−0.36	−0.115
n-C_4H_9	−0.39	−0.13
i-C_4H_9	−0.93	−0.125
neo-C_5H_{11}	−1.74	−0.165
$ClCH_2$	−0.24	+1.05
ICH_2	−0.37	+0.85
Cl_2CH	−1.54	+1.94
Cl_3C	−2.06	+2.65
CH_3OCH_2	−0.19	+0.52
$C_6H_5CH_2$	−0.38	+0.215
$C_6H_5CH_2CH_2$	−0.38	+0.08
$CH_3CH=CH$	−1.63	+0.36
C_6H_5	−2.55	+0.60

[a]*Source:* J. Shorter, *Quarter. Rev.* (London), 24:423 (1970).

B. OVERVIEW OF THE ENVIRONMENT

In this section, we will attempt to provide background information that sets the stage for an analysis of the occurrence of environmental organic reactions. Environmental milieux are highly variable in their temperatures, pressures, oxygen content, etc., and reactions that are important in one medium such as the gas phase of the atmosphere may be negligible in another, such as bottom sediments. The section will begin with an overview of the atmosphere and its organic constituents, followed by brief surveys of the organic environments of surface and groundwaters, and of soils and sediments. Emphasis will be on the characteristics of the matrices and the organic materials associated with them; details on the reaction pathways, followed by some of the organic constituents, will be discussed in later chapters.

1. The Troposphere and the Stratosphere

Today's atmosphere is the product of thousands of millions of years of evolution; it has changed almost completely since the primeval atmosphere first formed. The evolution of the atmosphere is, of course, continuing. Although the rates of change of major atmospheric constituents are very small, significant changes in some minor components are occurring now. For example, the activities of humans over the past century or so have caused a host of novel organic compounds to enter the atmosphere, and the combustion of huge quantities of fossil fuels has significantly increased its CO_2 content. Since the late 1960s, it has become clear that the entire troposphere is a transport system for trace organic compounds, as well as a gigantic reactor where chemical changes in those compounds are taking place under the influence of sunlight and in the presence of highly reactive intermediates.

a. The Thermal Structure of the Atmosphere

Virtually all (99.999%) of the mass of the earth's atmosphere lies within 80 km of its surface; 80% is within the lowest 12 km. The atmosphere is not a region of smoothly varying properties, but is distinctly stratified.

The traditional classification of atmospheric regions is based on temperature (Figure 1.4). Closest to the earth lies the **troposphere** (Greek *tropos*, change), a region of generally declining temperature in the upward direction. The cause of this thermal effect is the conversion of visible sunlight radiation at the earth's surface to heat (infrared) radiation. At about 10–12 km, a temperature minimum of approximately 210°K is reached; this is generally considered the top of the troposphere or **tropopause**. From 10 to about 50 km (the **stratosphere**) the temperature increases to about 270°K. Little organic chemistry occurs above 50 km. The region from 50 to 90 km (the **mesosphere**) is a region of generally declining temperatures, and finally, at still higher elevations (the **thermosphere**), the temperature increases sharply, to a maximum of about 2000°K at a few thousand kilometers.

The stratosphere is heated, principally, by a strongly exothermic, photochemically-driven reaction sequence;

$$O_2 + h\nu\ (<242\ \text{nm}) \rightarrow 2\ O\cdot \quad (1.44)$$

$$O_2 + O\cdot \rightarrow O_3 + 100\ \text{kJ} \quad (1.45)$$

$$O_3 + O\cdot \rightarrow 2\ O_2 + 400\ \text{kJ} \quad (1.46)$$

$$O_3 + h\nu\ (240\text{–}290\ \text{nm}) \rightarrow O_2 + O\cdot \quad (1.47)$$

The sequence is initiated by the homolysis of molecular oxygen. Light with wavelengths shorter than about 240 nm is virtually all absorbed by constituents of the upper regions of the atmosphere such as molecular N_2; however, at the top of the stratosphere the concentration of O_2 becomes sufficiently great for Reaction 1.44 to become important. The product of this reaction is two atoms of oxygen, which combine with molecular oxygen exothermically (in the presence of a third body, usually N_2), to produce ozone. (Ozone also reacts with oxygen atoms in a very exothermic reaction that returns two equivalents of O_2.) Molecular ozone, in turn, absorbs 240–290 nm actinic radiation to regenerate oxygen molecules and atoms.

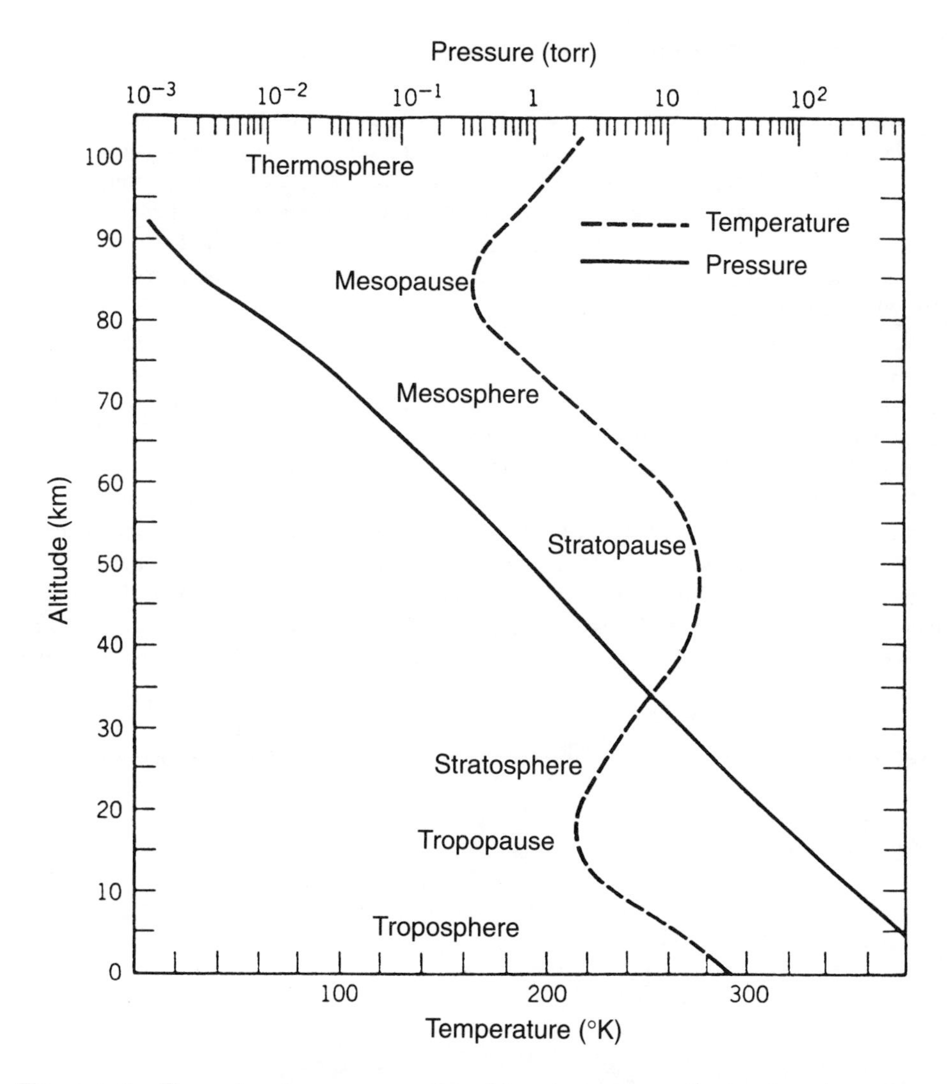

Figure 1.4. Thermal and pressure profiles of the atmosphere with elevation above the earth's surface. From B. J. Finlayson-Pitts and J. N. Pitts, Jr., *Atmospheric Chemistry*, John Wiley & Sons 1986. Reprinted by permission.

The concentration of ozone is at a maximum at about 25 km, roughly in the middle of the stratosphere.

The stratosphere, unlike the troposphere, undergoes little vertical mixing and few clouds are present; therefore, molecules that reach the stratosphere (SO_2 from volcanic eruptions, chlorofluorocarbons, etc.) tend to remain there for some time and are not "rained out" (deposited back to the earth's surface) as compounds in the troposphere can be.

b. Solar Energy Distribution

The sun emits radiation whose frequency distribution is closely similar to that predicted for a blackbody radiator at 5900°K (Figure 1.5). It essentially begins in the vacuum UV at about 120 nm, rises to a maximum at 400 nm at the beginning of the visible spectrum, and falls gradually through the visible and infrared regions. The total amount of solar energy received at the earth's surface is about 1370 J/m^2/sec, and a quantity approximately equal to this is either reflected back into space or absorbed by the atmosphere.

Diurnal and seasonal variations in solar intensity are, of course, of utmost importance to ecosystems. In the extreme polar regions there is no direct solar radiation at all for more than four months of the year, whereas near the equator the overall intensity of sunlight fluctuates less than 10% annually. The spectral energy distribution also varies with the season. For example, in July in the middle latitudes (ca. 40°), the fraction of shorter-wave UV (290–315 nm) in the total solar radiation is more than three times higher than it is in December, due to the shorter path these easily scattered wavelengths have to traverse through the atmosphere. For similar reasons, shortwave UV is more intense at high elevations, particularly in the tropics where stratospheric ozone is less concentrated (Caldwell et al., 1980).

Because several constituents of the atmosphere (O_3, O_2, CO_2, and water) have

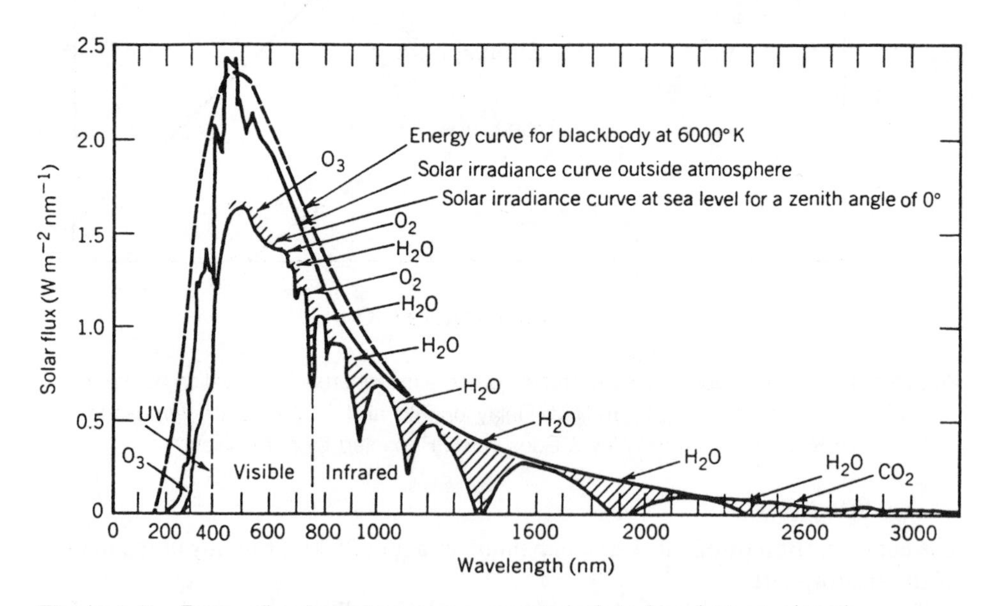

Figure 1.5. Energy flux (in watts per square meter) plotted against wavelength, calculated for a 6000°K blackbody radiator, and compared to the observed solar flux at sea level and outside the earth's atmosphere. Reprinted from Chapter 16, "Thermal Radiation," in *Handbook of Geophysics*, U.S. Air Force Cambridge Research Center, 1960.

important absorption maxima in various regions of the solar spectrum, the frequency distribution of radiation at the earth's surface differs significantly from that emitted by the sun. Water, in the form of cloud droplets, for example, absorbs energy in the IR region and also reflects and scatters other sunlight wavelengths, so that the earth's surface is cooled when it is present in high concentrations. Other particles in the atmosphere also absorb and scatter incoming light in a complex manner that depends on their size, shape, and composition.

Because tropospheric CO_2 levels have been increasing due to fossil fuel combustion over the past 100 or so years, it has been theorized that more and more infrared radiation is being absorbed, warming the troposphere (the so-called greenhouse effect). Other IR-absorbing atmospheric gases such as CH_4, nitrous oxide, and chlorofluorocarbons also appear to be important contributors to the hypothesized global warming (Ramanathan et al., 1985; Rowland, 1989; Lashof and Ahuja, 1990). Tropical deforestation also has been implicated in global CO_2 increases, since when trees are cut down, the succeeding grassy or weedy plant community can remove only a small fraction of atmospheric CO_2, compared to the forest. Furthermore, much of the wood obtained is burned, adding to the CO_2 burden directly.

In addition, stratospheric ozone concentrations appear to be declining, due to photolysis by shortwave UV of volatile chlorine-containing compounds such as $CFCl_3$ and its relatives (Reaction 1.48); the Cl atoms produced in this reaction scavenge O atoms (Reaction 1.49) and

$$CFCl_3 + h\nu\ (<250\ \text{nm}) \rightarrow {\cdot}CFCl_2 + {\cdot}Cl \qquad (1.48)$$

catalyze an efficient chain process (Reaction 1.6) that removes ozone molecules and converts them to oxygen molecules (Molina and Rowland, 1974; Rowland, 1989; Rowland, 1991):

$$Cl{\cdot} + O_3 \rightarrow ClO{\cdot} + O_2 \qquad (1.49)$$

$$ClO{\cdot} + O \rightarrow Cl{\cdot} + O_2 \qquad (1.50)$$

$$\text{(sum)} \qquad O_3 + O \rightarrow 2\ O_2$$

These compounds reach the stratosphere because they are so unreactive in all tropospheric processes, including reactions with HO· to which they are virtually inert. Measurements of atmospheric ozone in Antarctica (Stolarski et al., 1986), in the Arctic (Zurer, 1990), and even in the temperate latitudes (Watson et al., 1988) all point to a decrease in stratospheric ozone concentrations. Recent attempts to diminish the use of ozone-depleting compounds on a global basis appear to have been successful; however, only minor effects on the rate of ozone destruction are likely to be observed for many years to come.

c. Chemical Constituents and Their Reactions

The reactions leading to the transformation of organic compounds in the atmosphere are primarily oxidations; their rates are governed by the reactivities of the various classes of organic substances toward the oxidants (typically reactive oxygen-containing species) present. The majority of these oxygenated forms and the mechanisms of their reactions are discussed in detail in Chapter 4. Only an overview will be presented in this section.

Although a great assortment of organic molecules is present in the troposphere at all times, most occur only in trace concentrations. Because of the enormous extent of the atmosphere, however, it can be a sink for huge quantities of a given compound. One part in 10^9, over the whole of the atmosphere, would correspond to three million tons of a uniformly distributed substance.

A comprehensive list of organic compounds detected in both natural and polluted atmospheres has been compiled by Graedel (1978). The book consists principally of tables, listing more than 1,600 identified compounds by structural class, concentration, and atmospheric lifetime, if known. The tables include substances found in tobacco smoke, forest fires, volcanic eruptions, and industrial emissions.

Hydrocarbons. The simplest hydrocarbon, methane (CH_4), is also by far the most abundant organic species in the atmosphere. It is largely produced at or slightly below the earth's surface by biotic mechanisms such as the anaerobic microbial decomposition of organic matter in such places as marshes, lake bottoms, landfills, and the stomachs of ruminant animals. Other sources of CH_4 of probable lesser importance include natural gas seepages and incomplete combustion. Recent measurements of atmospheric methane concentration show that it has been increasing at about 1% per year, from typical levels of about 1.5 ppm measured 15 or so years ago, to about 1.75 ppm at present (Rasmussen and Khalil, 1983; Khalil and Rasmussen, 1990). The lifetime of methane in the atmosphere is almost completely determined by its reaction with hydroxyl radical; for details see Chapter 4.B.2. A possible decline in the global levels of atmospheric hydroxyl radicals has been postulated as a contributing reason for the increasing methane concentration (Khalil and Rasmussen, 1985), although these workers believe that increased emission from agricultural and industrial activities contribute the majority of the increased concentration observed.

Higher alkanes found in the atmosphere, sometimes included in a category called **nonmethane hydrocarbons** or NMHC, principally come from natural gas seepages, transportation, and industrial activities. Normally, either ethane or butane is the second most abundant aliphatic hydrocarbon in atmospheric samples [usually in the 50–100 ppb range, more than an order of magnitude lower in concentration than methane (Cavanagh et al., 1969; Altshuller, 1983)]. Higher alkanes (up to C_{40}) have been detected in aerosols and other atmospheric samples (Appel, 1981; Hutte et al., 1984; Sexton and Westberg, 1984; Harkov, 1986; Levsen et al., 1991). In clean environments, *n*-alkanes of odd carbon number from C_{23} to C_{33} predominate. The principal source of these compounds is the epicuticular wax layer of plants.

Branched-chain alkanes in the C_{24}–C_{30} range appear to be largely products of vehicular exhaust, and are apparently synthesized during fuel combustion (Boone and Macias, 1987).

Unsaturated aliphatic and cyclic hydrocarbons have numerous biological sources. Ethylene, $H_2C=CH_2$, the simplest of the series, is emitted by green plants in substantial quantities. It has hormonal activity and has been implicated in the control of many physiological processes in plants. The natural tropospheric concentration of ethylene is very low due to its high reactivity with ozone, •OH, and other atmospheric oxidants (Robinson and Robbins, 1968), but in polluted atmospheres its concentrations can be much higher. It is a product of combustion of wood, coal, oil, natural gas, and petroleum. Elevated levels of ethylene, such as may occur in homes where coal gas is used for cooking, can be very deleterious to plants.

Green plants also release substantial quantities of isoprene (**2**) as well as monoterpenoid hydrocarbons such as α-pinene (**3**), myrcene (**4**), and limonene (**5**). These compounds contribute significantly to atmospheric organic concentrations in remote or forested areas (Altshuller, 1983; Hutte et al., 1984). It has been estimated that the global output of these substances may equal about 28% of the annual output of methane (Rasmussen and Went, 1965).

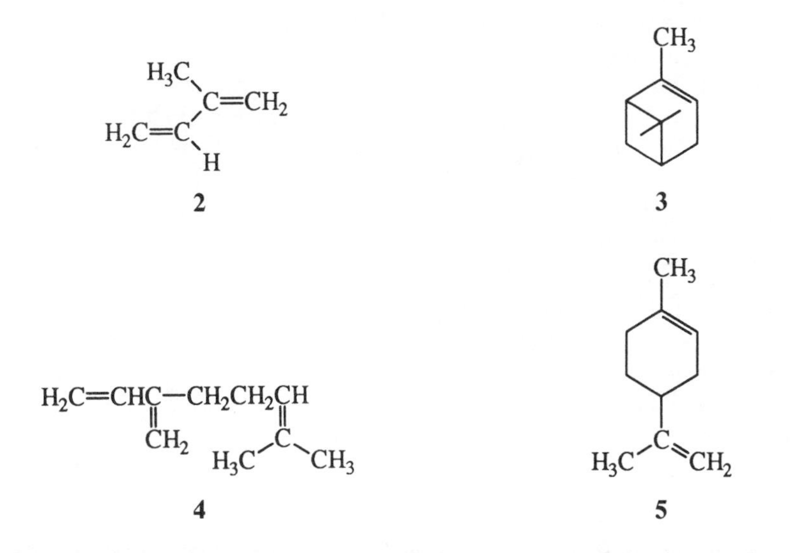

Aromatic hydrocarbons are produced anthropogenically and have been identified in urban air. Generally, toluene is the most abundant aromatic hydrocarbon; benzene concentrations range from 1/2 to 2/3 of that of toluene. Simple aromatic compounds such as xylenes and other alkylated benzenes have also been detected at much lower (0.001 to 0.010 ppm) concentrations (Wathne, 1983). The principal source of these compounds is probably gasoline fuel evaporation, but it has been shown that simple aromatic hydrocarbons can be produced by cracking higher condensed molecules. Diesel exhaust contains a high proportion of two-ring hydrocarbons such as naphthalenes, indans, and tetrahydronaphthalenes (Levins, 1972).

Polycyclic aromatic hydrocarbons (PAHs), defined as aromatic hydrocarbons with three or more rings, are also found in the atmosphere; they (especially the higher PAHs) are largely associated with particles. The mechanisms of formation of PAHs are discussed in Chapter 4. PAHs having fused ring numbers from two (naphthalene) to seven (coronene) have been observed (Lee et al., 1981; Hoff and Chan, 1987; Levsen et al., 1991). Concentration ranges reported are from less than 1 ng/m^3 to hundreds of ng/m^3. The combustion of almost any organic material produces PAHs (Table 1.6.) Benzo[a]pyrene (**6**), a particularly active carcinogenic PAH, has been identified repeatedly in urban air studies. Usually, PAH concen-

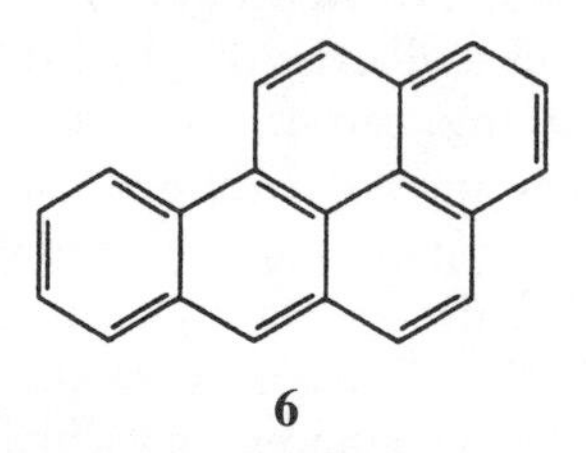

6

trations are highest in winter months, showing that heating systems are the most important contributors of atmospheric PAH; but relatively large quantities are also produced by automobiles, electricity-generating stations, and refuse burning. Tobacco smoke and "tar" have especially high PAH concentrations; a heavy smoker probably inhales up to 1 μg of benzo[a]pyrene daily (Wynder and Hoffmann, 1968). The atmospheric reactions of PAH include oxidations, usually by hydroxyl radical, to give rise to phenols and quinones (König et al., 1983) and also reactions with nitrogen oxides (usually surface-catalyzed) to produce nitro-PAH derivatives (Pitts et al., 1978; Arey et al., 1986; Pitts, 1987).

Fuel-rich conditions and relatively low temperatures permit the processes of PAH formation to continue for a long enough period so that roughly spherical aggregates of organic particles of colloidal dimensions (soot) are produced. In practical terms, a PAH with a molecular weight of about 2,000 can be considered to be a soot particle (Miller and Fisk, 1987). These high-carbon particles are long-lived and appear to be transported over very considerable distances.

Table 1.6. Formation of PAH During Pyrolysis of Various Organic Compounds

Material Pyrolyzed	Benzo[a]pyrene Produced (μg/g)
Glucose	48
Fructose	98
Cellulose	289
Stearic acid	1200
Dotriacontane	3130
β-Sitosterol	3750

Source: Schmeltz and Hoffmann (1976).

Hazes in the Arctic have been systematically observed since the late 1970s. Known to contain high concentrations of sulfur oxides as well as carbon, they form in the late autumn when weather and temperature patterns are favorable for transport of sooty particles from the industrial regions of North America and Europe (particularly the former Soviet Union). During the winter, such hazes may cover a region approximating the size of North America and extending vertically to at least 9,000 m. In the spring, increasing atmospheric moisture and rain combine to scavenge and "rain out" the aerosols. It is probable that their carbon constituents may absorb 10% or more of the incident solar energy over the Arctic ice cap in the late winter months when their extent is greatest. The effects of these hazes on climate or on the Arctic ecosystem are almost entirely unknown (Hileman, 1983; Novakov, 1984).

Atmospheric particles are produced in virtually all combustion processes and are also formed when volatile products of plant origin are irradiated in the presence of a light-absorbing species such as NO_2. The bluish atmospheric hazes common in forested mountainous areas have been attributed to the latter process. Traces of SO_2 also promote the formation of organic particulate material. These particles, entering the atmosphere, are the principal locus for PAHs.

Oxygen-containing organic compounds. Data on oxygenated species in the atmosphere are relatively scarce, principally due to the lack of suitable analytical procedures for the determination of polar constituents in the presence of a large excess of hydrocarbons. Interest in the area is increasing, however, because of evidence that many reactive oxygen-containing species (epoxides, lactones, and hydroperoxides, for example) may be important in the epidemiology of respiratory disorders and cancer.

Methanol and ethanol occur in crankcase emissions and have been detected in urban air, methanol at levels up to 0.1 ppm (Bellar and Sigsby, 1970). In addition, rather high concentrations of *n*-butanol (0.019 ppm) occurred in the air at Point Barrow, Alaska (Cavanagh et al., 1969); it is likely that biological activity was responsible. Phenol is present at a few thousandths of a ppm; nitrophenols have been detected in rain water (Nojima et al., 1976; Levsen et al., 1991). Alkylphenols and alkyl-substituted nitrophenols have been identified in urban aerosols and in the exhaust from internal combustion engines (Schuetzle et al., 1975; Kuwata et al., 1980).

Carbonyl compounds are probably present at low levels in all air samples (Fung and Grosjean, 1981). A great variety of combustion processes produce aldehydes; they are particularly abundant in wood smoke, and have also been identified in the products of combustion of natural gas and fuel oil, and in engine exhaust samples. It is likely that they are relatively long-lived intermediates in atmospheric photooxidation reactions. Formaldehyde (CH_2O), for example, is always an obligatory intermediate in the conversion of methane to CO_2. In nearly all air samples so far examined, formaldehyde has been the most abundant aldehyde, with acetaldehyde, propionaldehyde, acrolein, benzaldehyde, crotonaldehyde, and furfural sometimes present at much lower concentrations (Kuwata et al., 1979; Penkett, 1982; Zhou and

Mopper, 1990; Levsen et al., 1991). In Los Angeles air, higher aldehydes have been reported to be present at concentrations comparable to those of formaldehyde. In rain, fog, and mist samples taken in California, formaldehyde and acetaldehyde were always present, but surprisingly high concentrations of glyoxal and pyruvaldehyde were also detected. One fog sample contained 276 μM (16 mg/L) of glyoxal. It was suggested that the α-dicarbonyl compounds were present at such high concentrations in the water because they were either more water-soluble or reacted more slowly with water or sulfur oxides than the monoaldehydes (Steinberg and Kaplan, 1984; Igawa et al., 1989). In Caribbean air samples, a series of aliphatic aldehydes ranging in chain length from C_1 to C_{18} was identified (Zhou and Mopper, 1990). A forested atmosphere in Colorado was shown to contain (in addition to formaldehyde and acetaldehyde) several higher aldehydes typical of higher plants, including 2-hexenal (**7**) and *p*-hydroxybenzaldehyde (**8**: Nondek et al., 1992). Naphthalene- and alkylnaphthaldehydes, anthracene-9-carboxaldehyde (**9**) and phenanthrene-9-carboxaldehyde (**10**) were identified in diesel exhaust particle extracts, together with some of their alkylated homologs (Choudhury, 1982; Newton et al., 1982).

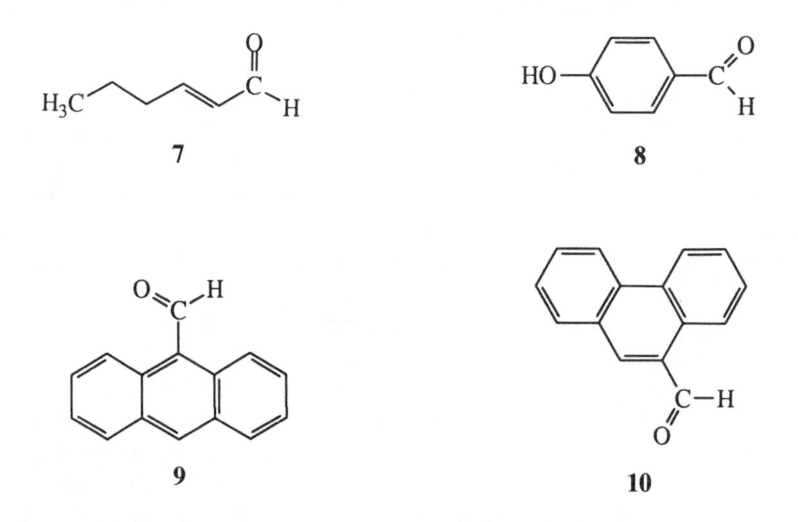

Not as much work has been reported on atmospheric ketone concentrations. Acetone was present at 0.001 ppm in the air at Point Barrow, Alaska (Cavanagh et al., 1969) and also, at half that concentration or less, in air over the Atlantic (Penkett, 1982; Zhou and Mopper, 1990); it is also present at higher elevations (Arnold et al., 1986). Phenyl and higher cyclic ketones, a few C_6–C_{18} aliphatic ketones, butanone, biacetyl and 4-methyl-2-pentanone have also been identified in ambient air (Kawamura and Kaplan, 1983; Ramdahl, 1983; Zhou and Mopper, 1990). Several quinones have been identified by Soxhlet extraction of atmospheric particulate material, including anthraquinone and more complex multi-ring compounds (Choudhury, 1982; Ramdahl et al., 1982). In addition, more than 40 other oxygenated PAHs including aldehydes, ketones, lactones, and anhydrides have also been characterized from airborne particle extracts (Choudhury, 1982; Newton et al.,

1982; Ramdahl et al., 1982; König et al., 1983; Helmig et al., 1992). Fluorenone (**11**) and its alkyl homologs were particularly abundant in diesel emission particulate material.

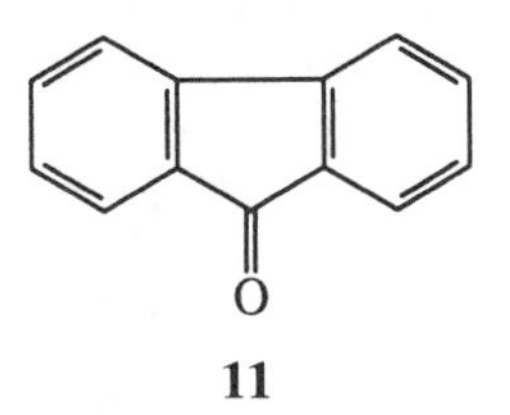

11

Carboxylic acids are produced in major quantities by the combustion of coal and wood. In fact, some time ago it was reported that far higher concentrations of organic acids than hydrocarbons were present in the volatile constituents from a large coal-fired power plant (Cuffe et al., 1964). However, the atmospheric fates of the acids are not completely understood. They do not appear to undergo important chemical reactions, except for keto acids, which photodegrade in a few hours (Grosjean, 1983; Berges and Warneck, 1992). Mostly, they are transferred back to the surface of the earth largely by dry deposition (if in the gas phase) and rainout (if in the particulate phase (Grosjean, 1989).

Formic acid is typically found at concentrations similar to those of the other one-carbon oxygenated compounds, methanol and formaldehyde. Other mono- and dicarboxylic acids are frequently identified in aerosols and other particles such as rain, fog, and mist; acetic acid is usually approximately as abundant as formic acid (Berkenbus, 1983; Kawamura and Kaplan, 1984; Kawamura et al., 1985; Steinberg et al., 1985; Harkov, 1986; Yokouchi and Ambe, 1986; Grosjean, 1989). Pyruvic acid, $CH_3C=O\text{-}COOH$, is produced during atmospheric oxidation of aromatic compounds such as *o*-cresol (Grosjean, 1984). It, like other keto acids, photodegrades further in sunlight to a mixture of products including formaldehyde, acetaldehyde, and PAN (Grosjean, 1983; Berges and Warneck, 1992).

Extracts of rain and snow from a southern Indiana location contained a long series of fatty acids, ranging in size from C_{12} to C_{28}. Concentrations of total acids ranged from 4 to 45 μg/L; palmitic and stearic acids (n-C_{16} and C_{18}, respectively) predominated. It was noted that concentrations of fatty acids were highest in the spring, when plant growth and pollen release were contributing organic matter to the atmosphere, and declined in the late autumn when deciduous plants died (Meyers and Hites, 1982). Analyses of atmospheric fallout samples in other areas have also revealed several long-chain fatty acids in the C_8–C_{34} range, including unsaturated and branched-chain acids. Phenolic acids were also identified from a Japanese sample, including *o*-, *m*-, and *p*-hydroxybenzoic acids and several hydroxycinnamic acids. It was concluded that volatilization from pollens and other plant parts probably contributed to the occurrence of these compounds in the atmosphere (Matsumoto and Hanya, 1980; Levsen et al., 1991).

Carboxylic esters are very rarely found in atmospheric samples, with the exception of phthalate esters (**12**). These materials are practically ubiquitous in air and water samples (Appel, 1981; Hoff and Chan, 1987). Dibutyl, dioctyl, and di(2-ethylhexyl) phthalates are produced in very large quantities for use as plasticizers, lubricants, and ingredients of inks, cosmetics, adhesives, and coatings. The compounds combine great oxidative and hydrolytic stability with low water solubility and relatively high vapor pressure, and thus enter and persist in the atmosphere.

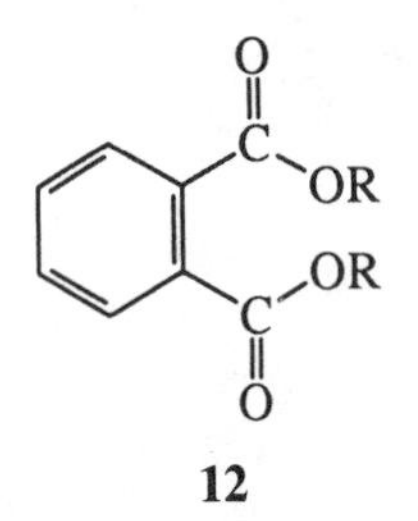

12

Novakov has suggested that "primary oxidants," hydrogen peroxide and organic hydroperoxides (as opposed to the "secondary oxidants" formed in photochemical smog episodes), may play important roles in atmospheric chemistry and may have significant biological effects. The intermediacy of peroxy radicals, of course, is well established (see Section 4.A.1) in atmospheric photooxidation reactions, but the role of relatively long-lived nonradical forms has not been fully explored. Both of these types of oxidants are formed when pure or mixed hydrocarbons are burned in internal combustion engines (Novakov, 1984). Hydroperoxides and peroxy acids such as peracetic acid, $CH_3C(O)OOH$ and hydroxymethyl hydroperoxide, $HOCH_2OOH$, have also been shown to be formed in laboratory studies from a variety of hydrocarbon and carbonyl compound precursors (Hanst and Gay, 1983; Hewitt and Kok, 1991). Methyl, hydroxymethyl, and hydroxyethyl ($HOCH_2CH_2OOH$) hydroperoxides have been detected in air (Heikes et al., 1987; Hellpointner and Gäb, 1989). The sum of the concentrations of methyl hydroperoxide, hydroxymethyl hydroperoxide, and three other organic peroxides, measured in ambient air sampling studies in Colorado, was found to approximate or slightly exceed that for hydrogen peroxide (Hewitt and Kok, 1991).

Other organic compounds. Volatile organic nitrogen compounds of natural origin include aliphatic amines such as methyl, dimethyl, ethyl, diethyl, *n*-propyl, isopropyl, *n*-butyl, and *sec*-butyl, which have been identified in cattle feedlot air (Mosier et al., 1973). Dimethylnitrosamine, which could be formed by reactions of secondary amines with nitrogen oxides in the air or have a terrestrial origin, has been detected in ambient air samples in urban areas (Epstein et al., 1976; Fine et al., 1976). A number of heterocyclic nitrogen compounds have been detected in the nanograms/m^3 (parts per trillion) range, including caffeine, isoquinoline, and carbazole (Lee et al., 1981; Kawamura and Kaplan, 1983). Three nitriles (cyano compounds) derived from PAHs have been reported from urban air particles; unidentified isomers of acenaphthene-, anthracene-, and pyrenecarbonitriles (Ramdahl et al., 1982).

Sulfur compounds include alkylthiols (products of anaerobic microbial activity) and dimethyl sulfide (Lovelock et al., 1972; Adams et al., 1981; Andrae and Raemdonck, 1983). Thiols are also by-products of pulp and paper bleaching operations. In polycyclic fractions, dibenzothiophene and a few related compounds have been identified (Lee et al., 1981). Several compounds containing both sulfur and oxygen have been reported, including monomethyl and dimethyl sulfate, methanesulfonic acid (Panter and Penzhorn, 1980) and *bis*-hydroxymethyl sulfone, $HOCH_2SO_2CH_2OH$ (Eatough and Hansen, 1984; Eatough et al., 1986). The latter compound was proposed to be formed by a series of reactions involving ozone, ethylene, and SO_2.

Atmospheric halogen compounds (Penkett, 1982) are of both natural and industrial origin. Probably the most abundant halocarbon in the troposphere is methyl chloride, CH_3Cl, which is present at a level of 0.6–2 parts in 10^9. It appears to be present in volcanic emissions, formed by microbial fermentation, by the combustion of vegetation (Lovelock, 1975), and by the S_N2 reaction of methyl iodide (a constituent of marine algae) with the large excess of chloride ion in seawater (Zafiriou, 1975). Methyl iodide is also present in the atmosphere in highly variable amounts, averaging about one part in 10^{12} ; much higher concentrations are present in seawater (Lovelock et al., 1973). It is likely that other halogenated organic compounds of natural origin are also present in the atmosphere; for example, a host of Cl-, Br- and I-containing haloforms and a variety of other halogen compounds have been identified as algal metabolites (Burreson et al., 1975; Gschwend et al., 1985). Organohalogen compounds appear to be much more common in nature than was previously thought; Swedish workers have shown that the concentrations of halogen-containing organic compounds in 135 lakes in southern Sweden were comparable to those of polluted rivers such as the Rhine (Asplund and Grimvall, 1991). These authors rejected the hypothesis of long-range atmospheric transport and concluded that enzymatic halogenation reactions were the source of the bulk of the observed haloorganic compounds.

Other halogenated compounds detected in low concentrations (usually less than 0.03 part in 10^9) include chloroform ($CHCl_3$), dichloromethane (CH_2Cl_2), 1,1-dichloroethane ($ClCH_2CH_2Cl$), 1,1-dibromoethane or EDB ($BrCH_2CH_2Br$), 1,1,1-trichloroethane (CH_3CCl_3), trichloroethylene ($ClCH=CHCl_2$) and tetrachloroethylene ($Cl_2C=CCl_2$). Virtually all of these compounds have an anthropogenic origin and, in many cases, their tropospheric concentrations have been increasing (Tsani-Bazaca et al., 1982; von Düszeln and Thiemann, 1985; Prinn et al., 1987). EDB and 1,1-dichloroethane are added to motor fuels because of their antiknock properties; trichloroethylene, trichloroethane, and tetrachloroethylene are used as solvents or in dry cleaning operations. It is likely that a significant fraction of the atmospheric chloroform is volatilized from drinking water, cooling waters, and wastewaters which have been treated with aqueous chlorine (see Section 5.A.3; Barcelona, 1979). Chloroform and 1,2-dichloroethane are also produced in automobile exhaust (Harsch et al., 1977; Clark et al., 1982).

Polychlorinated biphenyls (PCBs: **13**) as well as DDT (**14**) and related compounds have been detected in air and atmospheric particle samples (Harvey and Steinhauer, 1974; Bidleman and Olney, 1974). Indoor air may contain rather high levels of

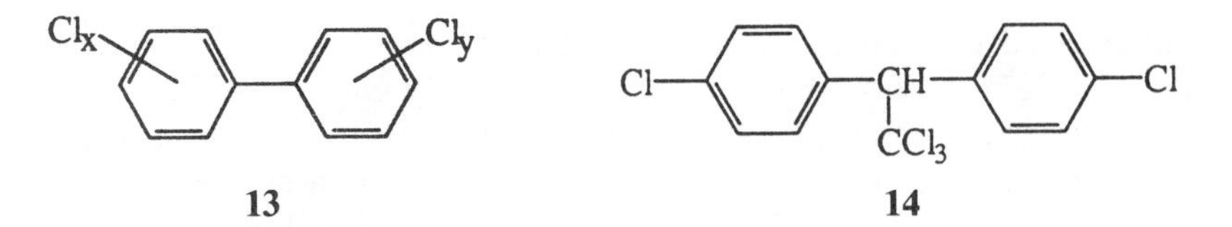

PCBs. Many products once contained PCBs and were used primarily inside buildings, such as printing ink, carbonless carbon paper, small electrical capacitors, and fluorescent light ballasts. MacLeod (1981) showed that indoor PCB concentrations in offices, private homes, and laboratories were (in the early 1980s) at least 10x and often 100x higher than outdoor levels. She also sampled an office in which a fluorescent light ballast had burned out and noted that immediately after the burnout, concentrations rose to more than 10 $\mu g/m^3$ and then fell gradually over a period of at least three months to more typical values of 200 ng/m^3.

Among the chlorinated heterocyclic compounds, chlorinated dibenzodioxins (**15**) and chlorinated dibenzofurans (**16**) have received much attention recently because of their high acute toxicity in some systems and their occurrence in the herbicide Agent Orange, in waste oils, and in bleached paper products. The compounds have been observed to be formed in low yields in many combustion-related processes such as municipal waste incineration (Rappe, 1984), and have been measured in ambient air samples by numerous investigators (see, e. g., Czuczwa and Hites, 1986; Marklund et al., 1986). Halogenated PAHs have been identified in flue gas effluents (Eklund and Stromberg, 1983).

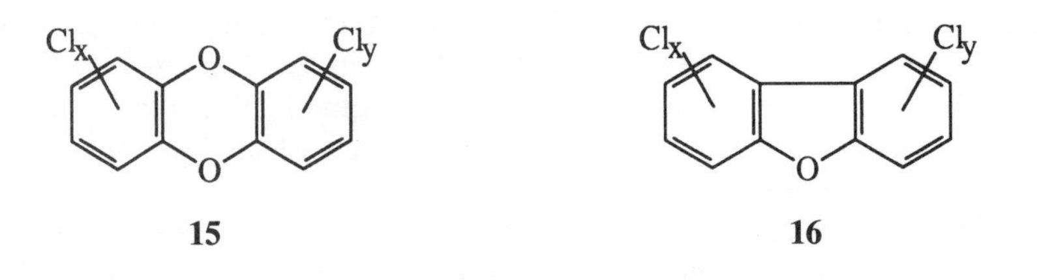

2. Natural Waters

The earth is unique in having a significant (ca. 70%) surface covering of liquid water. Table 1.7 summarizes the global distribution of surface and near-surface water. The oceans contain a very large majority of the terrestrial water; most of the remainder is glacial ice.

Almost all of the fresh liquid water is groundwater, with at least half of the total present at depths of greater than 1000 m. (Much less than 1% of the total is in the root zone of the soil.) Lakes and rivers, despite their extreme importance for humans, are a quantitatively insignificant component of the world's water supply. There is about as much water in the atmosphere as there is in all the world's rivers.

Evaporation and precipitation are in almost exact balance on a global basis. Of the rain and snow falling on land, about a third runs off and enters the ocean via river and lake drainage systems. The other two-thirds is retained temporarily by the

Table 1.7. Global Distribution of Water

	Volume (km^3)	%
Oceans	1.37×10^9	97.3
Ice caps, glaciers	3.0×10^7	2.1
Lakes	1.2×10^5	8×10^{-3}
Rivers	1.2×10^4	8×10^{-4}
Ground water	8.3×10^6	0.6
Soil moisture	2.5×10^4	1.8×10^{-3}
Atmosphere	1.3×10^4	9×10^{-4}
	1.41×10^9	100.0

soil and returned to the atmosphere directly (by evaporation) or indirectly (by transpiration from plants).

a. Water as Solvent and Reactant

Water is an angular molecule with an H–O–H bond angle of 104.5° and an O–H bond length of 0.96 Å. The energy of dissociation of the O–H bond is about 450 kJ/mol, much higher than that for typical C–C bonds which average around 350 kJ/mol.

Water has a very high latent heat of vaporization (2300 J/g, by far the highest of all common liquids), and accordingly, considerable energy must be supplied to the liquid to convert it to the vapor. Nevertheless, the atmosphere always contains water both in the vapor state and in the form of small aerosol droplets. At 20°C, up to 17 g water vapor/m^3 can be held in the air.

Liquid water. The structure of liquid water is still controversial. Some models suggest that it can be represented as a slightly disordered ice, although these representations do not account very well for the remarkable degree of supercooling (up to 40°C) that the liquid can exhibit. It would be expected that ice-like assemblages would be efficient nuclei for crystallization.

An early proposal by Pauling (1960) was that liquid water could be considered a self-clathrate (Figure 1.6), with a central, nonhydrogen-bonded "guest" molecule surrounded by a hydrogen-bonded framework of pentagonally and hexagonally-faced polyhedra. This model appears to be somewhat too "ordered" to account for many of the properties of water (Frank and Quist, 1961), but its central idea of a structure incorporating both "framework" (hydrogen-bonded) and "interstitial" (nonhydrogen-bonded) water molecules has been included in many of the more recent attempts to describe liquid water. The "flickering cluster" model (Nemethy and Scheraga, 1964) has many adherents; it posits the existence of compact, short-lived assemblages of hydrogen-bonded molecules in equilibrium with unassociated, "vapor-like" water. Cluster models can be expressed as an equilibrium between two

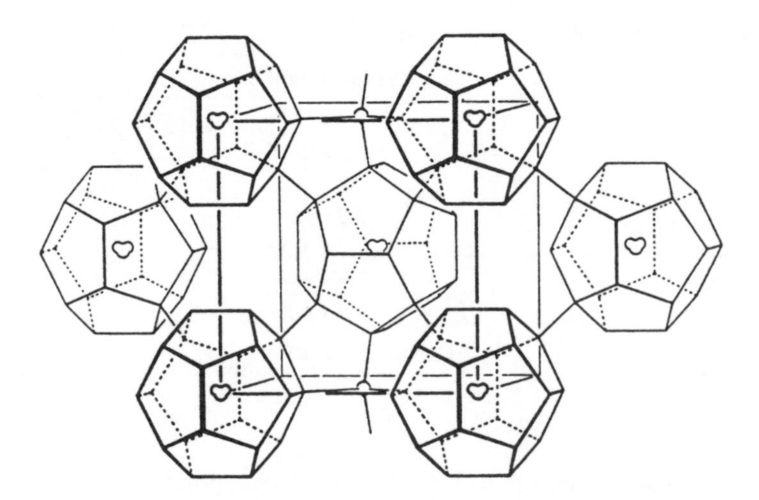

Figure 1.6. Pauling's host-guest model for liquid water structure. The central water molecule is envisioned as not taking part in hydrogen bonding with the surrounding icosahedral assemblage of 20 fully hydrogen-bonded molecules. From J. L. Kavanau (1964), *Water and Solute-Water Interactions.* Reprinted by permission of Holden-Day, Inc.

forms of water; "bulky" water $(H_2O)_b$ which is ice-like; that is, consisting of associated, open structures with considerable void space, and "dense" water $(H_2O)_d$ which is more closely packed and less hydrogen-bonded. Cooling such an assemblage would cause a shift in the equilibrium toward the bulky form, which would eventually crystallize, whereas heating it would increase the proportion of dense water which would escape as vapor.

Water as a solvent. Aqueous solutions can be thought of as rearrangements of the structure of liquid water in order to accommodate foreign molecules, which can interact with the solvent in more or less consequential ways. Two extremes can be imagined; a virtual noninteraction in which the water molecules merely circumvent the solute, arranging themselves around it; and an almost complete incorporation, in which the functional groups of the solute molecule placidly combine in the hydrogen-bonded lattice of the solvent. The former extreme is approximated in solutions of aliphatic hydrocarbons; the second, in solutions of sugars and sugar alcohols.

The molar solubility of aliphatic hydrocarbon gases in water shows a maximum at C_3–C_4. This appears to represent the optimum size for a clathrate-like structure in which the host water molecules arrange themselves around the guest hydrocarbon. This is a good example of the so-called "hydrophobic interaction" or "like-dissolves-like" phenomenon. Presumably, larger cages, such as would be required to surround a larger molecule, are less stable. The negative entropy change that occurs when a nonpolar gas is dissolved in water is quite considerable, suggesting that a more ordered, "ice-like" state results as the solute exerts a tendency for the water molecules in its vicinity to become more "bulky."

The properties of water in the presence of nonpolar solutes (those not having functional groups that can significantly interact with the water lattice) are to some degree similar to those of water containing nonpolar gases. Thus, for example, there is a roughly inverse linear relationship between the molar volume of a saturated hydrocarbon and its solubility. The aqueous solubilities of aliphatic hydrocarbons decrease sharply with increasing chain length, from about 5×10^{-4} M for *n*-pentane to 2.5×10^{-5} M for *n*-heptane and 1×10^{-7} M for *n*-dodecane (McAuliffe, 1966). It appears as if the formation of structured cavities having water molecules of sufficient size to accommodate hydrocarbons having more than a very few carbon atoms is energetically quite unfavorable.

Aromatic hydrocarbons also decrease in solubility with size, although they are far more soluble (approximately two orders of magnitude on a molar basis) than aliphatic hydrocarbons having the same number of carbon atoms (Mackay and Shiu, 1977; Pearlman et al., 1984). The solubilities of benzene, toluene, and naphthalene are, respectively, 1750 mg/L (0.022 M), 550 mg/L (6×10^{-3} M), and 32 mg/L (2.5×10^{-5} M). Even a rather large aromatic compound such as phenanthrene ($C_{14}H_{10}$) has a solubility of nearly 1μM. It is probable that there are significant charge-transfer interactions between water molecules and the pi electron clouds of aromatic species.

Striking increases in water solubility result when heteroatoms are introduced into hydrocarbon chains, due to the ability of the substituent to take part in hydrogen bonding and assume a position in the structural framework of the liquid. The solubility of *n*-hexane is only about 1×10^{-4} M, but that of di-*n*-propyl ether is about 2.5×10^{-2} M, and triethylamine is miscible with water. Introduction of polar or charged functional groups with exchangeable or ionizable hydrogens, such as -OH, -NH-, $-SO_3^-$, or $-COO^-$/-COOH, causes dramatic changes in solubility behavior; hydrophobic interactions become less important than the short-range, stronger interactions between the polar groups and water molecules, unless the polar group is contained in a molecule having extensive nonpolar regions (such as a long-chain alcohol). As an example, a short-chain alcohol such as *n*-butanol may participate in water clusters by forming up to three hydrogen bonds, using the hydrogen atom and two electron pairs on oxygen. The nonpolar end of the molecule is probably surrounded by a clathrate-like region similar to the solvation cages of *n*-alkanes. By analogy to hydrocarbons, aromatic -OH compounds (phenols) are much more soluble than the corresponding linear alcohols; the C_7 aliphatic alcohol *n*-heptanol has a solubility of 8×10^{-3} M, and the cresols (H_3C-Ph-OH) around 0.2 M.

As more OH groups are introduced into a molecule, its solubility characteristics depend to some extent on how well it fits into the framework of the structured component of water. It has been noted that the spacings between oxygen atoms of many very water-soluble compounds correspond rather closely to the first and second nearest-neighbor distances of the ice lattice (Berendsen, 1975). Seemingly minor changes in the stereochemistry of polyfunctional compounds often, however, lead to changes in solubility that are not well understood. These differences have been studied for some cyclic polyols and monosaccharides. For example, the oxygen atoms of β-D-glucose (**17**), which are all equatorially disposed, fit very well into an ice lattice; the water-water hydrogen bonds of the lattice are replaced by glucose-water bonds, and the solubility of glucose is extremely high (90 g/100 mL H_2O); but mannose (**18**), in which the 3-OH is axial, has a solubility of 250 g/100 mL, and β-D-galactose (**19**), with a 5-axial OH, has a solubility of 58 g/100 mL (Tait et al., 1973).

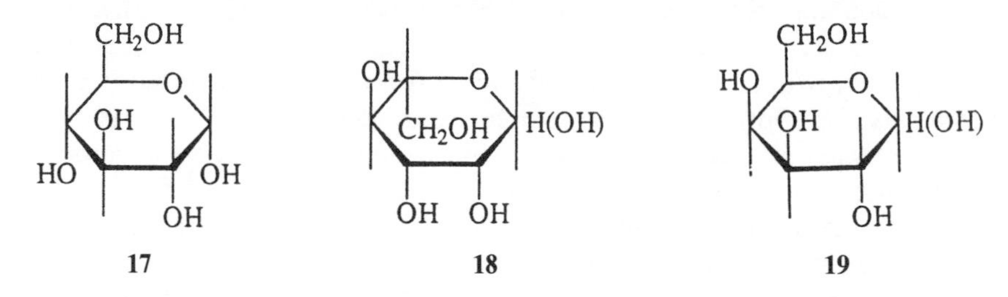

Many naturally occurring substances as well as compounds of anthropogenic origin are amphiphilic; that is, contain a highly polar region and a nonpolar region. Fatty acids and synthetic detergents are straightforward examples of this class of compounds, but polymeric dissolved organic species such as humic substances (see Sections 1.B.1f and 1.B.3a) also display these characteristics (Wershaw, 1986; Yonebayashi and Hattori, 1987). At very low concentrations, these substances may exist in true solution, but as their concentrations increase, interactions among their hydrophobic portions become dominant, and they produce oligomeric aggregates which eventually form particles of colloidal dimensions, called micelles (Figure 1.7).

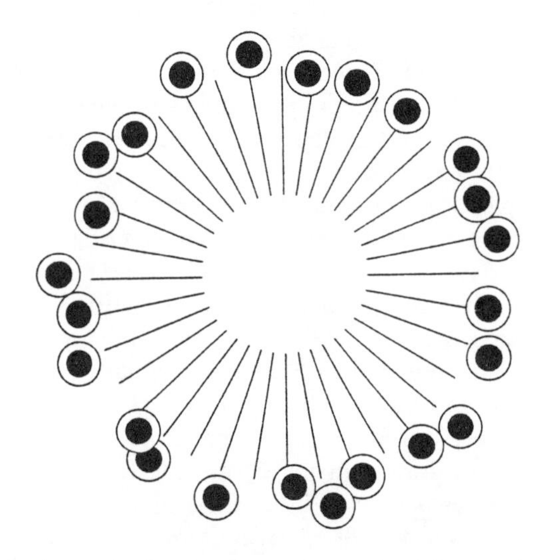

Figure 1.7. An idealized micelle, showing polar, hydrophilic head groups (circles) and hydrophobic chains (lines).

Humic macromolecules may not form true micelles, but because they appear to include distinct hydrophobic and hydrophilic regions (Wershaw, 1986), they might be classified as "self-micelles." As in the case of completely apolar solutes, the formation of micelles is favored by the reduced volume of water cavity space required to surround the micelle relative to its individual molecules, but in addition, the polar regions of amphiphiles are oriented at the surface of the micelle and provide additional solubilizing capacity through ionic interactions, especially if they are charged.

The aggregates of many surface-active compounds are capable of enhancing the solubility of nonpolar organic compounds by bringing them into association with the hydrophobic portion of the amphiphilic cluster (Kile and Chiou, 1989). Humic aggregates and micelles, in particular, may incorporate nonpolar materials such as hydrocarbons, a property that gives them the ability to act as soaps and detergents and transport hydrophobic substances into the aqueous phase (Carter and Suffet, 1982). In the environment, this property becomes important in spills of materials such as petroleum; some of the insoluble hydrocarbons can become incorporated into naturally occurring micelles. For example, dissolved organic matter from seawater was shown to enhance the solubility of *n*-alkanes, but not branched or aromatic hydrocarbons (Boehm and Quinn, 1973). Other investigators, however, have found that the solubility of aromatic hydrocarbons was increased by some samples of humic material (Gauthier et al., 1987); humic substances with higher aromatic carbon content were generally more efficient at binding pyrene. It is possible that dissolved organic matter may enhance solubility of some, especially aromatic, hydrocarbons and quinones by charge-transfer interactions (Kress and Ziechmann, 1977; Melcer et al., 1987). The solubilities of hydrophobic compounds such as chlorinated pesticides have also been reported by many investigators to be enhanced in the presence of humic and fulvic acids (see, e. g., Chiou et al., 1986), although the enhancements vary greatly, depending on the source and chemical characteristics of the humic material (Chiou et al., 1987).

Amphiphilic compounds are also surface-active; their differently polarized regions cause them to accumulate at interfacial zones in the environment. For example, at the air-water interface, amphiphiles tend to orient themselves in "surface microlayers" or "surface films" (see Section 1.B.2d), where the polar region of the molecule is associated with the water phase and the nonpolar region is forced out of solution and extends up into the air phase. Often these surface layers are visible by the damping effect they exert on wave action (Figure 1.8); they are apparent as smoother patches among the ripples on a lake or in the ocean.

b. Marine Waters and Estuaries

The oceans. The volume of water contained in the world's oceans (about 1.4×10^9 km^3, or 1.4×10^{21} liters) defies the imagination. A generalized cross-section of the ocean (Figure 1.9) illustrates that only a small fraction of its water occurs near the surface; approximately half of it is more than 4 km deep. The range of hydrostatic pressure (1–1000 bars) in the seas is significant because at high pressures water no

Figure 1.8. Aerial photograph of a marine surface layer (to right of picture) in the Gulf of Mexico near the Mississippi delta. Small square objects are oil drilling platforms. From *Oceans from Space*, by Badgley, Miloy, and Childs. Copyright 1969 by Gulf Publishing Company, Houston, TX. Used with permission. All rights reserved.

longer behaves as an ordinary fluid; it becomes supercritical. Under high pressure, volume effects, which can be neglected near the surface, provide a driving force (especially for reactions that produce or consume gases) that can alter the rates and equilibria in chemical reactions.

The oceans of the world by no means have uniform physical or chemical properties. There are major differences in near-surface salinity and temperature gradients among the major oceans, which depend on local meteorological conditions, ocean current regimes, and nearness to land or ice masses. However, a combination of tides, currents, waves, and upwellings tend to keep the waters of the oceans more or less well mixed, so that on the whole the ratios of the major dissolved ionic constituents of seawater are nearly constant (Table 1.8). To a crude approximation, seawater is a solution 0.5 M in NaCl and 0.04 M in $MgSO_4$. The mean pH of the ocean is about 8.2.

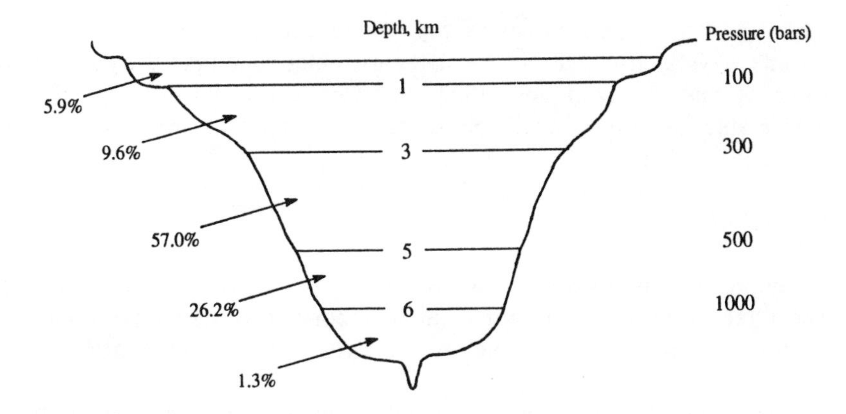

Figure 1.9. Generalized cross-section of the oceans.

Near shore, the interaction between the ocean and coastal rivers results in the formation of fresh-salt water interfaces, known as **estuaries**. Depending on coastal geology, they can assume various forms, such as fjords, drowned river valleys, or bar-built estuaries, characterized by barrier islands. Tidal effects, storms, and the mixing, or lack thereof, of salt water and freshwater strongly influence the biology

Table 1.8. Principal Ionic Constituents of Sea Water

Ion	Concentration (g/kg)	Molarity
Cl^-	19.4	0.55
Na^+	10.8	0.47
SO_4^{2-}	2.7	0.03
Mg^{2+}	1.3	0.05
Ca^{2+}	0.4	0.01
K^+	0.4	0.01
HCO_3^-	0.14	0.002
Br^-	0.07	0.001
BO_3^{3-}	0.024	4×10^{-4}
CO_3^{2-}	0.012	2×10^{-4}
Sr^{2+}	0.0088	1×10^{-4}
F^-	0.0013	7×10^{-5}
Al^{3+}	0.0011	4×10^{-5}
NO_3^-	0.0012	2×10^{-5}
Li^+	0.00014	2×10^{-5}
PO_4^{3-}	0.0014	1.5×10^{-5}

and chemistry of these transitional regions. The chemistry of estuaries is complex and in some ways intermediate between the extremes exemplified by marine and freshwater systems. For example, salting-out effects can result in the precipitation of a considerable fraction of the dissolved organic matter of a stream when it meets saline water in an estuary.

c. Lakes and Rivers

Lakes and rivers combined make up less than 0.01% of the global water supply. However, these waters are disproportionately important to the human population, as well as being ecologically valuable resources for innumerable animal and plant species.

Standing (*lentic*) waters include lakes, ponds, bogs, swamps, and many impoundments. There is no sharp gradation between lentic and *lotic* (running) waters, in which the presence of a unidirectional current, leading to more or less rapid turnover of the mass of water at a given point, is characteristic.

Freshwater is typically much more dilute in terms of inorganic solutes than marine water (Table 1.9). The total quantity of dissolved salts may be as little as 1 mg/L, but the average freshwater body contains approximately 120 mg/L (Gibbs, 1972). The sum (in terms of chemical equivalents) of inorganic cations normally exceeds the sum of anions, due to the occurrence of dissolved organic matter, whose anions are mostly carboxyl groups (see Section 1.C.3c).

Table 1.9. Mean Inorganic Composition of the World's Rivers

	mg/L	M	meq/L
Anions			
HCO_3^-	55.9	9.2×10^{-4}	0.92
SO_4^{2-}	10.6	1.1×10^{-4}	0.22
Cl^-	8.1	2.3×10^{-4}	0.23
NO_3^-	0.84	1.4×10^{-5}	0.01
			1.38
Cations			
Ca^{2+}	15.0	3.8×10^{-4}	0.76
Mg^{2+}	3.9	1.6×10^{-4}	0.32
Na^+	6.9	3.0×10^{-4}	0.30
K^+	2.1	5.4×10^{-5}	0.05
Fe^{3+}	0.41	7.3×10^{-6}	0.02
			1.45
Neutral			
SiO_2	13.1	2.2×10^{-4}	

The pH of freshwater may vary from ca. 0, equivalent to 1 M HCl, in acid mine drainages to as high as 12.3 in very productive and actively photosynthesizing lakes (Schütte and Ellsworth, 1954). However, 80% of the natural surface waters of the United States (and, presumably, elsewhere) have pHs between 6.0 and 8.4 (American Chemical Society, 1978). Some poorly buffered water bodies, however, are becoming increasingly acidic due to acid deposition, a phenomenon caused by combustion of sulfur- and nitrogen-containing fuels and transport of the sulfur and nitrogen oxides downwind.

Lakes. Lakes may be broadly categorized as *eutrophic* (nutrient-rich) and *oligotrophic* (nutrient-poor); eutrophic lakes are generally geologically older, higher in concentrations of inorganic salts, shallower, and higher in suspended solids and plankton (and thus less transparent) than oligotrophic lakes. The variable physical and chemical properties in lake waters profoundly affect their chemistry and biology. Lakes tend to evolve quite rapidly, in geological terms, after their formation. Most small lakes, in fact, fill in and disappear relatively quickly (a few tens of millennia is typical) through the process of **eutrophication**, defined as the enrichment of a lake by the provision of nutrients from the watershed and from the decomposition of aquatic plants and animals. As an oligotrophic lake becomes more nutrient-rich, bigger and bigger populations of algae and consumer organisms develop, die, and decompose, with the result that the lake basin becomes shallower. Concomitantly, plant communities invading from the land cause the shoreline to move closer to the center of the lake. Eventually, after the lake becomes choked with vegetation and detritus, it becomes a swamp or bog. The process of eutrophication can be greatly accelerated by human activity such as industrialization of a lake watershed or the construction of homes on its banks.

Chemical and biological processes in lakes, particularly redox phenomena, may be strongly affected by **stratification**. This is a process characteristic of deep (> 10 m) lakes in temperate climates; that is, with alternating warm and cold seasons. As such a lake warms in the spring, a surface layer of warm water (the *epilimnion*; *epi*, upon) develops, which becomes well-mixed and well-aerated by wind. Because warm water is less dense, it floats on a colder, deoxygenated layer, the *hypolimnion* (*hypo*, below), which extends to the bottom of the lake. (Normally, winds are not strong enough to mix a deep lake to homogeneity.) These layers are stable as long as the air temperature is high enough to keep the epilimnion a few degrees warmer than the hypolimnion. Because algae grow at or near the surface of the lake, they also contribute oxygen to the hypolimnion, with the results that dissolved oxygen concentrations will constantly be near saturation and that most chemical species will tend to be in their oxidized forms (CO_2, NO_3^-, SO_4^{2-}, etc.). The hypolimnion, by contrast, is a largely anoxic and cold zone in which most species are in their reduced forms (CH_4, NH_3, H_2S, etc.)

As the air temperature declines in the autumn, the temperature of the epilimnion also drops until it approaches that of the hypolimnion. At this point, there is no longer a density barrier between the two layers sufficient to maintain stratification. The entire lake tends to behave as a single hydrologic entity and becomes completely mixed over a short period of time. This phenomenon, known as the **overturn**, results in water with a nearly uniform temperature and oxygen concentration, and persists until the spring.

Streams. In limnological jargon, any flowing water is a *stream*. Streams form wherever precipitation exceeds evaporation. This simple notion was not recognized until the 18th century; before that it was usually assumed that large underground reservoirs were the source of stream flow. The English astronomer, Halley, made the first careful measurements of the rates of precipitation, evaporation, and discharge for the major rivers of the Mediterranean watershed, and concluded that the volume of water falling on the basin was more than sufficient to account for the total observed flow.

In the United States, the average annual precipitation is about 76 cm. About 53 cm of this amount is returned to the atmosphere by direct evaporation or evapotranspiration by plant tissues. Although part of the remaining 23 cm goes into groundwater recharge, most of it is used to maintain river flow (Leopold et al., 1964).

Streams are tremendously variable in size. A small mountain spring may emit only a few mL of water per second, but if the river system that it is part of were to be followed to its end, a discharge of hundreds of m^3/sec would not be unexpected. It has been difficult to agree on a simple classification scheme for streams; sometimes the concept of *stream order* is used, in which a stream is codified by the nature of its tributaries. A first-order stream has no tributaries; a second-order stream has first-order tributaries; and so on. The Mississippi River is a tenth-order stream in this system (Strahler, 1957). The useful concept of a "river continuum," (Vannote et al., 1980), although primarily a biological hypothesis, is also helpful when thinking about chemical processes in streams. The continuum idea views a stream as part of a whole river system from origins to mouth and sets out the philosophy that what happens in headwater streams may have important consequences for the larger downstream segments. It also views streams as integrally connected to their watersheds, so that changing the nature of a watershed, for example, by clearcutting it, may profoundly affect not only the stream that happens to immediately drain it, but the entire system.

d. The Air-Water Interface: The Surface Microlayer

Liquids exhibit a surface tension that is proportional to the degree of interaction between the molecules of the liquid. The surface tension of pure water (0.073 newtons/m) is, owing to hydrogen bonding, the highest of all pure liquids (with the exception of molten salts). Dissolved inorganic salts actually increase the surface tension of water somewhat, so that in absolutely clean seawater the value increases to 0.075 newtons/m. Organic liquids such as benzene, heptane, etc., where intermolecular interactions are minimal, display surface tensions in the .020 to .050 range. Most organic solutes in water decrease its surface tension markedly, and natural water samples ordinarily exhibit much lower surface tensions. A large number of studies over the past 35 years have demonstrated that the air-water interface is a zone of substantially enriched organic carbon concentration. In many instances, the organic material at the surface forms a virtually continuous film, or microlayer.

Two types of organic films can form on the surface of a body of water. Highly insoluble molecules whose densities are lower than that of water, such as aliphatic hydrocarbons, can simply form an overlying phase that does not significantly interact with the lower aqueous layer. Films of this sort are exemplified by petroleum oil spills, at least in their early stages before extensive oxidative weathering has developed.

Surface-active or amphiphilic molecules, in contrast to insoluble compounds, interact with the water phase. There are essentially two types of film-forming, surface-active compounds; **dry** surfactants, in which the major portion of the molecule extends out of the water (Figure 1.10.A), and **wet** surfactants, in which the bulk of the molecule is within the water phase and only small portions of the structure project into the air (MacIntyre, 1974: Figure 1.10.B). Synthetic detergents (**20**), long-chain aliphatic compounds with a polar benzenesulfonate "tail," are typical dry surfactants, whereas proteins, with many hydrophilic regions and only a limited

$$CH_3(CH_2)_nSO_3^{\ominus}$$

20

number of hydrophobic sites, epitomize the wet type. Dry surfactants could potentially form true monolayers of molecular dimensions (<20 Å), but wet surfactants, being polymeric, tend to form thicker films on the order of 1μM.

Visible slicks of surface-active material may occur on both marine and freshwater. Hutchinson (1957) noted that lakes sometimes develop oily patches ("taches d'huile") where there has been no history of human contamination. Circulation of water masses by the wind (Eisenreich et al., 1978) or by thermal gradients (Lewis, 1974) can cause slicks or foam masses to occur as surface-borne organic matter is skimmed off and concentrated at the margins of upwelling or downwelling regions.

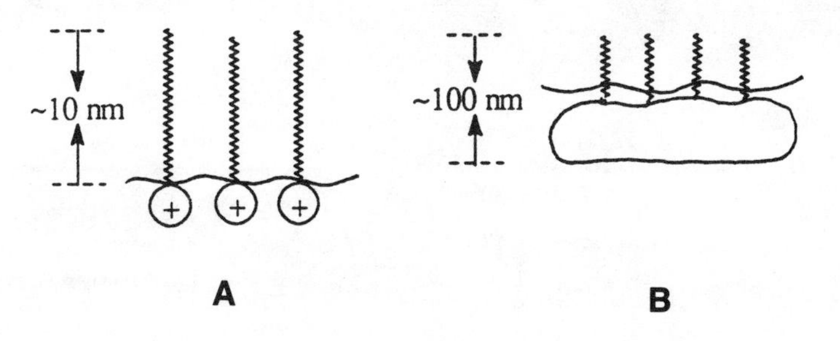

Figure 1.10. (A) A "dry" surfactant film, showing the hydrophilic head-groups dissolved in the aqueous phase and the hydrophobic chains extending into the gas phase. (B) A "wet" surfactant, showing the bulk of the molecule dissolved in water and a few hydrophobic moieties in the gas phase.

Very large masses of foam can be generated in organic-rich freshwaters flowing over dams or sills; a dramatic example from the Okefenokee Swamp, Georgia, is shown in Figure 1.11. Here, the outflow from a sill is further agitated by power boats; foam masses more than a meter high may form.

The occurrence of visible foam masses on freshwaters is usually blamed by onlookers on detergent discharges, but it often is due to natural organic material. Foaming in New England streams, for example, is prevalent during spring, when sap flows in watershed trees are at their peak. The formation of these foams is presumed to be due to saponins (steroidal glycosides) or related surface-active natural materials (Pojasek and Zajicek, 1978). In more southerly mid-Atlantic and midwestern U.S. states, however, foaming is most noticeable in the autumn leaf fall period.

Marine natural slicks are very often due to wax esters, storage lipids of many abundant zooplankton organisms. These substances are esters of long-chain (C_{18}–C_{36}) acids with alcohols of similar chain length. In some instances the compounds account for more than 75% of the weight of the animals (Sargent, 1978). When large populations suddenly die off in adverse conditions, immense slicks can result. In 1974, for example, a red slick covering more than 1,500 square kilometers of the

Figure 1.11. Foam masses in the Okefenokee Swamp, showing macroscopic, visible evidence of surface-active DOM. The water is agitated by passing over a sill and further mixed by boat engines. From an original photograph by Alan C. Graham. Reprinted by permission.

Pacific Ocean was reported; lipid made up 55% of the slick, and 80% of the total lipid was wax esters. The red color of the slick was caused by the carotenoid pigment, astaxanthin, also a constituent of the zooplankton (Lee and Williams, 1974). The same compounds are the principal constituents of a solid waxy material that accumulates during cold winters on the shores of Bute Inlet, a British Columbia fjord.

There have been many investigations of the composition of freshwater and marine surface microlayers, using a variety of collecting devices based on different principles. These samplers collect surface layers of varying thicknesses. It is not surprising that many different types of compounds have been identified, nor that there is great disagreement over the degrees of enrichment for given compound classes in the surface region.

Early work on surface film composition focused on the easily analyzable lipid fraction, particularly the fatty acids and their derivatives. Both free and combined fatty acids are usually found, with the even-carbon saturated acids from C_{12} to C_{22} usually predominant (Garrett, 1967; Zsolnay, 1977; Ehrhardt et al., 1980). In unpolluted environments, C_{15} and C_{17} branched-chain acids, characteristic of some bacteria and blue-green algae, are occasionally present. Oxidative reactions at this interface reduce the concentration of unsaturated fatty acids relative to their abundance in their source materials, the lipids of bacteria and algae, which are notable for their high levels of polyunsaturated fatty acids. Unsaturated fatty acids are unstable to solar ultraviolet radiation and decompose by complex mechanisms involving reactive oxygen species (see Chapters 4 and 6) to produce a variety of oxidation products including peroxides, alcohols, ketones, epoxides, hydrocarbons, polymers, and even some poorly characterized water-soluble compounds (Baker and Wilson, 1966).

The lipid fraction of many surface films, however, may make up only a minor fraction of the total organic carbon present. Proteinaceous and carbohydrate-containing material is abundant in samples taken from both marine and freshwater locations. For example, Baier et al. (1974), using internal reflectance infrared spectroscopy and other techniques, proposed that glycoprotein-like materials predominated in lake and ocean surface samples. Analyses of amino acids and carbohydrates isolated from near-shore marine microlayer samples indicated that more than half of the particulate organic matter consisted of hydrolyzable polypeptides and polysaccharides; the soluble organic matter was similar to that of subsurface waters, consisting largely of uncharacterized, probably humic, material (Henrichs and Williams, 1985). Wind-generated foams on Lake Mendota, Wisconsin, also were shown to contain roughly 40% protein (Eisenreich et al., 1978). A large fraction of the organic matter also appears to consist of humic polymers; absorption and infrared spectra, as well as polarographic responses, of freshwater surface microlayer samples were characteristic of humic or fulvic acids (Eisenreich et al., 1978; Pellenbarg, 1978; Cosovic et al., 1985). Enrichment of phenolic material has also been demonstrated to occur in the surface microlayer of coastal marine waters off the coast of Maine (Carlson and Mayer, 1980). The enrichments were shown to be greater in more highly saline waters, consistent with a "salting-out" effect of polyphenolic material of freshwater origin.

Many hydrophobic pollutants will tend to become more concentrated in surface layers, both those of the natural type and those caused by discharges of petroleum oil or other industrial fluids. In the first instance, only a limited fraction of the pollutant load is likely to occur in the microlayer because of its thinness and fragility; however, in chronic or acute spillage events a more important fraction may be concentrated at the surface. The presence of a coherent, organic-rich film at the surface of a water body may promote organic reactions by a number of mechanisms. For example, it may act as a hydrophobic, concentrating phase into which potential reactants may diffuse; or it may promote oxidation reactions because of the higher solubility of molecular oxygen in organic phases than in water.

Disturbance of the natural surface microlayer by chronic or adventitious spills of petroleum has become an unfortunately common occurrence in the last 30 years. In areas with heavy ship traffic, as much as 85% of the surface organic matter may be petroleum hydrocarbons (Morris, 1974). Both crude and refined oils have been repeatedly spilled in large accidents. Crude petroleum is a bewilderingly complex mixture of hydrocarbons, phenols, carboxylic acids, nitrogen and sulfur heterocycles, and other constituents. During the refining process, some classes of compounds may be almost entirely removed, but others may be produced; in addition, additives such as antioxidants may be introduced.

The fates of spilled petroleum are still not well understood despite two decades of investigation. Petroleum spilled on water undergoes a series of complex changes which, taken together, usually result in its relatively rapid disappearance from the surface. For example, rapid initial spreading of most oils to a thin film on the surface leads to a surprisingly rapid loss of its volatile constituents to the atmosphere by evaporation; certainly within a few days, at most, virtually all of the even slightly volatile hydrocarbons that have not been removed by other processes will evaporate (Stiver and Mackay, 1984; Stiver et al., 1989).

Emulsification of oil with water is promoted by wave action; the usual form is the "water-in-oil" type, or chocolate mousse, which consists of water droplets surrounded by an envelope of high-boiling hydrocarbons or more or less surface-active heterocycles. Mousse often sinks when it absorbs a sufficient quantity of particulate inorganic material to increase its density to greater than that of water. If surface-active material such as a detergent is deliberately added to the oil, a second type of emulsion, namely oil-in-water, may result. In this case, the nonpolar tails of the detergent molecules become incorporated into the oil and their polar head groups dissolve in the aqueous phase. The result is an emulsion that is readily dispersed into the water column.

Light-induced weathering of petroleum films can also lead to major changes in the types of organic compounds present. These reactions have been reviewed by Payne and Phillips (1985). Some hydrocarbons are readily converted by light and air to much more soluble oxygen-containing species such as phenols, ketones, acids, and hydroperoxides (Larson et al., 1979; see also Section 6.D.3e).

e. Groundwater

Subsurface water has been an important resource for human populations throughout recorded history. At present, about half of the drinking water in the United States is groundwater, as well as roughly a third of all water used for irrigation.

Groundwater-containing formations are usually porous rocks like sandstone or unconsolidated materials like gravels, sands, or subsurface soils. Permanent groundwater occurs in trapped zones referred to as **aquifers** (Figure 1.12), in which a layer of impervious material such as bedrock or fine-textured clay prevents further downward passage of water that has infiltrated into the soil. In such a situation, an equilibrium develops between input and output (downslope migration, evaporation, withdrawal, etc.), resulting in a sort of underground lake whose approximately planar surface is called the **water table**. The top of the water table is defined as that point where the pressure of groundwater is equal to atmospheric pressure and the soil is referred to as **saturated**. Between the soil surface and the top of the water table is a region of variable soil moisture called the **vadose** or **unsaturated** zone. Permanent streams and lakes occur where the water table intersects the land surface at a valley slope or canyon wall.

Since it is not exposed to the atmosphere, most groundwaters are low in dissolved oxygen, and accordingly ions and naturally occurring organic compounds tend to be in a reduced state. The populations of microorganisms in groundwater, formerly thought to be negligibly small, have been discovered to be large and abundantly active (Matthess, 1982). Methane, a product of the metabolic activities of microorganisms, is commonly found in subsurface waters, especially in deep wells. In an extensive survey of groundwater total organic (TOC) concentrations in 27 U.S. states, Leenheer et al. (1973) reported that the median concentration was 0.7 mg

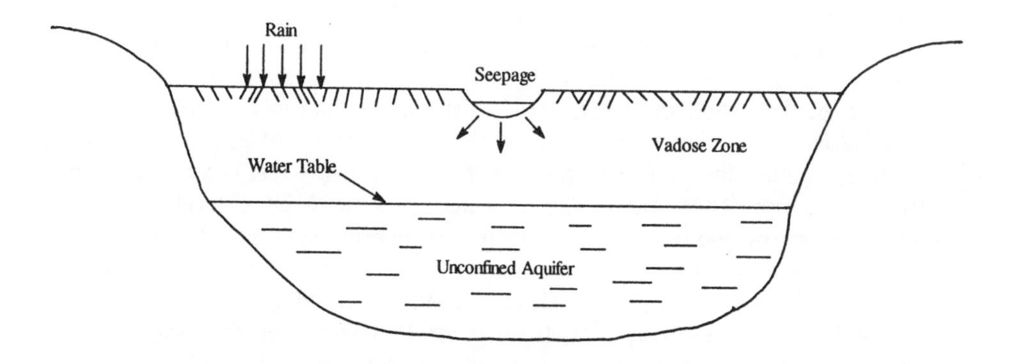

Figure 1.12. Groundwater. Water seeping through surface soil gathers below the surface in regions bounded by impermeable rock or clay layers. Adapted from David Todd, *Groundwater Hydrology*, John Wiley & Sons, New York, 1980. Reprinted by permission.

C/L; 85% of the 100 samples tested contained less than 2 mg/L. An extreme value from a south Florida limestone aquifer was 15 mg/L. This aquifer was apparently hydrologically connected to organic-rich surface water.

Potentially hazardous urban and industrial chemical wastes disposed of on land, as in landfilling, may percolate into and contaminate subsurface water. The passage of organic matter through the soil and into groundwater represents especially acute problems of groundwater pollution. Once contaminated, groundwater is very difficult or impossible to reclaim. Reported groundwater contaminants have included hydrocarbons, surfactants, insecticides and herbicides, and other halogenated compounds such as chlorinated ethylenes (Zoeteman et al., 1980; McCarty et al., 1981; Matthess, 1982).

In landfills, the presence of a compacted clay liner has been thought to be an adequate barrier for the transport of leachates out of the impounded area; however, organic solvents and petroleum products, which are typical constituents of landfills that have accepted industrial and chemical processing wastes, have been shown to permeate clays at much faster rates than do aqueous solutions (Brown and Thomas, 1984).

f. Organic Matter in Aquatic Environments

All natural waters contain some carbonaceous material. Even rainwater has been shown to include about 1 mg C/L (Thurman, 1985). Some of this organic material is in the form of "truly dissolved," soluble compounds, whereas other fractions consist of colloidal or particulate organic matter. The definitions of dissolved and particulate carbon that have been more or less established by convention among aquatic scientists are somewhat elastic and arbitrary; one criterion that is widely used sets the boundary between the two at 0.45 μM (the pore size of some common filter media). These demarcations tend to disguise the fact that there is a continuum of organic particles in water that ranges in relative size from molecules to microorganisms.

Among surface waters, mid-ocean marine samples contain the least dissolved organic carbon, or DOC; less than 1 mg C/L. (A related acronym, DOM, is used to refer to dissolved organic matter; the basic difference is that DOC is usually derived from an ultimate elemental analysis for carbon, whereas DOM normally alludes to the soluble organic molecules, which of course contain hydrogen, oxygen, etc. There is accordingly more DOM than DOC in water.) "Average" DOC levels for marine samples (near-shore and open-water) are on the order of 1 mg C/L (Williams, 1971), whereas freshwaters normally contain higher amounts with a mean in the range of 5–6 mg C/L (Moeller et al., 1979; Meybeck, 1981; Mulholland and Watts, 1982). (High values for both marine and freshwaters that are three or more times the "average" level have been reported, but most waters are probably within a factor of 2 or so of the "average." An exception needs to be made for waterlogged locales, such as bogs and swamps that derive most of their carbon from the continuous leaching of recently shed plant litter, exemplified by the Okefenokee Swamp in the southeastern U.S. and related environments. These waters typically contain approximately 20–50 mg DOC/L and are quite acidic (pH 2 to 4.5; Given, 1975; McKnight et al., 1985). Groundwater (Leenheer et al., 1973) has a median concentration of about 0.7 mg DOC/L. Polluted water organic content varies widely, but can be very high, as indicated by a DOC value of 300 mg/L for typical untreated sewage (Painter and Viney, 1959).

Small molecules. Although the bulk of the DOC in practically all natural surface waters is polymeric, approximately 10% or so consists of small, "identifiable" compounds. It is probable that in the open ocean the excretion of cellular metabolites from living and dead planktonic animals, plants, and bacteria whose home is the ocean contributes the great majority (> 99%) of dissolved organic matter; nearer to shore, or in landlocked bodies of water, sources that are land-derived (soil, leachates of land plants, etc.) become more predominant.

The largest fraction of identifiable dissolved organic compounds, in both marine and freshwaters, is comprised of carbohydrates. Polymeric forms (starch, cellulose, etc.) are invariably predominant over monomeric sugars and disaccharides (Johnson and Sieburth, 1977; Mopper et al., 1980; Sweet and Perdue, 1982). Several classic investigations on natural waters have clearly demonstrated that concentrations of most monosaccharides and amino acids are kept low by the highly active uptake of these nutrients by aquatic bacteria, fungi, and other heterotrophic organisms (Wright and Hobbie, 1965; Andrews and Williams, 1971). However, the flux of these smaller molecules may be very high and, therefore, over a long period of time they may account for most of the energy cycling in aquatic environments (Bada and Lee, 1977). In marine systems, the probable source of most of the carbohydrate is extracellular excretion by phytoplankton. It has been known for a long time that marine algae release some portion of their photosynthetically fixed carbon into the surrounding water, particularly during periods of rapid growth and high light intensities. Hellebust (1965), for example, reported that 22 species of algae excreted, on average, about 5% of the carbon they fixed; the released material included polysaccharides, polyols, lipids, proteins, and glycolic acid.

Individual free amino acids, also, exist only in low concentrations (seldom exceeding 10 μg/L for an individual compound or 50 μg/L for all free amino acids) in natural waters (for some of the many articles on the subject see, e.g., Riley and Segar [1970, seawater] and Lytle and Perdue, [1981, freshwater]); it has long been recognized that combined amino acids (in proteins, peptides, and mixed polymers such as humic-bound forms) greatly dominate the aquatic nitrogen budget (Bada and Lee, 1977). Scully et al. (1989) showed that the protein content of a lake was highly variable, increasing by a factor of ca. 10 during algal bloom periods, up to a maximum concentration of approximately 3 mg/L.

Carbonyl compounds, carboxylic acids, and esters of a wide variety of types have been isolated from surface waters. Aldehydes and ketones are produced during the photolysis of larger polymers such as humic materials; formaldehyde, acetaldehyde, and glyoxal appear to be the most abundant in seawater samples (Kleber and Mopper, 1990). Low molecular weight mono-, di-, and tricarboxylic acids such as acetic, formic, lactic, glycolic, malic, and citric acids have been reported by many investigators from marine and freshwaters. Glycolic acid is an especially important excretion product of many phytoplankton species (Hellebust, 1965; Wright and Shah, 1977). Longer-chain fatty acids (and their esters, such as triglycerides) tend to partition into the air-water interface (see Section 1.B.2d).

Hydrocarbons may enter the aquatic environment via *de novo* synthesis, accidental oil spillages, runoff from the land, leaching from sediments, or atmospheric fallout. Concentrations of total hydrocarbons in marine surface waters vary from 0.1 to 75 μg/L, with the majority of samples from unpolluted environments having values toward the lower end of the range. Alkanes (branched, cyclic, and linear), as well as many aromatic compounds, commonly occur (Zsolnay, 1977).

Aquatic humic materials. The humic substances of the soil will be described in Section 1.B.3d. These materials are polymeric, ill-defined substances which originate from the breakdown of plant litter via oxidation and condensation reactions between polyphenols, polysaccharides and polyamino acids of plant and microbial origin. Significant amounts of these substances of terrestrial origin have also been detected in many natural waters, especially freshwaters. Marine samples also contain "humic-like" materials, but they may differ significantly in origin, structure, and properties from inland humic substances. Various estimates concerning the fraction of DOC that can be considered humic in nature seem to center around 50%. Leenheer and Noyes (1984) have described a mobile sequential column-adsorption technique for separating aquatic DOM into "hydrophobic acid," (essentially, humic material), hydrophilic acid, basic, and neutral fractions.

The idea that there is a connection between soil humus and the colored organic matter in inland waters is of relatively recent lineage, dating back to an article by Wilson (1959) in which he stated that colored DOM was the same as the fulvic acid (see Section 1.B.3d) fraction of soil organic matter. Black and Christman (1963a, 1963b) were probably the first to investigate these materials using a variety of modern techniques including absorption and fluorescence spectroscopy. Aquatic humic materials, at least those of freshwaters, do appear to have many features in common with soil humic substances, particularly their fulvic acid fractions. Some dark-colored freshwaters, however, contain a significant fraction (approximately 10%) of materials with humic acid characteristics (Martin and Pierce, 1971; McKnight et al., 1985). Details of the characteristic physical and chemical properties of humic compounds will be deferred to Section 1.B.3d, and only an overview of their attributes will be given here, along with information that is specific to the aquatic environment.

Marine polymeric DOC has often been given the special name of "Gelbstoff" (Ger., "yellow material," after the name given it by Kalle; cf. his review of the subject in 1966); it was first recognized in the 1930s as the basis for the color of seawater. On the basis of a cursory similarity to the humic substances of soils and freshwaters, it has been called "marine humus," although it may actually bear practically no molecular resemblance to the other polymers of this type. In an early comparison of the two classes of material, Sieburth and Jensen (1968) concentrated both marine and freshwater Gelbstoff on nylon columns, eluted them with NaOH, and compared the eluates by two-dimensional paper chromatography and UV spectroscopy, finding significant differences. Other workers have also shown similarities in isotopic enrichment (^{13}C and ^{2}H) between the carbon and hydrogen of Gelbstoff and that of marine plankton, and differences in comparison to terrestrial carbon and hydrogen (Nissenbaum and Kaplan, 1972; Nissenbaum, 1974). Ertel and Hedges (1983) summarized several differences in the chemical and spectroscopic characteristics of humic compounds isolated from marine sediments and those of their terrestrial analogs:

1. The marine humic acids were higher in hydrogen and nitrogen concentrations.
2. The marine humic acids had fewer highly conjugated or aromatic structural units.
3. Marine polymers had greater levels of unchanged or slightly altered carbohydrate, protein, and pigment molecules.

NMR studies of marine polymeric DOC have also shown it to differ from terrestrial and freshwater humic material, as described below.

It has long been recognized that the humic material of oceanic water samples is more abundant in near-shore environments, where both terrestrial inputs and in situ production would be greatest. Gardner and Menzel (1974), Hedges and Parker (1976), and Ertel et al. (1984) have obtained evidence for the largely indigenous origin of marine polymeric material in studies of organic constituents of water and sediment samples from rivers or lakes and also from the adjacent continental shelf. Aromatic aldehydes obtained by nitrobenzene or CuO oxidation of sedimentary lignin were used as tracers for terrestrially derived organic matter. Organic carbon from sediment samples of inland locations contained up to 2% of a mixture of vanillin (**21**), *p*-hydroxybenzaldehyde (**8**), and syringaldehyde (**22**). These concentrations dropped rapidly toward the estuary mouth and were very much lower ($< 0.1\%$) in marine sediments. The data are consistent with the origin of the terrestrial humic materials from land plants, which contain lignin, a polyphenolic polymer that would be a plausible precursor for the phenolic aldehydes observed; marine algae do not contain lignin.

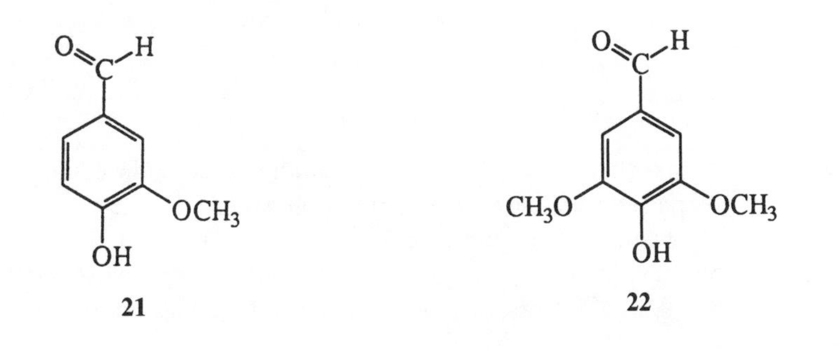

Nitrogen content in aquatic humic materials usually ranges from 0.5 to 2%, although in seawater DOM it has been reported to be as high as 6.5% (Thurman and Malcolm, 1989). Little is known about the nature of the combined nitrogenous structural types that are present, although hydrolytic studies have revealed some bound amino acids (Lytle and Perdue, 1981; Thurman and Malcolm, 1989).

The NMR spectra of aquatic humic materials show many of the characteristics also observed in their soil-derived counterparts (Figures 1.13 to 1.16; cf. Section 1.B.3d). As in the case of soil humic substances, both ^{1}H and ^{13}C NMR spectra have been obtained for a wide variety of samples. In general, ^{13}C NMR spectra are more informative, with cleaner and sharper signals that lend themselves to less equivocal interpretations. Furthermore, proton NMR spectra usually corroborate quite closely the inferences revealed by analyses of ^{13}C spectra.

Four major ^{13}C bands at ca. 30 ppm (aliphatic C), ca. 80 ppm (C bound to oxygen), ca. 130 ppm (aromatic C) and ca. 170 ppm (carboxyl C) dominate typical spectra. In addition, a small peak at around 200 ppm occurs in many aquatic humus samples; it has been assigned to carbonyl carbon. Leenheer et al. (1987) compared the ^{13}C NMR spectra of unaltered and chemically modified aquatic humic materials to suggest that aromatic α-ketones of the acetophenone type were the most abundant types of carbonyl structures in these polymers.

By use of spin echo techniques and pulse enhancement methods, Buddrus et al. (1989) estimated that for a groundwater-derived humic material, most of the aliphatic carbon was of the tertiary R_3CH type relative to methylene and methyl carbons. Methyl groups were found in surprisingly low amounts. In a ^{13}C-NMR study of dissolved organic matter from sediment pore waters, Orem and Hatcher (1987) proposed that two different forms of DOM, "carbohydrate-paraffinic" and "aromatic-paraffinic," predominated, depending on the redox properties of the sedimentary environment and the source of the sedimentary organic detritus. In sediments that were aerobic, DOM with decreased carbohydrate and enhanced aromatic

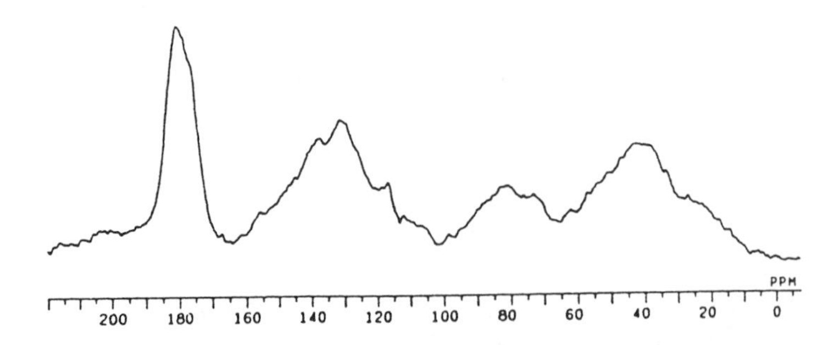

Figure 1.13. Spin-echo 100-MHz ^{13}C NMR spectrum of a groundwater fulvic acid. Reprinted with permission from Buddrus et al. "Quantitation of Partial Structures of Aquatic Humic Substances by One- and Two-Dimensional Solution ^{13}C Nuclear Magnetic Resonance Spectroscopy," *Anal. Chem.* 61, 628–631. Copyright 1989, American Chemical Society.

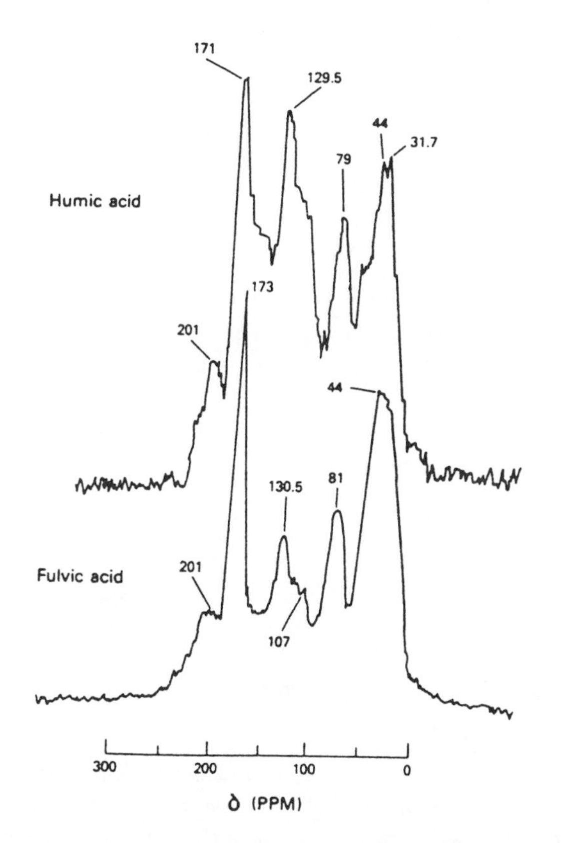

Figure 1.14. Solid-state 22.6 MHz Fourier-transform ^{13}C NMR spectrum of Suwanee River standard humic acid. Reprinted with permission from *Organic Geochemistry*, Vol. 11, Leenheer et al. "Presence and Potential Significance of Aromatic-Ketone Groups in Aquatic Humic Substances," Copyright 1987, Pergamon Press PLC.

Figure 1.15. Solid-state 22.6 MHz Fourier-transform ^{13}C NMR spectrum of Suwanee River standard fulvic acid. Reprinted with permission from *Organic Geochemistry*, Vol. 11, Leenheer et al. "Presence and Potential Significance of Aromatic-Ketone Groups in Aquatic Humic Substances," Copyright 1987, Pergamon Press PLC.

character was suggested to prevail, whereas in anoxic sediments the carbohydrate type was the principal form present.

Gillam and Wilson (1983) and Harvey et al. (1984) have performed model studies directed toward the synthesis of marine humic material from its assumed precursor compounds, metabolites from planktonic algae. Gillam and Wilson extracted lipid-soluble materials from a large quantity of the diatom *Phaeodactylum tricornutum* and compared the "humic" fraction of the extract to a "humic" material from coastal

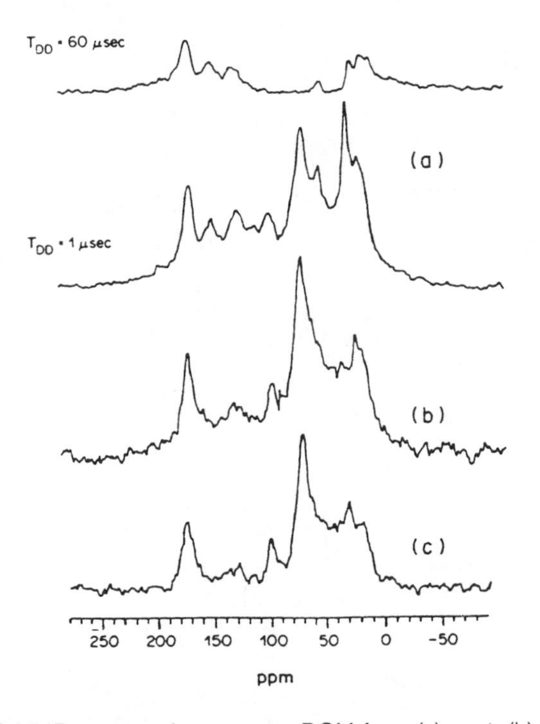

Figure 1.16. ^{13}C NMR spectra of pore water DOM from (a) peat, (b) Great Bay sediment, and (c) Mangrove Bay sapropel. Reprinted with permission from *Organic Geochemistry*, Vol. 11, Orem and Hatcher. "Solid State ^{13}C NMR Studies of Dissolved Organic Matter in Pore Waters from Different Depositional Environments." Copyright 1987, Pergamon Press PLC.

seawater. Both extracellular and intracellular fractions from the algae showed close similarities to the ^{13}C-NMR spectrum of the aqueous material (Figure 1.17). Harvey et al. allowed pure unsaturated lipids or freeze-dried cells of a marine diatom, *Skeletonema costatum,* to autooxidize in aerated, illuminated seawater; they likewise obtained pale yellow products whose ^{1}H–NMR, IR, and fluorescence spectra closely resembled that of a marine "fulvic acid." They hypothesized that marine humic substances were water-soluble, largely aliphatic acids that were formed by crosslinking and autooxidation of unsaturated lipids in seawater. A contrary suggestion that marine fulvic material may be derived predominantly from polysaccharides (Steelink et al., 1983) has received less support in the literature.

The fluorescence spectra of many freshwater humic materials have been found to be surprisingly similar to one another (Larson and Rockwell, 1980; Table 1.10). The fluorescence does not resemble that of other naturally occurring polymers, such as lignin, nor does it show the mirror-image absorption/fluorescence relationship characteristic of pure compounds. This suggests that the active fluorophores contribute only a small fraction of the light absorption (Visser, 1983). The quantum efficiency

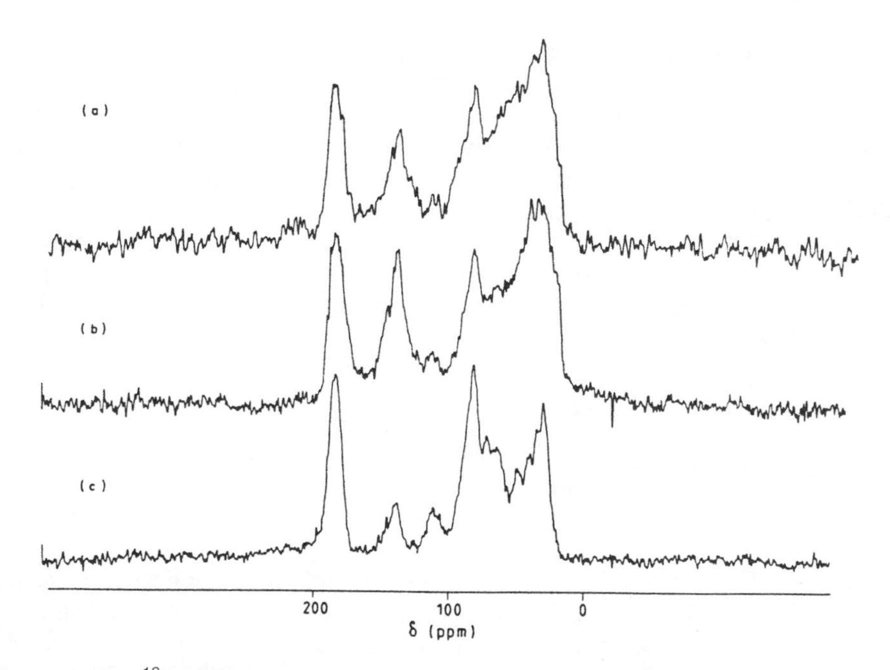

Figure 1.17. ^{13}C NMR spectra of (a) coastal seawater humic material; (b) extract of extracellular metabolites of the marine diatom, *Phaeodactylum cornutum*; (c) alkaline extract of cellular material from *P. cornutum*. From Gillam and Wilson (1983). Reprinted by permission of Lewis Publishers.

Table 1.10. Fluorescence of Freshwater Dissolved Organic Materials

Source	λ max Excitation	λ max Emission	Ref.
"Colored waters" (USA)	365	490	Black & Christman, 1963
"Colored waters" (USA)	361	460	Christman & Ghassemi, 1966
Lakes & rivers (USA)	360–370	450–460	Ghassemi & Christman, 1968
Lake (Netherlands)	365	470	de Haan (1972)
Lake (USA)	346–427	426–510	Hall & Lee (1974)
River (Sweden)	—	420–430	Almgren, et al. 1975
Feedlot runoff (Canada)	360	460	Lakshman, 1975
Various waters (UK)	340–350	410–460	Smart et al., 1976
Moor (Germany)	365	470	Müller-Wegener, 1977
River (USA)	375	465	Larson, 1978
Spring seep (USA)	344	432	Larson & Rockwell, 1980
River (Georgia)	350	455	Goldberg & Weiner, 1989
Rivers and lakes (Quebec)	330–340	410–440	Visser 1983
River (Florida)	350	450	Zepp & Schlotzhauer, 1981

of fluorescence was low and variable, with the highest yield of eleven substances tested being about 1% (Zepp and Schlotzhauer, 1981). The nature of the fluorophores is speculative, although it was shown that caffeic acid (**23**), a compound typical of degraded lignins, was converted to the highly fluorescent coumarin, esculetin (**24**) on standing in alkaline solution or upon exposure to light. Coumarins have fluorescence excitation maxima at about 363 nm and emission maxima at 447 nm, very similar to those reported for humic materials (Larson and Rockwell, 1980). Other fluorophores, such as Maillard or browning-reaction products, and the "age pigments" derived from oxidized fatty acid products and amines

(Chio et al., 1969) are possibilities, based on their published fluorescence characteristics. Goldberg and Weiner (1989) proposed that the fulvic acid from the Suwanee River, Georgia, contained two distinct fluorophores, as shown by phase-resolved fluorescence emission spectroscopy. The fluorophores had calculated lifetimes in water of about 1 and 6.5 nsec, respectively.

3. Solid Phases

a. Soil Structure

Soils have formed gradually over the eons; first, primordial rocks were fractured into particles by geological and meteorological forces; later, plants, animals, and microorganisms contributed their activities and additives. Physical effects such as freezing and thawing in rock fissures, or fragmentation by the abrasive action of ice or flowing water, account for most of the disintegration of large rocks into smaller debris. Further decomposition occurs when the exposed mineral faces are attacked by water, particularly if it contains sufficient dissolved CO_2 to be appreciably acidic; this leads to (relatively) rapid dissolution of some constituents, hydrolytic fragmentation of other linkages, and secondary reactions between the products. Soils may vary widely in pH; some aluminum-rich or tropical soils reach values as low as 2, whereas other carbonate-rich soils high in sodium attain pHs above 10 (Wolfe et al., 1990).

Soil stratification. The layers of a typical soil, which make up a "soil profile," are illustrated in a general way in Figure 1.18. It is clear that the upper layers of the profile are enriched in organic matter in various stages of decomposition. The topmost horizon, 01, consists of relatively unaltered organic litter; in the 02 horizon,

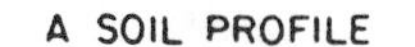

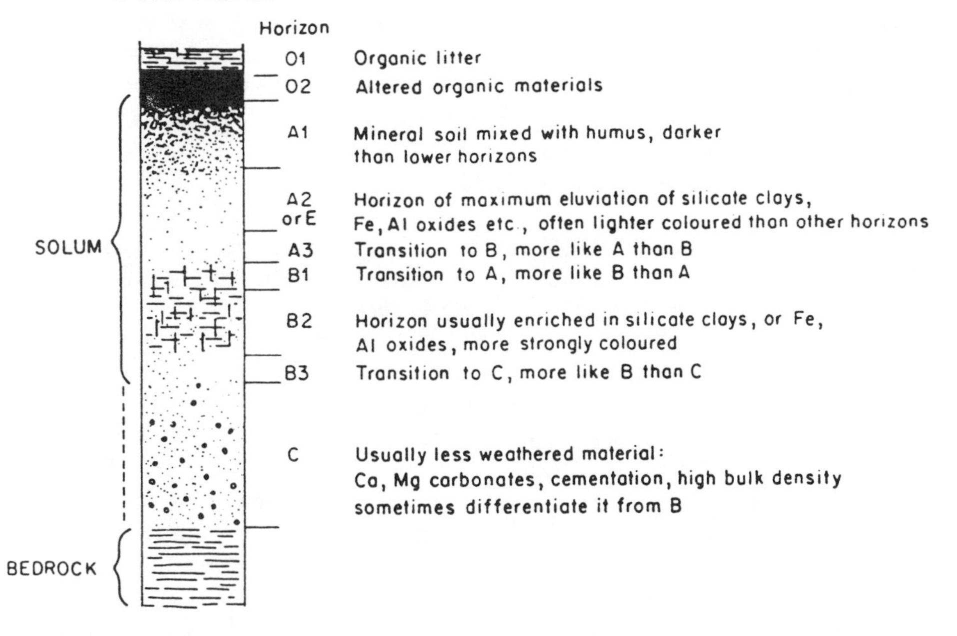

Figure 1.18. A soil profile, illustrating the traditional O, A, B, and C horizons and some further subdivisions of them. From D. J. Greenland and M. H. B. Hayes, Eds., *The Chemistry of Soil Constituents*, John Wiley & Sons, 1978. Reprinted by permission.

dark brown to black organic matter in a more or less advanced stage of decay predominates. Below these layers is a region of transition between the organic-rich upper horizons and the organic-poor lower horizons. This A1 region, roughly equivalent to what is usually called topsoil, is typically dark gray in color and contains a mixture of mineral and organic particles. Below these regions are the B horizon, where many of the constituents that have leached from the upper layers accumulate, and the C horizon, which consists of the underlying bedrock or other parent material for the soil.

It should be recognized that soils are extremely variable and that the example given is a general one. The thicknesses and even the existence of any of the horizons and sub-horizons mentioned can vary greatly depending on source material, topography, biological activity, pH, temperature, and rainfall patterns, drainage characteristics, etc. In fact, soil taxonomy is a thriving discipline; more than 10,000 "soil series" (analogous in some ways to biological species) have been described in the United States alone (Buol et al., 1973).

Clays. Probably the most abundant constituents of most soils are the clay minerals, a mixture of crystalline aluminosilicates of rather variable composition (Dixon and

Weed, 1977). These minerals typically take the form of extremely fine particles, often with diameters of less than 2 μM, and having a high relative surface area. Clays are highly organized, often forming stacked layers of parallel planes made up of silica tetrahedra and alumina octahedra. The ability of the planar arrays to move relative to one another, especially in the presence of water, accounts for the slipperiness and pliability of these materials.

In some clays, aluminum and phosphorus can partially replace silicon in the silicate layers, and magnesium, iron, and zinc can substitute for aluminum in the octahedral layer (Laszlo, 1987). The presence of higher-valent metals such as silicon with lower-valent atoms such as aluminum results in a net negative charge, almost always observed in these minerals. To achieve electrical neutrality, cations migrate from solution to the clay surface or to the interlayer spaces. In soils or waters with few alkali metal or alkaline earth ions, the cations may be protons, making it possible for the surface of the clay to be strongly acidic (Skujins, 1967). Kaolinite in a yeast culture experiment has been shown to have a surface pH of about 2 units lower (or a $[H^+]$ of 100 times higher) than the surrounding medium (McLaren, 1957). Such a difference can strongly affect not only the rate of certain chemical reactions, such as hydrolyses, but also enhance or inhibit the ionization of many compounds such as phenols or anilines. In addition to proton-donating (Brønsted) acidity, clays may also possess Lewis acid (electron-accepting) sites, by virtue of their content of reducible transition metal cations such as Fe^{3+}, Cu^{2+}, etc. Lewis acid sites, however, also attract water molecules, and may be deactivated in high-moisture environments such as sediment-water interfaces (Soma et al., 1983) For synthetic or catalytic applications, it is possible to exchange the clay's inherent cations for almost any desired cation; by this means clays can be made into "solid acids," made to bear catalytic metal ions for redox reactions, or made lipophilic by treating them with quaternary ammonium salts having long aliphatic chains.

Interactions between clay minerals and organic compounds have been reviewed by Theng (1974) and Lagaly (1984). The surface chemistry of these minerals (and of their associated impurities such as hydrous oxides of transition metals), governs many of the chemical reactions and equilibria (organic as well as inorganic) which occur in soils. Clay minerals can interact with organic compounds by adsorption, intercalation (in which the molecules enter the interlayer space and deform the silicate layers), and ion-exchange processes. Depending on the ionic composition of the clay, redox, displacement, and other reactions can be observed. These types of reactions will be discussed in the appropriate sections.

Other inorganic constituents. Many soils, especially in dry climates, are rich in forms of calcium carbonate such as limestone; some soils may contain over 50% $CaCO_3$. This mineral is quite alkaline and contributes to elevated pH values which may exceed 8.0. Metal oxides in varying amorphous and crystalline forms may also be important constituents of some soils; oxisols, for example, may contain as much as 80% Fe_2O_3 (Wolfe et al., 1990). Metal oxides may interact with clays and also with organic constituents of soils.

Water in soils may exist either in the spaces between soil particles (displacing air from the soil), within the pores of the soil particles, or adsorbed to the surfaces of the particles. The moisture content of the soil is constantly in flux due to changes in temperature, humidity, and precipitation levels. At high water content, a separate liquid phase of the soil ("soil solution") is present, an environment of relatively high solute concentrations. Reactions that are sensitive to acid-base catalysis, salt effects, or organic matter concentrations may be strongly promoted or retarded in such environments. Even the mere evaporation of soil water significantly concentrates dissolved substances such as pesticides, affecting the kinetics of their reactions. Due to differential evaporation and extraction of soil constituents, the composition of the soil solution as well as its pH and Eh (see Section 4.A) may change significantly with depth.

A chemically important fraction of the soil is its air space; about 30% of the volume of a typical soil consists of voids. Although air penetrates readily into well-drained soils, it does not exchange rapidly with the atmosphere, and may be enriched in CO_2 and depleted of O_2 by the respiratory activity of soil organisms. Because of this phenomenon, the acidity of water within soils may be significantly higher than in nearby environments.

b. Aquatic Sediments

Solid phases which are more or less permanently overlain by water are referred to as sediments. Sediments are normally made up of heterogeneous mixtures of organic and inorganic particles of a wide range of sizes. The constituents of the sediment are variously derived from weathering of primordial rock, terrestrial runoff, and detritus derived from minerals and living organisms in the water column. Water percolates around and into the sediment grains, leading to virtually complete saturation of these particles. At the sediment-water interface, microbiological activity is usually intense due to the enrichment of this site with organic matter that descends from the water column.

The pH values for sediments are affected by biological activity, the nature of the sedimentary mineral components, and the ionic strength of the water in the water column; however, they are usually rather close to that of the overlying water. In addition, depending on the redox characteristics of the interface (Eh value), either oxidative or reductive pathways can be favored. In a study of the biological transformation of cellulose, chitin, and protein at a lake bottom-water interface, for example, Brewer and Pfaender (1979) showed that cellulose was more resistant to degradation than the other polymers; cellulose is quite susceptible to oxidative degradation and was apparently stabilized in this particular sediment.

Because sediments are so high in organic matter, they tend to act as a sink or trap for hydrophobic compounds, such as many typical environmental pollutants, which partition strongly into the organic-rich region from the water column. For example, petroleum hydrocarbons are quite common constituents of polluted sediments (Hargrave and Phillips, 1975; Shaw and Baker, 1978). Once an organic compound enters the sedimentary phase, it may find itself in a highly reducing environment; then, depending on its molecular characteristics, it may either be preserved or rapidly reduced, as mentioned above.

The water held inside the sediment particles is referred to as pore water. These waters contain much higher DOC levels than the overlying waters. For example, in nearshore marine sediments where the seawater DOC concentration is about 1 mg/L, pore waters have been found to contain as much as 400 mg/L (Lyons et al., 1979). It is likely that most of this DOC is produced by the activities of microorganisms such as bacteria that are intimately associated in organic metabolism at the particle-water interface (Barcelona, 1980). Although polymeric DOC such as humic substances prevails in these waters, as it does in other natural waters, there are indications from the literature that a somewhat larger fraction (on the order of 20–30%) of the total DOC is made up of small molecules such as carbohydrates (Mopper et al., 1980), carboxylic acids (Barcelona, 1980), and individual amino acids (Henrichs and Farrington, 1979; Gardner and Hanson, 1979).

c. *Soil Organic Matter*

Almost all soils are enriched with organic matter only near their surfaces. Plant litter and the activities of most large organisms occur only at the surface, and the great majority of all organic matter is deposited there. By the time one descends a few meters below the surface of the earth, the fraction of organic matter normally falls to much less than 1%; at the surface it can be as high as 6% or even more.

Small molecules. Almost all the organic matter in soils is either polymeric or strongly bound to soil particles. Most readily soluble molecules would rapidly be leached or degraded by soil microorganisms, and only a very small fraction of such molecules are present at equilibrium. Nevertheless, because trace constituents of simple structure are readily identified, a fairly large number of small molecules have been reported from soils.

Sugars and polysaccharides in the soil may be synthesized by microorganisms or may derive from plant litter (Cheshire et al., 1973). About 10% of the organic matter in soil is carbohydrate (Flaig, 1971), and virtually all of it is polymeric, but some free monosaccharides have been identified; glucose, galactose, fructose, xylose, arabinose and ribose, for example, were detected in cold water extracts of a Norwegian pine-forest soil. (Grov, 1963). Isotopic labeling studies by Cheshire (1977) implied that most of the hexose sugars in soil originated through microbial synthesis. However, the pentose sugars in soil polysaccharides appeared to be almost entirely derived from plant residues or gums. Sugar phosphates, particularly phytic acid (inositol hexaphosphate, **25**) are common in many soils; as much as 50% of the organic P in soil may be esterified to sugar alcohols. Several different stereoisomers of inositol phosphates have been identified from various soils. It is probable that several of the isomers are microbially synthesized; the remainder may be of plant origin (Caldwell and Black, 1958; Martin and Wicken, 1966).

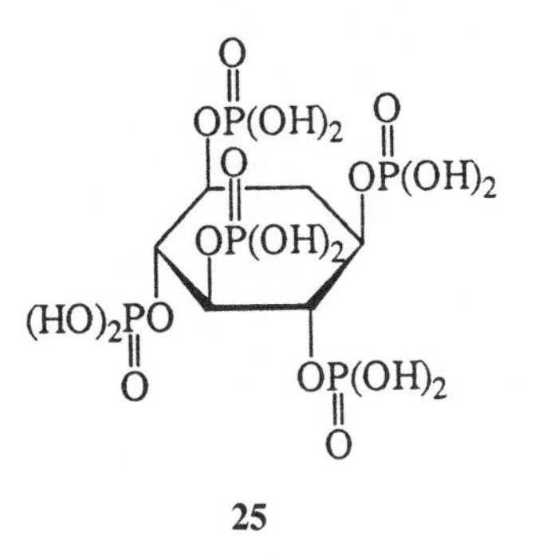

25

The degradation of lignin and other plant polyphenols releases relatively large quantities of free phenolic acids to the soil. These compounds are potentially important because of their pronounced antifungal and phytotoxic activity. The five most common phenolic acids, all related to the lignins of grasses, conifers and deciduous trees, are *p*-hydroxybenzoic acid, vanillic acid (**26**), ferulic acid (**27**), syringic acid (**28**), and *p*-coumaric acid (**29**). Quantitative studies are rare, but the most abundant

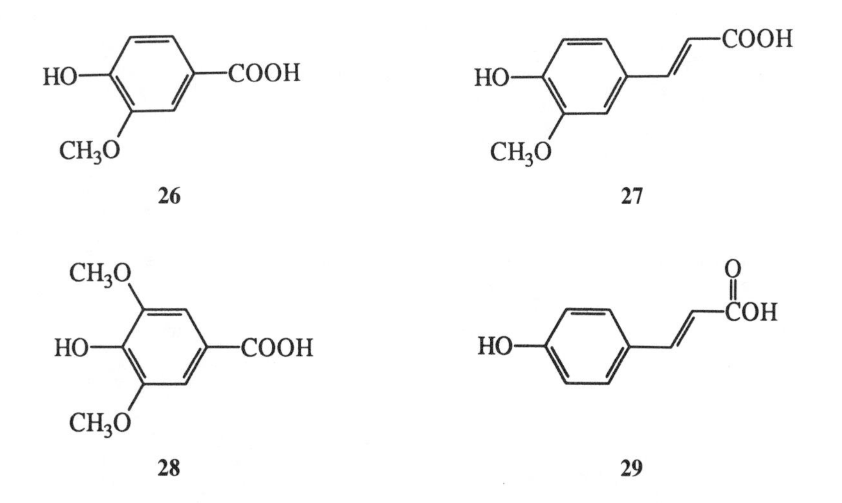

26 **27** **28** **29**

acids (usually *p*-coumaric and *p*-hydroxybenzoic acids) have been reported at about 10 ppm (ca. 1×10^{-5} M) in soil. Formic and acetic acids are the most abundant aliphatic acids in nearly all soils so far examined (Stevenson, 1969; Flaig, 1971). For example, in 10 soils studied by Batistic (1974), acetic acid constituted an average of 57% of all the volatile (<6 C atoms) acids present; formic acid contributed another 36%. Few simple esters have been isolated from soil, but one of the more interesting classes of compounds that has been reported to date is a group of *omega*-feruloyloxy acids (**30**), comprising a major fraction of the polar lipids of an acidic *Sphagnum* peat from eastern France (Riess-Kautt et al., 1988). These compounds are plausibly derived from partially decomposed suberin (see structure **37**).

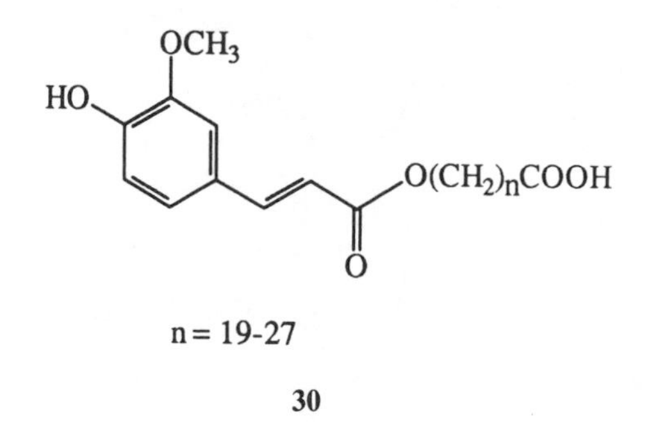

n= 19-27

30

There have been occasional reports of simple aldehydes in soils, such as methylglyoxal (**31**: Enders and Sigurdsson, 1947), vanillin (**21**), *p*-hydroxybenzaldehyde (**8**), and cinnamaldehyde (**32**: Terentev et al., 1968). However, most of the stable, simple carbonyl compounds reported from soils so far are quinones (Matsui and Kumada, 1974).

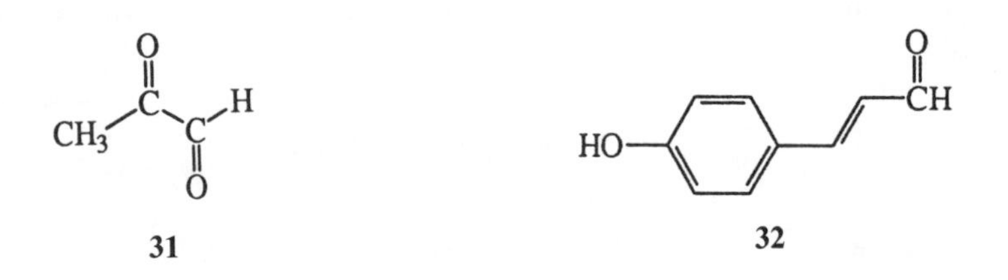

31 **32**

For example, chrysotalunin (**33**), an anthraquinone dimer, is a red pigment which occurs in many soils worldwide (McGrath, 1967; Foo and Tate, 1977; Fujitake

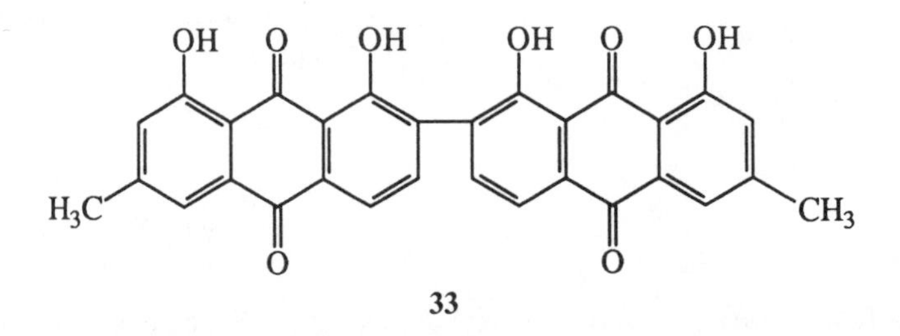

33

et al., 1991). Probably the most abundant anthraquinone derivative in most soils, it may be of fungal origin. So-called “green soils” or “P-type humic acids” have been found in many locations around the world; they also appear to contain dark, variously colored quinonoid pigments having characteristic visible absorption maxima. A minor constituent of a Japanese green soil was shown to be 4,9-dihydroxyperylene-3,10-quinone (**34**: Sato, 1976). Extracts of some Southern Hemisphere green soil contained chloroquinones (Cameron and Sidell, 1976); one purple pigment was assigned the structure **35**.

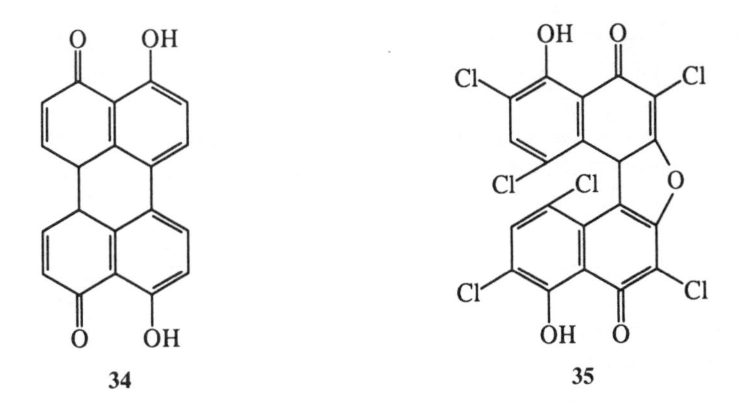

The mode of introduction of chlorine, and even the ultimate source of the material, is unknown.

The organic nitrogen content of soil varies widely, from a few hundredths of a percent to perhaps 3% in organic-rich peat bog soils. Most of the nitrogen appears to be associated with highly polymeric material, but a small fraction is low in molecular weight. In most investigations alanine, glycine, leucine, isoleucine, serine, threonine and lysine are found to be the most abundant amino acids (Flaig, 1971). Amino sugars in combined form make up a significant fraction of the total soil organic nitrogen. Glucosamine and galactosamine are by far the most abundant; their nitrogen may constitute up to 10% or even more of the organic N in some soils (Chen et al., 1977). Both of these hexosamines probably originate from soil organisms.

Humic materials. The so-called "humic substances" are probably the most abundant polymers in nature. The standing crop of humic materials in the soil has been estimated to be approximately $2\text{–}3 \times 10^{12}$ tons (Bazilevitch, 1974). They are brown, acidic materials which bear no immediately apparent resemblance to conceivable precursor materials, and appear to be formed by oxidation and condensation reactions between polyphenols, polysaccharides, and polyamino acids of plant and microbial origin. As Hurst and Burges (1967) pointed out, "This situation is undoubtedly the most complicated and variable existing in nature." Despite nearly two centuries of investigation, environmental chemists still cannot agree on a rational structure for humic substances or on mechanisms for their formation.

Humic materials of soil are intimately bound to its mineral constituents. They are polyelectrolytes, bearing strong negative charges, and are associated with metallic cations such as those found in clays. No single organic solvent is useful for the extraction of more than a small fraction of soil carbon. Water is likewise a poor extractant, although some polymeric material having osmometrically determined mean molecular weights of a few hundreds can be removed. The most commonly used efficient extractant is aqueous alkali; typically soil is shaken under N_2 with 0.5% NaOH, the suspension is centrifuged, and the aqueous layer is decanted. In the traditional workup of soil organic matter extracts, the alkaline fraction is next acidified, usually causing the precipitation of a large fraction of the soluble material. This base-soluble, acid-insoluble humic fraction is called *humic acid*. Organic material remaining in solution after acidification is referred to as *fulvic acid*. (Whether the two fractions differ sufficiently in their chemistry to warrant separate classifying names is doubtful.) Further distinctions are sometimes made; for example, organic carbon which cannot be extracted from soil by alkali has been called *humin*.

No universally accepted technique has materialized that permits a reproducible subfractionation of humic and fulvic acids. Humic acid polymers can often be largely freed of small molecular contaminants, both salts and organic impurities, by dialysis. Fulvic acids, however, appear to consist of smaller molecules or aggregates, and attempts to dialyze them result in large losses of organic carbon to the medium. Adsorption, partition, and ion-exchange chromatography techniques have been employed to some extent. In general, ion-exchange techniques have not been widely used because of the problem of contamination by impurities from the resin, and because it is quite difficult to achieve complete desorption of the humic material from the resin.

Similar difficulties have plagued attempts to fractionate humic materials by adsorption chromatography. Charcoal, alumina, and other adsorbents have been tried with limited success. The most widely used method of fractionation for soil humic materials has been gel permeation chromatography. Although some work has been done using controlled-pore glass and polyacrylamide beads, most investigators have used Sephadex® dextran gels. These materials are available with a wide range of molecular exclusion properties (different degrees of cross-linking) based on pore size; ideally, molecules whose dimensions are smaller than the pore sizes of the gel are held up within the pores and elute from the column later than large molecules which are excluded from the pores. In practice, the method works well for relatively homogeneous mixtures such as closely related proteins or polysaccharides, but mixtures of polydisperse, chemically different substances such as humic materials also exhibit different electrostatic interactions with the dextran gel material. In particular, polyphenolic materials display strong hydrogen bonding and are eluted much later than would otherwise be predicted (Brook and Munday, 1970). Thus, a significant fraction of a humic mixture is likely to display an apparent molecular size much smaller than the "true" value. Fractionation of humic materials by dextran gel filtration may be useful for comparative purposes, but probably not for the assignment of molecular weight distribution without careful consideration of other variables. Many other methods of molecular weight determination have been applied to soil humic preparations, but none has been entirely successful (Wershaw and Aiken, 1985).

Ultimate analyses of many soil humic substances have been reported. The dominant elements are carbon (40–65%) and oxygen (30–50%); a selection of values is given in Table 1.11. In general, humic acids are somewhat higher in carbon and lower in oxygen than fulvic acids. Nitrogen content almost always ranges from 1% to 6%, hydrogen from 3% to 6%. Few determinations of other elements have been

Table 1.11. Mean Elemental Composition of Selected Humic Materials (in %)

	C	H	N	O	S
Soil humic acid	57.3	5.0	2.8	34.4	0.6
Coal humic acid	64.8	4.1	1.2	28.7	1.2
Soil fulvic acid	47.0	4.4	1.5	46.4	0.7
Water fulvic acid	46.2	5.9	2.6	45.3	—

Source: Schnitzer and Khan (1972).

performed; sulfur and phosphorus appear usually to be present to the extent of a few tenths of a percent, at most.

Stevenson and Butler wrote (1969), "A variety of functional groups, including COOH, phenolic OH, enolic OH, quinone, hydroxyquinone, lactone, ether and alcoholic OH, have all been reported in humic substances, but there is considerable disagreement as to the amounts present; in some cases even proof of existence is lacking." Matters have not improved significantly in the intervening couple of decades. There is a paucity of analytical methods which are sufficiently precise and free from interferences for reliable determination of organic functionalities in complex polymers such as humic materials. Published infrared and ultraviolet spectra of humic substances are generally of little or no value for this purpose because of their lack of resolution or fine structure, as would be expected for complex mixtures.

Probably ^{13}C and ^{1}H NMR spectroscopy have provided the most useful information about the general characteristics of the structures of humic substances. After a brief summary of the findings from these spectra, we will turn our attention to other methods, such as functional group analyses and chemical degradation studies, and finally give a selection of type structures that have been proposed by various researchers over the past 30 years.

NMR techniques, when applied to polymers, allow an observer to estimate the number or fraction of protons or carbon atoms in different magnetic environments. This is highly useful, because a carbon attached to an ether oxygen, for example, has a quite different and characteristic NMR spectral position relative to one attached to a carbonyl oxygen or an alkyl carbon. Chemical shift assignments for groups typical of humic samples are listed in Tables 1.12 and 1.13. Thorn (1989) and Snape et al. (1989), however, have pointed out that it is difficult to obtain quantitatively accurate ^{13}C NMR spectra; that is, spectra in which the signal intensity is exactly proportional to the number of nuclei causing the signal, because of the presence of paramagnetic impurities, differential spin-lattice relaxation times, and other problems of molecular and nuclear motion. These problems appear to be particularly acute with aromatic carbon atoms and can lead to underestimates of the fraction of aromatic carbon in a sample. Therefore, extrapolation of ^{13}C NMR data to quantify carbon types in humic material should be done with discretion.

Since the earliest reported NMR investigations of humic materials, it has appeared that aromatic protons and carbon atoms cannot play as important a role in most of their structures as many early hypotheses had suggested. These primeval structures

Table 1.12. ^{1}H-Resonances from Humic Materials

Chemical Shift (δ) ppm	Assignment
0.8→1.0	terminal methyl groups of methylene chains
1.0→1.4	methylene of methylene chains (δ = 1.25); CH_2,CH at least two carbons or further from aromatic rings, or polar functional groups
1.4→1.7	methylene of alicyclic compounds
1.7→2.0	protons of methyl and methylene groups α to aromatic rings; β protons of indanes and tetralins
2.0→3.3	protons of methyl groups and methylene groups α to aromatic rings; protons α to carboxylic acid groups; α protons of indanes and tetralins
3.3→5.0	protons α to carbon attached to oxygen groups; sugars of carbohydrates
5.0→6.5	olefins
6.5→8.1	aromatic protons including quinones, phenols, oxygen containing heteroaromatics aromatics
8.1→9.0	sterically hindered protons of aromatics; nitrogen heteroaromatics; formate
Variable	acidic protons of phenols and carboxylic acids, confirmed by D_2O exchange

were based on polyphenols and highly condensed aromatic ring systems. For example, Lüdemann et al. (1973) showed that ^{1}H–NMR spectra of "synthetic" humic substances prepared by autooxidative polymerization of catechol or hydroquinone, which had long been thought to resemble natural humic substances, contained very much higher concentrations of aromatic protons (64% in the case of catechol-derived polymers; 78% from hydroquinone) than did authentic humic acids extracted from black soil (ca. 20%) or a podzol (44%). The most important proton types in the latter materials were aliphatic protons and hydrogens attached to carbon atoms bound to electron-withdrawing groups (H–C–O and H–C–C=O types, e.g.). High frequency (270 MHz) Fourier transform proton NMR experiments confirmed the relative importance of aliphatic and aromatic proton types in a humic material extracted from the A horizon of a New Zealand soil. Aliphatic protons were revealed by strong signals at 0.8, 1.2, 1.9, and 2.1 ppm. In addition, a strong, broad band around 3.6 ppm was assigned to H–C–O protons; these appeared to constitute 27% of the total (excluding exchangeable) number of protons. Possibly these signals arose from sugar-like constituents or other polyhydroxylated materials. Aromatic protons were present, but only to the extent of ca. 15% (Wilson et al., 1978). Similar

Table 1.13. ^{13}C-Resonances in Humic Materials

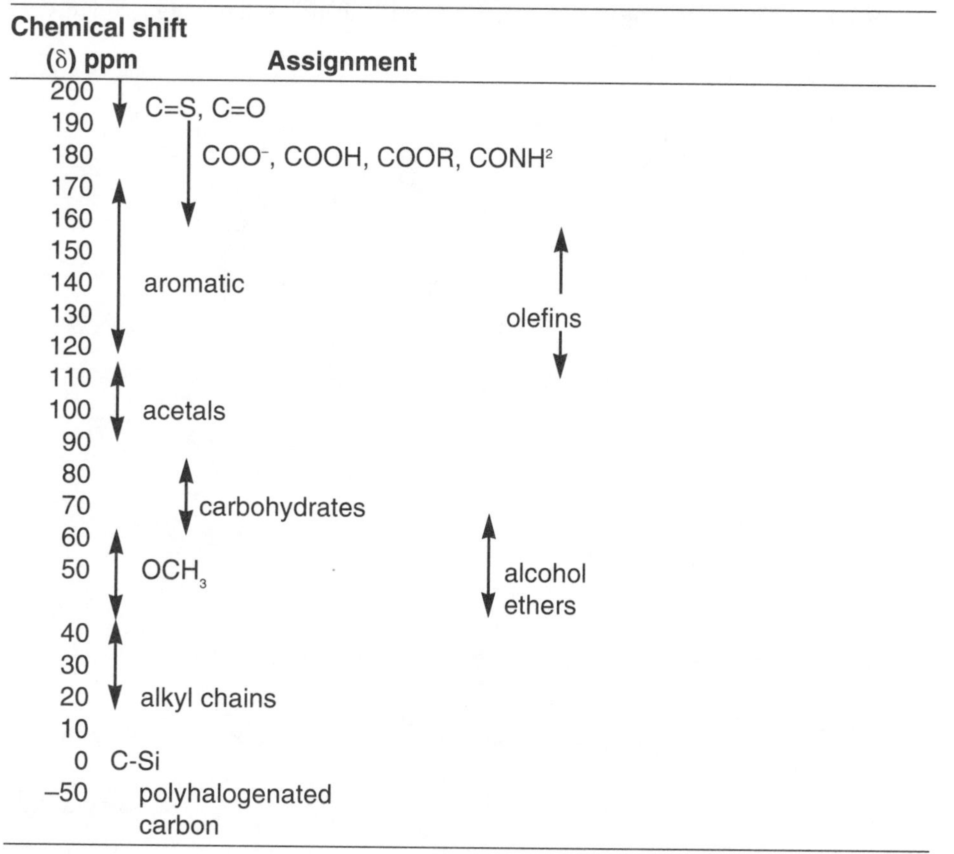

Source: Wilson (1981).

data have been obtained for a variety of other soil humic materials (see, e.g., Wilson et al., 1983). Hatcher et al. (1981a, b) examined a group of humic materials from soils and from terrestrial and aquatic sediments and calculated that in the sedimentary materials, aromatic protons and carbon atoms contributed from 25% to 45% of the total. The soil humic substances had generally higher aromatic contents, around 50%, with extreme values of 70–92% in a few Japanese soils. These authors proposed that terrestrial humic materials contained a mixture of lignin-derived (aromatic) and lipid-derived (aliphatic) structural elements. They suggested that phenolic residues, which are prominent in many older model structures for humic materials, were practically absent as evidenced by the lack of absorption at about 150 ppm in the ^{13}C-NMR spectra of soil humic and fulvic acids.

An interesting hypothesis for the origin of aliphatic structures in humic materials has been put forth by Tegelaar et al. (1989). These authors have suggested that polymers such as cutin (**36**) and suberin (**37**), polyester derivatives that play an important role in protecting external and internal organs of many higher plants, as

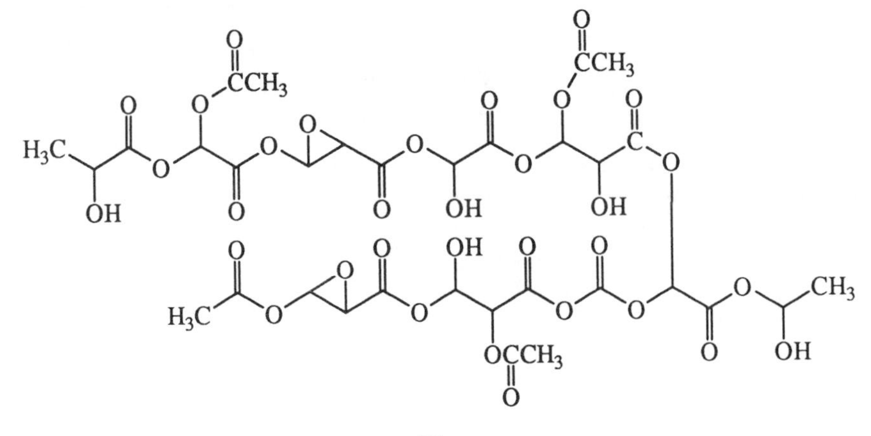

36

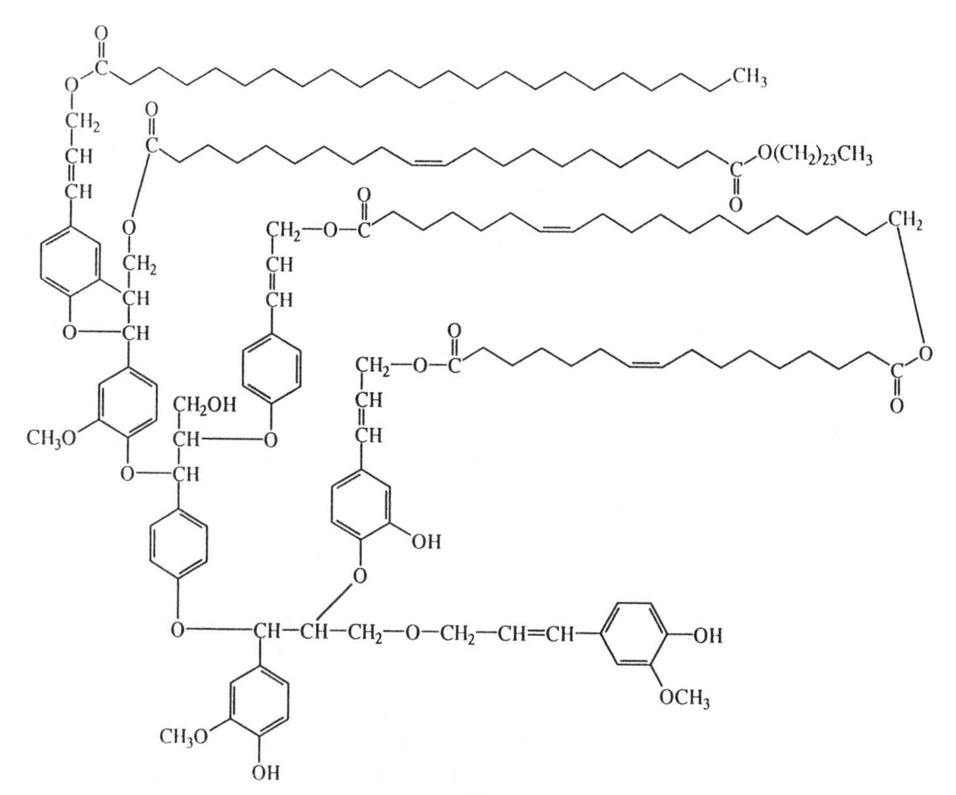

37

well as other poorly characterized aliphatic biopolymers, might be important precursors because of their abundance and resistance to degradation. They also called attention to the occurrence of poorly characterized, highly aliphatic substances characteristic of the cell walls of some algae. Goni and Hedges (1990) demonstrated that copper oxide oxidation products of cutin, long-chain hydroxy fatty acids, were similar to those observed when sediment-derived humic substances were oxidized by the same reagent.

The distribution of OH groups in humic materials has been examined by Mikita et al. (1981). Using ^{13}C-enriched methylating agents, first diazomethane and then sodium hydride–methyl iodide, they synthesized "permethylated" humic and fulvic acid from a soil and a groundwater, respectively (Figure 1.19). The data demonstrate that aliphatic and aromatic carboxyl groups were the most important functional groups

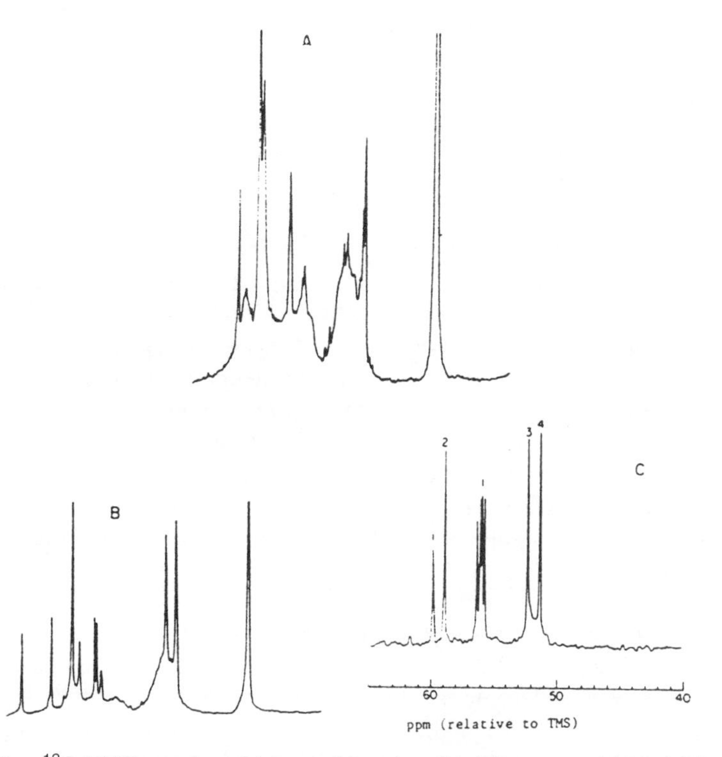

Figure 1.19. ^{13}C NMR spectra of (a) a soil humic acid; (b) a groundwater fulvic acid; (c) a mixture of ^{13}C-methylated reference compounds: (1) pentamethylquercetin; (2) glycerol dimethyl ether; (3) methyl benzoate; (4) methyl pentanoate. Reprinted with permission from Mikita et al., "Carbon-13 Enriched Nuclear Magnetic Resonance Method for the Determination of Hydroxyl Functionality in Humic Substances," *Anal. Chem.* 53, 1715–1717. Copyright 1981, American Chemical Society.

Table 1.14. Estimated Abundance of Hydroxyl Groups as Percentages of Total Hydroxyl Functionality in Two Humic Materials

Chemical Shift of Methoxyl Band, ppm	Assignments	% of Total OCH_3 Area of Spectrum	
		Fulvic Acid[a]	Humic Acid[b]
51.2	Aliphatic carboxyl	15	9
52.5	Aromatic carboxyl	24	20
55–56	Phenol	12	14
57.5	Aliphatic OH	11	9
58–60	Aliphatic or carbohydrate OH	28	38
61–62	Aliphatic or carbohydrate OH	10	10
45.6[c]	Amino nitrogen	20[c]	44[c]

Source: Mikita et al. (1981).
[a]Biscayne aquifer fulvic acid.
[b]Mollisol humic acid.
[c]As a percent of the total OCH_3 peak area.

susceptible to methylation by their procedure in the fulvic acid, with aliphatic OH closely behind (Table 1.14). In the humic acid, aliphatic OH was more abundant than carboxyl OH. In both samples, phenolic OH was of lesser importance. Using a similar procedure, the presence of alkylatable carbon atoms was discovered by Thorn (1989). Apparently, in the hydride–methyl iodide step in the methylation procedure, carbons α to carbonyl groups or phenolic hydroxyls become methylated (Figure 1.20). These signals are seen as broad resonances at about 20–26 ppm.

Whole soils have been examined by ^{13}C NMR using the technique of magic-angle spinning (Barron and Wilson, 1981; Preston and Ripmeester, 1982). The soils studied appeared to contain organic matter low in aromatic carbon and high in aliphatic carbon atoms.

In view of the fact that lignin is often suggested as a precursor for soil humic materials, it is interesting that its ^{13}C NMR spectrum (Figure 1.21) shows little or no alkyl chain signal, a prominent $-OCH_3$ peak at about 56 ppm, intense peaks around 150 ppm that are due to aromatic ether carbon, and strong aromatic signals in the 110–135 ppm range (Bartuska et al., 1980). Accordingly, it is likely that lignin would have to be greatly modified if its carbon atoms were to make up a significant fraction of the structural framework of humic polymers. In fairness, it has been demonstrated that partial lignin decomposition by fungi does lead to great changes in its elemental analysis and other chemical properties (Kirk and Chang, 1974).

Functional groups. The acidic nature of humic polymers has been recognized for many decades; they can readily be titrated with basic reagents such as barium

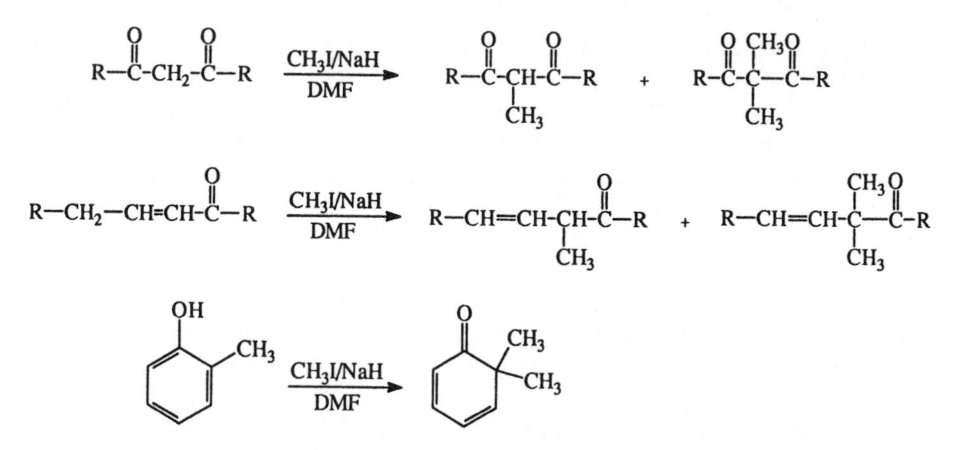

Figure 1.20. Selected C-methylation reactions that occur when humic materials are treated with methyl iodide-sodium hydride. Reprinted from R. C. Averett et al., "Humic Substances in the Suwanee River, Georgia: Interactions, Properties, and Proposed Structures," U.S. Geological Survey report #87–557.

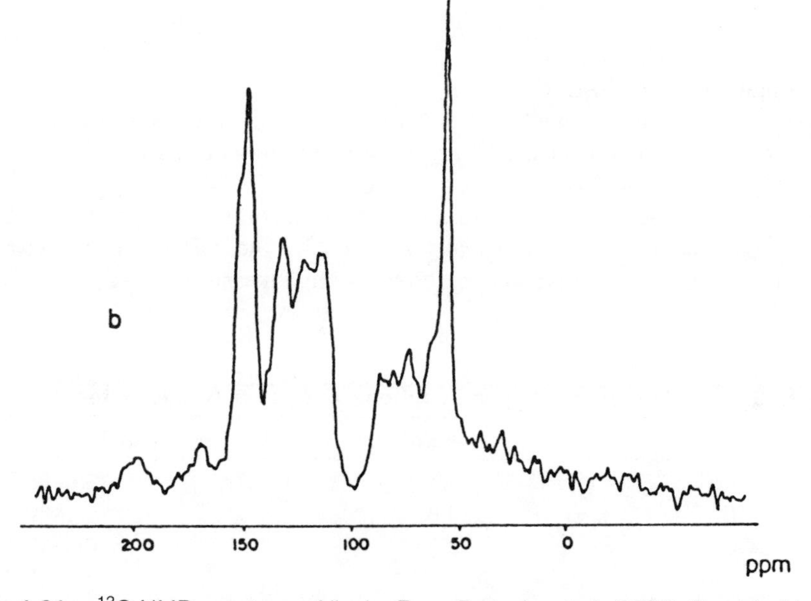

Figure 1.21. ^{13}C NMR spectrum of lignin. From Bartuska et al. (1980). Reprinted by permission of Walter de Gruyter & Co.

hydroxide and readily reduced by B_2H_6 (Martin et al., 1963). As Table 1.15 indicates, values obtained for carboxyl group concentrations in soil humic substances range from 2-9 meq/g. That at least some of the carboxyl groups are on carbon atoms close to one another is indicated by the work of Wood et al. (1961), who showed that when a lignite humic acid was heated in solution or treated with acetic anhydride, new bands in the IR spectrum characteristic of 5- or 7-ring anhydrides were observed. These were attributed to partial structures of types **38** and **39** occurring in the parent humic acid and were estimated to account for 30–50% of the total carboxyl content. Similar results were reported by Arai and Kumada (1983), who measured the formation of a fluorescent chromophore after the reaction of resorcinol with 1,2-dicarboxylate groups in a soil humic acid; they estimated that 1,2-dicarboxylates made up 20–50% of the total acidity. Based on the fluorescence characteristics of model compounds, the authors suggested that 1,2-benzenedicarboxylate (phthalate) groups were present in actual humic materials.

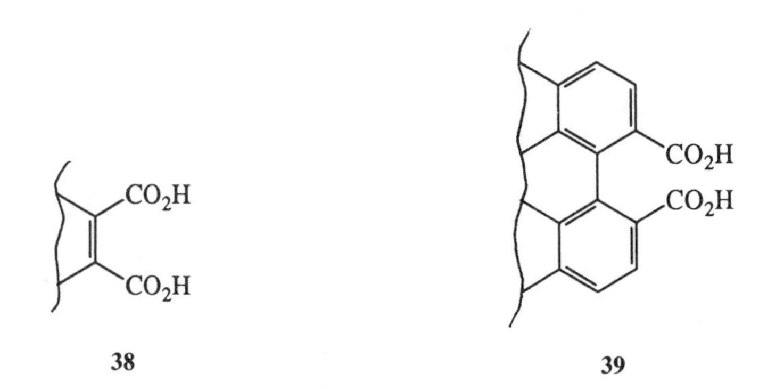

Although widely ranging values for carboxyl content in humic substances have been reported, there is a general trend for fulvic acid fractions to have higher concentrations than humic acids. This is consistent with their higher oxygen content and greater water solubility.

Few good methods exist for determination of phenolic groups in humic substances. Their concentration is usually assessed by the difference between total acidity measured by titration and carboxyl acidity measured by calcium acetate

Table 1.15. Distribution of Oxygen Functional Groups in Humic Materials

Group	meq/g	% of Total O (range)
-COOH	5.1	13.6–65.0
Phenolic -OH	3.6	9.1–38.0
Aliphatic -OH	3.0	0.9–16.4
C=O	3.2	4.1–28.7
$-OCH_3$	0.5	1.0–9.4

Source: Mean of 7–11 determinations reported by Schnitzer and Khan (1972).

exchange (Schnitzer and Gupta, 1965). In the few instances when an attempt has been made to estimate phenolic content by an independent method, the values obtained are almost invariably far lower than those measured by difference (Tsutsuki and Kuwatsawa, 1978). Representative phenolic hydroxyl data for soil humic materials are summarized in Table 1.15. Perdue et al. (1980) and Bonn and Fish (1991) have clearly demonstrated that titration methods have severe limitations when applied to polyacids with continuous pKa distributions. It is perhaps appropriate to mention that some enolizable ketones may have significant acidity (and would thus be titratable), but may not be readily oxidized by the colorimetric reagents such as the Folin reagent usually used for the determination of phenols. Black and Christman (1963b), who attempted to measure enols in aquatic fulvic acids by use of a brominative method, appear to be practically the only authors to have obtained data relating to this possibility. All fulvic acid samples tested were reported to give positive results. Low yields of phenolic acids of the lignin type have been found among the hydrolysis products of some humic acids (Tate and Anderson, 1978).

It seems certain that another considerable fraction of the oxygenated functional groups in most humic substances is made up of aliphatic alcoholic hydroxyl groups (Clapp, 1957; Tan and Clark, 1969). Two groups have described fractionation procedures for soil fulvic acids incorporating differential elution from charcoal (Anderson et al., 1977) or polyamide (Sequi et al., 1975) columns. In both cases, the investigators reported that it was possible to selectively elute a fraction that was very high in polysaccharide carbon. Other investigators have shown that humus fractions derived almost exclusively from carbohydrates, with very little geochemical alteration, can be separated from some soils (Wilson et al., 1986). The nature of the carbohydrate fraction of aquatic fulvic acids was investigated by enzymatic hydrolysis, which revealed that about 20% of the glucose in these materials was starch-like and another 10% was cellulose-like (Bertino et al., 1987). About half of the glucose that could be obtained through acid hydrolysis of the fulvic acid was not accessible to any of the enzymes tested.

A wide range of values for soil organic matter carbonyl content has been reported by various research groups (Table 1.10). Not much can be said concerning the types of carbonyl groups present; presumably, many aldehydes would not be likely to survive for long periods of exposure to oxygen, but quinones, dialkyl or aryl ketones, diketones, amides and esters cannot be ruled out. In a study in which a fulvic acid was derivatized with ^{15}N-hydroxylamine, signals characteristic of hydroxamic acids were observed; these signals presumably arose from the reaction of the derivatizing agent with naturally occurring ester groups (Thorn, 1989).

There has been considerable controversy over the occurrence of quinone groups in humic substances. The presence of quinones might be expected if theories of the formation of humic acids from degraded lignins are correct. Infrared spectroscopic data are inconclusive because of the many interfering functional groups in the 1630–1670 cm^{-1} region characteristic of quinones. Probably the best evidence for quinones in humic material is provided by ESR spectroscopy; signals characteristic of semiquinone ion radicals have been observed from several humic preparations (Atherton et al., 1967).

Not much is known about the structural characteristics of nitrogen compounds associated with soil humic material (Hayes and Swift, 1978). Roughly half of the nitrogen may be more or less loosely bound as attached amino acids or amino sugars, either singly or as part of polymeric structures. Heterocyclic nitrogen appears to constitute a minor fraction of the total, perhaps in the form of pyrrole or acridine moieties (Stevenson, 1982; Schnitzer, 1985). It is known that aniline derivatives can be oxidatively coupled with humic acids in the presence of free radical-generating systems such as enzymes (Bollag, 1983), but there is little evidence for the occurrence of these types of linkages in native humic materials. Electron spin resonance data suggest that some of the nitrogen in soil humic acid occurs in the form of porphyrins. High resolution ESR spectra showed splittings for a copper-containing fraction which were nearly identical to that reported for a copper-tetraphenylporphyrin chelate (Goodman and Cheshire, 1973).

Models for the structure of humic materials. Leenheer et al. (1989) have pointed out that the notion of a "general," "average," or "type" structure for humic substances is controversial, with some investigators believing that such mixtures are so innately diverse and heterogeneous that structural models have little meaning. On the other hand, other workers believe that attempts to synthesize the available information into approximate or figurative representations are of value in helping to predict the physical, biological, and chemical consequences of the presence of humic materials in the environment. In any event, a few investigators have attempted to synthesize the information available about humic substances into general or type structures. These structures display a wide range of chemical complexity and sensitivity to mechanistic concepts. They tend to fall into two general categories: first, "uniformist" models that try to explain all of the properties of humic compounds in terms of a very simple series of reactions or structures; and secondly, "eclectic" models characterized by the incorporation of all kinds of subunits of very diverse chemical nature. We present in this section a very limited selection of the wide variety of proposed structures available in the literature to give some insight into the thought processes that have gone into the models.

A very early model by Fuchs (Figure 1.22; see Swain, 1963) featured a highly condensed aliphatic and aromatic ring system substituted by carboxyl and phenolic hydroxyl groups. The structure appears to have been influenced by the flat, polychair hexagonal structure of graphite and also by early models for coal which depicted it as largely made up of planar polyaromatic hydrocarbons. It is not entirely clear how such a structure could arise from naturally occurring precursors. This model was influential for a time because it illustrated a central "core" region with highly conjugated double bonds, which would account for the color of humic materials, and allowed for flexible modes of attachment of chains of other functionalities, which could be chosen more or less randomly according to which properties of the humic structure the proposer wished to emphasize.

Another type of uniformist model is that of Flaig (1950, Figure 1.23) which considered humic materials to be polyphenol ethers. This sort of structure appears

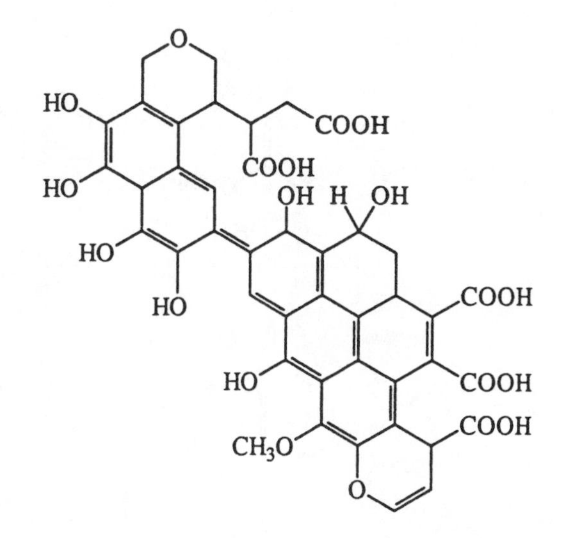

Figure 1.22. Fuchs' model for humic acid structure. From Swain (1963). Reprinted by permission of Pergamon Press.

to be inspired by the indisputable fact that solutions of polyphenols such as pyrogallol and hydroquinone turn brown on standing, especially in the presence of alkali.

The apparent source materials for forming such a hypothetical humic polymer would be lignin degradation products or plant polyphenols such as flavonoids. Because it has little relationship with the physical or chemical characteristics of actual humic materials, this type of model has fallen out of favor almost completely, except for ambiguous statements, still occasionally encountered, that the color of natural waters is due to "tannins."

Schnitzer and Khan (1972) proposed a structure for fulvic acid that pictures it as an assemblage of small aromatic molecules, held together by hydrogen bonds (Figure 1.24). The model is derived from results of oxidative degradation studies of humic materials in which the largest group of identifiable products was a number of aromatic polycarboxylic (3 to 6 -COOHs per benzene ring) acids (see, e.g., Schnitzer and Ortiz de Serra, 1973). Although this formulation looks simplistic by today's

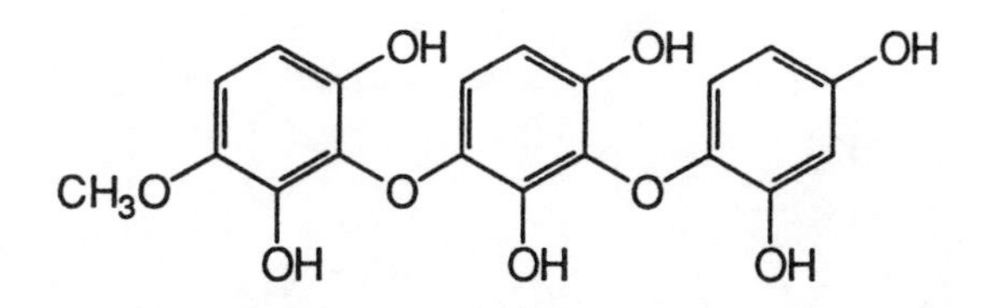

Figure 1.23. Flaig's model for humic acid structure. From Swain (1963). Reprinted by permission of Pergamon Press.

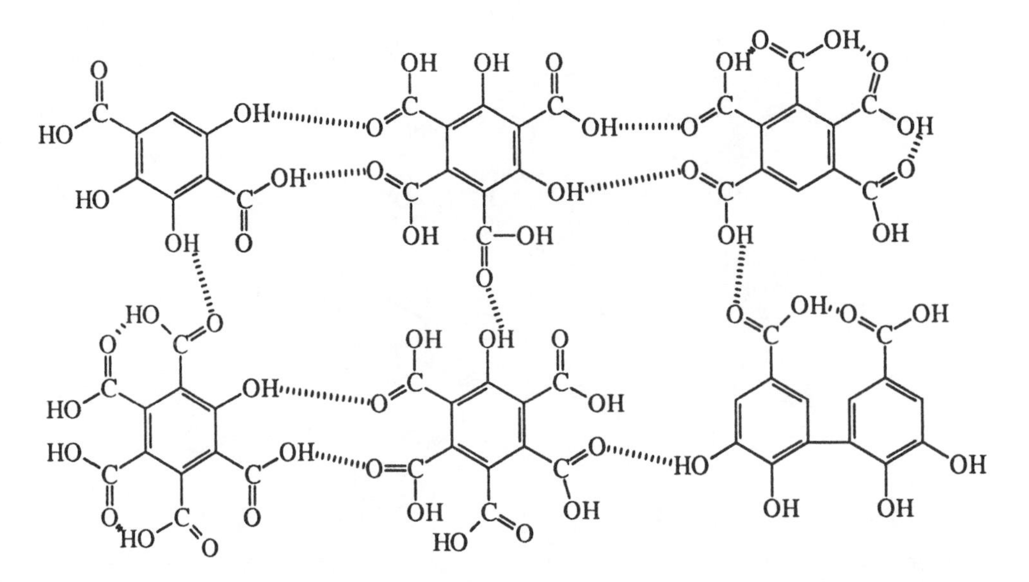

Figure 1.24. Schnitzer and Kahn's (1972) model for fulvic acid structure. Reprinted by permission of Marcel Dekker, Inc.

standards, it is difficult to better reconcile the observed degradation products from humic materials with rational precursors for them that might be incorporated in a larger type structure. It has, however, been demonstrated that benzenecarboxylic acids are formed even when apparent structurally unrelated polymers, such as polysaccharides, are subjected to oxidizing conditions (Cheshire et al., 1968; Almendros and Leal, 1990). The mechanisms of formation of aromatic compounds from carbohydrate precursors are uncertain, although simple sugars have been reported to yield

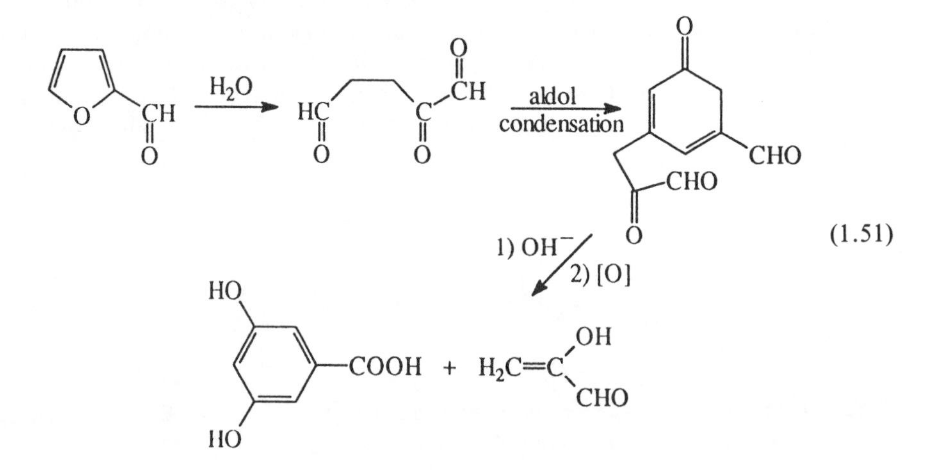

aromatic products when heated under mildly acidic conditions (Popoff and Theander, 1972, 1976). Cheshire et al. (1968) proposed that aldol condensations between two molecules of a five-carbon keto aldehyde derived from furfural hydrolysis (Reaction 1.51) could give rise to phenolic acids.

A uniformist structure for marine humic substances (Figure 1.25) was proposed by Harvey et al. (1983). This model takes as its starting point the polyunsaturated lipids of marine planktonic organisms. As the figure shows, a polyunsaturated triglyceride can be hypothetically converted via free-radical reactions to form inter-strand linkages that might become partially aromatized by oxidation and dehydra-

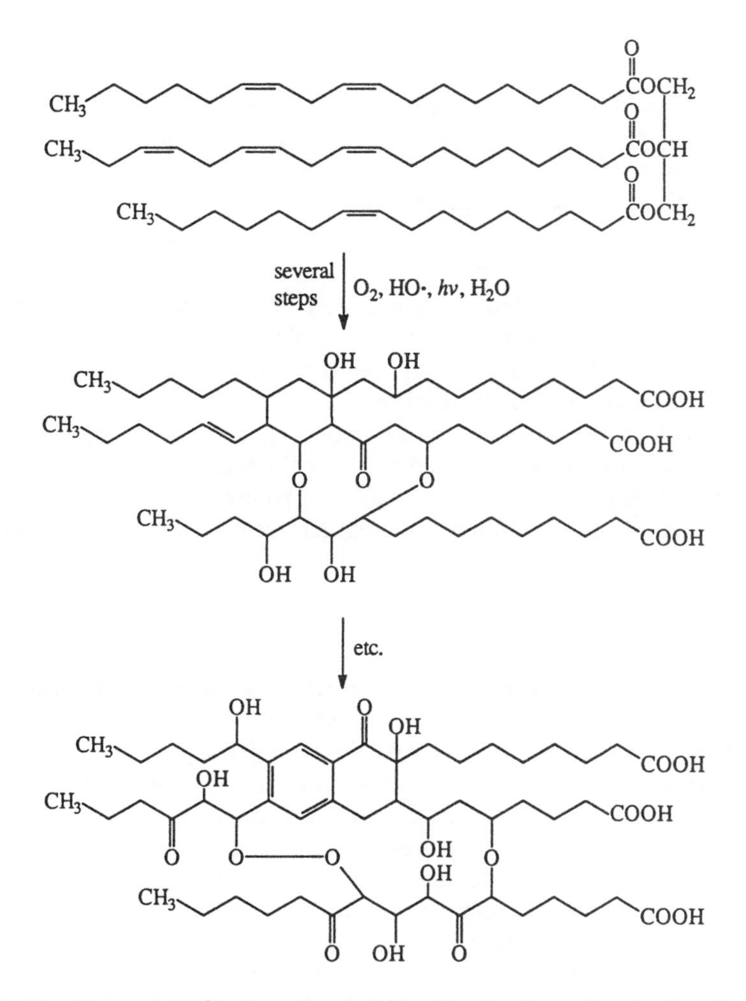

Figure 1.25. Harvey and Stanhauer's model for the structure of marine humic substances. From Harvey and Stanhauer (1983). Reprinted by permission of Elsevier Scientific Publishers.

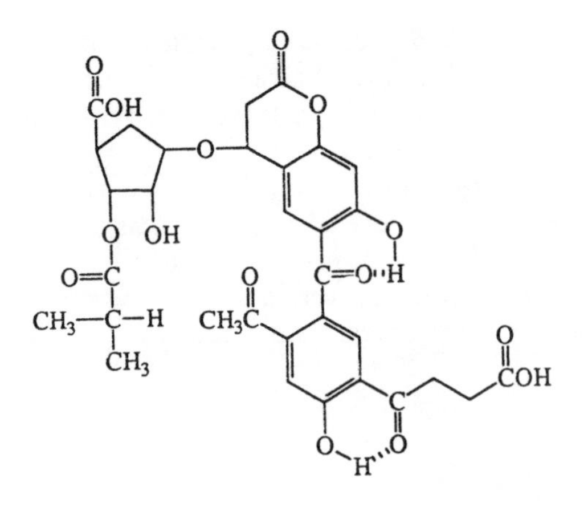

Figure 1.26. Leenheer's "average structural model" for freshwater fulvic acid. Reprinted from R. C. Averett et al., "Humic Substances in the Suwanee River, Georgia: Interactions, Properties, and Proposed Structures," U.S. Geological Survey report #87–557,

tion. After ester hydrolysis, the type structure is revealed as a largely aliphatically substituted aromatic (1-tetralone) structure containing carboxyl, carbonyl, alcohol, and peroxide groups. The structure proposed is intriguing and ought to lend itself to testing through model compound reactions.

Perhaps the ultimate uniformist model was proposed by Anderson and Russell (1976), who stated that fulvic acid had many of the properties of a synthetic polymer, polymaleic acid. This polymer could be represented as $[\text{-CHCOOH-}]_n$, by analogy with polyethylene, although other structures such as anhydrides, cyclopentanone units, and keto olefinic structures have been suggested to be present. The polymer was shown to differ in many important respects such as its UV, fluorescence, and ^{13}C NMR characteristics from an authentic fulvic acid (Spiteller and Schnitzer, 1983).

Leenheer et al. (1989) proposed several semieclectic "average structural models" for a fulvic acid isolated from a highly colored river water. The parameters chosen for the models were designed to agree with the extensive spectroscopic and chemical data available for this material, and were also based on hypothetical reactions that could have transformed known plant-derived compounds into the fulvic acid. The average molecular formula for the models was taken to be $C_{33}H_{32}O_{19}$, based on molecular weight and elemental composition analyses. Nitrogen, sulfur, and phosphorus were not incorporated into the model structures due to their low abundance in the fulvic acid. One of these structures is shown in Figure 1.26.

One example of extreme eclecticism in a humic model is Gjessing's (1976; Figure 1.27); there are many other examples like this, mostly sharing the same apparent reluctance to discriminate among all possible environmental reactants.

To conclude, it is apparent that despite more than a century of work on humic

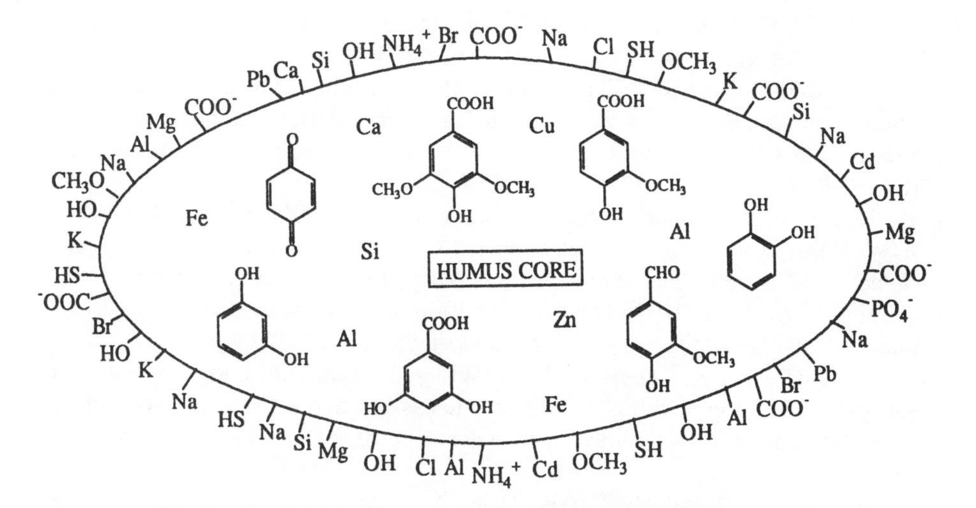

Figure 1.27. Gjessing's (1976) "eclectic" concept of humic material. Reprinted by permission of Lewis Publishers.

materials, we still have only a very approximate and incomplete understanding of their molecular characteristics. It is to be hoped that in future years we will have a much better conception of the chemical nature of these important polymers.

REFERENCES

Adams, D. F., S. O. Farwell, E. Robinson, M. R. Pack, and W. L. Bamesberger. 1981. Biogenic sulfur source strengths. *Environ. Sci. Technol.* 15: 1493–1498.

Almendros, G. and J. A. Leal. 1990. An evaluation of some oxidative degradation methods of humic substances applied to carbohydrate-derived humic-like polymers. *J. Soil Sci.* 41: 51–59.

Almgren, T. B., B. Josefsson, and G. Nyquist. 1975. A fluorescence method for studies of spent sulfite liquor and humic substances in sea water. *Anal. Chim. Acta* 78: 411–421.

Altshuller, A. P. 1983. Natural volatile substances and their effect on air quality in the United States. *Atmos. Environ.* 17: 2131-2165.

American Chemical Society. 1978. Cleaning our environment—a chemical perspective. Washington, DC.

Anderson, H. A. and J. D. Russell. 1976. Possible relationship between soil fulvic acid and polymaleic acid. *Nature* 260: 597.

Anderson, H. A., A. R. Fraser, A. Hepburn, and J. D. Russell. 1977. Chemical and infrared spectroscopic studies of fulvic acid fractions from a podzol. *J. Soil Sci.* 28: 623–633.

Andrae, M. O. and H. Raemdonck. 1983. Dimethyl sulfide in the surface ocean and marine atmosphere: a global view. *Science* 221: 744–747.

Andrews, P. and P. J. LeB. Williams. 1971. Heterotrophic utilization of dissolved organic compounds in the sea. III. Measurement of the oxidation rates and concentrations of glucose and amino acids in sea water. *J. Mar. Biol. Assoc. U.K.* 51: 111–125.

Appel, B. R. 1981. Characterization of carbonaceous materials in atmospheric aerosols by high-resolution mass spectrometric thermal analysis. In G. M. Hidy, ed., *The aerosol characterization experiment*. Wiley-Interscience, New York.

Arai, S. and K. Kumada. 1983. The determination of 1,2-dicarboxylate structures in humic acid by fluorescein formation. *Geoderma* 31: 151–162.

Arey, J., B. Zielinska, R. Atkinson, A. M. Winer, T. Ramdahl, and J. N. Pitts, Jr. 1986. The formation of nitro-PAH from the gas-phase reactions of fluoranthene and pyrene with the OH radical in the presence of NO_x. *Atmos. Environ.* 20: 2339–2345.

Arnold, F., G. Knop, and H. Ziereis. 1986. Acetone measurements in the upper troposphere and lower stratosphere—implications for hydroxyl radical abundances. *Nature* 321: 505–507.

Asplund, G. and A. Grimvall. 1991. Organohalogens in nature. *Environ. Sci. Technol.* 25: 1346–1350.

Atherton, N. M., P. A. Cranwell, A. J. Floyd, and R. A. Haworth. 1967. ESR spectra of humic acids. *Tetrahedron* 23: 1653–1667.

Bada, J. L. and C. Lee. 1977. Decomposition and alteration of organic compounds dissolved in seawater. *Mar. Chem.* 5: 523–534.

Baier, R. E., W. D. Goupil, S. Perlmutter, and R. King. 1974. Dominant chemical composition of the sea surface films, natural slicks and foams. *J. Res. Atmos.* 8: 571–600.

Baker, N. and L. Wilson. 1966. Water-soluble products of UV-irradiated, autoxidized linoleic and linolenic acids. *J. Lipid Res.* 7: 341–348.

Barcelona, M. J. 1979. Human exposure to chloroform in a coastal urban environment. *J. Environ. Sci. Health* A14: 267–283.

Barcelona, M. J. 1980. Dissolved organic carbon and volatile fatty acids in marine sediment pore waters. *Geochim. Cosmochim. Acta* 44: 1977–1984.

Barron, P. F. and M. A. Wilson. 1981. Humic soil and coal structure study with magic-angle spinning ^{13}C CP-NMR. *Nature* 289: 275–276.

Bartuska, V. I., G. E. Maciel, H. I. Bolker, and B. I. Fleming. 1980. Structural studies of lignin isolation procedures by ^{13}C NMR. *Holzforschung* 34: 214–217.

Batistic, L. 1974. Les acides aromatiques et volatils dans les sols. *Plant Soil* 41: 73–80.

Bazilevitch, N. I. 1974. Soil-forming role of substance and energy exchange in the soil-plant system. *Trans. 10th Internat. Congr. Soil Sci.*, Moscow, 6: 17–27.

Bellar, T. A. and J. E. Sigsby. 1970. Direct gas chromatographic analysis of low molecular weight substituted organic compounds in emissions. *Environ. Sci. Technol.* 4: 150–156.

Berendsen, H. J. C. 1975. Specific interactions of water with biopolymers. In F.

Franks, ed., *Water: a comprehensive treatise*. Vol. 5. Plenum, New York. pp. 293–330.

Berges, M. G. M. and P. Warneck. 1992. Product quantum yields for the 350-nm photodecomposition of pyruvic acid in air. *Ber. Bunsen Ges. Phys. Chem.* 96: 413–416.

Berkenbus, B. D. 1983. Methodology for the extraction and analysis of hydrocarbons and carboxylic acids in atmospheric particulate matter. *Atmos. Environ.* 17: 1537–1543.

Bertino, D. J., P. W. Albro, and J. R. Haas. 1987. Enzymatic hydrolysis of carbohydrates in aquatic fulvic acid. *Environ. Sci. Technol.* 21: 859–863.

Bidleman, T. F. and C. E. Olney. 1974. Chlorinated hydrocarbons in the Sargasso Sea atmosphere and surface water. *Science* 183: 516–518.

Black, A. P. and R. F. Christman. 1963a. Characteristics of colored surface waters. *J. Amer. Water Works Assoc.* 55: 753–769.

Black, A. P. and R. F. Christman. 1963b. Chemical characteristics of fulvic acids. *J. Amer. Water Works Assoc.* 55: 897–912.

Boehm, P. D. and J. G. Quinn. 1973. Solubilization of hydrocarbons by the dissolved organic matter in sea water. *Geochim. Cosmochim. Acta* 34: 2459–2477.

Bolin, B. 1979. *The global carbon cycle*. John Wiley, Chichester, UK.

Bolin, B. and R. B. Cook, eds. 1983. *Major biogeochemical cycles and their interactions*. John Wiley, Chichester, UK.

Bollag, J.-M. 1983. Cross-coupling of humus constituents and xenobiotic substances. In R. F. Christman and E. T. Gjessing, eds., *Aquatic and terrestrial humic materials*. Ann Arbor Sci. Pub., Ann Arbor, MI, pp. 127–141.

Bonn, B. A. and W. Fish. 1991. Variability in the measurement of humic carboxyl content. *Environ. Sci. Technol.* 25: 232–240.

Boone, P. M. and E. S. Macias. 1987. Methyl alkanes in atmospheric aerosols. *Environ. Sci. Technol.* 21: 903–909.

Brewer, W. S. and F. K. Pfaender. 1979. The distribution of selected organic molecules in freshwater sediment. *Water Res.* 13: 237–240.

Brezonik, P. L. 1990. "Principles of Linear Free-Energy and Structure-Activity Relationships and Their Applications to the Fate of Chemicals in Aquatic Systems," in *Aquatic Chemical Kinetics*, Stumm, W. Ed., John Wiley, New York, pp. 113–144.

Brook, A. J. W. and K. C. Munday. 1970. The effect of temperature on the interaction of phenols with Sephadex gels. *J. Chromatogr.* 51: 307–310.

Brown, K. W. and J. C. Thomas. 1984. Conductivity of three commercially available clays to petroleum products and organic solvents. *Hazardous Waste* 1: 545–553.

Brunner, S., E. Hornung, H. Santi, E. Wolff, O. G. Piringer, J. Altschuh, and R. Brüggeman. 1990. Henry's law constants for polychlorinated biphenyls: experimental determination and structure-property relationships. *Environ. Sci. Technol.* 24: 1751–1754.

Buddrus, J., P. Burba, H. Herzog, and J. Lambert. 1989. Quantitation of partial structures of aquatic humic substances by one- and two-dimensional solution ^{13}C nuclear magnetic resonance spectroscopy. *Anal. Chem.* 61: 628–631.

Buol, S. W., F. D. Hole, and R. J. McCracken. 1973. *Soil genesis and classification*. Iowa State Univ. Press, Ames, IA.

Burkhard, L. P., D. E. Armstrong, and A. W. Andren. 1985. Henry's law constants for the polychlorinated biphenyls. *Environ. Sci. Technol.* 19: 590–596.

Burreson, B. J., R. E. Moore, and P. Roller. 1975. Haloforms in the essential oil of the alga *Asparagopsis taxiformis* (Rhodophyta). *Tetrahedron Lett.* 473–476.

Caldwell, A. D. and C. A. Black. 1958. Inositol hexaphosphate. I. Quantitative determination in extracts of soils and manures. *Proc. Soil Sci. Soc. Amer.* 22: 290–293.

Caldwell, M. M., R. Robberecht, and W. D. Billings. 1980. A steep latitudinal gradient of solar ultraviolet-B radiation in the arctic-alpine life zone. *Ecology* 61: 600–611.

Cameron, D. W. and M. D. Sidell. 1976. A polychloroquinone from green soils. *J. Chem. Soc. Chem. Commun.* 252–253.

Carlson, D. J. and L. M. Mayer. 1980. Enrichment of dissolved phenolic material in the surface microlayer of coastal waters. *Nature* 286: 482–483.

Carter, C. W. and I. H. Suffet. 1982. Binding of DDT to dissolved humic materials. *Environ. Sci. Technol.* 16: 735–740.

Cavanagh, L. A., C. F. Schadt, and E. Robinson. 1969. Atmospheric hydrocarbon and carbon monoxide measurements at Point Barrow, Alaska. *Environ. Sci. Technol.* 3: 251–257.

Chapman, N. B. and J. Shorter, eds. 1970. *Advances in Linear Free Energy Relationships,* London: Plenum, Foreword.

Chen, Y., F. J. Sowden, and M. Schnitzer. 1977. Nitrogen in Mediterranean soils. *Agrochimica* 21: 7–14.

Cheshire, M. V. 1977. Origins and stability of soil polysaccharide. *J. Soil Sci.* 28: 1–10.

Cheshire, M. V., P. A. Cranwell, and R. D. Haworth. 1968. Humic acid—III. *Tetrahedron* 24: 5155–5167.

Cheshire, M. V., C. M. Mundie, and H. Shepherd. 1973. The origin of soil polysaccharide: transformation of sugars during the decomposition in soil of plant material labelled with ^{14}C. *J. Soil Sci.* 24: 54–68.

Chio, K. S., U. Reiss, B. Fletcher, and A. L. Tappel. 1969. Peroxidation of subcellular organelles: formation of lipofuscinlike fluorescent pigments. *Science* 166: 1535–1536.

Chiou, C. T., R. L. Malcolm, T. I. Brinton, and D. E. Kile. 1986. Water solubility enhancement of some organic pollutants and pesticides by dissolved humic and fulvic acids. *Environ. Sci. Technol.* 20: 502–508.

Chiou, C. T., D. E. Kile, T. I. Brinton, R. L. Malcolm, J. A. Leenheer, and P. MacCarthy. 1987. A comparison of water solubility enhancements of organic solutes by aquatic humic materials and commercial humic acids. *Environ. Sci. Technol.* 21: 1231–1234.

Choudhury, D. R. 1982. Characterization of polycyclic ketones and quinones in diesel emission particulates by gas chromatography/mass spectrometry. *Environ. Sci. Technol.* 16: 102–106.

Christman, R. F. and M. Ghassemi. 1966. Chemical nature of organic color in water. *J. Amer. Water Works Assoc.* 58: 723–741.

Clapp, C. E., Jr. 1957. High molecular weight water-soluble muck: isolation and determination of constituent sugars of a borate complex-forming polysaccharide employing electrophoretic techniques. *Diss. Abstr.* 17: 963-964.

Clark, A. L., A. E. McIntyre, J. N. Lester, and R. Perry. 1982. Evaluation of a Tenax GC sampling procedure for collection and analysis of vehicle-related aromatic and halogenated hydrocarbons in ambient air. *J. Chromatogr.* 252: 147–157.

Cordes, E. H. and W. P. Jencks. 1963. The mechanism of hydrolysis of Schiff bases derived from aliphatic amines. *J. Am. Chem. Soc.*, 85: 2843-2848.

Cosovic, B., V. Vojvodic, and T. Plese. 1985. Electrochemical determination and characterization of surface active substances in freshwaters. *Water Res.* 19: 175–183.

Cuffe, S. T., R. W. Gerstle, A. A. Orning, and C. H. Schwartz. 1964. Air pollutants from coal-fired power plants; report no. 1. *J. Air Pollut. Contr. Assoc.* 14: 353–362.

Czuczwa, J. M. and R. A. Hites. 1986. Airborne dioxins and dibenzofurans: sources and fates. *Environ. Sci. Technol.* 20: 195–200.

Dixon, J. B. and S. B. Weed, eds. 1977. Minerals in soil environments. *Soil Sci. Soc. Amer.*, Madison, WI.

Eatough, D. J. and L. D. Hansen. 1984. Bis-hydroxymethyl sulfone: a major aerosol product of the atmospheric reactions of $SO_2(g)$. *Sci. Total Environ.* 36: 319–328.

Eatough, D. J., V. F. White, L. D. Hansen, N. L. Eatough, and J. L. Cheney. 1986. Identification of gas-phase dimethyl sulfate and monomethyl hydrogen sulfate in the Los Angeles atmosphere. *Environ. Sci. Technol.* 20: 867–878.

Ehrhardt, M., C. Osterroht, and G. Petrick. 1980. Fatty-acid methyl esters dissolved in seawater and associated with suspended particulate material. *Mar. Chem.* 10: 67–76.

Eisenreich, S. J., A. W. Elzerman, and D. E. Armstrong. 1978. Enrichment of micronutrients, heavy metals, and chlorinated hydrocarbons in wind-generated lake foam. *Environ. Sci. Technol.* 12: 413–417.

Eklund, G. and B. Stromberg. 1983. Detection of polychlorinated polynuclear aromatics in flue gases from coal combustion and refuse incinerators. *Chemosphere* 12: 657–660.

Enders, C. and S. Sigurdsson. 1947. Chemie der Huminsäurebildung unter physiologischen Bedingungen. *Biochem. Z.* 318: 44–46.

Epstein, S. S., D. H. Fine, D. P. Rounbehler, and N. M. Belcher. 1976. N-nitroso compounds: detection in ambient air. *Science* 192: 1328–1330.

Ertel, J. R. and J. I. Hedges. 1983. Bulk chemical and spectroscopic properties of marine and terrestrial humic acids, melanoidins and catechol-based synthetic polymers. In R. F. Christman and E. T. Gjessing, eds., *Aquatic and terrestrial humic materials*. Ann Arbor Sci. Pub., Ann Arbor, MI, pp. 143–163.

Ertel, J. R., J. I. Hedges, and E. M. Perdue. 1984. Lignin signature of aquatic humic substances. *Science* 223: 485–487.

Fine, D. H., D. P. Rounbehler, E. Sawacki, K. Krost, and G. A. DeMarrais. 1976. N-nitroso compounds in the ambient community air of Baltimore, Maryland. *Anal. Lett.* 9: 595–604.

Flaig, W. 1950. Zur Kenntnis der Huminsäuren. I. Mitteilung. Zur chemischen Konstitution der Huminsäuren. *Z. Pflanzenernähr. Düng. Bodenkunde* 51: 193–212.

Flaig, W. 1971. Organic compounds in soil. *Soil Sci.* 111: 19–33.

Foo, L. Y. and K. R. Tate. 1977. Isolation of chrysotalunin, a red pigment from a New Zealand soil. *Experientia* 33: 1271.

Frank, H. S. and A. S. Quist. 1961. Pauling's model and the thermodynamic properties of water. *J. Chem. Phys.* 34: 604–611.

Fujitake, N., J. Azuma, T. Hamasaki, H. Nakajima, and K. Saiki. 1991. Determination of chrysotalunin, a predominant soil anthraquinone pigment, by high-performance liquid chromatography. *Geoderma* 48: 83–91.

Fung, K. and D. Grosjean. 1981. Determination of nanogram amounts of carbonyls as 2,4-dinitrophenylhydrazones by high performance liquid chromatography. *Anal. Chem.* 53: 168–171.

Gardner, W. S. and D. W. Menzel. 1974. Phenolic aldehydes as indicators of terrestrially derived organic matter in the sea. *Geochim. Cosmochim. Acta* 38: 813–822.

Gardner, W. S. and R. B. Hanson. 1979. Dissolved free amino acids in interstitial waters of Georgia salt marsh soils. *Estuaries* 2: 113–116.

Garrett, W. D. 1967. The organic chemical composition of the sea surface. *Deep-Sea Res.* 14: 221–227.

Gauthier, T. D., W. R. Seitz, and C. L. Grant. 1987. Effects of structural and compositional variations of dissolved humic materials on pyrene K_{oc} values. *Environ. Sci. Technol.* 21: 243–248.

Ghassemi, M. and Christman, R. F. 1968. Properties of the yellow organic acids of natural waters. *Limnol. Oceanogr.* 13: 583–597.

Gibbs, R. J. 1972. Water chemistry of the Amazon River. *Geochim. Cosmochim. Acta* 36: 1061–1066.

Gillam, A. H. and M. A. Wilson. 1983. Application of ^{13}C-nmr spectroscopy to the structural elucidation of dissolved marine humic substances and their phytoplanktonic precursors. In R. F. Christman and E. T. Gjessing, eds., *Aquatic and terrestrial humic materials*. Ann Arbor Sci. Pub., Ann Arbor, MI, pp. 25–35.

Given, P. H. 1975. Environmental organic chemistry of bogs, marshes, and swamps. In G. Eglinton, ed., *Environmental chemistry*. Ann. Rep. Chem. Soc., London, pp. 55–80.

Gjessing, E. T. 1976. *Physical and chemical characteristics of aquatic humus*. Ann Arbor Sci. Pub., Ann Arbor, MI.

Goldberg, M. C. and E. R. Weiner. 1989. Fluorescence measurements of the volume, shape, and fluorophore composition of fulvic acid from the Suwanee River. In R. C. Averett, J. A. Leenheer, D. M. McKnight, and K. A. Thorn, eds. *Humic substances in the Suwanee River, Georgia: interactions, properties, and proposed structures*. USGS Report 87-557, Denver, CO, pp. 183-204.

Goni, M. A. and J. I. Hedges. 1990. Cutin-derived CuO reaction products from purified cuticles and tree leaves. *Geochim. Cosmochim. Acta* 54: 3065-3072.

Goodman, B. A. and M. V. Cheshire. 1973. Electron paramagnetic resonance evidence that copper is complexed in humic acid by the nitrogen of porphyrin groups. *Nature New Biol.* 244: 158-159.

Gossett, J. M. 1987. Measurement of Henry's law constants for C_1 and C_2 chlorinated hydrocarbons. *Environ. Sci. Technol.* 21: 202-208.

Graedel, T. E. 1978. *Chemical compounds in the atmosphere*. Academic Press, New York.

Grosjean, D. 1983. Atmospheric reactions of pyruvic acid. *Atmos. Environ.* 17: 2379-2382.

Grosjean, D. 1984. Atmospheric reactions of ortho cresol: gas phase and aerosol products. *Atmos. Environ.* 18: 1641-1652.

Grosjean, D. 1989. Organic acids in southern California air: ambient concentrations, mobile source emissions, in situ formation and removal processes. *Environ. Sci. Technol.* 23: 1506-1514.

Grov, A. 1963. Carbohydrates in cold water extracts of a pine forest soil. *Acta Chem. Scand.* 17: 2301-2306.

Gschwend, P., J. MacFarlane, and K. Newman. 1985. Volatile halogenated organic compounds released to seawater from temperate marine macrolayer. *Science* 227: 1033-1035.

de Haan, H. 1972. Some structural and ecological studies of soluble humic compounds from Tjeukemeer. *Verh. Int. Verein. Limnol.* 18: 685-695.

Hall, K. J. and G. F. Lee. 1974. Molecular size and spectral characterization of organic matter in a meromictic lake. *Water Res.* 8: 239-251.

Hammett, L. P. 1937. The effect of structure upon the reactions of organic compounds. benzene derivatives, *J. Am. Chem. Soc.*, 59:96-103.

Hammett, L. P. 1970. *Physical Organic Chemistry,* 2nd edition, McGraw-Hill, New York, p. 347.

Hanst, P. L. and B. W. Gay, Jr. 1983. Atmospheric oxidation of hydrocarbons: formation of hydroperoxides and peroxyacids. *Atmos. Environ.* 17: 2259-2265.

Hargrave, B. T. and G. A. Phillips. 1975. Estimates of oil in aquatic sediments by fluorescence spectroscopy. *Environ. Pollut.* 8: 193-215.

Harkov, R. 1986. Semivolatile organic compounds in the atmosphere: a review. *J. Environ. Sci. Health* A21: 409-433.

Harsch, D. E., R. A. Rasmussen, and D. Pierotti. 1977. Identification of a potential source of chloroform in urban air. *Chemosphere* 11: 769-775.

Harvey, G. A. and W. G. Steinhauer. 1974. Atmospheric transport of polychlorinated biphenyls to the North Atlantic. *Atmos. Environ.* 8: 777-782.

Harvey, G. R., D. A. Boran, L. A. Chesal, and J. M. Tokar. 1983. The structure of marine fulvic and humic acids. *Mar. Chem.* 12: 119-132.

Harvey, G. R., D. A. Boran, S. R. Piotrowicz, and C. P. Weisel. 1984. Synthesis of marine humic substances from unsaturated lipids. *Nature* 309: 244–246.

Hassett, J. J., J. C. Means, W. L. Banwart, and S. G. Wood. 1980. *Sorption properties of sediments and energy-related pollutants.* EPA Report #EPA-600/3-80-041, U.S. Environmental Protection Agency, Washington, DC.

Hatcher, P. G., G. E. Maciel, and L. W. Dennis. 1981a. Aliphatic structure of humic acids: a clue to their origin. *Org. Geochem.* 3: 43–48.

Hatcher, P. G., M. Schnitzer, L. W. Dennis, and G. E. Maciel. 1981b. Aromaticity of humic substances in soils. *Soil Sci. Soc. Amer. J.* 45: 1089–1094.

Hawker, D. W. 1989. Vapor pressures and Henry's law constants of polychlorinated biphenyls. *Environ. Sci. Technol.* 23: 1250–1253.

Hayes, M. H. B. and R. S. Swift. 1978. The chemistry of soil organic colloids. In D. J. Greenland and M. H. B. Hayes, eds. *The chemistry of soil constituents.* Wiley, Chichester, UK, pp. 179–320.

Hedges, J. I. and P. L. Parker. 1976. Land-derived organic matter in surface sediments from the Gulf of Mexico. *Geochim. Cosmochim. Acta* 40: 1019–1029.

Heikes, B. G., G. L. Kok, J. G. Walega, and A. L. Lazrus. 1987. H_2O_2, O_3, and SO_2 measurements in the lower troposphere over the eastern United States during fall. *J. Geophys. Res.* 92: 915–931.

Hellebust, J. A. 1965. Excretion of some organic compounds by marine phytoplankton. *Limnol. Oceanogr.* 10: 192–206.

Hellpointner, E. and S. Gäb. 1989. Detection of methyl, hydroxymethyl and hydroxyethyl hydroperoxides in air and precipitation. *Nature* 337: 631–634.

Helmig, D., J. W. Arey, W. P. Harger, and J. Lopez-Cancio. 1992. Formation of mutagenic nitrobenzopyranones and their occurrence in ambient air. *Environ. Sci. Technol.* 26: 622–624.

Henrichs, S. M. and J. W. Farrington. 1979. Amino acids in interstitial waters of marine sediments. *Nature* 279: 319–322.

Henrichs, S. M. and P. M. Williams. 1985. Dissolved and particulate amino acids and carbohydrates in the sea surface microlayer. *Mar. Chem.* 17: 141–163.

Hewitt, C. N. and G. L. Kok. 1991. Formation and occurrence of organic hydroperoxides in the troposphere: laboratory and field observations. *J. Atmos. Chem.* 12: 181–194.

Hileman, B. 1983. Arctic haze. *Environ. Sci. Technol.* 17: 232A-236A.

Hoff, R. M. and K.-W. Chan. 1987. Measurement of polycyclic aromatic hydrocarbons in the air along the Niagara River. *Environ. Sci. Technol.* 21: 556–561.

Hurst, H. M. and N. A. Burges. 1967. Lignin and humic acids. In A. D. McLaren and G. H. Peterson, eds., *Soil biochemistry*, Vol. 1. Marcel Dekker, New York. pp. 260–286.

Hutchinson, G. E. 1957. *A treatise on limnology.* John Wiley, New York, p. 360.

Hutte, R. S., E. J. Williams, J. Staehelin, S. B. Hawthorne, R. M. Barkley, and R. E. Sievers. 1984. Chromatographic analysis of organic compounds in the atmosphere. *J. Chromatogr.* 302: 173–179.

Igawa, M., J. W. Munger, and M. R. Hoffmann. 1989. Analysis of aldehydes in cloud- and fogwater samples by HPLC with a postcolumn reaction detector. *Environ. Sci. Technol.* 23: 556–561.

Johnson, K. M. and J. McN. Sieburth. 1977. Dissolved carbohydrates in seawater. 1. A precise spectrometric analysis for monosaccharides. *Mar. Chem.* 5: 1–13.

Kalle, K. 1966. The problem of the Gelbstoff in the sea. *Mar. Biol. Ann. Rev.* 4: 91–104.

Karickhoff, S. W. 1984. Organic pollutant sorption in aquatic systems. *J. Hydraul. Eng.* 110: 707–735.

Karickhoff, S. W. and K. R. Morris. 1985. Sorption dynamics of hydrophobic pollutants in sediment suspensions. *Environ. Toxicol. Chem.* 4: 469–479.

Kawamura, K. and I. R. Kaplan. 1983. Organic compounds in the rainwater of Los Angeles. *Environ. Sci. Technol.* 17: 497–501.

Kawamura, K. and I. R. Kaplan. 1984. Capillary gas chromatography determination of volatile organic acids in rain and fog samples. *Anal. Chem.* 56: 1616–1620.

Kawamura, K., S. Steinberg, and I. R. Kaplan. 1985. Capillary GC determination of short-chain dicarboxylic acids in rain, fog, and mist. *Internat. J. Environ. Anal. Chem.* 19: 175–188.

Khalil, M. A. K. and R. A. Rasmussen. 1985. Causes of increasing atmospheric methane: depletion of hydroxyl radicals and the rise of emissions. *Atmos. Environ.* 19: 397–407.

Khalil, M. A. K. and R. A. Rasmussen. 1990. Atmospheric methane: recent global trends. *Environ. Sci. Technol.* 24: 549–553.

Kile, D. E. and C. T. Chiou. 1989. Water solubility enhancements of DDT and trichlorobenzene by some surfactants above and below the critical micelle concentration. *Environ. Sci. Technol.* 23: 832–838.

Kirk, T. K. and H.-M. Chang. 1974. Decomposition of lignin by white-rot fungi. II. Characterization of heavily degraded lignins from decayed spruce. *Holzforschung* 29: 56–64.

Kleber, R. J. and K. Mopper. 1990. Determination of picomolar concentrations of carbonyl compounds in natural waters, including seawater, by liquid chromatography. *Environ. Sci. Technol.* 24: 1477–1481.

König, J., E. Balfanz, W. Funcke, and K. Romanowski. 1983. Determination of oxygenated polycyclic organic matter in airborne particulate material by capillary gas chromatography and gas chromatography–mass spectrometry. *Anal. Chem.* 55: 599–603.

Kress, B. M. and W. Ziechmann. 1977. Wechselwirkung zwischen Humusstoffen und aromatischen Kohlenwasserstoffen. *Chem. Erde.* 36: 209–217.

Kuwata, K., M. Uerobi, and Y. Yamasaki. 1979. Determination of aliphatic and aromatic aldehydes in polluted airs as their 2,4-dinitrophenylhydrazones by high performance liquid chromatography. *J. Chromatogr. Sci.* 17: 264–268.

Kuwata, K., M. Uerobi, and Y. Yamasaki. 1980. Determination of phenol in polluted air as p-nitrobenzeneazophenol derivative by reversed phase high performance liquid chromatography. *Anal. Chem.* 52: 857–860.

Lagaly, G. 1984. Clay-organic interactions. *Phil. Trans. Roy. Soc. Lond.* A311: 315–332.

Lakshman, G. 1975. Monitoring agricultural pollution using natural fluorescence. *Water Resources Res.* 11: 705–708.

Larson, R. A. 1978. Dissolved organic matter of a low-coloured stream. *Freshwat. Biol.* 8: 91–104.

Larson, R. A., T. L. Bott, L. L. Hunt, and K. Rogenmuser. 1979. Photooxidation products of a fuel oil and their antimicrobial activity. *Environ. Sci. Technol.* 13: 965–969.

Larson, R. A. and A. L. Rockwell. 1980. Fluorescence spectra of water-soluble humic materials and some potential precursors. *Arch. Hydrobiol.* 89: 416–425.

Lashof, D. A. and D. R. Ahuja. 1990. Relative contributions of greenhouse gas emissions to global warming. *Nature* 344: 529–531.

Laszlo, P. 1987. Chemical reactions on clays. *Science* 235: 1473–1477.

Lee, M. L., M. V. Novotny, and K. D. Bartle. 1981. *Analytical chemistry of polycyclic aromatic compounds*. Academic Press, New York.

Lee, R. F. and P. M. Williams. 1974. Copepod "slick" in the northwest Pacific Ocean. *Naturwissenschaften* 61: 505–506.

Leenheer, J. A. and T. I. Noyes. 1984. A filtration and column-adsorption system for onsite concentration and fractionation of organic substances from large volumes of water. USGS Water-Supply Paper #2230. 16 pp.

Leenheer, J. A., R. L. Malcolm, P. W. McKinley, and L. A. Eccles. 1973. Occurrence of dissolved organic carbon in selected ground-water samples in the United States. *J. Res. U. S. Geol. Surv.* 2: 361–369.

Leenheer, J. A., M. A. Wilson, and R. L. Malcolm. 1987. Presence and potential significance of aromatic-ketone groups in aquatic humic substances. *Org. Geochem.* 11: 273–280.

Leenheer, J. A., D. M. McKnight, E. M. Thurman, and P. MacCarthy. 1989. Structural components and proposed structural models of fulvic acid from the Suwanee River. In R. C. Averett, J. A. Leenheer, D. M. McKnight, and K. A. Thorn, eds. *Humic substances in the Suwanee River, Georgia: interactions, properties, and proposed structures*. USGS Report 87–557, Denver, CO, pp. 335–359.

Leopold, L. B., M. G. Wolman, and J. P. Miller. 1964. *Fluvial processes in geomorphology*. Freeman, San Francisco, p. 51.

Lerman, A. 1971. Time to chemical steady states in lakes and oceans. In J. D. Hem, ed. *Nonequilibrium systems in natural water chemistry*. Advan. Chem. Ser. # 106, American Chemical Society, Washington, DC, pp. 30–76.

Levins, P. L. 1972. Analysis of the odorous compounds in diesel engine exhaust. EPA Report #EPA-R2-73-275. 129 pp.

Levsen, K., S. Behnert, and H. D. Winkeler. 1991. Organic compounds in precipitation. *Fresenius J. Anal. Chem.* 340: 665–671.

Lewis, W. M. 1974. An analysis of surface slicks in a reservoir receiving heated effluent. *Arch. Hydrobiol.* 74: 304–315.

Liss, P. S. and P. G. Slater. 1974. Flux of gases across the air-sea interface. *Nature* 247: 181–184.

Lovelock, J. E. 1975. Natural halocarbons in the air and in the sea. *Nature* 256: 193–194.

Lovelock, J. E., R. J. Maggs, and R. A. Rasmussen. 1972. Atmospheric dimethyl sulfide and the natural sulphur cycle. *Nature* 237: 452–453.

Lovelock, J. E., R. J. Maggs, and R. J. Wade. 1973. Halogenated hydrocarbons in and over the Atlantic. *Nature* 241: 194–196.

Lüdemann, H.-D., H. Lentz, and W. Ziechmann. 1973. Protonenresonanz-spektroskopie von Ligninen und Huminsären bei 100 Megahertz. *Erdöl Kohle* 26: 506–509.

Lundquist, K., B. Josefsson, and G. Nyquist. 1978. Analysis of lignin products by fluorescence spectroscopy. *Holzforschung* 32: 27–32.

Lyman, W. J. 1990a. Octanol-water partition coefficient. In W. J. Lyman, W. F. Reehl, and D. H. Rosenblatt, eds. *Handbook of chemical property estimation methods*. American Chemical Society, Washington, DC, pp. 1-1 – 1-51.

Lyman, W. J. 1990b. Adsorption coefficient for soils and sediments. In W. J. Lyman, W. F. Reehl, and D. H. Rosenblatt, eds. *Handbook of chemical property estimation methods*. American Chemical Society, Washington, DC, pp. 4-3 to 4-33.

Lyons, W. B., H. E. Gaudette, and A. D. Hewitt. 1979. Dissolved organic matter in pore water of carbonate sediments from Bermuda. *Geochim. Cosmochim. Acta* 43: 433–437.

Lytle, C. R. and E. M. Perdue. 1981. Free, proteinaceous, and humic-bound amino acids in river water containing high concentrations of aquatic humus. *Environ. Sci. Technol.* 15: 224–228.

MacIntyre, F. 1974. Nonlipid-related possibilities for chemical fractionation in bubble film caps. *J. Rech. Atmos.* 8: 515–527.

Mackay, D. and P. J. Leinonen. 1975. Rate of evaporation of low solubility contaminants from water bodies to the atmosphere. *Environ. Sci. Technol.* 9: 1178–1180.

Mackay, D. and W. Y. Shiu. 1977. Aqueous solubility of polynuclear aromatic hydrocarbons. *J. Chem. Eng. Data* 22: 399–402.

Mackay, D. and W. Y. Shiu. 1981. A critical review of Henry's law constants for chemicals of environmental interest. *J. Phys. Chem. Ref. Data* 10: 1175–1199.

Mackay, D., A. Bobra, and W. Y. Shiu. 1980. Relationships between aqueous solubility and octanol-water partition coefficients. *Chemosphere* 9: 701–711.

MacLeod, K. E. 1981. Polychlorinated biphenyls in indoor air. *Environ. Sci. Technol.* 15: 926–928.

Marklund, S., L.-O. Kjeller, M. Hansson, M. Tysklind, C. Rappe, C. Ryan, H. Collazo, and R. Dougherty. 1986. Determination of PCDDs and PCDFs in incineration samples and pyrolytic products. In C. Rappe, G. Choudhary, and L. H. Keith, eds. *Chlorinated dioxins and dibenzofurans in perspective*. Lewis Publishers, Chelsea, MI. pp. 79–92.

Martin, D. F. and R. H. Pierce Jr. 1971. A convenient method of analysis of humic acid in fresh water. *Anal. Lett.* 49–52.

Martin, F. E., P. Dubach, N. C. Mehta, and H. Deuel. 1963. Bestimmung der funktionellen Gruppen von Huminstoffen. *Z. Pflanzenernähr. Düng. Bodenk.* 103: 29–39.

Martin, J. K. and A. J. Wicken. 1966. Fractionation of organic phosphorus in alkaline soil extracts and the identification of inositol phosphates. *New Zealand J. Agric. Res.* 9: 529–535.

Matsui, Y. and K. Kumada. 1974. Hydroxyanthroquinones in soil. *Soil Sci. Plant Nutr.* 20: 333–341.

Matsumoto, G. and T. Hanya. 1980. Organic constituents in atmospheric fallout in the Tokyo area. *Atmos. Environ.* 14: 1409–1419.

Matthess, G. 1982. *The properties of groundwater*. Wiley-Interscience, New York.

McAuliffe, C. 1966. Solubility in water of paraffin, cycloparaffin, olefin, acetylene, cycloolefin, and aromatic hydrocarbons. *J. Phys. Chem.* 70: 1267–1275.

McCarty, P. L., M. Reinhard, and B. E. Rittmann. 1981. Trace organics in groundwater. *Environ. Sci. Technol.* 15: 40–51.

McGrath, D. 1967. Nature and distribution in Irish soils of a new soil pigment. *Nature* 215: 1414.

McKnight, D. M., E. M. Thurman, R. L. Wershaw, and H. H. Hemond. 1985. Biogeochemistry of aquatic humic substances in Thoreau's Bog, Concord, Massachusetts. *Ecology* 66: 1339–1352.

McLaren, A. D. 1957. Concerning the pH dependence of enzyme reactions on cells, particulates and in solution. *Science* 125: 697.

Melcer, M. E., M. S. Zalewski, M. A. Brisk, and J. P. Hassett. 1987. Evidence for a charge-transfer interaction between dissolved humic materials and organic molecules: study of the binding interaction between humic materials and chloranil. *Chemosphere* 16: 1115–1121.

Meybeck, M. 1981. Carbon dioxide effects research and assessment program: flux of organic carbon by rivers to the ocean. In *Flux of organic carbon by rivers to the oceans*, USDOE Report CONF-8009140, pp. 219–269.

Meyers, P. A. and R. A. Hites. 1982. Extractable organic compounds in midwest rain and snow. *Atmos. Environ.* 16: 2169–2175.

Mikita, M. A., C. Steelink, and R. L. Wershaw. 1981. Carbon-13 enriched nuclear magnetic resonance method for the determination of hydroxyl functionality in humic substances. *Anal. Chem.* 53: 1715–1717.

Miller, J. A. and G. A. Fisk. 1987. Combustion chemistry. *Chem. Eng. News.* (Aug. 31). pp. 22–46.

Miller, M. M., S. Ghodbane, S. P. Wasik, Y. B. Tewari, and D. E. Martire. 1984. Aqueous solubilities, octanol/water partition coefficients, and entropies of melting of chlorinated benzenes and biphenyls. *J. Chem. Eng. Data* 29: 184–190.

Moeller, J. R., G. W. Minshall, K. W. Cummins, R. C. Petersen, C. E. Cushing, J. R. Sedell, R. A. Larson, and R. L. Vannote. 1979. Transport of dissolved organic carbon in streams of differing physiographic characteristics. *Org. Geochem.* 1: 139–150.

Molina, M. J. and F. S. Rowland. 1974. Stratospheric sink for chlorofluoromethanes: chlorine atom-catalysed destruction of ozone. *Nature* 249: 810–812.

Mopper, K., R. Dawson, G. Leibezelt, and V. Ittekott. 1980. The monosaccharide spectra of natural waters. *Mar. Chem.* 10: 55–56.

Morris, R. J. 1974. Lipid composition of surface films and zooplankton from the eastern Mediterranean. *Mar. Pollut. Bull.* 5: 105–109.

Mosier, A. R., C. E. Andre, and F. G. Viets, Jr. 1973. Identification of aliphatic amines volatilized from cattle feedyard. *Environ. Sci. Technol.* 7: 642–644.

Mulholland, P. J. and J. A. Watts. 1982. Transport of organic carbon to the oceans by rivers of North America: a synthesis of existing data. *Tellus* 34: 176–186.

Müller-Wegener, U. 1977. Fluoreszenzspektroskopische Untersuchungen an Huminsäuren. *Z. Pflanzenernähr. Bodenkunde* 32: 27–32.

Murphy, T. J., M. D. Mullin, and J. A. Meyer. 1987. Equilibration of polychlorinated biphenyls and toxaphene with air and water. *Environ. Sci. Technol.* 21: 155–162.

Nemethy, G. and H. Scheraga. 1964. Structure of water and hydrophobic bonding in proteins. IV. The thermodynamic properties of liquid deuterium oxide. *J. Chem. Phys.* 41: 680–689.

Newton, D. L., M. D. Erickson, K. B. Tomer, E. D. Pellizzari, P. Gentry, and R. B. Zweidinger. 1982. Identification of nitroaromatics in diesel exhaust particulates using gas chromatography/negative ion chemical ionization mass spectrometry and other techniques. *Environ. Sci. Technol.* 16: 206–213.

Nissenbaum, A. 1974. Deuterium content of humic acids from marine and submarine environments. *Mar. Chem.* 2: 59–63.

Nissenbaum, A. and I. R. Kaplan. 1972. Chemical and isotopic evidence for the in situ origin of marine humic substances. *Limnol. Oceanogr.* 17: 570–582.

Nojima, K., K. Fukuya, S. Fukui, S. Kanno, S. Nishiyama, and Y. Wada. 1976. Formation of nitrophenols by the photochemical reaction of toluene in the presence of nitrogen monoxide and nitrophenols in rain. *Chemosphere* 5: 25–30.

Nondek, L., D. R. Rodler, and J. W. Birks. 1992. Measurement of sub-ppbv concentrations of aldehydes in a forest atmosphere using a new HPLC technique. *Environ. Sci. Technol.* 26: 1174–1178.

Novakov, T. 1984. The role of soot and primary oxidants in atmospheric chemistry. *Sci. Total Environ.* 36: 1–10.

Orem, W. H. and P. G. Hatcher. 1987. Solid-state ^{13}C NMR studies of dissolved organic matter in pore waters from different depositional environments. *Org. Geochem.* 11: 73–82.

Painter, H. A. and M. Viney. 1959. Composition of a domestic sewage. *J. Biochem. Microbiol. Technol. Eng.* 1: 143–162

Panter, R. and R.-D. Penzhorn. 1980. Alkyl sulfonic acids in the atmosphere. *Atmos. Environ.* 14: 149–151.

Pauling, L. 1960. *The nature of the chemical bond.* 3rd ed. Cornell Univ. Press, Ithaca, NY, pp. 469–473.

Pavelich, W. A. and R. W. Taft. 1957. The evaluation of inductive and steric effects on reactivity. The methoxide ion-catalyzed rates of methanolysis of *l*-menthyl esters in methanol, *J. Am. Chem. Soc.*, 79:4935–4940.

Payne, J. R. and C. R. Phillips. 1985. Photochemistry of petroleum in water. *Environ. Sci. Technol.* 19: 569–579.

Pearlman, R. S., S. H. Yalkowsky, and S. Banerjee. 1984. Water solubilities of polynuclear aromatic and heteroaromatic compounds. *J. Phys. Chem. Ref. Data* 13: 555–562.

Pellenbarg, R. E. 1978. *Spartina alterniflora* litter and the aqueous surface microlayer in the salt marsh. *Estuar. Coast. Mar. Sci.* 6: 187–195.

Penkett, S. A. 1982. Non-methane organics in the remote troposphere. In E. D. Goldberg, ed., *Atmospheric chemistry*, Springer, Berlin. pp. 329–355.

Perdue, E. M., J. H. Reuter, and M. Ghosal. 1980. The operational nature of acidic functional group analyses and its impact on mathematical descriptions of acid-base equilibria in humic substances. *Geochim. Cosmochim. Acta* 44: 1841–1851.

Pitts, J. N., Jr. 1987. Nitraton of gaseous polycyclic aromatic hydrocarbons in simulated and ambient urban atmospheres: a source of mutagenic nitroarenes. *Atmos. Environ.* 21: 2531–2547.

Pitts, J. N., Jr., K. A. Van Cauwenberghe, D. Grosjean, J. P. Schmid, D. R. Fitz, W. L. Belser Jr., G. B. Knudson, and P. M. Hynds. 1978. Atmospheric reactions of polycyclic aromatic hydrocarbons: Facile formation of mutagenic nitro derivatives. *Science* 202: 515–519.

Pojasek, R. B. and O. T. Zajicek. 1978. Surface microlayers and foams—source and metal transport in aquatic systems. *Water Res.* 12: 7–10.

Popoff, T. and Theander, O. 1972. Formation of aromatic compounds from carbohydrates. I. Reaction of D-glucuronic acid, D-galacturonic acid, D-xylose, and L-arabinose in slightly acidic, aqueous solution. *Carbohydrate Res.* 22: 135–149.

Popoff, T. and Theander, O. 1976. Formation of aromatic compounds from carbohydrates. III. Reaction of D-glucose and D-fructose in slightly acidic, aqueous solution. *Acta Chem. Scand.* 30: 397–402.

Preston, C. M. and J. A. Ripmeester. 1982. Application of solution and solid-state ^{13}C NMR to four organic soils, their humic acids, fulvic acids, humins and hydrolysis residues. *Can. J. Spectrosc.* 27: 99–105.

Prinn, R., D. Cunnold, R. Rasmussen, P. Simmonds, F. Alyea, A. Crawford, P. Fraser, and R. Rosen. 1987. Atmospheric trends in methylchloroform and the global average for the hydroxyl radical. *Science* 238: 945–950.

Ramanathan, V., R. J. Cicerone, H. B. Singh, and T. Kiehl. 1985. Trace gas trends and their potential role in climate change. *J. Geophys. Res.* 90: 5547–5566.

Ramdahl, T., G. Becher, and A. Bjørseth. 1982. Nitrated polycyclic aromatic hydrocarbons in urban air particles. *Environ. Sci. Technol.* 16: 861–865.

Ramdahl, T. 1983. Polycyclic aromatic ketones in environmental samples. *Environ. Sci. Technol.* 17: 660–670.

Rappe, C. 1984. Analysis of polychlorinated dioxins and furans. *Environ. Sci. Technol.* 18: 78A-90A.

Rasmussen, R. A. and M. A. K. Khalil. 1983. Global production of methane by termites. *Nature* 301: 700–702.

Rasmussen, R. A. and F. W. Went. 1965. Volatile organic material of plant origin in the atmosphere. *Proc. Nat. Acad. Sci.* U.S. 53: 215–220.

Riess-Kautt, M., J. P. Kintzinger, and P. Albrecht. 1988. *Omega*-feruloyloxyacids, a novel class of polar lipids in peat soil. *Naturwissenschaften* 75: 305–307.

Riley, J. P. and D. A. Segar. 1970. The seasonal variation of the free and combined dissolved amino acids in the Irish Sea. *J. Mar. Biol. Assoc.* U.K. 50: 713–720.

Robinson, E. and R. C. Robbins. 1968. *Sources, abundance, and fate of gaseous atmospheric pollutants.* Stanford Research Institute Project Report #PR-6755, Huntsville, AL. 123 pp.

Rowland, F. S. 1989. Chlorofluorocarbons and the depletion of stratospheric ozone. *Amer. Scientist* 77: 36–45.

Rowland, F. S. 1991. Stratospheric ozone in the 21st century: the chlorofluorocarbon problem. *Environ. Sci. Technol.* 25: 622–628.

Sargent, J. R. 1978. Marine wax esters. *Sci. Progr.* (Oxf.) 65: 437–458.

Sato, O. 1976. The chemical nature of components of PG (green fraction of P type humic acid). *Soil Sci. Plant Nutr.* 22: 485–488.

Schmeltz, I. and D. Hoffman. 1976. Formation of polynuclear aromatic hydrocarbons from combustion of organic matter. In R. Freudenthal and P. W. Jones, eds. Carcinogenesis: a comprehensive survey. Vol. 1. *Polynuclear aromatic hydrocarbons: chemistry, metabolism, and carcinogenesis.* Raven Press, New York.

Schnitzer, M. 1985. Nature of nitrogen in humic substances. In G. R. Aiken, D. M. McKnight, R. L. Wershaw, and P. MacCarthy, eds. *Humic substances in soil, sediment, and water.* Wiley, New York, pp. 303–325.

Schnitzer, M. and U. C. Gupta. 1965. Determination of acidity in soil organic matter. *Soil Sci. Soc. Amer. Proc.* 29: 274–277.

Schnitzer, M. and S. U. Khan. 1972. *Humic substances in the environment.* Marcel Dekker, New York.

Schnitzer, M. and M. I. Ortiz de Serra. 1973. The chemical degradation of a humic acid. *Can. J. Chem.* 51: 1554–1556.

Schuetzle, D., D. Cronn, A. L. Crittenden, and R. J. Charlson. 1975. Molecular composition of secondary aerosol and its possible origin. *Environ. Sci. Technol.* 9: 838–845.

Schütte, K. H. and J. F. Ellsworth. 1954. The significance of large pH fluctuations observed in some South African vleis. *J. Ecol.* 42: 148–150.

Scully, F. E., Jr., G. D. Howell, R. Kravitz, J. T. Jewell, V. Hahn, and M. Speed. 1989. Proteins in natural waters and their relation to the formation of chlorinated organics during water disinfection. *Environ. Sci. Technol.* 22: 537–542.

Sequi, P., G. Guidi, and G. Petruzzelli. 1975. Distribution of amino acid and carbohydrate components in fulvic acid fractionated on polyamide. *Can. J. Soil Sci.* 55: 439–445.

Sexton, K. and H. Westberg. 1984. Nonmethane hydrocarbon composition of urban and rural atmospheres. *Atmos. Environ.* 18: 1125–1132.

Shaw, D. G. and B. A. Baker. 1978. Hydrocarbons in the marine environment of Port Valdez, Alaska. *Environ. Sci. Technol.* 13: 1200–1205.

Sieburth, J. McN. and A. Jensen. 1968. Studies on algal substances in the sea. 1. Gelbstoff (humic material) in terrestrial and marine waters. *J. Exp. Mar. Biol. Ecol.* 2: 174–189.

Skujins, J. J. 1967. Enzymes in soil. In A. D. McLaren and G. H. Peterson, eds., *Soil biochemistry*, Vol. 1. Marcel Dekker, New York. pp. 371–416.

Smart, P. L., B. L. Finlayson, W. D. Rylands, and C. M. Ball. 1976. The relationship of fluorescence to dissolved organic carbon in surface waters. *Water Res.* 10: 805–811.

Snape, C. E., D. E. Axelson, R. E. Botto, J. J. Delpeuch, P. Tekely, B. C. Gerstein, M. Pruski, G. E. Maciel, and M. A. Wilson. 1989. Quantitative reliability of aromaticity and related measurements on coals by ^{13}C n. m. r. A debate. *Fuel* 68: 547–560.

Soma, Y., M. Soma, and I. Harada. 1983. Raman spectroscopic evidence of formation of *p*-dimethoxybenzene cation on Cu- and Ru-montmorillonite. *Chem. Phys. Lett.* 94: 475–478.

Spiteller, M. and M. Schnitzer. 1983. A comparison of the structural characteristics of polymaleic acid and a soil fulvic acid. *J. Soil Sci.* 34: 525–537.

Steelink, C., M. A. Mikita, and K. A. Thorn. 1983. Magnetic resonance studies of humates and related model compounds. In R. F. Christman and E. T. Gjessing, eds., *Aquatic and terrestrial humic materials*. Ann Arbor Sci. Pub., Ann Arbor, MI, pp. 83–105.

Steinberg, S. and I. R. Kaplan. 1984. The determination of low molecular weight aldehydes in rain, fog and mist by reversed phase liquid chromatography of the 2,4-dinitrophenylhydrazone derivatives. *Int. J. Environ. Anal. Chem.* 18: 253–266.

Steinberg, S., K. Kawamura, and I. R. Kaplan. 1985. The determination of α-keto acids and oxalic acid in rain, fog and mist by HPLC. *Int. J. Environ. Anal. Chem.* 19: 251–260.

Stevenson, F. J. 1969. Organic acids in soil. In A. D. McLaren and G. H. Peterson, eds., *Soil biochemistry*, Vol. 1. Marcel Dekker, New York. pp. 119–146.

Stevenson, F. J. 1982. *Humus chemistry: genesis, composition, reactions*. Wiley-Interscience, New York.

Stevenson, F. J. and J. H. A. Butler. 1969. Chemistry of humic acids and related pigments. In G. Eglinton and M. J. Murphy, eds. *Organic geochemistry*. Springer Verlag, New York. pp. 534–557.

Stiver, W. and D. Mackay. 1984. Evaporation rate of spills of hydrocarbons and petroleum mixtures. *Environ. Sci. Technol.* 18: 834–840.

Stiver, W., W. Y. Shiu, and D. Mackay. 1989. Evaporation times and rates of specific hydrocarbons in oil spills. *Environ. Sci. Technol.* 23: 101–105.

Stolarski, R. S. 1986. Nimbus 7 satellite measurements of the springtime Antarctic ozone decrease. *Nature* 322: 808–812.

Strahler, A. N. 1957. Quantitative analysis of watershed geomorphology. *Trans. Amer. Geophys. Union* 83: 913–920.

Stumm, W. and J. J. Morgan. 1981. *Aquatic chemistry.* 2nd ed., Wiley-Interscience, New York.

Swain, F. M. 1963. Geochemistry of humus. In I. A. Breger, ed. *Organic geochemistry*. Pergamon, New York, pp. 81–147.

Sweet, M. S. and E. M. Perdue. 1982. Concentration and speciation of dissolved sugars in river water. *Environ. Sci. Technol.* 16: 692–698.

Taft, R. W., Jr. 1956. Separation of polar, steric, and resonance effects in reactivity, in *Steric Effects in Organic Chemistry*, M. S. Newman, Ed., John Wiley, New York, pp. 556–675.

Tait, M. J., A. Suggett, F. Franks, S. Ablett, and P. A. Quikenden. 1973. Hydration of monosaccharides: a study by dielectric and nuclear magnetic relaxation. *J. Solution Chem.* 1: 131–151.

Tan, K. H. and F. E. Clark. 1969. Polysaccharide constituents in fulvic and humic acids extracted from soil. *Geoderma* 2: 245–255.

Tate, K. R. and H. A. Anderson. 1978. Phenolic hydrolysis products from gel chromatographic fractions of humic acids. *J. Soil Sci.* 29: 76–83.

Tegelaar, E. W., J. W. de Leeuw, and C. Saiz-Jimenez. 1989. Possible origin of aliphatic moieties in humic substances. *Sci. Total Environ.* 81/82: 1–17.

Terentev, V. M., R. I. Tsareva, and O. V. Shchutskaya. 1968. *Isotopes and radiation in soil organic-matter studies*. Intl. Atomic Energy Agency, Vienna, p. 421.

Tewari, Y. B., M. M. Miller, S. P. Wasik, and D. E. Martire. 1982. Aqueous solubility and octanol/water partition coefficient of organic compounds at 25.0°C. *J. Chem. Eng. Data* 27: 451–454.

Theng, B. K. G. 1974. *The chemistry of clay-organic reactions*. Wiley, New York.

Thibodeaux, L. J. 1979. *Chemodynamics*. Wiley, New York.

Thomas, R. G. 1990. Volatilization from water. In W. J. Lyman, W. F. Reehl, and D. H. Rosenblatt, eds. *Handbook of chemical property estimation methods*. American Chemical Society, Washington, DC, pp. 15-1 – 15-34.

Thorn, K. A. 1989. Nuclear-magnetic-resonance spectrometry investigations of fulvic and humic acids from the Suwanee River. In Thurman, E. M. and R. L. Malcolm. 1989. Nitrogen and amino acids in humic and fulvic acids from the Suwanee River. USGS Report 87-557, Denver, CO, pp. 251–309.

Thurman, E. M. 1985. *Organic geochemistry of natural waters*. Martinus Nijhoff/ Dr. W. Junk Pub., Dordrecht, Netherlands.

Tsani-Bazaca, E., A. McIntyre, J. Lester, and R. Perry. 1982. Ambient concentrations and correlations of hydrocarbons and halocarbons in the vicinity of an airport. *Chemosphere* 11: 11–23.

Tsutsuki, K. and S. Kuwatsawa. 1978. Composition of oxygen-containing functional groups of humic acids. *Soil Sci. Plant Nutr.* 24: 547–570.

Vannote, R. L., G. W. Minshall, K. W. Cummins, J. R. Sedell, and C. E. Cushing. 1980. The river continuum concept. *Can. J. Fish. Aquat. Sci.* 37: 130–137.

Verschueren, K. 1983. *Handbook of environmental data on organic chemicals*. 2nd ed. Van Nostrand-Reinhold, New York.

Visser, S. A. 1983. Fluorescence phenomena of humic matter of aquatic origin and microbial cultures. In R. F. Christman and E. T. Gjessing, eds., *Aquatic and terrestrial humic materials*. Ann Arbor Sci. Pub., Ann Arbor, MI, pp. 183–202.

von Düszeln, J. and W. Thiemann. 1985. Volatile chlorinated hydrocarbons in a coastal urban atmosphere. *Sci. Total Environ.* 41: 187–194.

Walker, A. and D. V. Crawford. 1968. The role of organic matter in adsorption of the triazine herbicides by soils. *Proc. Sympos. Radiation in Soil Org. Matter Studies*, pp. 91–108.

Wathne, B. 1983. Measurements of benzene, toluene and xylenes in urban air. *Atmos. Environ*. 17: 1713–1722.

Watson, R. T., M. J. Prather, and M. J. Kurylo. 1988. Present state of knowledge of the upper atmosphere 1988: an assessment report. NASA Ref. Publ. #1208, Washington, DC.

Weber, W. J., Jr., P. M. McGinley, and L. E. Katz. 1991. Sorption phenomena in subsurface systems: concepts, models and effects on contaminant fate and transport. *Water Res*. 25: 499–528.

Wershaw, R. L. 1986. A new model for humic materials and their interactions with hydrophobic organic chemicals in soil-water or sediment-water systems. *J. Contam. Hydrol*. 1: 29–45.

Wershaw, R. L. and G. R. Aiken. 1985. Molecular size and weight measurements of humic substances. In G. R. Aiken, D. M. McKnight, R. L. Wershaw, and P. MacCarthy, eds. *Humic substances in soil, sediment, and water*. Wiley, New York, pp. 477–492.

Williams, P. M. 1971. The distribution and cycling of organic matter in the ocean. In S. J. Faust and J. V. Hunter, eds. *Organic compounds in aquatic environments*. Marcel Dekker, New York. pp. 145–163.

Wilson, A. L. 1959. Determination of fulvic acids in water. *J. Appl. Chem*. 9: 501–510.

Wilson, M. A. 1981. Applications of NMR spectroscopy to the study of the structure of soil organic matter. *J. Soil Sci*. 32: 167–186.

Wilson, M. A., A. J. Jones, and B. Williamson. 1978. NMR spectroscopy of humic materials. *Nature* 276: 487–489.

Wilson, M. A., P. J. Collin, and K. R. Tate. 1983. ^{1}H-nuclear magnetic resonance study of a soil humic acid. *J. Soil Sci*. 34: 297–304.

Wilson, M. A., K. M. Goh, P. J. Collin, and L. G. Greenfield. 1986. *Origin of humus variation*. Org. Geochem. 9: 225–231.

Wolfe, N. L., L. A. Burns and W. C. Steen. 1980a. Use of linear free energy relationships and an evaluation model to assess the fate and transport of phthalate esters in the aquatic environment, *Chemosphere*, 9:393–402.

Wolfe, N. L., W. C. Steen and L. A. Burns. 1980b. "Phthalate Ester Hydrolysis: Linear Free Energy Relationships," *Chemosphere*, 9:403–408.

Wolfe, N. L., U. Mingelgrin, and G. C. Miller. 1990. Abiotic transformations in water, sediments, and soil. In Pesticides in the soil environment; SSSA Book Series #2. Soil Sci. Soc. Amer., Madison, WI. pp. 103–168.

Wood, J. C., S. E. Moschopedis, and W. den Hertog. 1961. Studies in humic acid chemistry. II. Humic anhydrides. *Fuel* 40: 491–502.

Woodwell, G. M. and E. V. Pecan, eds. 1973. *Carbon and the biosphere*. 24th Brookhaven Symposium in Biology. 1973.

Wright, R. T. and J. E. Hobbie. 1965. The uptake of organic solutes in lake water. *Limnol. Oceanogr.* 10: 22–28.

Wright, R. T. and N. M. Shah. 1977. The trophic role of glycolic acid in coastal seawater. II. Seasonal changes in concentration and heterotrophic use in Ipswich Bay, Massachusetts, USA. *Marine Biol.* 43: 257–263.

Wynder, E. L. and D. Hoffmann, eds. 1968. *Tobacco and tobacco smoke: studies in experimental carcinogenesis.* Academic Press, New York.

Yokouchi, Y. and Y. Ambe. 1986. Characterization of polar organics in airborne particulate matter. *Atmos. Environ.* 20: 1727–1734.

Yonebayashi, K. and T. Hattori. 1987. Surface active properties of soil humic acids. *Sci. Total Environ.* 62: 55–64.

Zafiriou, O. C. 1975. Reaction of methyl halides with seawater and marine aerosols. *J. Mar. Res.* 33: 75–81.

Zepp, R. G. and P. F. Schlotzhauer. 1981. Comparison of photochemical behavior of various humic substances in water. III. Spectroscopic properties of humic substances. *Chemosphere* 10: 479–486.

Zhou, X. and K. Mopper. 1990. Measurement of sub-parts-per-billion levels of carbonyl compounds in marine air by a simple cartridge trapping procedure followed by liquid chromatography. *Environ. Sci. Technol.* 24: 1482–1485.

Zoeteman, B. C. J., K. Harmsen, J. B. H. J. Linderts, C. F. H. Morra, and W. Slooff. 1980. Persistent organic pollutants in river water and ground water of the Netherlands. *Chemosphere* 9: 231–249.

Zsolnay, A. 1977. Inventory of non-volatile fatty acids and hydrocarbons in the oceans. *Mar. Chem.* 5: 465–475.

Zurer, P. S. 1990. Chlorine eroding arctic as well as antarctic ozone, scientists confirm. *Chem. Eng. News* (Mar. 15), 22–23.

CHAPTER 2

HYDROLYSIS

A. INTRODUCTION

In general terms, hydrolysis is defined as a chemical transformation in which an organic molecule, RX, reacts with water, resulting in the formation of a new covalent bond with OH and cleavage of the covalent bond with X (the leaving group) in the original molecule. The net reaction is the displacement of X by OH^- (Harris, 1981; Mill and Mabey, 1988)

$$RX + H_2O \rightarrow ROH + X^- + H^+ \tag{2.1}$$

Hydrolysis is an example of a larger class of reactions referred to as nucleophilic displacement reactions in which a nucleophile (an electron-rich species containing an unshared pair of electrons) attacks an electrophilic atom (an electron-deficient reaction center). Hydrolytic processes encompass several types of reaction mechanisms that can be defined by the type of reaction center (i.e., the atom bearing the leaving group, X) where hydrolysis occurs. The reaction mechanisms encountered most often are direct and indirect nucleophilic substitution and nucleophilic addition-elimination.

The chemical structures of hydrolyzable functional groups that are prevalent in environmental chemicals, and the reaction products resulting from their hydrolysis, are listed in Table 2.1. For these classes of chemicals, hydrolysis may be the dominant pathway for their transformation in aquatic ecosystems. Hydrolytic processes are not limited to the bodies of water such as rivers, streams, lakes, and oceans usually associated with the term aquatic ecosystems. Hydrolysis of organic chemi-

Table 2.1. Examples of Hydrolyzable Functional Groups

1. Halogenated Aliphatics

Nucleophilic substitution

$$RCH_2X \xrightarrow{H_2O,\ OH^-} RCH_2OH + HX$$

Elimination

$$-\overset{X}{\underset{|}{\overset{|}{C}}}-\overset{|}{\underset{H}{\underset{|}{C}}}- \xrightarrow{H_2O,\ OH^-} >C{=}C< + HX$$

2. Epoxides

$$\text{epoxide } (C-C \text{ bridged by } O) \xrightarrow{H^+,\ OH^-} -\overset{OH}{\underset{|}{\overset{|}{C}}}-\overset{OH}{\underset{|}{\overset{|}{C}}}-$$

3. Organophosphorus Esters

$$R_1O-\overset{X}{\overset{\|}{P}}(OR_2)-OCH_2R_3 \xrightarrow{H_2O,\ OH^-} R_1OH + {}^-O\overset{X}{\overset{\|}{P}}(OR_2)OCH_2R_3 \quad \text{or} \quad R_1O\overset{X}{\overset{\|}{P}}(OR_2)O^- + HOCH_2R_3$$

X=O,S

4. Carboxylic Acid Esters

$$R_1C(=O)OR_2 \xrightarrow{H^+,\ OH^-} R_1C(=O)O^- + HOR_2$$

5. Anhydrides

$$R_1C(=O)OC(=O)R_2 \xrightarrow{H^+,\ OH^-} R_1C(=O)O^- + {}^-OC(=O)R_2$$

6. Amides

$$R_1C(=O)N(H)R_2 \xrightarrow{H^+,\ OH^-} R_1C(=O)O^- + H_2NR_2$$

7. Carbamates

$$R_1N(H)C(=O)OR_2 \xrightarrow{H^+,\ OH^-} R_1NH_2 + CO_2 + HOR_2$$

8. Ureas

$$R_1N(H)C(=O)N(H)R_2 \xrightarrow{H^+,\ OH^-} R_1NH_2 + CO_2 + H_2NR_2$$

cals also can occur in fogwater, biological systems, groundwater systems, and the aqueous microenvironment associated with soils and sediments.

B. HYDROLYSIS KINETICS

1. Specific Acid and Base Catalysis

In addition to nucleophilic attack by H_2O (neutral hydrolysis), we find that hydrolytic reactions are sensitive to specific acid and specific base catalysis (i.e., catalysis by hydronium ion, H^+, and hydroxide ion, OH^-, respectively). Accordingly, hydrolysis kinetics must take into account the potential for H_2O to dissociate. Even at pH 7.0, where the concentration of H^+ and OH^- is only 10^{-7} M, specific acid and specific base catalysis can significantly accelerate hydrolysis kinetics. Specific acid and base catalysis occurs because the hydronium ion and hydroxide ions provide an alternative mechanism for hydrolysis that is energetically more favorable. In specific acid catalysis, hydronium ion is thought to provide a reaction pathway of lower energy by withdrawing electron density from the atom bearing the leaving group, X, thus making it more susceptible to nucleophilic attack by H_2O. Specific base catalysis occurs because OH^- is a much more reactive nucleophile than H_2O (typically by a factor of about 10^4: Streitwieser, 1962). Accordingly, a hydrolysis reaction involving nucleophilic attack by OH^- will occur at a faster rate than the pathway involving nucleophilic attack by H_2O alone.

Taking into consideration specific acid and base catalysis (the concentration of a catalyst is reflected in the rate law but is not reflected in the equilibrium constant), the hydrolysis rate term for hydrolyzable chemicals can be described by:

$$d[RX]/dt = k_{hyd}[RX] = k_a[H^+][RX] + k_n[RX] + k_b[OH^-][RX] \quad (2.2)$$

where [RX] is the concentration of the hydrolyzable compound, k_{hyd} is the observed or measured hydrolysis rate constant, and k_a, k_n, and k_b are the rate constants for the acid-catalyzed, neutral, and base-catalyzed processes. Assuming that the individual rate processes for the acid, neutral, and base hydrolyses obey first-order kinetics with respect to the hydrolyzable chemical, RX, it is possible to write the following equation for k_{hyd}:

$$k_{hyd} = k_a[H^+] + k_n + k_b[OH^-] \quad (2.3)$$

From the equilibrium term for the ionization of water, K_w:

$$K_w = [OH^-][H^+] = 1 \times 10^{-14} \quad (2.4)$$

it is possible to substitute for $[OH^-]$ in Equation 2.4, giving:

$$k_{hyd} = k_a[H^+] + k_n + k_b(K_w/[H^+]) \tag{2.5}$$

Because k_{hyd} is a pseudo first-order rate constant at a fixed pH (i.e., the hydrolysis is independent of RX concentration), the half-life for hydrolysis can be calculated from Equation 2.6.

$$t_{1/2} = \ln 2/k_{hyd} \tag{2.6}$$

2. pH Dependence

It is apparent from Equation 2.5 that the overall rate constant for hydrolysis, k_{hyd}, depends on pH and the magnitude of the rate constants for the individual processes. Plots of log k_{hyd} versus pH are very useful for determining the contribution of the acid, neutral, and base terms for hydrolysis of a compound of interest at a specific pH. Figure 2.1 illustrates the log k_{hyd} versus pH plot for several chemicals of interest. These data demonstrate that the relationship between hydrolysis kinetics and pH is dependent on the nature of the hydrolyzable functional group. For example, over typical environmental pH values (4 to 8), the neutral hydrolysis rate term for ethylene oxide and methyl chloride will dominate. Only below pH 4 will the acid-catalyzed rate term for ethylene oxide contribute to the overall hydrolysis rate term. Likewise, base-catalyzed hydrolysis of ethylene oxide and methyl chloride will not

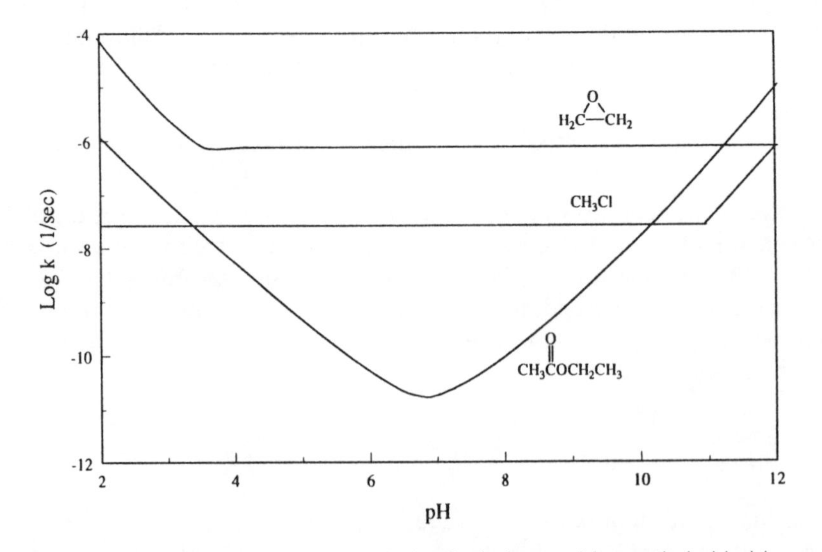

Figure 2.1. pH-rate profiles for the hydrolysis of ethylene oxide, methyl chloride and ethyl acetate (Data taken from Mabey and Mill, 1978).

compete with neutral hydrolysis until pH 11, which is well above the pH range of natural aquatic ecosystems. By contrast, the neutral hydrolysis of ethyl acetate occurs over a much narrower pH range (6 to 7). Either below or above this pH range, the rate terms for acid- or base-catalyzed hydrolysis will dominate.

C. HYDROLYSIS REACTION MECHANISMS

Reaction mechanisms for hydrolysis can be classified according to the type of reaction center involved. The primary distinction is made between reaction at saturated and unsaturated centers. With respect to carbon-centered functional groups, which will be the primary focus of this chapter, hydrolysis involves reactions at sp^3 (saturated) or sp^2 (unsaturated) hybridized carbons. Nucleophilic reactions at sp^3 carbons are termed nucleophilic substitution (2.7). The reaction at sp^2 carbons is termed nucleophilic addition-elimination or acyl substitution (2.8).

$$\mathrm{RX} + \mathrm{Y:} \longrightarrow \mathrm{RY} + \mathrm{X:} \qquad (2.7)$$

$$\mathrm{R\overset{\overset{\displaystyle O}{\|}}{C}X} + \mathrm{Y:} \longrightarrow \mathrm{R\overset{\overset{\displaystyle O}{\|}}{C}Y} + \mathrm{X:} \qquad (2.8)$$

Because both the nucleophile, Y:, and the leaving group, X:, are Lewis bases, these are examples of Lewis acid-base reactions in which one Lewis base replaces another in the Lewis acid-base adduct (Jensen, 1978).

1. Nucleophilic Substitution

The limiting cases of nucleophilic substitution have been described as the ionization mechanism (S_N1, substitution-nucleophilic-unimolecular) and the direct displacement mechanism (S_N2, substitution-nucleophilic-bimolecular: Gleave et al., 1935). The S_N1 and S_N2 mechanisms describe the extremes in nucleophilic substitution reactions. Pure S_N1 and S_N2 reaction mechanisms, however, are rarely observed. More often a mix of these reaction mechanisms are occurring simultaneously.

S_N1 Mechanism

The S_N1 mechanism begins by a rate-determining heterolytic dissociation of the substrate to an sp^2 hybridized carbocation (commonly referred to as a carbonium ion) and the leaving group (2.9):

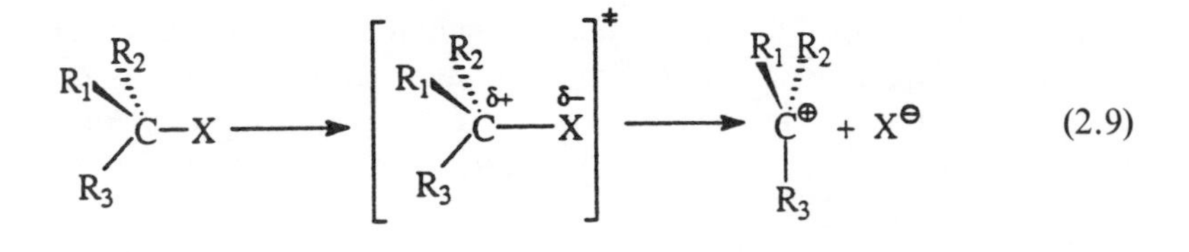

(2.9)

$$\text{Rate} = k_1[\text{RX}] \tag{2.10}$$

where $[R - X]^{\ddagger}$ represents a transition state, an energy maximum in the dissociation of RX to the R^+ and X^- ions. Because formation of the carbocation is rate-determining, the reaction kinetics for the S_N1 mechanism will exhibit first-order kinetics overall (i.e., the rate of S_N1 reactions is independent of the concentration or the nature of the attacking nucleophile). Accordingly, factors that stabilize the transition state leading to the formation of the carbocation, and stabilization of the cation itself, will increase reactivity. Thus, the reactivity of S_N1 reactions is affected primarily by electronic factors such as resonance or inductive effects.

In aqueous systems, formation of the highly reactive carbonium ion by the S_N1 mechanism is usually followed by reaction with water, resulting in the formation of an alcohol (2.11); however, prior to reaction with water, rearrangement may occur through migration of substituents, usually protons or alkyl groups, resulting in formation of a more stable carbonium ion (2.12). Also, elimination of a proton on the carbon adjacent to the carbocation, which results in the formation of alkenes, may occur (2.13). The observation of reaction products resulting from these processes can serve as evidence for the intermediacy of a carbonium ion.

$$\text{H–C(R}_1\text{)(R}_2\text{)–C}^{\oplus}\text{(R}_3\text{)(H)} \xrightarrow{H_2O} \text{H–C(R}_1\text{)(R}_2\text{)–C(R}_3\text{)(H)–}\overset{\oplus}{\text{O}}\text{H}_2 \xrightarrow{-H^+} \text{H–C(R}_1\text{)(R}_2\text{)–C(R}_3\text{)(H)–OH} \tag{2.11}$$

$$\text{H–C(R}_1\text{)(R}_2\text{)–C}^{\oplus}\text{(R}_3\text{)(H)} \xrightarrow[\text{-1,2-proton migration}]{\text{~H}} {}^{\oplus}\text{C(R}_1\text{)(R}_2\text{)–C(R}_3\text{)(H)–H} \xrightarrow{H_2O} \text{HO–C(R}_1\text{)(R}_2\text{)–C(R}_3\text{)(H)–H} \tag{2.12}$$

$$\text{H–C(R}_1\text{)(R}_2\text{)–C}^{\oplus}\text{(R}_3\text{)(H)} \xrightarrow[\text{proton elimination}]{-H^+} \text{(R}_1\text{)(R}_2\text{)C=C(R}_3\text{)(H)} \tag{2.13}$$

S_N2 Mechanism

The direct displacement or S_N2 mechanism occurs through a transition state in which bond breaking and bond making occur simultaneously.

$$X{-}C(R_1)(R_2)(R_3) + :Y \longrightarrow \left[\overset{\delta-}{X} \cdots C(R_1)(R_2)(R_3) \cdots \overset{\delta-}{Y} \right]^{\ddagger} \longrightarrow X: + (R_1)(R_2)(R_3)C{-}Y \quad (2.14)$$

$$\text{Rate} = k_2[\text{R-X}][:\text{Y}] \quad 2.15$$

Because both the substrate (or electrophile) and the attacking nucleophile appear in the transition state, the concentrations of both of these species will appear in the rate term. Also, because the nucleophile approaches the substrate at 180° from the leaving group (termed backside attack), inversion of configuration will occur at carbons that are chiral (i.e., carbon atoms bearing four nonidentical substituents). Due to the geometry of the transition state for the S_N2 mechanism, the reactivity of S_N2 reactions is largely dependent on steric factors and to a lesser degree on electronic factors. The carbon at which nucleophilic attack occurs is bonded to five substituents in the transition state of the S_N2 reaction. The result is a crowded transition state. Increasing the bulk of these substituents will increase the nonbonded interactions in the transition state, raising the energy of the transition state and slowing the rate of the reaction.

Classes of organic pollutants that hydrolyze via nucleophilic substitution reactions include the halogenated hydrocarbons, epoxides, and phosphorus esters. Further discussion of the factors affecting the reactivity of nucleophilic substitution reactions will be made as the hydrolysis mechanisms of these chemicals are examined in greater detail.

2. Functional Group Transformation by Nucleophilic Substitution Reactions

a. Halogenated Aliphatics

The widespread contamination of surface and groundwaters by halogenated aliphatics has prompted efforts to determine pathways for the degradation of these environmental pollutants. Hydrolysis half-lives for halogenated aliphatics range from months to years; however, because residence times for these pollutants in groundwater systems are often measured in years, hydrolysis will be an environmentally significant transformation pathway for these compounds.

Simple halogenated aliphatics. Hydrolysis of the simple halogenated aliphatics (halogen substitution at one saturated carbon atom) by nucleophilic substitution with H_2O is generally pH-independent, resulting in the formation of alcohols (2.16).

$$RX + H_2O \rightarrow ROH + HX \quad (2.16)$$

Although a number of the simple halogenated aliphatics are susceptible to base-catalyzed hydrolysis, the rate term for the base-catalyzed process will have a negligible effect on the overall hydrolysis rate under environmental conditions. The data listed in Table 2.2 summarize hydrolysis half-lives at 25°C and pH 7 for a series of simple halogenated aliphatics (Mabey and Mill, 1978).

The measured half-lives range from 19 sec to 7000 yr, suggesting that structure variation can have significant effects on hydrolysis rates. The reactivity of these chemicals can be rationalized in terms of the limiting mechanisms presented for nucleophilic substitution. It is apparent from the data in Table 2.2 that the fluorinated aliphatics are much more stable than the chlorinated aliphatics, which in turn are more stable than the brominated aliphatics. This trend in reactivity reflects the strength of the carbon-halogen bond, which follows the order F>Cl>Br, that is broken in the nucleophilic substitution reaction.

Of the simple halogenated aliphatics in Table 2.2, the halogenated methanes are of the greatest environmental concern because of their ubiquitous nature as groundwater pollutants. The halogenated methanes, except for the trihalomethanes, hydrolyze by direct nucleophilic displacement by water (S_N2 mechanism). Hydrolysis of the trihalomethanes, CHX_3, is thought to occur initially by proton abstraction (E1cB mechanism) to form the trichloromethyl carbanion (2.17). This carbanion subsequently undergoes chloride ion elimination to form dichlorocarbene (2.18),

Table 2.2. Hydrolysis Half-Lives for Simple Halogenated Aliphatics at pH 7 at 25°C[a]

RX	X = F	x = Cl	x = Br
CH_3X	30 y	0.93 y	20 d
CH_2X_2	–	704 y	183 y
CHX_3	–	3500 y	686 y
CX_4	–	7000 y	–
CH_3CH_2X	–	38 d	30 d
$(CH_3)_2CHX$	–	38 d	2.1 d
$(CH_3)_3CX$	50 d	23 s	–
$CH_2{=}CHX$	–	NR[b]	
PhX	–	NR[b]	
$CH_2{=}CH{-}CH_2X$	–	69 d	12 h
$C_6H_5CH_2X$	–	15 h	–
$C_6H_5CHX_2$	–	0.1 h	–
$C_6H_5CX_3$	–	19 s	–
CH_3OCH_2X	–	1.99 min[c]	–

[a]Data taken from Mabey and Mill (1978) except where noted.
[b]No significant hydrolysis reported at 100°C after 10 d. (Hill et al., 1976).
[c]Data taken from Van Duuren et al. (1972).

which reacts with OH^- to form carbon monoxide and chloride ion (2.19) (Hine et al., 1956):

$$CHCl_3 + OH^- \rightarrow {}^-CCl_3 + H_2O \qquad (2.17)$$

$${}^-CCl_3 \rightarrow :CCl_2 + Cl^- \qquad (2.18)$$

$$:CCl_2 + 2OH^- \rightarrow CO + 2Cl^- + H_2O \qquad (2.19)$$

For the mono-, di-, and tetrahalomethanes, an increase in the number of halogen substituents on carbon increases the hydrolysis half-life because of the greater steric hindrance about the site of nucleophilic attack. By contrast, however, as the steric bulk about the carbon atom bearing the halogen is increased by the addition of methyl groups [CH_3X, CH_3CH_2X, $(CH_3)_2CHX$] a significant decrease in reactivity (decreasing half-lives) is not observed, as might be predicted, for nucleophilic displacement reactions occurring by an S_N2 mechanism. The significant increase in reactivity of $(CH_3)_2CHBr$ compared to CH_3CH_2Br reflects the contribution of the S_N1 mechanism to hydrolysis of the isopropyl derivative. Similarly, the dramatic increase in the reactivity of *t*-butyl chloride compared to that of isopropyl chloride signals a shift in reaction mechanism from one that is a mix of S_N2 and S_N1 mechanisms for isopropyl chloride to primarily a S_N1 mechanism for *t*-butyl chloride. The allylic ($CH_2{=}CH{-}CH_2{-}X$), benzylic ($PhCH_2X$) and alkoxymethyl halides ($ROCH_2X$) are other examples of halogenated organics that hydrolyze by a S_N1 mechanism. With the exception of allyl chloride, the hydrolysis half-lives for these chemicals are on the order of minutes to hours. The dramatic increase in reactivity is due to the structural features of these compounds that allow for delocalization, and thus stabilization, of the carbonium ion intermediate. Resonance structures can be drawn for the allylic and benzylic carbonium ions that delocalize the positive charge over several carbon atoms:

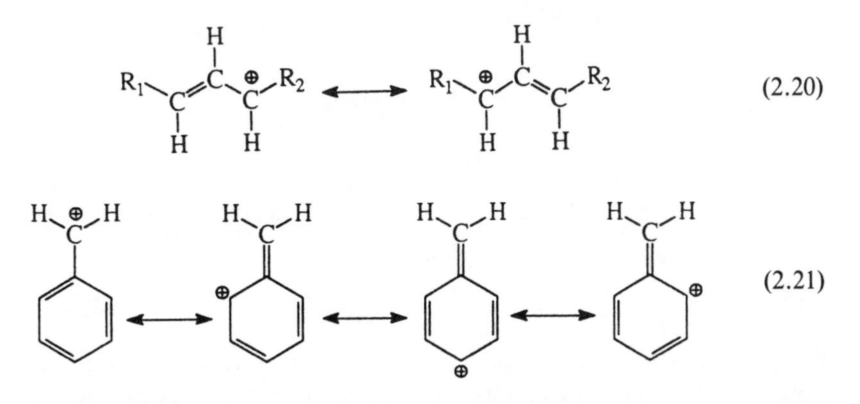

(2.20)

(2.21)

Resonance effects can also be used to explain the facile hydrolysis of 2-chloromethyl ether (Table 2.2). Although the oxygen atom, which is bonded directly to the carbon bearing the chlorine in 2-chloromethyl ether, is electron-withdrawing by the inductive effect (i.e., electron-withdrawing and -donating effects that have their origin in bond dipoles), the electron-donating resonance effect is dominant:

$$R-\ddot{\underset{\cdot\cdot}{O}}-\overset{\oplus}{C}H_2 \longleftrightarrow R-\overset{\oplus}{\underset{\cdot\cdot}{O}}=CH_2 \tag{2.22}$$

Similar rate enhancements are expected for aliphatic hydrocarbons containing nitrogen and sulfur hetero-atoms bonded directly to the halogen-bearing carbon. It is readily apparent that chemicals containing structural features that allow for the delocalization of carbonium ion intermediates will not be persistent in aquatic environments.

The lack of reactivity observed for vinyl chloride and chlorobenzene are a reflection of the high energy pathway required for nucleophilic substitution at vinylic and aromatic carbons. Vinyl and phenyl carbonium ions have been observed, but their formation requires very reactive leaving groups, such as trifluoromethanesulfonate.

Polyhalogenated aliphatics. The hydrolysis kinetics for the polyhalogenated ethanes and propanes (e.g., 1,1,2,2-tetrachloroethane and 1,2-dibromo-3-chloropropane) are somewhat more complex than the simple halogenated aliphatics. In addition to nucleophilic substitution reactions, degradation of these compounds can occur through the base-catalyzed loss of HX. Depending on structure type, elimination (or dehydrohalogenation) may be the dominant reaction pathway at environmentally significant pHs. Although this type of reaction does not fit the definition for hydrolysis presented earlier (i.e., a covalent bond with -OH is not formed), nucleophilic substitution and elimination reactions for halogenated organics are typically combined under the general term hydrolysis. The loss of HX occurs through a bimolecular elimination (E2) reaction in which abstraction of the hydrogen on the β carbon occurs simultaneously with cleavage of the C–X bond:

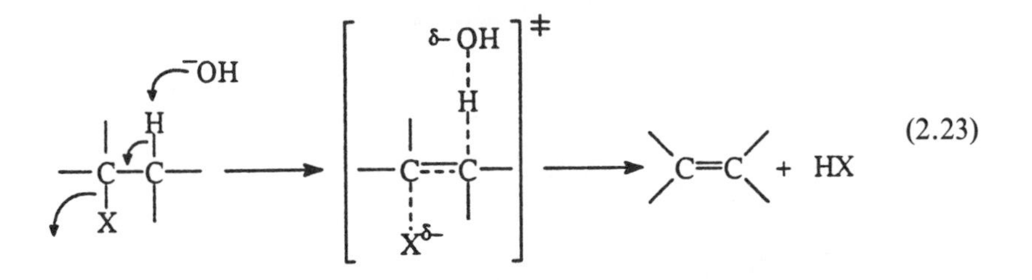

This process often results in the formation of halogenated alkenes that are generally more hazardous and more persistent than are the products of substitution reactions (Bolt et al., 1982). Clearly, understanding these competing reaction processes is of considerable environmental significance.

Nucleophilic Substitution versus Elimination. The distribution of reaction products resulting from the hydrolytic degradation of the polyhalogenated ethanes and propanes will depend on the relative rates for the nucleophilic substitution and dehydrohalogenation reaction pathways. Because strong bases favor elimination over nucleophilic substitution, base-catalyzed reactions (i.e., reaction with the strong base, OH^-) of polyhalogenated aliphatics are expected to be dominated by elimination reactions (March, 1985). On the other hand, neutral hydrolysis (i.e., reaction with the weaker base, H_2O) is expected to occur primarily by nucleophilic substitution reactions. As a result, reaction product distribution will be pH dependent. Comparison of the hydrolysis kinetics and reaction pathways of the insecticides, DDT and methoxychlor, illustrate this point (Wolfe et al., 1977). The pH half-life profile for methoxychlor and DDT is shown in Figure 2.2.

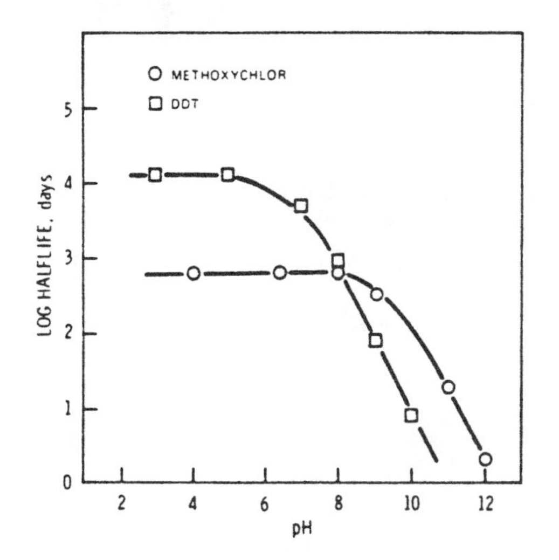

Figure 2.2. Calculated pH half-life profile for methoxychlor and DDT hydrolytic degradation in water at 27°C (Reprinted by permission of the American Chemical Society).

At environmentally relevant pHs (5 to 8) the hydrolysis of methoxychlor is pH independent. Over this same pH range, the hydrolysis of DDT is pH dependent. Product studies conducted at pH 9 show that the base-catalyzed reaction pathway for DDT and methoxychlor is dehydrochlorination, resulting in the formation of DDE and DMDE, respectively (Reactions 2.24 and 2.25). In the case of DDT, the elimination pathway will predominate in most natural waters; thus, DDE, which is more stable than its parent compound, DDT, is expected to be the major reaction product. On the other hand, neutral hydrolysis of methoxychlor predominates at environmental pHs. Product studies of methoxychlor conducted at pH 7 show that hydrolysis proceeds through formation of a carbonium ion (S_N1 mechanism), which can either eliminate a proton to give DMDE (minor pathway) or rearrange (1,2-migration of the phenyl moiety) to give the more stable carbonium ion, which subsequently reacts with water to give the alcohol (Figure 2.3). Facile hydrolysis of this "intermediate" occurs (S_N1 mechanism) to give anisoin. Depending on reaction conditions, anisoin may be oxidized to anisil. These results demonstrate that to predict reaction product formation, a detailed understanding of the hydrolysis kinetics as a function of pH may be required.

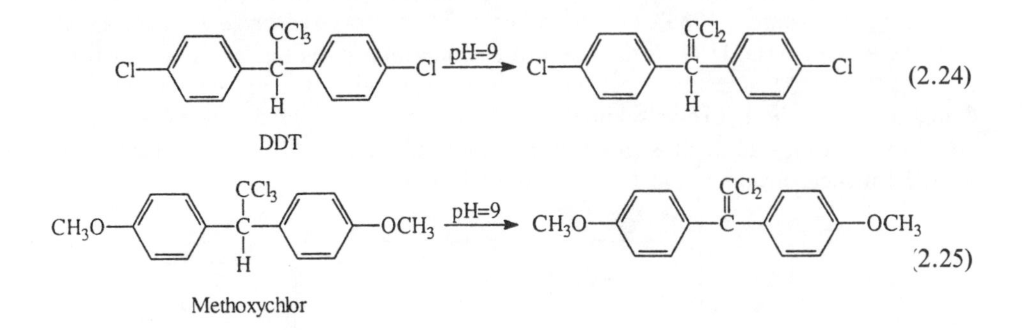

To gain insight into the structural features of halogenated aliphatics that affect the relative rates of the neutral and base-catalyzed hydrolysis pathways, Jeffers et al. (1989) measured k_n and k_b for a series of polychlorinated ethanes, propanes and

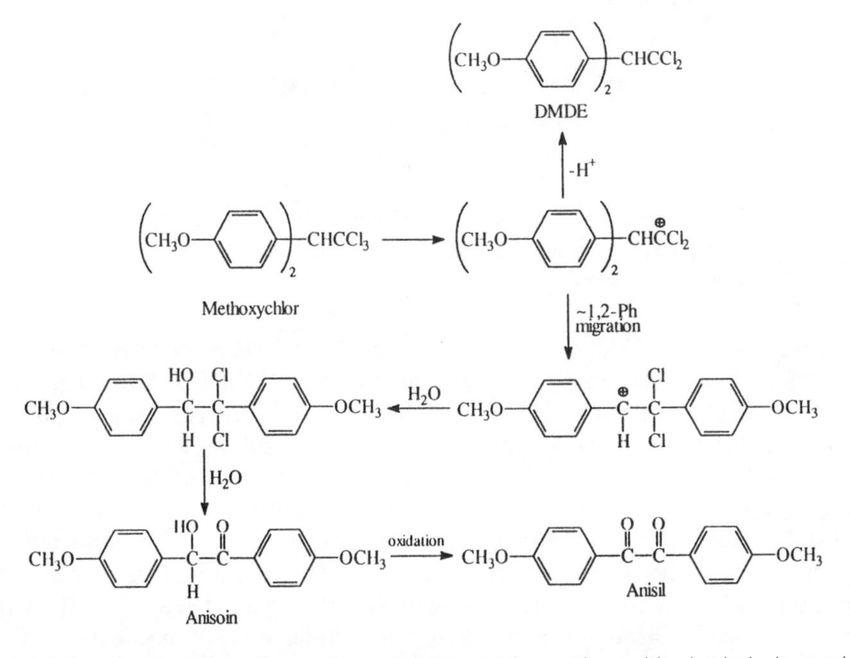

Figure 2.3. Proposed reaction pathway to account for methoxychlor hydrolysis products (Adapted from Wolfe et al., 1977).

ethenes. These data are summarized in Table 2.3. A wide range of reactivity is observed for these compounds; the environmental hydrolysis half-lives range over seven orders of magnitude. For a number of the polyhalogenated aliphatics it is apparent that both neutral and base-catalyzed hydrolysis will occur at environmental pHs, and that the relative contributions of these processes will be dependent on the degree and pattern of halogen substitution.

Trends in neutral hydrolysis reactivity for the chlorinated ethanes and propanes seem to be dependent on a combination of factors including C-Cl bond strength and steric bulk at the site of nucleophilic substitution. For example, the degree of chlorination has significant effects on neutral hydrolysis rates. Comparison of the hydrolysis half-lives for chloroethane (Table 2.2), 1,1-dichloroethane and 1,1,1-trichloroethane (Table 2.3) indicates that the addition of a second chlorine substituent (compare reactivity of chloroethane and 1,1-dichloroethane) significantly retards neutral hydrolysis, suggesting that the steric bulk of the chlorine atom is of greater significance than its inductive (electron-withdrawing) effect. The addition of the third chlorine substituent (1,1,1-trichloroethane), however, greatly accelerates neutral hydrolysis. The observed increase in reactivity for 1,1,1-trichloroethane may be explained by invoking an S_N1 mechanism for hydrolysis. Formation of the carbonium ion intermediate is stabilized by the resonance electron-donating effect of the

Table 2.3. Hydrolysis Kinetics of Chlorinated Ethanes, Ethenes, and Propanes at 25°C[a]

Compound	k_n (min^{-1})	k_b(pH7) (min^{-1})	k_{hyd} min^{-1})	$t_{1/2}$(y)
$Cl_2HC\text{-}CH_3$	2.15×10^{-8}	7.20×10^{-14}	2.15×10^{-8}	61.3
$ClH_2C\text{-}CH_2Cl$	1.83×10^{-8}	1.04×10^{-11}	1.83×10^{-8}	72.0
$Cl_2HC\text{-}CH_2Cl$	5.19×10^{-11}	9.42×10^{-9}	9.47×10^{-9}	139.2
$Cl_3C\text{-}CH_3$	1.24×10^{-6}	0.00	1.24×10^{-6}	1.1
$Cl_2HC\text{-}CHCl_2$	9.70×10^{-9}	3.02×10^{-6}	3.03×10^{-6}	0.4
$Cl_3C\text{-}CH_2Cl_1$	2.60×10^{-8}	2.15×10^{-9}	2.82×10^{-8}	46.8
$Cl_3C\text{-}CHCl_2$	4.93×10^{-8}	1.31×10^{-4}	1.31×10^{-4}	0.010
$Cl_3C\text{-}CCl_3$	0.00	7.18×10^{-16}	7.18×10^{-16}	1.8×10^{9}
$ClH_2C\text{-}CH_2\text{-}CH_2Cl$	5.87×10^{-7}	1.67×10^{-12}	5.87×10^{-7}	2.2
$H_3C\text{-}CCl_2\text{-}CH_3$	3.18×10^{-4}	0.00	3.18×10^{-4}	0.004
$Cl_2HC\text{-}CHCl\text{-}CH_2Cl$	0.00	7.84×10^{-5}	7.84×10^{-5}	0.017
$Cl_2HC\text{-}CHCl\text{-}CHCl_2$	4.71×10^{-7}	9.81×10^{-5}	9.85×10^{-5}	0.013
$Cl_2C=CH_2$	0.00	1.09×10^{-4}	1.09×10^{-14}	1.2×10^{8}
$ClHC=CHCl$	0.00	6.32×10^{-17}	6.32×10^{-17}	2.1×10^{10}
$Cl_2C=CHCl$	0.00	1.07×10^{-12}	1.07×10^{-12}	1.3×10^{6}
$Cl_2C=CCl_2$	0.00	1.37×10^{-15}	1.37×10^{-15}	9.9×10^{8}

[a]Data taken from Jeffers et al. (1989).

chlorine substituents. Likewise, the hydrolysis of 2,2-dichloropropane, which also contains a perchlorinated carbon (i.e., all of the hydrogen atoms bonded directly to the carbon atom have been replaced with chlorine) and undergoes facile neutral hydrolysis, may also be occurring through a S_N1 mechanism.

Base-catalyzed elimination becomes important relative to neutral hydrolysis (i.e., nucleophilic substitution) as the degree of chlorination increases, as long as adjacent carbons contain at least one chlorine substituent. This trend is a reflection of the increased acidity of protons that are bonded to the same carbon containing the electron-withdrawing chlorine substituents. Over the series 1,2-dichloroethane, 1,1,2-trichloroethane, 1,1,2,2-tetrachloroethane, and pentachloroethane, an increase in 5 orders of magnitude is observed for k_b (2.26). As expected, 1,1,2,3-tetrachloropropane and 1,1,2,3,3-pentachloropropane also exhibit enhanced reactivity toward base-catalyzed elimination. The stability of 1,1,1-trichloroethane and 2,2-dichloropropane, suggests that perhalogenated carbon atoms are unreactive toward nucleophilic substitution by OH^-.

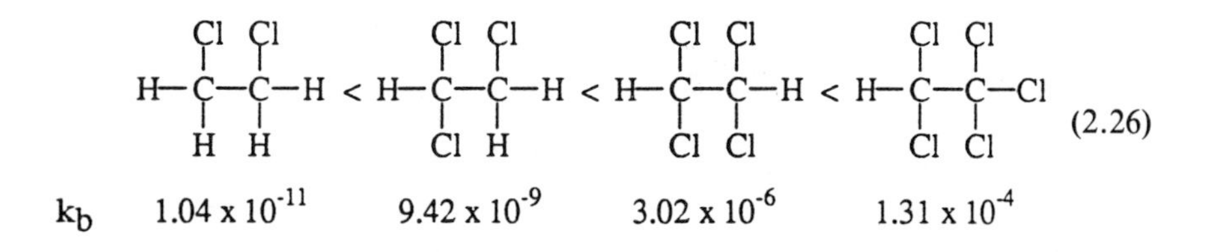

As with nucleophilic substitution reactions, rates of dehydrohalogenation reactions will be dependent on the strength of the C-X bond being broken in the elimination process. Accordingly, it is expected that the ease of elimination of X will follow the series Br > Cl > F. The relative reactivities of Br and Cl toward elimination is evident from the hydrolysis product studies of 1,2-dibromo-3-chloropropane (DBCP: Burlinson et al., 1982). DBCP has been used widely in this country as a soil fumigant for nematode control and has been detected in groundwaters (Mason et al., 1981) and subsoils (Nelson, et al., 1981). Hydrolysis kinetic studies demonstrated that the hydrolysis of DBCP is first order both in DBCP and hydroxide ion concentration above pH 7. Below pH 7, hydrolysis occurs via neutral hydrolysis; however, the base-catalyzed reaction will contribute to the overall rate of hydrolysis as low as pH 5. Product studies performed at pH 9 indicate that transformation of DBCP occurs initially by E2 elimination of HBr and HCl (Figure 2.4).

The product ratio of the two intermediate alkenes, 2-bromo-3-chloropropene (BCP) and 2,3-dibromopropene (DBP), demonstrates that the elimination of HBr occurs at a rate that is approximately 20 times faster than elimination of HCl. It is interesting to note that BCP and DBP hydrolyzed at rates that were 100 times faster and 16 times faster, respectively, than the parent compound DBCP, resulting in the formation of the stable alkene, BAA. Thus, accumulation of BCP and DBP in the environment would not be expected to occur. The hydrolysis kinetics of BCP and

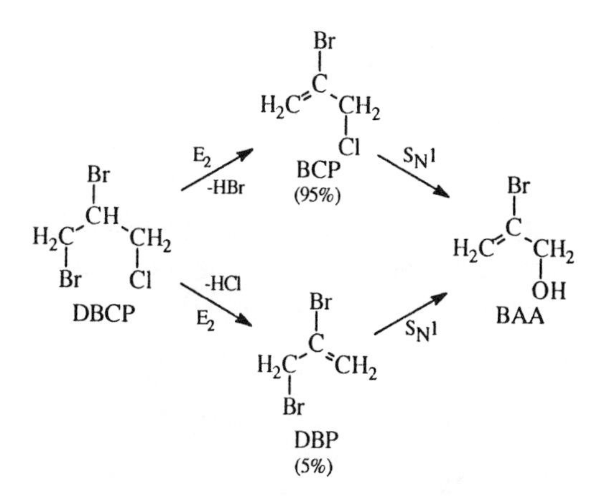

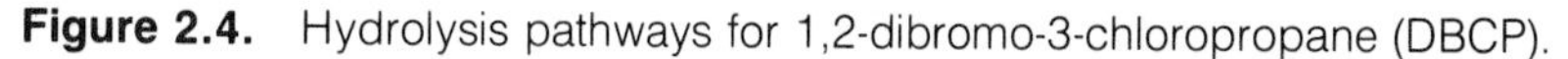
Figure 2.4. Hydrolysis pathways for 1,2-dibromo-3-chloropropane (DBCP).

DBP were independent of hydroxide concentration, which is consistent with a S_N1 mechanism and is typical for allylic halides.

b. Epoxides

Much of our understanding concerning the hydrolysis mechanisms of epoxides results from the interest in the carcinogenic bay-region diol epoxides, such as the 7,8-diol-9,10-epoxide of benzo[a]pyrene (2.27) (Jerina et al., 1978; Okamoto et al., 1978). The epoxides are also important intermediates used in the preparation of many industrial chemicals, especially polymers, and represent several high volume agrochemicals.

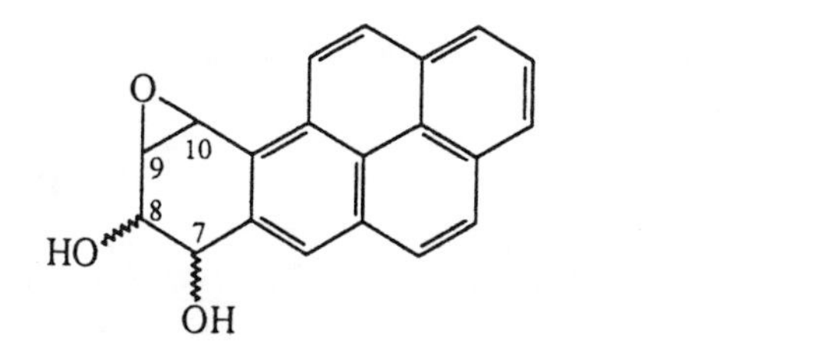

(2.27)

The hydrolysis of epoxides is pH dependent and can occur through acid, neutral, or base-promoted processes. Because the acid and neutral processes dominate over environmentally significant pH ranges, the base-catalyzed process can often be ignored. The reaction products resulting from epoxide hydrolysis are diols, and to a lesser extent, carbonyl products, which result from the rearrangement of carbonium ion intermediates.

$$\text{epoxide } [\text{H, R}_1]\text{C}{-}\text{O}{-}\text{C}[\text{R}_2, \text{R}_3] \longrightarrow \text{H}{-}\text{C}(\text{OH})(\text{R}_1){-}\text{C}(\text{OH})(\text{R}_2){-}\text{R}_3 + \text{H}{-}\text{C}(\text{R}_3)(\text{R}_1){-}\text{C}({=}\text{O}){-}\text{R}_2 \qquad (2.28)$$

The acid-catalyzed hydrolysis of epoxides proceeds by either heterolytic cleavage of the carbon-oxygen bond to form a carbonium ion intermediate that subsequently reacts with water to form the diol (S_N1 mechanism):

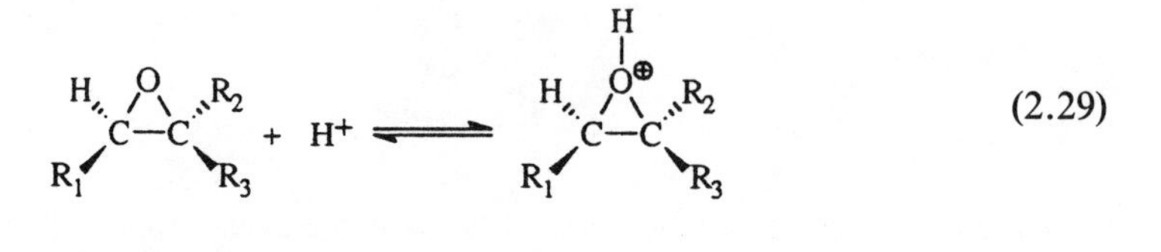

(2.29)

$$\text{protonated epoxide } [\text{H, R}_1]\text{C}{-}\text{O}^{\oplus}\text{H}{-}\text{C}[\text{R}_2, \text{R}_3] \xrightleftharpoons{\text{slow}} [\text{H, R}_1]\text{C}^{\oplus}{-}\text{C}(\text{OH})(\text{R}_3){-}\text{R}_2 \qquad (2.30)$$

$$[\text{H, R}_1]\text{C}^{\oplus}{-}\text{C}(\text{OH})(\text{R}_3){-}\text{R}_2 + \text{H}_2\text{O} \xrightleftharpoons{\text{fast}} \text{H}{-}\text{C}(\text{H}_2\overset{\oplus}{\text{O}})(\text{R}_1){-}\text{C}(\text{OH})(\text{R}_3){-}\text{R}_2 \qquad (2.31)$$

$$\text{H}{-}\text{C}(\text{H}_2\overset{\oplus}{\text{O}})(\text{R}_1){-}\text{C}(\text{OH})(\text{R}_3){-}\text{R}_2 \rightleftharpoons \text{H}{-}\text{C}(\text{HO})(\text{R}_1){-}\text{C}(\text{OH})(\text{R}_3){-}\text{R}_2 + \text{H}^+ \qquad (2.32)$$

or by nucleophilic ring opening of the protonated species, followed by formation of the diol (S_N2 mechanism):

$$\text{epoxide } [\text{H, R}_1]\text{C}{-}\text{O}{-}\text{C}[\text{R}_2, \text{R}_3] + \text{H}^+ \rightleftharpoons [\text{H, R}_1]\text{C}{-}\text{O}^{\oplus}\text{H}{-}\text{C}[\text{R}_2, \text{R}_3] \qquad (2.33)$$

$$[\text{H, R}_1]\text{C}{-}\text{O}^{\oplus}\text{H}{-}\text{C}[\text{R}_2, \text{R}_3] + \text{H}_2\text{O}\!: \xrightleftharpoons{\text{slow}} \text{H}{-}\text{C}(\text{H}_2\overset{\oplus}{\text{O}})(\text{R}_1){-}\text{C}(\text{OH})(\text{R}_3){-}\text{R}_2 \qquad (2.34)$$

$$\mathrm{H{-}\underset{R_1}{\overset{H_2\overset{\oplus}{O}}{C}}{-}\underset{R_3}{\overset{OH}{C}}{-}R_2 \rightleftharpoons H{-}\underset{R_1}{\overset{HO}{C}}{-}\underset{R_3}{\overset{OH}{C}}{-}R_2 + H^+} \tag{2.35}$$

The nucleophilic ring opening of the protonated epoxide is often referred to as a "borderline S_N2" reaction because bond breaking in the transition state is more nearly complete than bond making (Parker and Isaacs, 1959). Consequently, electron-donating substituent groups will stabilize positive charge formation in the transition state in the same way as for the S_N1 mechanism. Unlike the S_N1 reactions, however, borderline S_N2 reactions will be subject to steric hindrance from alkyl groups, though to a smaller extent than an ordinary S_N2 reaction.

Analogous S_N1 and S_N2 mechanisms can be written for the neutral hydrolysis of epoxides. Neutral hydrolysis through an S_N1 mechanism involves initial formation of a carbonium ion, followed by either nucleophilic attack of water to give the diol:

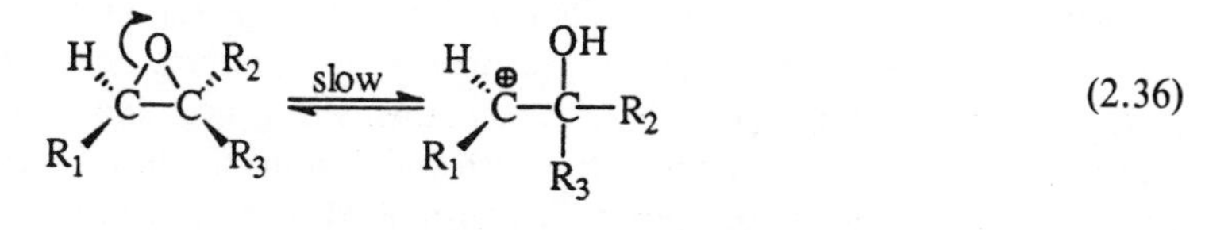

(2.36)

$$\mathrm{H(R_1)\overset{\oplus}{C}{-}\underset{R_3}{\overset{OH}{C}}{-}R_2 + H_2O \xrightleftharpoons{\text{fast}} H{-}\underset{R_1}{\overset{H_2\overset{\oplus}{O}}{C}}{-}\underset{R_3}{\overset{OH}{C}}{-}R_2} \tag{2.37}$$

$$\mathrm{H{-}\underset{R_1}{\overset{H_2\overset{\oplus}{O}}{C}}{-}\underset{R_3}{\overset{OH}{C}}{-}R_2 \rightleftharpoons H{-}\underset{R_1}{\overset{HO}{C}}{-}\underset{R_3}{\overset{OH}{C}}{-}R_2 + H^+} \tag{2.38}$$

or migration of a proton or alkyl group to give the carbonyl rearrangement product (Ukachukwu et al., 1986):

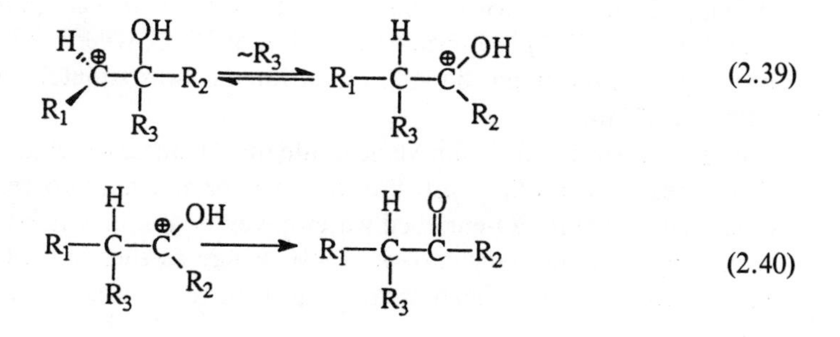

(2.39)

(2.40)

The S_N2 mechanism for neutral hydrolysis involves direct nucleophilic attack by water.

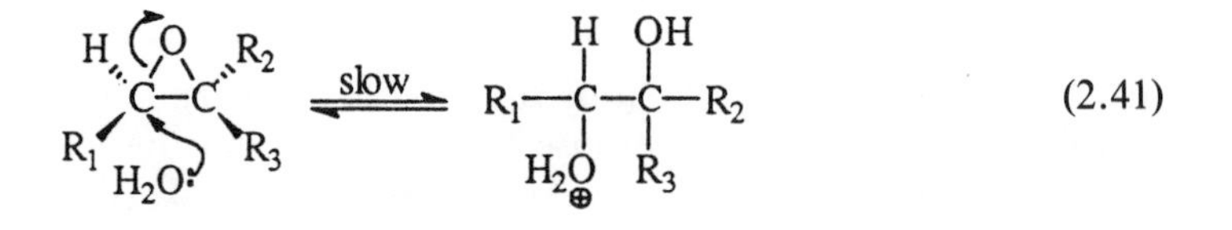

(2.41)

$$R_1-\underset{H_2\overset{\oplus}{O}}{\overset{H}{C}}-\underset{R_3}{\overset{OH}{C}}-R_2 \rightleftharpoons R_1-\underset{OH}{\overset{H}{C}}-\underset{R_3}{\overset{OH}{C}}-R_2 + H^+ \quad (2.42)$$

Reaction rates and product distributions for epoxide hydrolysis, which are strongly dependent on structural features of the epoxide, can be rationalized in terms of the limiting S_N1 and S_N2 mechanisms outlined above. For example, the rates of hydrolysis of epoxides by both acid catalysis and neutral processes are greatly enhanced when the epoxide is conjugated by aryl groups and carbon-carbon double bonds, suggesting the intermediacy of a carbonium ion (S_N1 mechanism) (Ross et al., 1982). The acid-catalyzed hydrolysis of 1,2,3,4-tetrahydronaphthalene-1,2-oxide (epoxide conjugated to the phenyl ring) occurs approximately 6×10^4 times faster than the isomeric 1,2,3,4-tetrahydronaphthalene-2,3-oxide (epoxide not conjugated to the phenyl ring) (Becker et al., 1979).

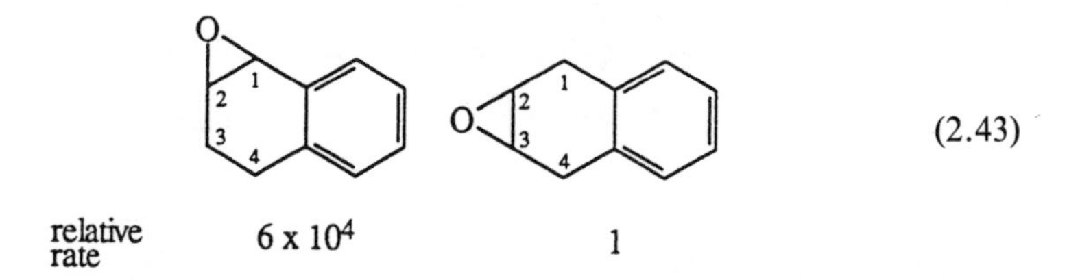

(2.43)

Likewise, the half-life for the neutral hydrolysis of the *cis*-7,8-diol-9,10-epoxide of benzo[a]pyrene (2.27) is approximately 30 s at 25°C, whereas the neutral hydrolysis of propylene oxide under the same conditions is approximately two weeks (Long and Pritchard, 1956).

Hydrolysis studies of isobutylene oxide in ^{18}O enriched water provide an example of how reaction product distributions provide insight into reaction mechanisms. Isobutylene oxide in ^{18}O-enriched water gave the diol in which the label was located exclusively at the primary carbon (2.44), where as the acid-catalyzed process gave the diol with the label located exclusively at the tertiary carbon (2.45) (Pritchard and Long, 1956).

$$H_2C\overset{O}{\frown}C(CH_3)_2 + H_2O^{18} \xrightleftharpoons{pH = 7} H-\underset{H}{\overset{HO^{18}}{C}}-\underset{CH_3}{\overset{OH}{C}}-CH_3 \tag{2.44}$$

$$H_2C\overset{O}{\frown}C(CH_3)_2 + H_2O^{18} \xrightleftharpoons{pH = 3} H-\underset{H}{\overset{HO}{C}}-\underset{CH_3}{\overset{^{18}OH}{C}}-CH_3 \tag{2.45}$$

These investigators concluded that the reaction at pH 7 proceeds by S_N2 reaction with water at the less sterically hindered primary carbon and that the acid-catalyzed reaction involves formation of a carbonium ion intermediate centered at the tertiary carbon (stabilization of the carbonium ion is provided by the electron-donating alkyl groups).

The stereoisomers, dieldrin and endrin, are two examples of agrochemicals that contain epoxide moieties. Both of these chemicals hydrolyze by S_N2 reaction with H_2O and OH^-, resulting in the formation of diols.

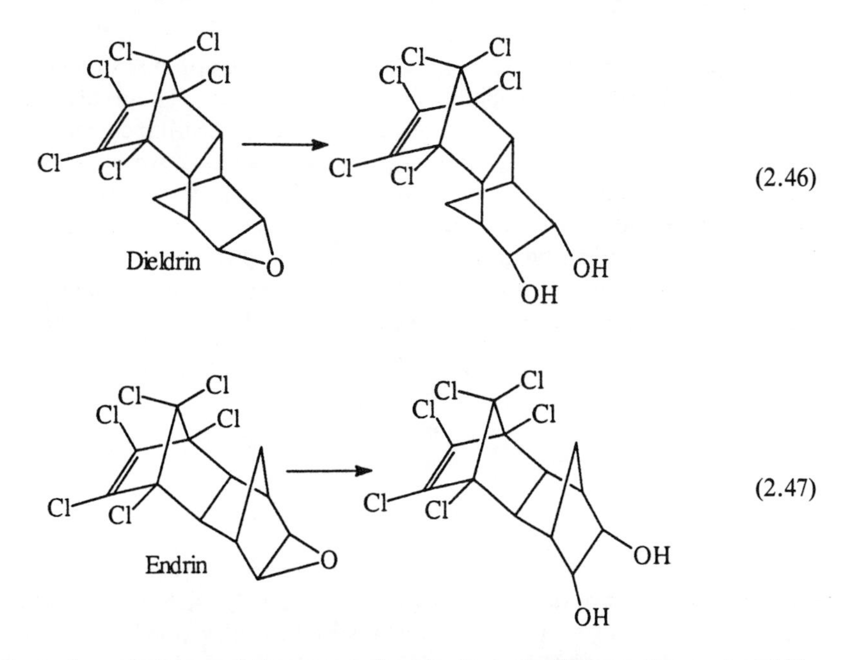

The bicyclic carbon skeleton of these agrochemicals sterically impedes nucleophilic attack by H_2O and OH^-. As a result, the hydrolysis half-lives for both of these compounds are quite long. Because of their persistence in aquatic ecosystems, these agrochemicals have been banned for use in the U.S.; however, they are still in use as soil insecticides for many crops outside of the U.S.

Epichlorohydrin is another example of an important industrial chemical containing an epoxide moiety. Epichlorohydrin is used primarily for the manufacture of glycerol and epoxy resins. Unlike the dieldrin and aldrin, epichlorohydrin is quite susceptible to hydrolysis. The calculated half-life for epichlorohydrin in distilled water at pH 7 at 20°C is 8.0 days (Santodonato, et al., 1980). Hydrolysis occurs initially by opening of the epoxide to form the chlorinated diol, which under basic conditions, leads to formation of glycerol through the epoxide intermediate.

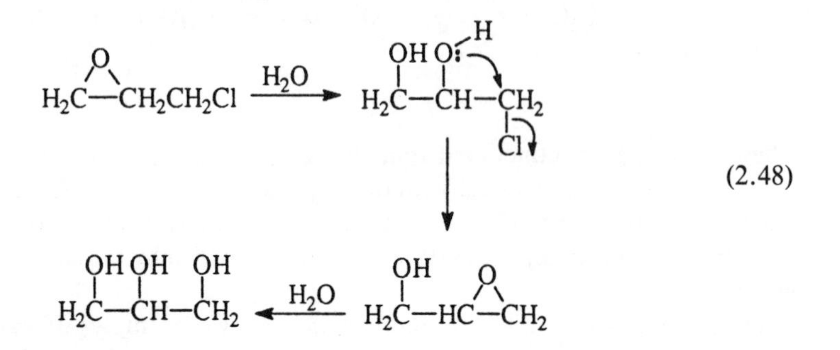

(2.48)

c. *Organophosphorus Esters*

The organophosphorus esters represent another class of environmental chemicals that are hydrolyzed by nucleophilic substitution reactions. These chemicals comprise one of the most important classes of agrochemicals with insecticidal activity (Fest and Schmidt, 1983). The organophosphorus esters also have important industrial uses such as oil additives, flame retardants, and plasticizers (Muir, 1988). The wide range of biological activity that this class of chemicals exhibits is due to the variety of substituents that can be attached to the central phosphorus atom.

The organophosphorus esters that have the greatest environmental significance because of their extensive use as insecticides have the general structure:

(2.49)

where R_1 and R_2 are alkyl groups (methyl or ethyl), X is O or S, and R_3 is an electron-withdrawing group. The electron-withdrawing group is required for biological activity (Fest and Schmidt, 1983). An interesting feature of the hydrolytic degradation of the organophosphate insecticides is that hydrolysis may occur by either nucleophilic substitution at the central phosphorous atom or at carbon resulting in cleavage of C–O and C–S bonds. It is generally observed that base-catalyzed hydrolysis favors P–O cleavage (2.50) and that neutral and acid-catalyzed hydrolysis favors C–O or C–S cleavage (2.51).

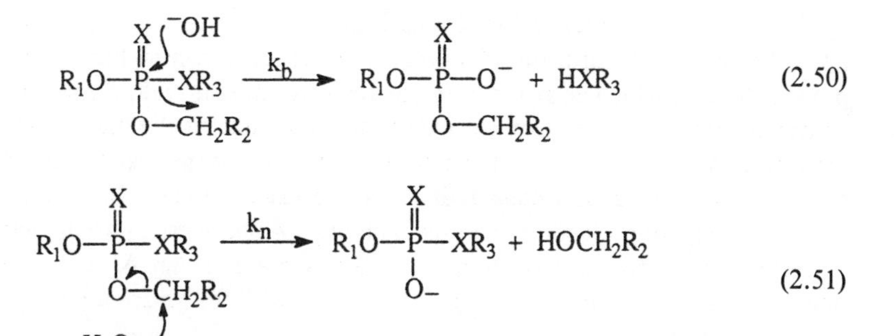

As a result, hydrolysis mechanisms and product distribution for the organophosphorus esters will be pH dependent. The diesters, the initial products formed from the hydrolysis of the triesters (2.50 and 2.51) will be further hydrolyzed to monoesters, but at a much slower rate than the parent compound. Little information exists to determine if formation of the more persistent diesters has deleterious effects on the health of environmental systems.

The structural features of organophosphorus esters, which are very similar to those of the carboxylic acid esters, suggest that nucleophilic displacement at the central phosphorus atom will occur by an addition-elimination mechanism. Mechanistic studies, however, have demonstrated that hydrolysis occurs through direct nucleophilic displacement at phosphorus and does not involve formation of a pentavalent intermediate with H_2O or OH^- (Hudson, 1965; Kirby and Warren, 1967). Accordingly, hydrolysis rates for phosphorus esters will be sensitive to electronic factors that alter the electrophilicity of the central phosphorus atom and steric interactions that impede nucleophilic attack.

The phosphorylating properties (phosphorylation is the mechanism by which the target enzyme, acetylcholinesterase, is inhibited) of these chemicals, like their hydrolysis properties, are governed by the electrophilicity of the central phosphorus atom. As a result, a balance between biological activity and hydrolytic stability has to be sought. Structure activity relationships indicate that phosphate esters with -XR substituents whose conjugate acids have pKa's between 6 and 8, have sufficient phosphorylating properties, and have sufficient hydrolytic stability to reach their biological target, but will not persist in aquatic environments (Schmidt, 1975; Schmidt and Fest, 1982). Phosphate esters with -XR substituents that have pK_a values below 6 (weak electron withdrawing group) are biologically inactive and too persistent in the environment. Phosphate esters with pKa values above 8 (strong electron withdrawing group) are biologically inactive because hydrolysis will occur before the substrate reaches the biological target.

The hydrolytic stability, as well as the biological activity, also can be affected by modifying other substituents at the phosphorus atom. For example, substitution of sulfur (P = S) for oxygen (P = O) in the ester moiety will reduce the electrophilicity of the phosphorus center because of the weaker electron withdrawing effect of sulfur. Accordingly, phosphorothioate esters (P = S) will exhibit greater stability toward neutral and base-catalyzed hydrolysis than their respective O-substituted counterparts. Likewise, substitution of alkyl groups with electron withdrawing groups (e.g., p-nitrophenyl), which activates P toward nucleophilic attack, dramatically increases hydrolysis rates. Comparison of hydrolysis rate data for the methyl and ethyl phosphoroate esters shows that increasing the steric bulk at the reactive center retards the rates of both neutral and base hydrolysis. Replacing one of the alkoxy groups with an alkyl group will retard hydrolysis. The phosphonate esters are in general a factor of 100 or more less reactive than the phosphoroate esters. Also, in contrast to the phosphoroate esters, base-catalyzed hydrolysis will be the dominant process for the phosphonate esters at pH 7.

The structures and hydrolysis half-lives at pH 7 at 20°C for a number of organophosphorus esters that are important agrochemicals are shown in Figure 2.5. It is evident from these data that hydrolysis will be an important transformation pathway for these chemicals in aquatic ecosystems.

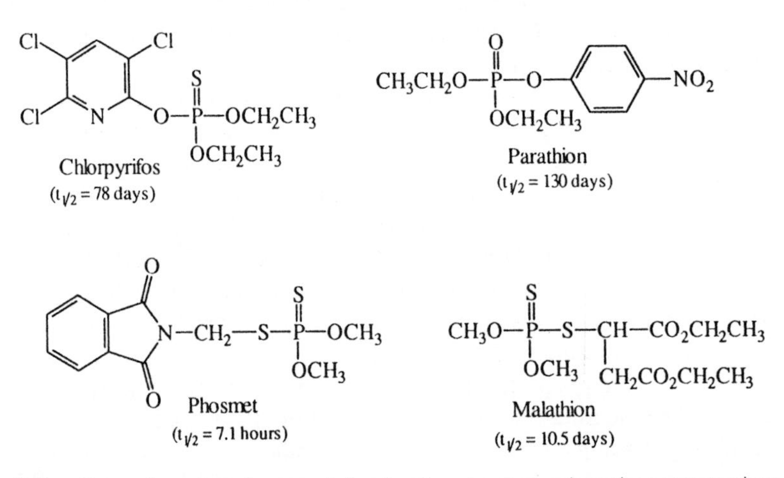

Figure 2.5. Examples of environmentally significant organophosphorus agrochemicals. Hydrolysis half-lives are given in parentheses.

3. Nucleophilic Acyl Substitution

a. Addition-Elimination Mechanism

Hydrolysis at unsaturated carbon occurs by a two-step process: (1) nucleophilic addition at the acyl group [RC(O)] to give a tetrahedral intermediate (2.52) and (2) elimination of the leaving group (2.53). This reaction mechanism is referred to as the addition-elimination mechanism or nucleophilic acyl substitution.

$$R-\overset{\overset{\displaystyle O}{\|}}{C}-X + Y: \rightleftharpoons R-\underset{\underset{\displaystyle Y}{|}}{\overset{\overset{\displaystyle O^-}{|}}{C}}-X \tag{2.52}$$

$$R-\underset{\underset{\displaystyle Y}{|}}{\overset{\overset{\displaystyle O^-}{|}}{C}}-X \rightleftharpoons R-\overset{\overset{\displaystyle O}{\|}}{C}-Y + X: \tag{2.53}$$

In general, nucleophilic displacement takes place much more readily at unsaturated or acyl carbons than at saturated carbons. Two factors account for this increase in reactivity of acyl substrates: (1) oxygen is more electronegative than carbon; the electron withdrawing power of the oxygen imparts a partial positive charge at the carbonyl group rendering it more electrophilic and (2) the transition state leading to the formation of the tetrahedral intermediate is relatively unhindered; a trigonal substrate becoming a tetrahedral intermediate.

The carboxylic acid derivatives (RC(O)OX) and carbonic acid derivatives (ROC(O)OX) represent two classes of environmental chemicals that hydrolyze through nucleophilic acyl substitution reactions. The general structural features of representative functional groups in these chemical classes are illustrated in Figure 2.6.

Further discussion will focus on detailed hydrolysis mechanisms for select functional groups shown in Figure 2.6.

4. Functional Group Transformation by Nucleophilic Acyl Substitution Reactions

a. Carboxylic Acid Derivatives

Detailed hydrolysis studies of the carboxylic acid derivatives by Bender (1960) and others (Samuel and Silver, 1965) provided the initial evidence for formation of the tetrahedral intermediate. These investigators found that when hydrolysis studies were conducted in ^{18}O-labeled water, incorporation of ^{18}O into unhydrolyzed substrate occurred.

$$R-\overset{\overset{\displaystyle O}{\|}}{C}-X + H_2O^{18} \rightleftharpoons R-\underset{\underset{\displaystyle ^{18}OH}{|}}{\overset{\overset{\displaystyle OH}{|}}{C}}-X \tag{2.54}$$

Carboxylic Acid Derivatives

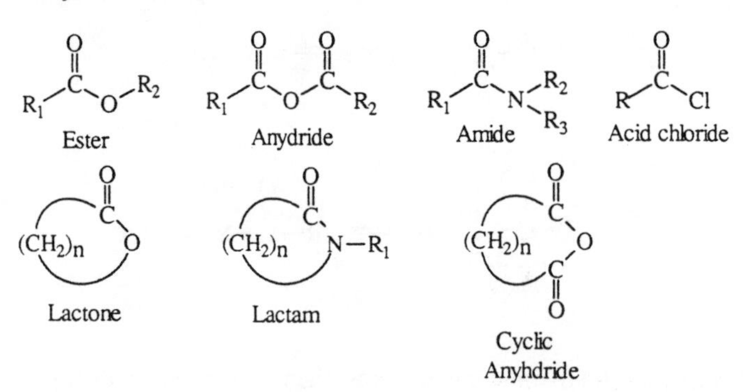

Carbonic Acid Derivatives

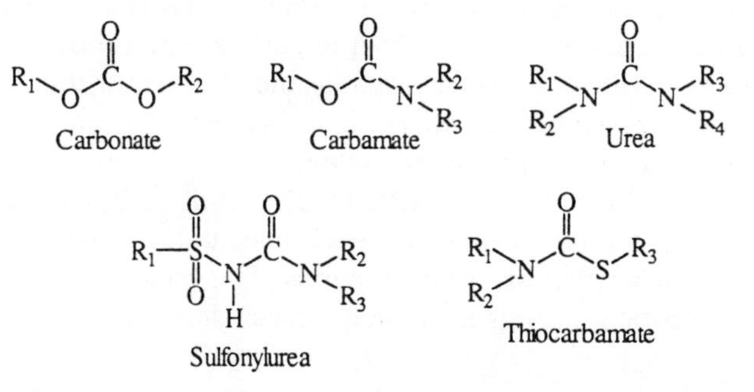

Figure 2.6. Basic chemical structures of carboxylic acid and carbonic acid derivatives.

$$\mathrm{R{-}C(OH)(^{18}OH){-}X} \rightleftharpoons \mathrm{R{-}C({=}O){-}X} + \mathrm{H_2O^{18}} \tag{2.55}$$

$$\mathrm{R{-}C(OH)(^{18}OH){-}X} \rightleftharpoons \mathrm{R{-}C({=}^{18}O){-}X} + \mathrm{H_2O} \tag{2.56}$$

$$\mathrm{R{-}C(OH)(^{18}OH){-}X} \rightleftharpoons \mathrm{R{-}C({=}^{18}O){-}OH} + \mathrm{X{:}} \tag{2.57}$$

The exchange of label could be rationalized through the tetrahedral intermediate if expulsion of water is competitive with expulsion of the leaving group, X:. The amount of ^{18}O associated with the unhydrolyzed acyl derivative provides a measure of the stability of the leaving group, X:; the better the leaving group, the less likely that reversal of the tetrahedral intermediate will occur. The amount of ^{18}O incorporation decreased in the following order:

$$RC(=O)NH_2 < R_1C(=O)OR_2 < R_1C(=O)OC(=O)R_2 < RC(=O)Cl \tag{2.58}$$

which reflects the decreasing basicity of the leaving group (NH_2^-, RO^-, $RC(O)O^-$, Cl^-, respectively). The weaker the base, the better the leaving group and the more susceptible the carboxylic acid derivative will be to hydrolysis. Because Cl^- is such a weak base, acid chlorides are almost instantly hydrolyzed to the parent carboxylic acid in aquatic environments.

Carboxylic Acid Esters. The hydrolysis mechanisms of the carboxylic acid esters have been thoroughly investigated. Although nine distinct mechanisms have been proposed (Ingold, 1969), our comments will be limited to the two most common mechanisms involving acyl-oxygen bond cleavage by acid catalysis ($A_{AC}2$) and base catalysis ($B_{AC}2$). Hydrolysis via the $A_{AC}2$ mechanism involves protonation of the carbonyl oxygen initially. Protonation polarizes the carbonyl group, removing electron density from the carbon atom making it more electrophilic and thus more susceptible to nucleophilic addition by water.

$A_{AC}2$ Mechanism

$$R_1-C(=O)-OR_2 + H^+ \rightleftharpoons R_1-C(=\overset{\oplus}{O}H)-OR_2 \tag{2.59}$$

$$R_1-C(=\overset{\oplus}{O}H)-OR_2 + H_2O \rightleftharpoons R_1-C(OH)(\overset{\oplus}{O}H_2)-OR_2 \tag{2.60}$$

$$R_1-C(OH)(\overset{\oplus}{O}H_2)-OR_2 \overset{\sim H^+}{\rightleftharpoons} R_1-C(OH)(HO)-\overset{\oplus}{O}(H)R_2 \tag{2.61}$$

$$R_1-C(HO)(OH)-\overset{\oplus}{O}(H)R_2 \rightleftharpoons R_1-C(=O)-OH + R_2OH + H^+ \tag{2.62}$$

The base-catalyzed mechanism ($B_{AC}2$) proceeds via the direct nucleophilic addition of OH^- to the carbonyl group. Base catalysis occurs because hydroxide ion is a stronger nucleophile than water.

$B_{AC}2$ Mechanism

$$R_1-\overset{\overset{\displaystyle O}{\|}}{C}-OR_2 + OH^- \rightleftharpoons R_1-\underset{\underset{\displaystyle OH}{|}}{\overset{\overset{\displaystyle O^-}{|}}{C}}-OR_2 \qquad (2.63)$$

$$R_1-\underset{\underset{\displaystyle OH}{|}}{\overset{\overset{\displaystyle O^-}{|}}{C}}-OR_2 \rightleftharpoons R_1-\overset{\overset{\displaystyle O}{\|}}{C}-OH + R_2O^- \qquad (2.64)$$

$$R_1-\overset{\overset{\displaystyle O}{\|}}{C}-OH + R_2O^- \rightleftharpoons R_1-\overset{\overset{\displaystyle O}{\|}}{C}-O^- + R_2OH \qquad (2.65)$$

Although neutral hydrolysis of carboxylic acid esters does occur, the base-catalyzed reaction will be the dominant pathway in most natural waters. Generally, acid hydrolysis will dominate in acidic waters with pHs below 4.

Both electronic and steric effects can significantly alter the reactivity of carboxylic acid esters. Because the acid-catalyzed process for both aliphatic and aromatic esters is relatively insensitive to structural changes, observed changes in the magnitude of k_{hyd} with structure are due primarily to changes of k_b, and to a lesser extent, k_n. Based on the mechanism for base catalysis ($B_{AC}2$), we would expect electronic factors that enhance the electrophilicity of the carbon atom of the carbonyl group would make it more susceptible to nucleophilic attack by OH^-. The electron-withdrawing groups can be substituents of either the acyl group (RC(O)) or the alcohol portion of the ester. The Hammett plot in Figure 2.7 illustrates the inductive effect of phenyl ring substituents on the hydrolysis of a series of ethyl benzoates (Tinsley, 1979). The positive ρ value of 2.56 indicates that electron-withdrawing substituents increase the rate of hydrolysis.

The hydrolysis kinetics of aliphatic esters (R_1 and R_2 = alkyl groups) are also sensitive to electronic effects. The hydrolysis data for a series of aliphatic esters are summarized in Table 2.4. The addition of chloride substituents to the R_1 group drastically increases the neutral and base hydrolysis rate constants. These data also demonstrate that as the steric bulk of R_2 increases, there is a significant decrease in k_b (compare ethyl acetate to *t*-butyl acetate).

Figure 2.8 illustrates hydrolytic reaction pathways for a number of environmentally significant carboxylic acid esters.

Bis(2-ethylhexyl)phthalate, a large volume plasticizer, is fairly resistant to hydrol-

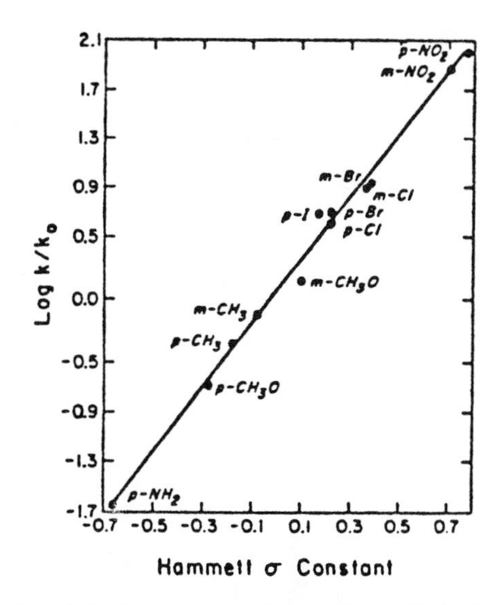

Figure 2.7. A plot of log k/k_0 for a series of phenyl substituted ethyl benzoates versus the Hammett sigma constant of the phenyl substituent groups. From Tinsley (1979). Reprinted by permission of John Wiley and Sons.

ysis ($t_{1/2}$ = 100 y at pH 8), suggesting that it will be fairly persistent in aquatic environments if hydrolysis is the primary reaction pathway (Wolfe et al., 1980). Fenpropathrin, which is a synthetic pyrethroid with high insecticidal activity and low mammalian toxicity, undergoes hydrolysis to give fenvaleric acid and the cyanohydrin, which rapidly loses the cyano group to give 3-phenoxybenzaldehyde (Takahashi et al., 1985a). Fenpropathrin is susceptible to neutral and base-catalyzed hydrolysis. Over the pH range of 5 to 9, the hydrolysis half-life for fenpropathrin will range from 8520 d to 11.3 d. The hydrolysis kinetics for fenpropathrin are typical of the synthetic pyrethroids (Camilleri, 1984 and Takahashi et al., 1985b). Because of their degradability and extreme potency, the synthetic pyrethroids have

Table 2.4. Hydrolysis of Aliphatic Esters at pH 7 at 25°C[a]

R_1	R_2	$k_a[H^+](s^{-1})$	$k_n(s^{-1})$	$k_b[OH^-](s^{-1})$	$k_{hyd}(s^{-1})$	$t_{1/2}$
Me	Et	1.1×10^{-11}	1.5×10^{-10}	1.1×10^{-8}	1.1×10^{-8}	2.0y
$ClCH_2$	Me	8.5×10^{-12}	2.1×10^{-7}	1.4×10^{-5}	1.4×10^{-5}	14h
Cl_2CH	Me	2.3×10^{-11}	1.5×10^{-5}	2.8×10^{-4}	3.0×10^{-4}	38m
Cl_3C	Me	—	$\geq7.7\times10^{-4}$	—	$\geq7.7\times10^{-4}$	$\leq$15m
Me	*i*-Pr	6.0×10^{-12}	—	2.6×10^{-9}	2.6×10^{-9}	8.4y
Me	*t*-Bu	1.3×10^{-11}	—	1.5×10^{-10}	1.6×10^{-10}	140y

[a]Data taken from Mabey and Mill (1978).

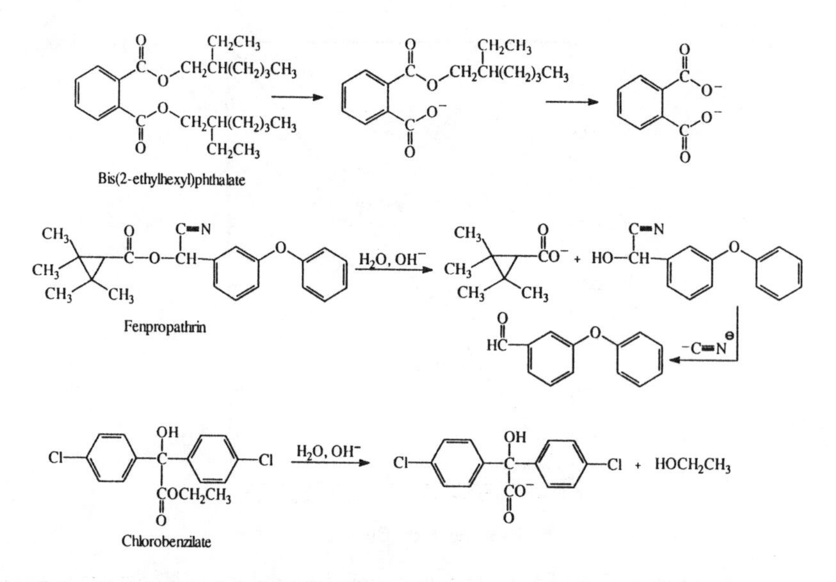

Figure 2.8. Examples of carboxylic acid ester hydrolysis.

become a popular alternative for the more environmentally persistent chlorohydrocarbons. The hydrolysis kinetics of chlorobenzilate, which is an acaracide (chemical agents used for the control of mites), have not been reported in the literature. Based on the hydrolytic behavior of other similar carboxylic acid esters, the hydrolysis half-life for chlorobenzilate at pH 7 at 25°C is expected to be 1 to 2 years.

Amides. In general, amides are much less reactive toward hydrolysis than esters. Typically, half-lives for amides at environmentally significant conditions are measured in hundreds to thousands of years (Mabey and Mill, 1978). This observation can be explained by the ground-state stabilization of the carbonyl group by the electron donating properties of the nitrogen atom (2.66).

$$R_1-\overset{\overset{\displaystyle O}{\|}}{C}-\ddot{N}\begin{matrix}R_2\\R_3\end{matrix} \longleftrightarrow R_1-\overset{\overset{\displaystyle O^-}{|}}{C}=\overset{+}{N}\begin{matrix}R_2\\R_3\end{matrix} \tag{2.66}$$

This stabilization is lost in the transition state leading to the formation of the tetrahedral intermediate. The result is that the hydrolysis of amides generally requires base or acid catalysis, both of which can compete at neutral pH. The basic hydrolysis of amides occurs through a $B_{AC}2$-type mechanism (Bender and Thomas, 1961).

$B_{AC}2$-type Mechanism

$$R_1-\overset{\overset{\displaystyle O}{\|}}{C}-NHR_2 + OH^- \rightleftharpoons R_1-\underset{\underset{\displaystyle OH}{|}}{\overset{\overset{\displaystyle O^-}{|}}{C}}-NHR_2 \tag{2.67}$$

$$R_1-\underset{\underset{\displaystyle OH}{|}}{\overset{\overset{\displaystyle O^-}{|}}{C}}-NHR_2 + H^+ \rightleftharpoons R_1-\underset{\underset{\displaystyle OH}{|}}{\overset{\overset{\displaystyle O^-}{|}}{C}}-\overset{+}{N}H_2R_2 \tag{2.68}$$

$$R_1-\underset{\underset{\displaystyle OH}{|}}{\overset{\overset{\displaystyle O^-}{|}}{C}}-\overset{+}{N}H_2R_2 \rightleftharpoons R_1-\overset{\overset{\displaystyle O}{\|}}{C}-OH + H_2NR_2 \tag{2.69}$$

$$R_1-\overset{\overset{\displaystyle O}{\|}}{C}-OH + OH^- \rightleftharpoons R_1-\overset{\overset{\displaystyle O}{\|}}{C}-O^- + H_2O \tag{2.70}$$

The primary difference, however, is that because of the poor leaving ability of the amide ion, protonation at nitrogen is usually required prior to breakdown of the tetrahedral intermediate. As a result, reversal of the tetrahedral intermediate to reactants is faster than its breakdown to the carboxylic acid and amine; thus, there is a substantial amount of exchange between the carbonyl oxygen and water (Samuel and Silver, 1965).

Acid hydrolysis of amides occurs through an $A_{AC}2$-type mechanism.

$A_{AC}2$-type Mechanism

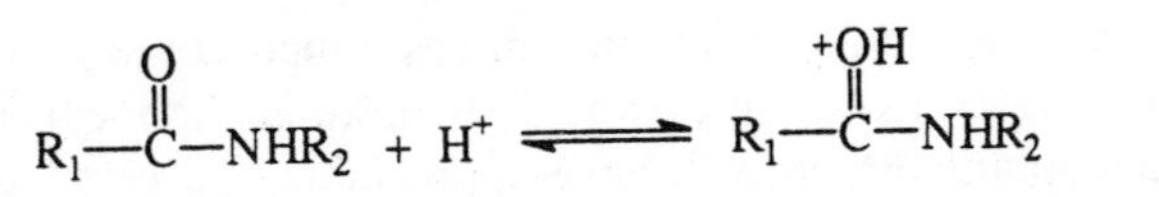

$$R_1-\overset{\overset{\displaystyle O}{\|}}{C}-NHR_2 + H^+ \rightleftharpoons R_1-\overset{\overset{\displaystyle ^+OH}{\|}}{C}-NHR_2 \tag{2.71}$$

$$R_1-\overset{\overset{\displaystyle ^+OH}{\|}}{C}-NHR_2 + H_2O \rightleftharpoons R_1-\underset{\underset{\displaystyle ^+OH_2}{|}}{\overset{\overset{\displaystyle OH}{|}}{C}}-NHR_2 \tag{2.72}$$

$$R_1-\underset{\underset{\displaystyle ^+OH_2}{|}}{\overset{\overset{\displaystyle OH}{|}}{C}}-NHR_2 \rightleftharpoons R_1-\underset{\underset{\displaystyle OH}{|}}{\overset{\overset{\displaystyle OH}{|}}{C}}-\overset{+}{N}H_2R_2 \tag{2.73}$$

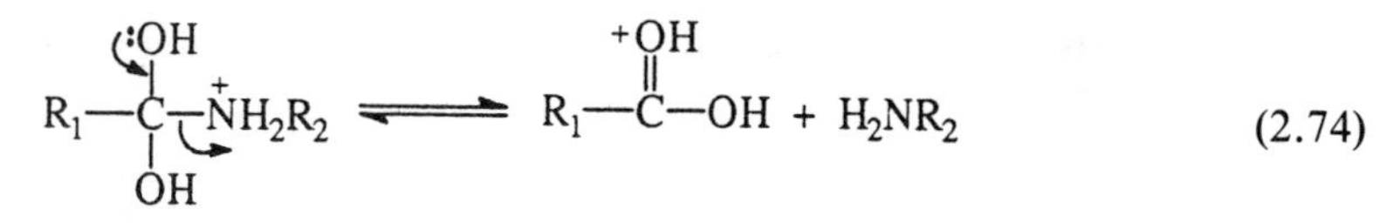

(2.74)

$$R_1{-}\overset{\overset{+OH}{\parallel}}{C}{-}OH + H_2NR_2 \rightleftharpoons R_1{-}\overset{\overset{O}{\parallel}}{C}{-}OH + H_3\overset{+}{N}R_2 \quad (2.75)$$

Although there are two possible sites for protonation of the amide moiety, the carbonyl oxygen atom and the nitrogen atom, protonation of the oxygen atom is preferred because of the resulting delocalization of the positive charge over the π-orbitals of the O-C-N linkage. Stabilization of the positive charge on nitrogen in this manner is not possible. Formation of the tetrahedral intermediate is followed by protonation of the nitrogen substituent, which is the most basic site in the tetrahedral intermediate. Protonation of the nitrogen substituent results in expulsion of the neutral amine, which is preferred to the loss of hydroxide ion. This mechanism is consistent with the observation that there is almost no exchange of carbonyl oxygen with water during acid-catalyzed hydrolysis of amides.

Table 2.5 summarizes the hydrolysis kinetic data for a number of aliphatic amides.

Table 2.5. Hydrolysis Kinetics for Amides at pH 7 at 25°C[a]

R_1	R_2	R_3	$k_a(M^{-1}s^{-1})$	$k_b(M^{-1}s^{-1})$	k_{hyd}	$t_{1/2}(y)$
CH_3	H	H	8.36×10^{-6}	4.71×10^{-5}	5.55×10^{-12}	3,950
$ClCH_2$	H	H	1.1×10^{-5}	1.5×10^{-1}	1.5×10^{-8}	1.46
Cl_2CH	H	H	—	3.0×10^{-1}	3.0×10^{-8}	0.73
CH_3	CH_3	H	3.2×10^{-7}	5.46×10^{-6}	5.76×10^{-13}	38,000
CH_3	CH_3CH_2	H	9.36×10^{-8}	3.10×10^{-6}	3.10×10^{-13}	70,000

[a]Data taken from Mabey and Mill (1978).

It is apparent that only those aliphatic amides which contain substituents that withdraw electron density from the carbonyl group (e.g., halides), making it more susceptible to nucleophilic attack, will have appreciable hydrolysis rates at pH 7 at 25°C. Alkyl substituents on nitrogen will increase hydrolysis half-lives due to steric effects (compare $t_{1/2}$ for acetamide to *N*-methyl- and *N*-ethylacetamide).

The benzamides are also quite stable to hydrolysis. For example, pronamide, which is a large volume herbicide, has a hydrolysis half-life of >700 y (Eq. 2.76).

b. Carbonic Acid Derivatives

Carbamates. Of the carbonic acid derivatives illustrated in Figure 2.6, the carbamates are of the greatest environmental significance because of their immense im-

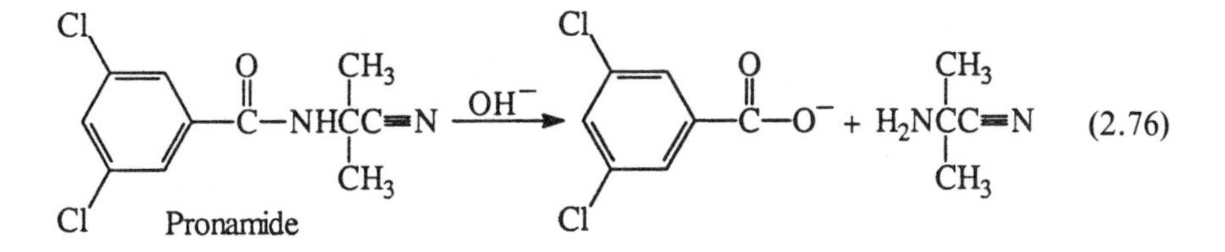

portance to the agrochemical industry. The carbamates exhibit a wide range of insecticidal and herbicidal activity. Because of their significance in the agrochemical industry, considerable effort has been made to understand their environmental fate pathways. A carbamate is hydrolyzed to an alcohol, carbon dioxide, and an amine (2.77).

$$R_1R_2N{-}\overset{O}{\overset{\|}{C}}{-}OR_3 \xrightarrow{OH^-} R_1R_2NH + CO_2 + HOR_3 \qquad (2.77)$$

They are susceptible to acid, neutral, and base hydrolysis, though in most cases, base hydrolysis will dominate at environmentally significant conditions. Hydrolysis data tabulated by Mabey and Mill (1978) for the carbamates indicate that hydrolysis half-lives at pH 7 at 25°C range from seconds to hundreds of thousands of years (Table 2.6).

In a manner analogous to carboxylic acid ester and amide hydrolysis, electron-withdrawing substituents on either oxygen or nitrogen will accelerate the rate of hydrolysis. The most dramatic differences in reactivities are observed when the hydrolysis half-lives of primary ($R_1 = H$, $R_2 =$ alkyl) and secondary ($R_1 =$ alkyl, $R_2 =$ alkyl) carbamates are compared (Table 2.6). It is evident that the primary carbamates hydrolyze at rates that are approximately 6 to 7 orders of magnitude faster than the corresponding secondary carbamates. These differences in reactivity can be explained by comparing their hydrolysis mechanisms that are outlined in Figure 2.9.

Table 2.6. Hydrolysis Kinetics of Primary and Secondary Carbamates at pH 7 at 25°C

R_3	R_2	R_1	k_h	$t_{1/2}$
C_6H_5	C_6H_5	H	5.4×10^{-6}	1.5 d
C_6H_5	C_6H_5	CH_3	4.2×10^{-12}	5,200 y
p-$NO_2C_6H_4$	C_6H_5	H	2.7×10^{-2}	26 s
p-$NO_2C_6H_4$	C_6H_5	CH_3	8.0×10^{-11}	2,700 y
1-$C_{10}H_9$	CH_3	H	9.4×10^{-7}	8.5 d
1-$C_{10}H_9$	CH_3	CH_3	1.8×10^{-11}	1,200 y

[a]Data taken from Mabey and Mill (1978).

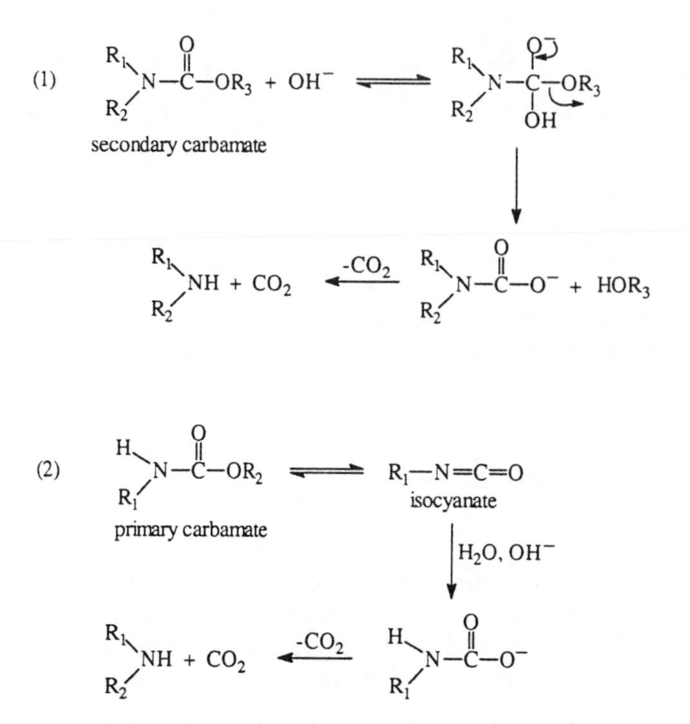

Figure 2.9. Hydrolysis pathways for primary and secondary carbamates.

Hydrolysis of secondary carbamates occurs by an analogous mechanisms previously described for the carboxylic acid derivatives. Nucleophilic attack of OH^- occurs at the carbonyl moiety ($B_{AC}2$ mechanism) to form the tetrahedral intermediate that breaks down by loss of alkoxide ion (RO^-) to give the carboxylated amine, which loses CO_2 rapidly to give the free amine. On the other hand, hydrolysis of the primary carbamates may proceed through an elimination mechanism (E_1cB) involving facile loss of proton on the nitrogen atom and subsequent formation of the isocyanate (Bergon et al., 1985; Hegarty and Frost, 1973; Williams, 1972). The isocyanates typically do not persist and react with H_2O or OH^- to afford the free amine after loss of CO_2.

Wolfe et al. (1978) have developed structure-activity relationships for the hydrolysis of a number of *N*-substituted and *N,N*-disubstituted carbamates. An excellent correlation was found between the second-order alkaline hydrolysis rate constant (k_b) for the carbamates and the pKa of the alcohol formed upon hydrolysis. The linear free energy relationships for the *N*-phenyl and *N*-methyl-*N*-phenyl carbamates are illustrated in Figure 2.10. These data indicate that only the *N*-phenyl carbamates will hydrolyze under environmental conditions. The half-life for the hydrolysis of these chemicals will be less than 6 months at pH 8 at 25°C.

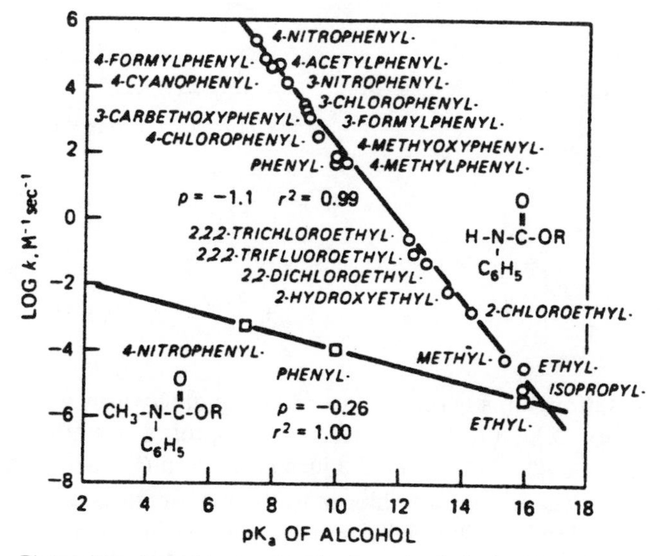

Figure 2.10. Plots of the log of second-order base hydrolysis rate constants of N-phenyl carbamates versus pK_a of the resulting alcohol at 25°C in water. From Wolfe et al. (1978). (Reprinted by permission of Pergamon Press Ltd.)

Sulfonylureas. The sulfonylureas have environmental significance because of their growing importance as herbicides. The sulfonyureas possess unprecedented levels of herbicidal activity and exhibit very low mammalian toxicity. The hydrolysis kinetics of sulfonylureas are quite interesting. They provide an example of how ionizable functional groups can significantly alter the hydrolysis kinetics of organic chemicals. Contrary to the normal behavior observed for carbonate derivatives, which typically exhibit base catalysis, the sulfonylureas are quite stable to hydrolysis under basic conditions but are readily hydrolyzed at acidic pHs (Brown, 1990: Hay, 1990). This behavior is a result of the sulfonyl group, which significantly enhances the acidity of the proton on the adjacent nitrogen atom. Typically, the sulfonylureas have pKa values ranging from 3.3 to 5.2, indicating that they will be significantly deprotonated even at neutral pH values (Beyer et al., 1987). Deprotonation of the sulfonylurea markedly deactivates the group toward hydrolysis because the resulting negative charge is distributed throughout the sulfonylurea moiety, reducing the electrophilicity of the carbonyl group (2.78).

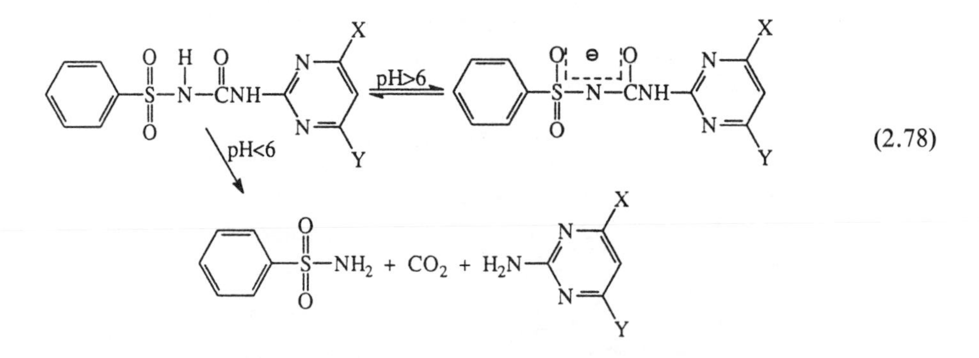

(2.78)

As a result, the sulfonylurea herbicides are more rapidly hydrolyzed in acidic waters and soils. Figure 2.11 illustrates the pH-rate profile for the hydrolysis of chlorimuron-ethyl at 45°C in buffered aqueous solution (Brown, 1990).

The hydrolysis rate constant for chlorimuron-ethyl decreases over 150-fold as the pH increases from 4 to 7. Additionally, ionization affects water solubility. The water solubility of chlorimuron-ethyl increases from 11 to 1200 mg/L when the pH increases from 5 to 7 (Hay, 1990). Consequently, the sulfonylureas are expected to be much more persistent and more readily transported in basic aquatic ecosystems.

D. OTHER NUCLEOPHILIC SUBSTITUTION REACTIONS

1. Reactions with Naturally Occurring Nucleophiles

Thus far, discussion of nucleophilic substitution processes has been limited to reactions with the oxygenated nucleophiles, OH^- and H_2O. Recently, however, there has been a great deal of interest in determining the contribution of other naturally occurring nucleophiles to the abiotic transformation of organic pollutants; namely, the halogenated aliphatics. Much of this interest comes from field studies that have detected organic sulfur compounds in groundwaters that presumably were formed from nucleophilic substitution reactions of bisulfide anion, HS^-, with haloaliphatic compounds (Schwarzenbach et al., 1985; Jackson et al., 1985; Weintraub and Moye, 1987; Barbash and Reinhard, 1989; Roberts et al., 1992). Laboratory studies have demonstrated the reactivity of HS^-, as well as other sulfur nucleophiles, for halogenated aliphatics (Schwarzenbach et al., 1985; Barbash and Reinhard, 1989a and 1989b; Haag and Mill., 1988b). In addition to having the capability to correctly predict the concentrations of halogenated compounds in aquatic ecosystems, there is concern over the formation of reaction products (e.g., thiols) that may be more persistent and hazardous than the parent compounds or their hydrolysis products.

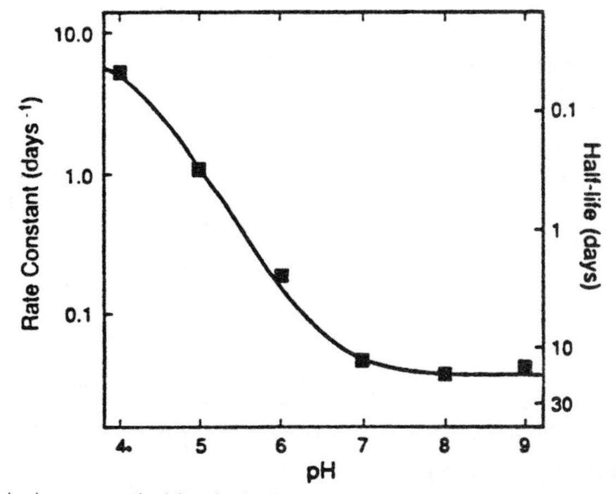

Figure 2.11. Chlorimuron-ethyl hydrolysis rate constant versus pH at 45°C in buffered aqueous solution. From Hay (1990). (Reprinted by permission of the Society of Chemical Industry).

a. Nucleophilic Reactivity

To assess the importance of nucleophilic substitution reactions of naturally occurring nucleophiles it is necessary to have some measure of their reactivity, relative to OH^- and H_2O. A number of properties of nucleophiles, all of which are some measure of the nucleophile's ability to donate electrons to an electrophile, have been used to correlate nucleophilic reactivity. These closely related properties include basicity, oxidation potential, polarizability, ionization potential, electronegativity, energy of the highest filled molecular orbital, covalent bond strength, and size (Jencks, 1987).

Hard-soft acid base model. Pearson (1963) has developed an empirical model for nucleophilic reactivity that has the ability to predict reactivity trends among different nucleophiles and electrophilic substrates. The model, known as Pearson's principle of hard and soft acids and bases (commonly abbreviated HSAB principle) classifies Lewis bases (nucleophiles) and Lewis acids (electrophiles) as either "hard" or "soft." Hard bases and acids are relatively small, of high electronegativity, and of low polarizability. Soft acids and bases are relatively large, of low electronegativity, and of high polarizability. Hardness is associated with a relatively large amount of ionic character and softness with a large amount of covalent character in a bond or transition state. On the whole it is found that hard acids form stronger bonds and react faster with hard bases, and soft acids form stronger bonds and react faster with soft bases. The HSAB principle summarizes this observation: *Hard acids prefer to bind to hard bases and soft acids tend to bind to soft bases.* Because nucleophilic-electrophilic reactions are an example of a Lewis base-Lewis acid reaction, the HSAB principle applies equally to the reaction of nucleophiles with electrophiles.

Environmentally significant nucleophiles are classified, according to Pearson's HSAB principle, as either hard, soft, or borderline (possessing intermediate hard-soft character) as follows:

Hard	**Borderline**	**Soft**
OH^-, $H_2PO_4^-$, HCO_3^-	H_2O, NO_2^-, SO_3^{2-}	HS^-, RS^-, PhS^-
NO_3^-, SO_4^{2-}, Cl^-, F^-	Br^-, $C_6H_5NH_2$	$S_2O_3^-$, I^-, CN^-
NH_3, CH_3COO^-		

The HSAB principle has been criticized because of the difficulty in defining the terms "hard" and "soft" quantitatively. More recently, Pearson (1983) has attempted to quantify the term hardness by defining hardness operationally as half the difference between ionization potential and electron affinity.

Because of the qualitative nature of the HSAB model, it tells us nothing about the absolute value of kinetic rate constants. The usefulness of the HSAB model arises from its ability to predict the relative reactivities of nucleophiles with various substrates. For example, C–O cleavage of dimethyl phosphates results from nucleophilic attack at carbon, a soft electrophilic site (carbonium ions are soft electrophiles), whereas cleavage of the P–O bond results from nucleophilic attack at the phosphorus atom (P–O), a hard electrophilic site. Accordingly, reaction of dimethyl phosphate with a soft nucleophile, such as sulfide, would expect to occur mainly at the soft carbon center, resulting in C–O cleavage (2.79). On the other hand, nucleophilic attack by a hard nucleophile, such as hydroxide ion, is expected to occur predominantly at the hard phosphorus center (2.80).

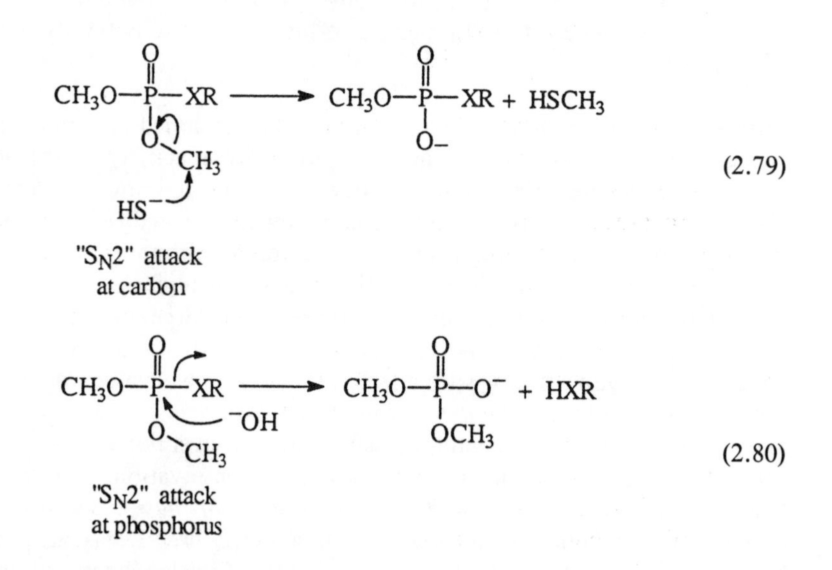

These predictions are consistent with experimental data on the competition between dealkylation and hydrolysis for the reactions of HS^- and OH^- with dimethyl phosphate (Schmidt, 1975).

The HSAB model also provides a rationale for the relative reactivities of HS^-, H_2O and OH^- toward halogenated aliphatics. Laboratory studies have demonstrated that HS^- reacts with 1,2-dibromoethane predominantly through nucleophilic substitution (2.81) (Haag and Mill, 1988b; Barbash and Reinhard, 1989a). By contrast, reaction of OH^- with 1,2-dibromoethane occurs predominantly through elimination (2.82) (Hine and Langford, 1956; Vogel and Reinhard, 1986; Weintraub, et al., 1986; Haag and Mill, 1988b; Weintraub and Moye, 1987). Reaction of water with 1,2-bromoethane favors nucleophilic substitution; however, a significant amount of elimination is also observed (2.83) (Junglaus and Cohen, 1986; Pignatello, 1986; Vogel and Reinhard, 1986; Weintraub et al., 1986; and Haag and Mill, 1988b).

$$\text{BrCH}_2\text{—CH}_2\text{Br} \xrightarrow[\text{"S}_\text{N}\text{2"}]{\text{HS}^-} \text{HSCH}_2\text{—CH}_2\text{SH} \tag{2.81}$$

$$\text{BrCH}_2\text{—CH}_2\text{Br} \xrightarrow[\text{"E2"}]{\text{OH}^-} \text{H}_2\text{C=C(Br)H} \tag{2.82}$$

$$\text{BrCH}_2\text{—CH}_2\text{Br} \xrightarrow[\text{"S}_\text{N}\text{2" + "E2"}]{\text{H}_2\text{O}} \text{HOCH}_2\text{—CH}_2\text{OH} + \text{H}_2\text{C=C(Br)H} \tag{2.83}$$

These results are consistent with the HSAB model if we consider that bimolecular elimination is a hard-hard interaction resulting from attack of the hard nucleophile, OH^-, at the proton, a hard electrophilic center. Whereas, nucleophilic substitution occurs by attack of the soft nucleophile, HS^-, at saturated carbon, a soft electrophilic center. The softness of H_2O is intermediate between that of OH^- and HS^-, resulting in a mixture of substitution and elimination products.

Correlation equations. Duboc (1978) has critically reviewed the equations used for correlating nucleophilic reactivity. The correlation equation that has been used to the greatest success is a two-term equation (Edwards, 1954 and 1956):

$$\log k/k_0 = \alpha E_n + \beta H \tag{2.84}$$

where E_n is a measure of the nucleophile's polarizability, H is a measure of its basicity, and α and β are the measures of susceptibility of the substrate to E_n and H, respectively. E_n has been defined in terms of the molar refractivity of the nucleophile. The limitations of this correlation result from the fact that it requires information that may not be generally available.

A simpler correlation that allows for direct comparison of reactivity for a series of nucleophiles for a particular substrate (RX) is that of Swain and Scott (1953), which is given by:

$$\log (k_{nuc}/k_{H_2O}) = ns \tag{2.85}$$

where k_{nuc} is the second-order rate constant for the reaction of the nucleophile of interest with RX, k_{H_2O} is the second-order rate constant for the reaction of RX with water, n is the nucleophilicity constant, and s is a parameter for the sensitivity of the substrate to nucleophilic reactions. Nucleophiles that are stronger than H_2O will have values for n that are positive (the larger the value of n, the stronger the nucleophile). If the substrate constant, s, is known for a particular substrate, the rate constant for the reaction of any nucleophile with the same substrate can be predicted using n values from a reference reaction. The Swain-Scott equation has been used successfully to correlate rates of reactions with similar transition states. The correlation is found to fail if reactions with fundamentally different transitions states (e.g., nucleophilic reactions at saturated and unsaturated carbons) are compared.

Values of n are determined by setting $n = 1$ for H_2O and $s = 1$ for a reference reaction, typically the S_N2 reaction of CH_3Br.

$$CH_3Br + H_2O \rightarrow CH_3OH + HBr \tag{2.86}$$

$$CH_3Br + Y: \rightarrow CH_3Y + Br^- \tag{2.87}$$

Table 2.7 lists nucleophilicity constants in water for a number of naturally occurring nucleophiles, values of rate constants for nucleophilic substitution (k_Y) relative to H_2O, and the calculated concentrations, $[Nuc]_{50\%}$, at which the rate constant for the S_N2 displacement by the nucleophile of interest is equal to the rate of S_N2 displacement by H_2O ($k_{Nuc}[Nuc]_{50\%} = k_{H_2O}[H_2O]$). It is readily apparent that as the magnitude for the value of n increases, the concentration of nucleophile needed to compete with hydrolysis decreases.

b. Reactions of Sulfur-Based Nucleophiles with Halogenated Aliphatics

The data in Table 2.8 demonstrate that the sulfur-based nucleophiles will have the greatest reactivity toward halogenated substrates (largest n values). In general, the concentrations of the naturally occurring nucleophiles typically found in uncontaminated, freshwater ecosystems are several orders of magnitude below their concentrations. At the concentrations of hydrogen sulfide, polysulfides, sulfite, and thiosulfite found in saltmarshes, anoxic surface waters and contaminated groundwaters, however, nucleophilic reactions with these nucleophiles can compete with hydrolysis (Haag and Mill, 1988b; Barbash and Reinhard, 1989b; Roberts et al.,

Table 2.7. Relative Reactivities of Naturally Occurring Nucleophiles

Nucleophile	n	k_{rel}	$[Nuc]_{50\%}$
H_2O	0.0	1.0	0.0
NO_3^-	1.0		~6
SO_4^{2-}	2.5	3×10^2	2×10^{-1}
Cl^-	3.0	1×10^3	6×10^{-2}
HCO_3^-	3.8	6×10^3	9×10^{-3}
HPO_4^{2-}	3.8	6×10^3	9×10^{-3}
Br^-	3.9	8×10^3	7×10^{-3}
OH^-	4.2	2×10^3	3×10^{-3}
I^-	5.0	1×10^5	6×10^{-4}
HS^-	5.1	1×10^5	6×10^{-4}
SO_3^{2-}	5.1	1×10^5	6×10^{-4}
S_4^{2-}	7.2		
S_5^{2-}	7.2		

1992). Table 2.8 compares hydrolysis half-lives with half-lives for reaction with sulfur-based nucleophiles for several halogenated aliphatics. These data show that the environmental half-lives for substrates such as 1-bromohexane and 1,2-dibromoethane can be substantially reduced in the presence of HS^- and polysulfides. Enhanced degradation of 2-bromopropane and 1,1,1-trichloroethane, as well as chloroform and carbon tetrachloride (results not shown) was not observed, suggesting that steric hindrance significantly impedes reaction with the sulfur based nucleophiles (Haag and Mill, 1988a).

Table 2.8. Comparison of Hydrolysis Half-Lives (days) with Half-Lives for the S_N2 Reaction of Sulfur-Based Nucleophiles with Halogenated Aliphatics at 25°C[a]

Nucleophile[b]	hexBr[c]	DBE[c]	2BP[c]	TrCE[c]
H_2O	20	1,000	2	350
5.0 mM HS^-	1.0	3.9	9.4	>3,200
0.07 mM S_x^{2-}	1.1			
0.5 mM $S_2O_3^{2-}$	17	170	240	>160,000
0.2 mM SO_3^{2-}	22	170		

[a]Data taken from Haag and Mill (1988b).

[b]Assuming complete ionization of sulfur-based nucleophiles.

[c]hexBr = 1-bromohexane; DBE = 1,2-dibromoethane; 2BP = 2-bromopropane; TrCE = 1,1,1 = trichloroethane.

In addition to accelerating degradation rates for halogenated aliphatics, reaction with sulfur nucleophiles will have significant consequences with respect to reaction product distributions. Schwarzenbach et al. (1985) have observed the formation of thiols and dialkyl sulfides from the S_N2 reactions of primary alkyl bromides with HS^- (Figure 2.12).

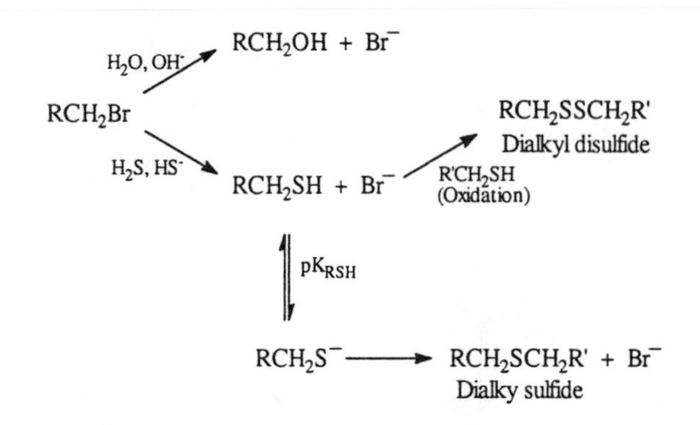

Figure 2.12. Pathways for the reaction of alkyl bromides with HO^- and HS^-.

Based on laboratory studies, Schwarzenbach et al. (1985) calculated that the reaction product distribution for primary alkyl bromides at an initial concentration of 1×10^{-5} M in an aquatic environment containing a constant concentration of total H_2S at 5×10^{-5} M (at pH 8.0) will result in the distribution of alcohol, thiol and dialkyl sulfide of 75%, 25%, and 0.2%, respectively. Formation of even small amounts of the dialkyl sulfides is of concern because of their persistence in groundwater systems. Schwarzenbach et al. (1985) have identified 24 dialkyl sulfides in groundwater samples from the site of a chemical plant that manufactures alkyl halides.

More recently Roberts et al. (1992) have studied the reactions of dihalomethanes with HS^-. The dihalomethanes are of environmental significance because of their wide use as organic solvents, degreasing agents, and chemical intermediates. They have been detected in industrial waste waters (Keith and Telliard, 1979) and groundwater and surface water samples (Page, 1981). Kinetic studies of CH_2Br_2, CH_2BrCl, and CH_2Cl_2 indicated that nucleophilic substitution reactions with HS^- exceed hydrolysis rates at HS^- concentrations greater than 2–17 μM. Rigorous product studies indicate that poly(thiomethylene) is the major product from the reaction of HS^- with dihalomethanes. The formation of minor amounts of dithiomethane was also observed. Figure 2.13 summarizes the proposed pathways for the formation of these products. Initially, the α-halo thiomethane is thought to form by the rate determining nucleophilic attack (S_N2 mechanism), resulting in the displacement of halide ion by HS^-. The α-halo thiomethane rapidly loses the second halide ion, presumably through a S_N1 mechanism, to give thioformaldehyde.

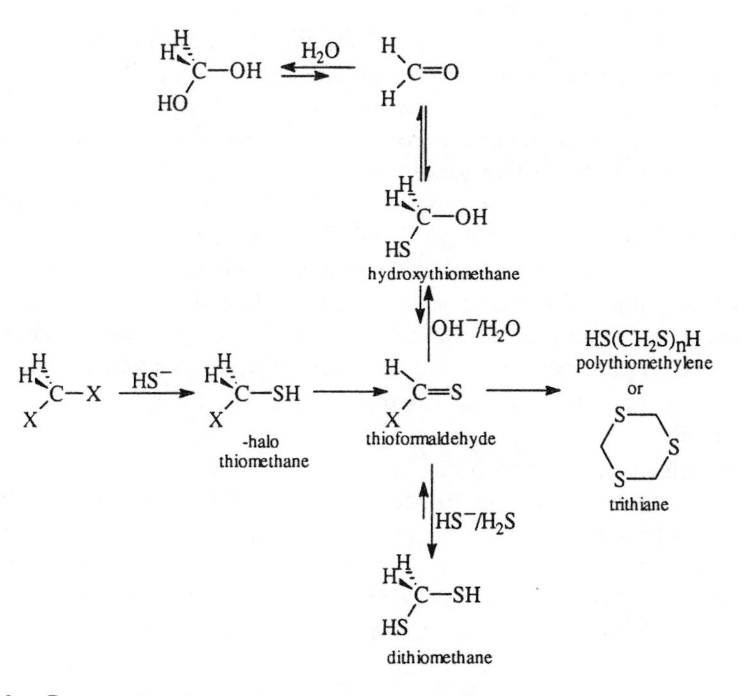

Figure 2.13. Proposed pathways for reactions of dihalomethanes and formaldehyde initiated by S_N2 reaction with HS^- (adapted from Roberts et al., 1992).

2. Neighboring Group Participation (Intramolecular Nucleophilic Displacement)

Neighboring group participation is a unique case of nucleophilic displacement that occurs when the attacking nucleophile adjacent to an atom bearing a leaving group acts as an internal nucleophile to displace the leaving group (2.88).

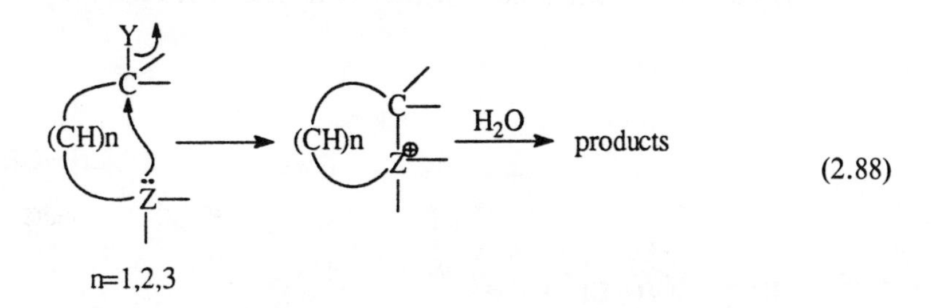

(2.88)

This is an example of an intramolecular nucleophilic substitution reaction. A cyclic intermediate is formed that undergoes attack by an external nucleophile, typically H_2O, to form products. These reactions typically involve 3-, 5-, or 6-membered ring intermediates.

Groups known to effect these reactions include sulfides, amines, alkoxides, halides, and carboxylates (Streitwieser, 1962). These nucleophilic groups are able to compete with H_2O or OH^-, which are typically present at much higher concentrations, because (1) the neighboring group is always in the immediate vicinity of the reactive site, thus its "effective concentration" is extremely high compared to external nucleophiles and (2) the electronic reorganization required to reach the transition state for nucleophilic displacement is minimal. As a result, hydrolysis rates for reactions involving neighboring group participation are often accelerated by several orders of magnitude compared to ordinary nucleophilic displacement by H_2O. When neighboring group participation results in a hydrolysis rate enhancement, the group is said to provide *anchimeric assistance* (Capon and McManus, 1976). Because one process almost always involves the other, these terms are often used interchangeably.

A comparison of the hydrolysis rate for 2-ethylthio-1-chloroethane ($EtSCH_2$-CH_2Cl) and 2-hydroxy-1-chloroethane ($HOCH_2CH_2Cl$) illustrates the dramatic effect that neighboring group participation can have on reactivity. Hydrolysis of 2-ethylthio-1-chloroethane occurs at a rate 10^4 times faster than 2-hydroxy-1-chloroethane. Hydrolysis of $EtSCH_2CH_2Cl$ is thought to occur through the sulfonium ion, which subsequently reacts with water to form the thio alcohol (2.89).

$$\text{EtS:}\,\text{CH}_2\text{CH}_2\text{—Cl} \longrightarrow \text{Cl}^{\ominus}\ \text{Et}\overset{\oplus}{\text{S}}\!\!<\!\!\begin{array}{c}\text{CH}_2\\|\\\text{CH}_2\end{array} \xrightarrow{\text{H}_2\text{O}} \text{EtSCH}_2\text{CH}_2\text{OH} \qquad (2.89)$$

Because the more electronegative oxygen atom does not readily share its lone pair of electrons, hydrolysis of $HOCH_2CH_2Cl$ occurs through ordinary S_N2 nucleophilic attack by water.

Evidence for the formation of the cyclic sulfonium ion is demonstrated from hydrolysis studies of 2-ethylthio-2-methyl-1-chloroethane (2.90) (Sykes, 1986).

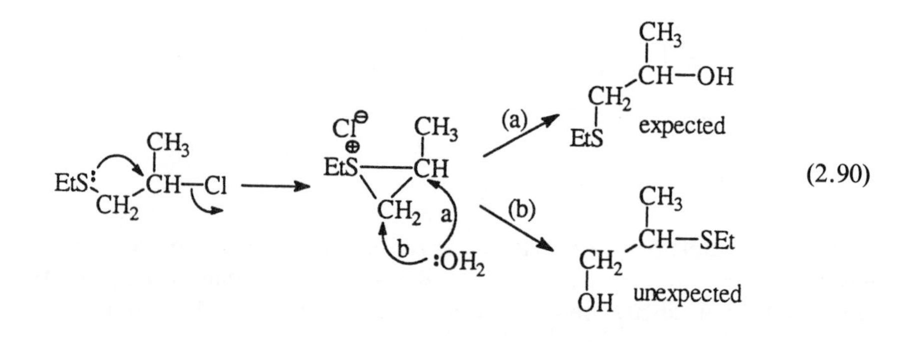

Formation of two alcohols indicates the intermediacy of an unsymmetrical sulfonium ion, which H_2O preferentially attacks at the less sterically hindered carbon, giving the primary alcohol as the major product.

Examples of environmentally significant chemicals whose hydrolyses occur with neighboring group participation include mustard gas and the related nitrogen mustards (2.91).

$$\underset{\text{Mustard gas}}{ClCH_2CH_2SCH_2CH_2Cl} \qquad \underset{\text{Nitrogen mustard}}{CH_3N(CH_2CH_2Cl)_2} \tag{2.91}$$

A thorough understanding of the environmental chemistry of these chemical warfare agents is taking on a new importance as the world community negotiates the destruction of millions of tons of these acutely toxic chemicals. The hydrolysis kinetics of mustard gas have shown the reaction to be first-order and pH-independent, consistent with a mechanism involving formation of the sulfonium ion in the rate determining step (Bartlett and Swain, 1949). The hydrolysis of the nitrogen mustards is much slower than the sulfur analogues because of the greater stability of the cyclic ammonium ion formed.

Intramolecular nucleophilic substitution reactions also are thought to compete with hydrolysis (i.e., S_N2 reaction with H_2O) in the hydrolysis of several organophosphorus esters. For example, the C-S cleavage of thiometen ($R = H$) and disulfoton ($R = CH_3$) can occur by intramolecular nucleophilic substitution by the sulfur atom of the electron-withdrawing group, which results in the formation of the cyclic sulfonium ion which subsequently reacts with water to give 2-(ethylthio)ethanol (Wanner et al., 1989).

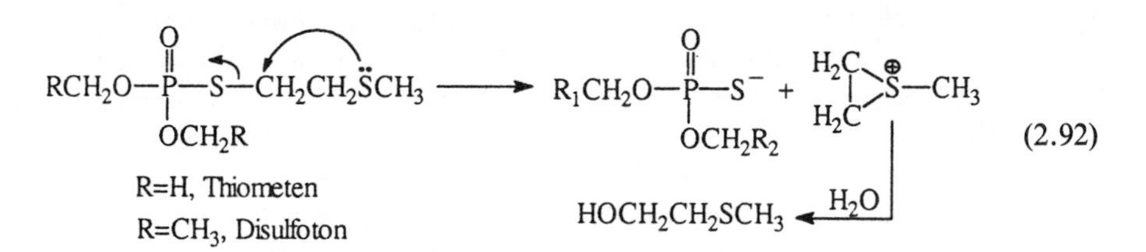

(2.92)

E. CATALYSIS OF HYDROLYTIC REACTIONS IN NATURAL AQUATIC ECOSYSTEMS

The complexity of natural aquatic ecosystems necessitates that we examine the ability of naturally occurring species and surfaces to alter the hydrolysis rates of organic pollutants. Failure to realize the potential for catalysis to occur in natural systems can lead to underestimations of hydrolysis rates when extrapolating from laboratory studies. Although this discussion will focus primarily on hydrolytic catalysis in natural water systems, several examples will be presented which demonstrate that components of natural systems also can impede hydrolysis.

Often there is confusion as to the definition of the term catalysis. A reaction is catalyzed if its rate is accelerated relative to the noncatalyzed pathway. A catalyst serves to alter the reaction mechanism of a chemical process by providing a new pathway with a lower potential-energy barrier. The Union of Pure and Applied Chemistry (IUPAC, 1981) definition of the term catalyst states "that a catalyst is a substance that increases the rate of a reaction without modifying the overall standard Gibbs energy change in the reaction; the process is called catalysis, and a reaction in which a catalyst is involved is known as a catalyzed reaction."

In the following discussion, examples of both homogeneous and heterogeneous catalysis will be presented. Homogeneous catalysis implies that both the substrate and the catalyst are in the dissolved state. Examples of homogeneous catalysis include specific-acid and -base catalysis, metal ion catalysis, and catalysis by dissolved organic matter. Heterogeneous catalysis refers to a process that takes place at the interface of two phases. Our discussion will be limited to the interface of water and solid surface such as a sediment, soil, or mineral oxides. In most examples of heterogeneous catalysis, the catalyst is associated with the solid phase and the reactants are dissolved in the aqueous phase (Hoffmann, 1990).

1. General Acid and Base Catalysis

We have already discussed the sensitivity of various hydrolyzable functional groups to specific-acid and specific-base catalysis (i.e., reaction with H^+ and OH^-, respectively). We must also consider the contribution of general acid and general base catalysis, or catalysis by all Brönsted acids and/or bases in the reaction system of interest. This phenomenon is generally referred to as buffer catalysis. A wide variety of weak acids and bases can be found in aquatic ecosystems that can potentially enhance hydrolysis rates of organic pollutants. The world average concentrations of these constituents in river water, seawater and interstitial water are summarized in Table 2.9. Additionally, buffer catalysis may be observed in laboratory studies where buffer salts are commonly used to control pH. Perdue and Wolfe (1983) developed a mathematical model, based on the application of the Brönsted equations for general acid-base catalysis to assess the contribution of general acid-base catalysis to hydrolysis reactions. They concluded that general acid-base catalysis is probably insignificant in most natural waters because of the very low concentrations of acids and bases found in these systems (Table 2.9). For example, a maximum 2% enhancement of k_{hyd} by buffer catalysis is predicted for the world average river, for which carbonic acid is the major buffer catalyst. The model does predict, however, that buffer catalysis may be significant in laboratory studies that employ buffers at concentrations greater than 0.001 M. The buffer catalysis of organohalides by phosphate, commonly used to control constant pH, is thought to occur through nucleophile displacement of the halide by phosphate ion (2.93) (Mabey and Mill, 1978; Junglaus and Cohen, 1986). The phosphate ester is then readily hydrolyzed to provide the alcohol (2.94).

Table 2.9. Concentration of Selected Organic and Inorganic Species in Selected Aquatic Environments[a]

Species	World Avg. River[b]	World Avg. Seawater[c]	World Interstitial Water[d]
Ammonia	—	—	2.00×10^{-4}
Carbonate	9.78×10^{-4}	2.33×10^{-3}	9.09×10^{-3}
Fulvic acid	1.00×10^{-4}	5.00×10^{-6}	2.00×10^{-4}
Phosphate	—	2.84×10^{-6}	1.80×10^{-5}
Silicate	2.18×10^{-4}	1.03×10^{-4}	—
Borate	—	4.10×10^{-4}	—

[a]mol/L.
[b]Data taken from Livingston (1963).
[c]Data taken from Turekian (1969).
[d]Data taken from Emerson (1976).

$$RX + HPO_4^{2-} \rightarrow ROPO_3H^- + X^- \quad (2.93)$$

$$ROPO_3H^- + H_2O \rightarrow ROH + H_2PO_4^- \quad (2.94)$$

2. Metal Ion Catalysis

Metal ion-catalyzed hydrolysis occurs primarily through two types of mechanisms: (1) the metal coordinates the hydrolyzable functional group, making it more electrophilic and thus more susceptible to nucleophilic attack (direct polarization mechanism) and (2) in situ generation of a reactive metal hydroxo species. The direct polarization mechanism is analogous to the acid-catalyzed hydrolysis of esters. The metal acts as Lewis acid, coordinating with the lone pair electrons of the oxygen atom, which results in the polarization of the carbonyl group, making it more susceptible to attack by H_2O or OH^-:

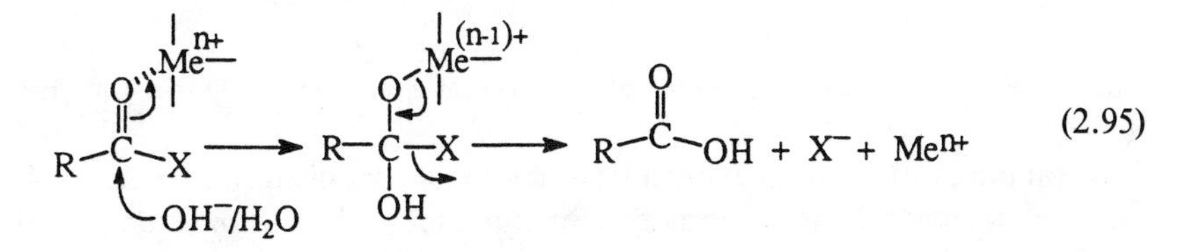

(2.95)

The electrophilicity of the reactive center also can be increased by the metal coordination of the leaving group, X, as in the case of phosphate ester substrates.

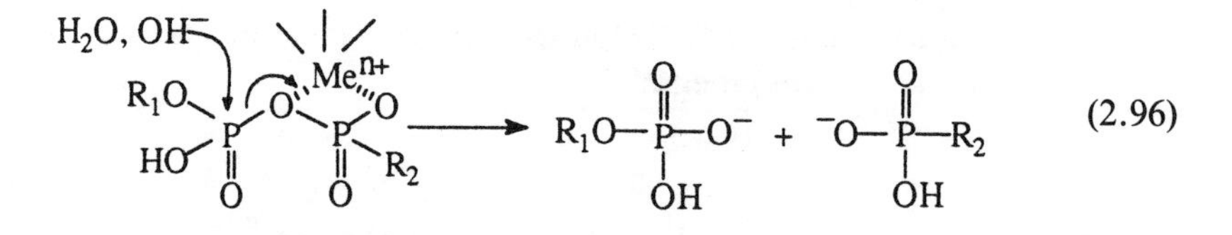

(2.96)

The reaction mechanism and the rate equations for metal ion catalysis by direct polarization have been presented by Plastourgou and Hoffmann (1984):

$$Me^{n+} + RCOX \leftrightarrow (MeRCOX)^{n+} \qquad K = k_1/k_2 \tag{2.97}$$

$$(MeRCOX)^{n+} + Y{:} \rightarrow RCOY + Me^{n+} + X{:} \tag{2.98}$$

$$d[RCOY]/dt = k_n[MeRCOX^{n+}][Y{:}] \tag{2.99}$$

$$d[RCOY]/dt = k_nK[Me^{n+}][Y{:}][RCOX] \tag{2.100}$$

$$d[RCOY]/dt = (k_nKMe_T[RCOX][Y{:}])/(1 + K[RCOX]) \tag{2.101}$$

where k_n is the second-order rate constant for nucleophilic substitution and K is the formation constant for the metal ion-carbonyl complex. It is evident that as the magnitude for K increases, the rate for metal ion catalysis increases. In general, complexation to more than a single ligand-donor group of the hydrolyzable chemical is necessary for metal ion catalysis to be observed (Hay, 1987). Thus, metal ion catalysis by the coordination mechanism will be limited primarily to those chemicals that possess auxiliary donor groups (e.g., carboxy, amino, hydroxy, etc.) near the hydrolyzable functional group that are capable of forming bidentate complexes (Houghton, 1979):

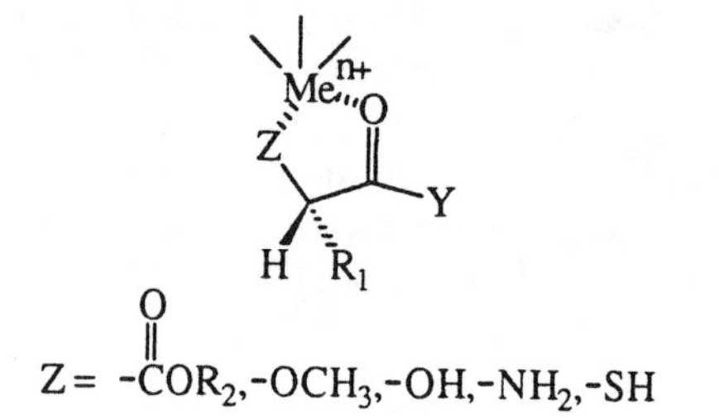

(2.102)

The stability of the bidentate complex provides greater polarization of the carbonyl moiety.

Metal ion catalysis by the direct polarization mechanism can accelerate hydrolysis rates by factors of 10^4 or greater (Buckingham, 1977). For example, the Cu^{2+}-catalyzed hydrolysis of α-amino acid esters occurs at a rate that is six orders of magnitude faster than the uncatalyzed process (Bender and Brubacher, 1973; Hay and Morris, 1976). The metal ion chelate is thought to have the following structure:

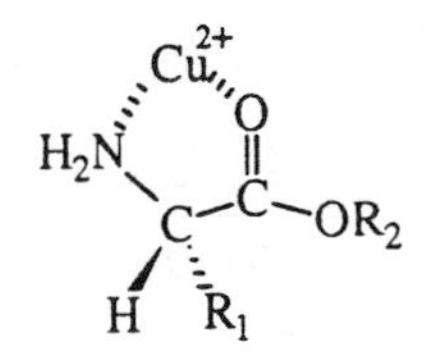

(2.103)

The organophosphothionate esters are also able to form bidentate complexes. For example, chlorpyrifos contains a pyridyl nitrogen capable of forming a six-membered chelate ring with metal ions.

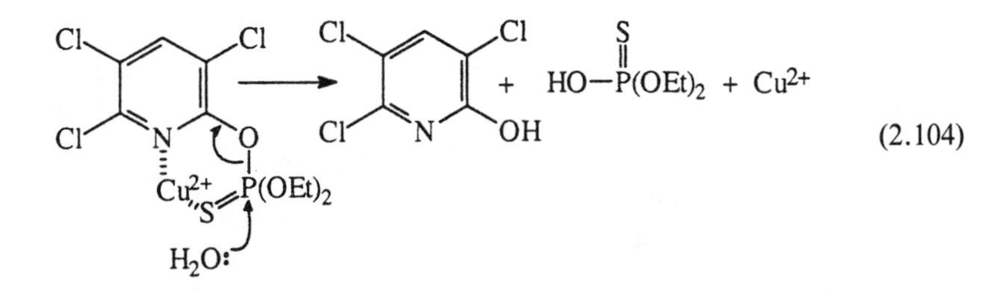

(2.104)

In an analogous manner to carbonyl-containing substrates, coordination increases the electrophilicity of the phosphorus atom enhancing its susceptibility to nucleophilic attack. Several metals, including Ca^{2+}, Mg^{2+}, Fe^{2+}, and Cu^{2+}, have been observed to catalyze the hydrolysis of chlorpyrifos (Blanchet and St. George, 1982). The greatest rate enhancement was measured for Cu^{2+}. Even so, Mill and Mabey (1988) calculated that the Cu^{2+} concentration would have to exceed 10^{-5} M to make a significant contribution to the hydrolysis rate of chlorpyrifos.

The metal catalyzed hydrolysis of parathion has been reported by Ketalaar et al. (1956). The calcium and copper catalyzed hydrolysis of parathion was found to compete with the uncatalyzed process at metal ion concentrations as low as 10^{-8} M. The enhanced rates observed for hydrolysis of parathion in the presence of the metal ions may be accounted for by formation of a complex in which the metal ion is chelated with the sulfur and alkoxy oxygen atoms (Mill and Mabey, 1988).

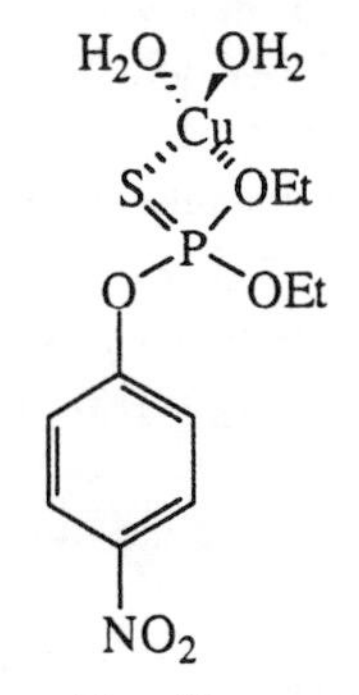

(2.105)

Parathion-Cu complex

Methyl parathion hydrolysis has also been observed to be catalyzed by Cu^{2+} (Mabey et al., 1984). Additionally, the Cu^{2+} affected the reaction product distribution. As expected, the neutral (pH < 8) hydrolysis of methyl parathion favors cleavage of the C–O bond (soft-soft interaction) (pathway a, 2.106), whereas the base-catalyzed (pH > 8) hydrolysis favors cleavage of the P–O bond (hard-hard interaction (pathway b, 2.106). In the presence of Cu^{2+}, however, heterolytic cleavage of the P–O bond resulting in the formation of nitrophenol is the dominant process even at low pH (pathway b, 2.106).

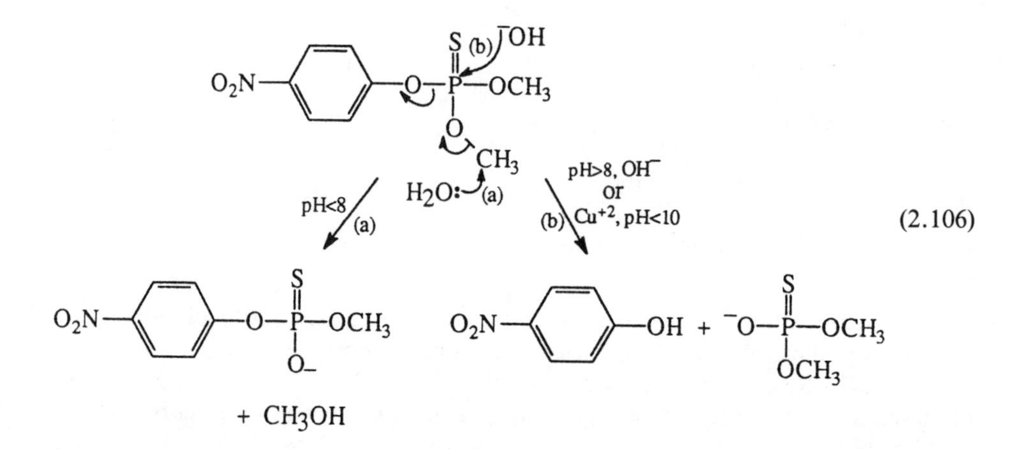

Formation of the metal ion chelate analogous to that for parathion and Cu^{2+} (2.105) simultaneously enhances the electrophilic character of the phosphorus center and weakens the aromatic P–O bond, facilitating the cleavage of this bond by S_N2 reaction with water.

Another example of metal ion-catalyzed hydrolysis reactions is the Cu^{2+}-promoted degradation of aldicarb, a widely used systemic pesticide (Bank and Tyrrell, 1985). The Cu^{2+} degradation of aldicarb is unusual in that complexation to the lone pair of electrons on nitrogen promotes two kinds of processes: proton removal (pathway a, Figure 2.14) and nucleophilic addition of water to the carbon-nitrogen double bond (pathway b, Figure 2.14) depending on the geometry of the carbon-nitrogen double bond. Complexation of Cu^{2+} with the trans isomer of aldicarb results in proton removal, and complexation with the cis isomer results in nucleophilic attack.

At pH 4.0, proton elimination, which leads to formation of the nitrile, is the dominant reaction process. These reaction products are the same as those found for the acid-catalyzed degradation of aldicarb (Bank and Tyrrell, 1984). Nucleophilic addition of water across the carbon-nitrogen double bond (pathway b) results in the formation of an unstable complex, which decomposes readily by cleavage of the carbon-nitrogen bond to give the aldehyde, methylamine, and carbon dioxide.

The second mechanism for metal ion-catalyzed hydrolysis involves formation of a

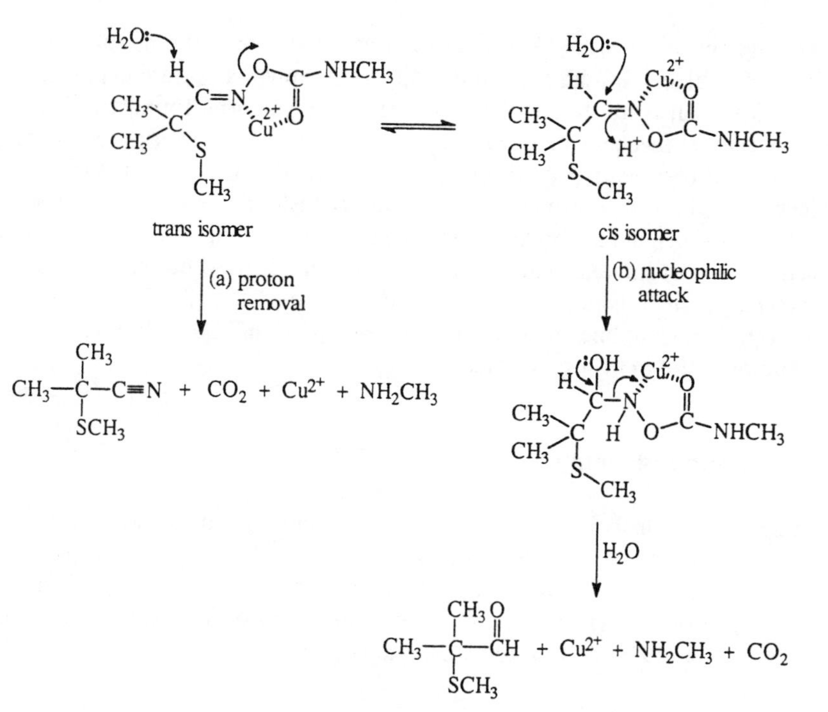

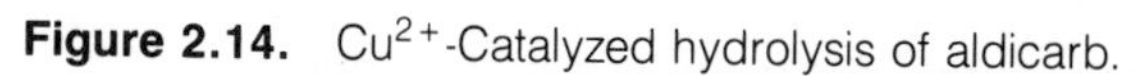

Figure 2.14. Cu^{2+}-Catalyzed hydrolysis of aldicarb.

metal-coordinated nucleophile that is considerably more reactive than the free nucleophile (Plastourgou and Hoffmann, 1984).

$$M(H_2O)_m{}^{n+} \leftrightarrow M(H_2O)_{m-1}OH^{(n+1)+} + H^+ \tag{2.107}$$

$$M(H_2O)_{m-1}OH^{(n+1)+} + RCOX \rightarrow (H_2O)_mM\text{-}O\text{-}CR^{n+} + X: \tag{2.108}$$

By effectively increasing the acidity of water molecules, metal coordination results in the in situ generation of OH^-. Both intramolecular and intermolecular catalysis have been proposed (Buckingham, 1977).

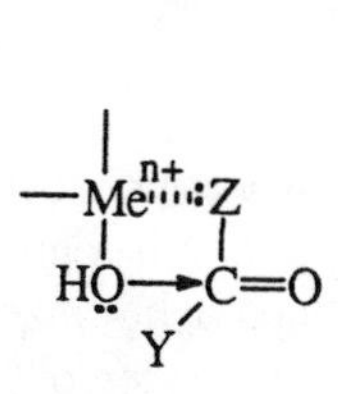

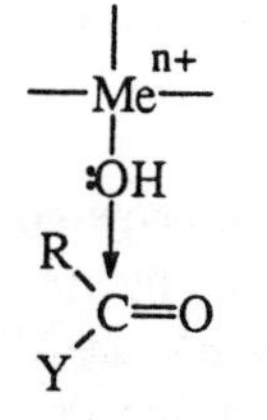

(2.109)

Buckingham and Clark (1981 and 1982) concluded that the metal-catalyzed hydrolysis of 4-nitro-, 2,4-dinitro-, and 2,4,6-trinitrophenyl acetate occurred through the reaction with metal-bound OH^-. Product analysis was found to be consistent with this type of mechanism and furthermore, a linear free energy relationship was found to exist between log k_n and pK_{a1} of the bound water molecule (Figure 2.15).

The observed linear free energy relationship (LFER) suggests that metal-catalyzed hydrolysis was occurring through similar pathways for all of the metal complexes studied (i.e., coordinated metal-nucleophile mechanism). The relative insensitivity of k_n to pK_{a1}, which is indicated by the small slope of the LFER, suggests that the metal-bound nucleophile is much more reactive than the unbound or free nucleophile over the pH range studied.

3. Surface-Bound Metals

Catalysis of hydrolytic reactions may also occur by surface-bound metals. Although there has been a greater focus on the study of metal catalysis by dissolved metal ions, there is increasing evidence to suggest that catalysis by surface-bound metals may be of greater importance in environmental systems such as groundwater aquifers. Stone (1989) postulates 3 mechanisms for catalysis at the mineral-water

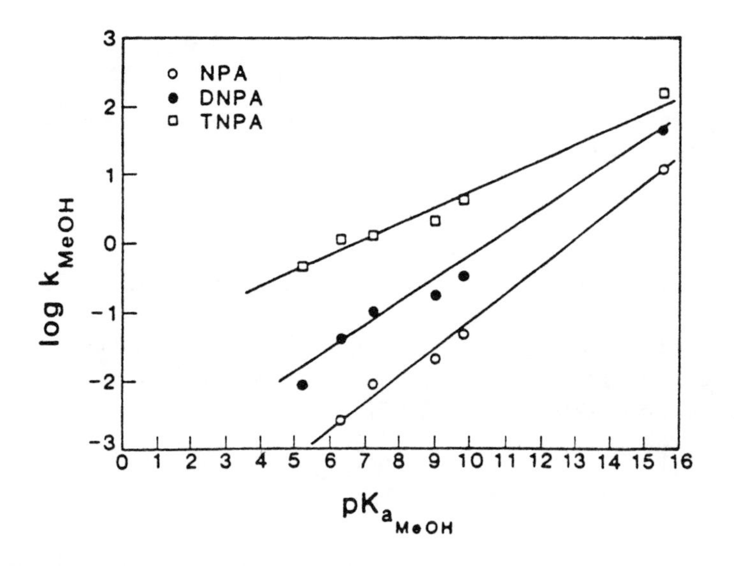

Figure 2.15. A linear free-energy relationship between the logarithm of rate constant for metal-catalyzed hydrolysis (log k_{MeOH}) of *p*-nitrophenyl acetate (NPA), dinitrophenyl acetate (DNPA), and trinitrophenyl acetate (TNPA) and the hydrolysis constant $pK_{a\ MeOH}$ for the formation of the corresponding hydroxy complex. From Buckingham and Clark (1982). Reprinted by permission of CSIRO Publications.

interface: (1) specific adsorption to metals of the mineral lattice resulting in the polarization of the hydrolyzable functional group, (2) metal hydroxo groups on the mineral surface act as nucleophiles, and (3) electrostatic interactions enhance the OH^- concentration relative to the bulk aqueous phase. Mechanisms 1 and 2 are analogous to the homogeneous metal ion catalysis previously discussed.

These mechanisms have been used to rationalize the effect of mineral oxides on the hydrolysis rates of carboxylic acid esters. For example, suspensions of TiO_2 and FeOOH dramatically accelerated the hydrolysis of phenyl picolinate (PHP) (Torrents and Stone, 1991). Suspensions of Al_2O_3 and SiO_2 had negligible effects on hydrolysis rates (Figure 2.16).

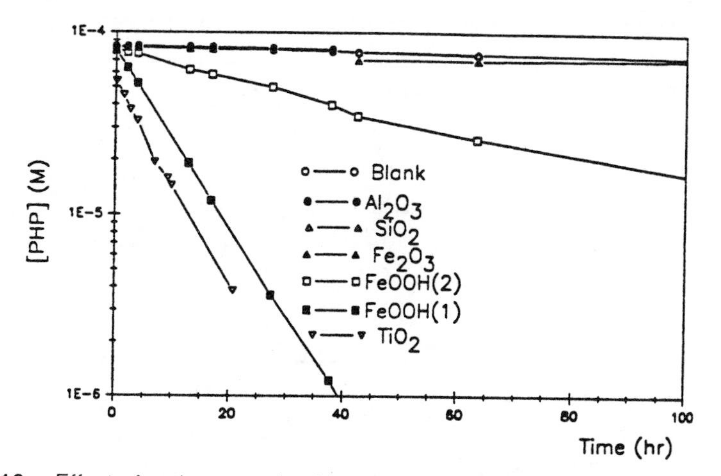

Figure 2.16. Effect of various metal oxides on the loss of phenyl picolinate (PHP) from solution via hydrolysis. All suspensions contained 10 g/L oxide, 1×10^{-3} M acetate buffer (pH 5.0) and 5×10^{-2} M NaCl. From Torrents and Stone (1991). (Reprinted by permission from the American Chemical Society).

The investigators proposed that the rate enhancement observed in TiO_2 and FeOOH suspensions was due to chelation of the carbonyl oxygen and pyridineal nitrogen of PHP by surface-bound metals (Mechanism 2.110).

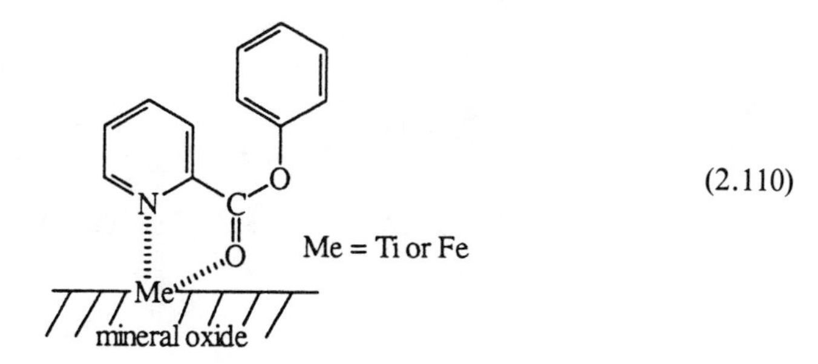

(2.110)

TiO_2 and FeOOH suspensions also suppressed the dependence of PHP hydrolysis kinetics on pH. In particle-free solution, the hydrolysis of PHP is pH dependent, exhibiting base-catalysis. In the presence of the metal oxides, however, hydrolysis was found to be pH independent, suggesting that H_2O could compete with the stronger nucleophile, OH^-, for the surface activated carbonyl group.

Mechanism 2 for catalysis at the mineral-water interface is analogous to the metal ion-catalyzed hydrolysis involving in situ generation of a reactive metal hydroxo species that was discussed previously, the difference being that the reactive metal hydroxo group of the mineral oxide is surface bound. The hydrolysis of *p*-nitrophenyl acetate in the presence of metal oxides provides an excellent example of this type of hydrolysis mechanism. The surface-catalyzed hydrolysis of *p*-nitrophenyl acetate has been proposed to occur through a surface complex in which nucleophilic hydroxyl groups on the surface of the metal oxide react directly with the carbonyl group of the ester linkage (Gonzalez and Hoffmann, 1989).

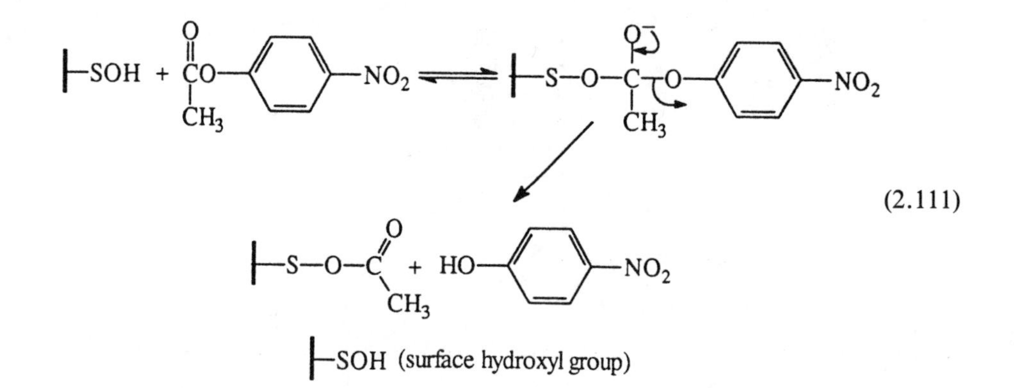

(2.111)

Favorable electrostatic interactions at the mineral-water interface can also result in the catalyzed hydrolysis of substrates (Mechanism 3). For example, hydrolysis rates of monophenyl terephthalate (MPT^-) in aluminum oxide suspensions were an order of magnitude greater than rates measured in homogeneous solutions (2.112) (Stone, 1989).

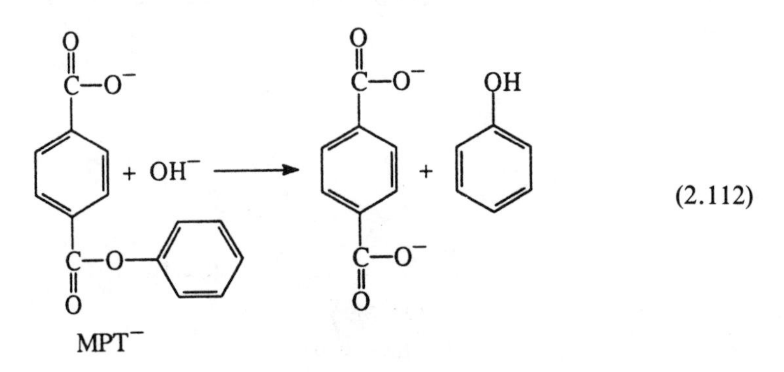

(2.112)

Stone suggests that catalysis occurs because the ionized carboxylate group of MPT^- is able to specifically sorb to the positively charged aluminum oxide surface where subsequent attack of hydroxide ions in the diffuse layer occurs.

$$>AlOH_2^+ + MPT^- \underset{k_{-1}}{\overset{k_1}{\rightleftharpoons}} >Al\text{—}MPT + H_2O \tag{2.113}$$

$$>Al\text{—}MPT + OH^- \xrightarrow{k_2} \text{products} \tag{2.114}$$

Consistent with this proposed mechanism is the observation that the addition of natural organic matter (NOM) to the Al_2O_3 suspensions inhibited the surface-catalyzed hydrolysis of MPT^- (Stone, 1988). Inhibition was thought to result from the competitive adsorption of DOM and a decrease in the positive charge of the Al_2O_3 surface.

4. Clays and Clay Minerals

The catalytic effects of clays on hydrolysis processes is generally associated with the acidic pH values measured at clay mineral surfaces. Numerous studies have demonstrated that the surface pH of clay minerals can be as much as 2 to 3 units lower than the bulk solution (Mortland, 1970; Bailey et al., 1968; Frenkel, 1974; Karickhoff and Bailey, 1976). The Brönsted acidity of clays arises primarily from the dissociation of water coordinated to exchangeable cations (2.115).

$$[M(H_2O)_x]^{n+} \rightarrow [M(OH)H_2O)_{x-1}]^{n-1+} + H^+ \tag{2.115}$$

The surface acidity, and thus catalytic activity, of a clay mineral will depend primarily on the nature of the exchangeable cation and moisture content. Generally, the higher the charge and the smaller the radius of the exchangeable cation, the greater its polarizing power and, therefore, the greater the degree of dissociation of the adsorbed water. Accordingly, the surface acidity of homoionic montmorillonites saturated with Na, Mg, or Al decreases in the order $Al > Mg > Na$. Moisture content can have a dramatic effect on surface acidity. In general, surface acidity will decrease with increasing water content. Typically, when the water content exceeds the sorbed or bound water, catalytic activity is markedly decreased (Voudrias and Reinhard, 1986).

El-Amamy and Mill (1984) measured the effect of the surface acidity of montmorillonite and kaolinite on the hydrolysis rate constants for a number of chemicals containing hydrolyzable functional groups that exhibit acid-catalyzed, base-catalyzed, and neutral hydrolysis. The chemicals that were studied included ethyl acetate, cyclohexene oxide, isopropyl bromide, 1-(4-methoxyphenyl)-2,3-epoxypropane, and N-methyl-*p*-tolyl carbamate (MTC). Aqueous suspensions of montmorillonite or kaolinite had minimal effects on the hydrolysis rates of these chemicals. The rate of hydrolysis of the epoxide, however, was increased by a factor of 10 in oven-dried clays up to the limits of sorbed water. When the moisture content of the clay exceeded the limit of sorbed water, the hydrolysis rate constant for the epoxide was found to decrease by a factor of 4. The dependence of the rate of hydrolysis of 1-(4-methoxyphenyl)-2,3-epoxypropane on montmorillonite saturated with various cations is illustrated in Figure 2.17.

It is apparent that at low moisture content (<10% for the Na-saturated clay mineral and <5% for the Ca-, or Mg-saturated clay mineral), where water is not available for hydrolysis, hydrolysis does not occur. This low moisture content corresponds with the saturation of the cation's first hydration shell. As the moisture content is increased to the upper limit of bound water (50% moisture content), a significant enhancement of the hydrolysis of the epoxide is observed. When the moisture content exceeds the upper limit of bound water (>50%), the rate constant for the hydrolysis of the epoxide was reduced by a factor of 4. It was concluded that water in excess of sorbed water diminishes the catalytic activity of clay surfaces by reducing the concentration gradient across the double layer, effectively raising the surface pH closer to that of the bulk water. In similar studies with MTC, the addition of water to oven-dried Na-montmorillonite and Na-kaolinite retarded the hydrolysis rate of the carbamate. This observation is consistent with the fact that MTC exhibits only neutral base-catalyzed hydrolysis.

Moisture content was also found to have a significant effect on the hydrolysis kinetics of parathion and methyl parathion on kaolinite. For example, the hydrolysis rate constant for parathion and methyl parathion on kaolinite increased by 2 orders of magnitude when the moisture content was increased to the limit of sorbed water (11% moisture content) (Saltzman et al. 1976). At moisture contents above the limit of sorbed water, however, a significant decrease in the hydrolysis rate constant was observed. Mingelgrin et al. (1977) concluded that the hydrolysis of parathion,

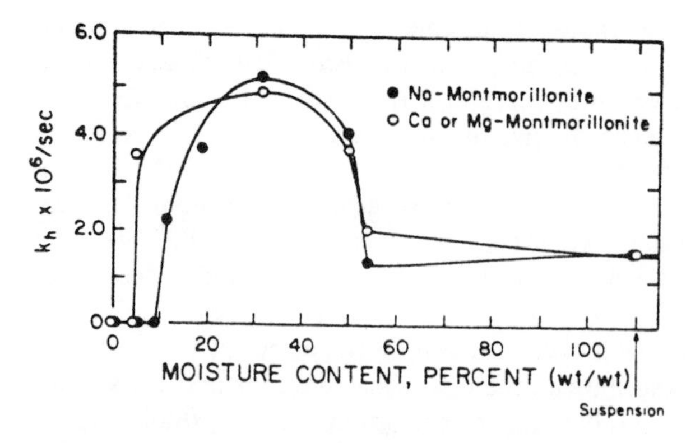

Figure 2.17. Rate of hydrolysis of 1-(4-methoxyphenyl)-2,3-epoxypropane on montmorillonite saturated with Na, Ca or Mg ions. From El-Amamy and Mill (1984). Reprinted by permission of the Clay Minerals Society.

and other organophosphorus esters, in general, occurs through the attack of a ligand water molecule of an exchange cation on the phosphorus-oxygen bond. Direct attack of the ligand water on the organophosphorus ester is enhanced by the cation-ligand interactions that polarize the water and weaken the HO–H bond.

Hydrolysis mechanisms in clay-water systems also can be dependent on the nature of the exchangeable cation. For example, Pusino et al. (1988) studied the catalytic hydrolysis of quinalphos on homoionic bentonite clays. On the Na- and K-clays, deethylation occurred, resulting in the formation of O-ethyl O-quinoxalin-2-yl thiophosphoric acid, whereas 2-hydroxyquinoxaline is the main reaction product on the Cu-, Fe- and Al-clays.

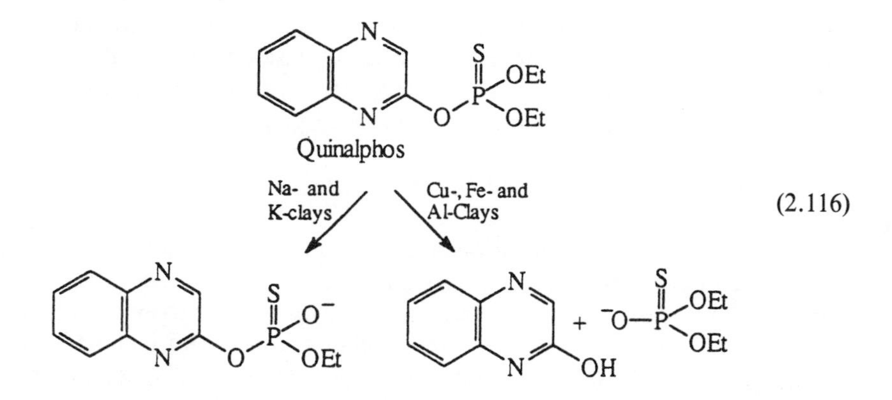

(2.116)

A 6-membered bidentate complex involving the ring nitrogen and the phosphorothioate sulfur of quinalphos was proposed to account for the formation of 2-hydroxyquinoxaline. IR and X-ray data suggest that formation of the chelate occurs through specific adsorption of quinalphos in the interlayer of the bentonite clays.

5. Natural Organic Matter

a. Dissolved Organic Matter

The ubiquitous nature of dissolved organic matter (DOM) in natural waters necessitates that we understand the role that DOM may play in altering the hydrolysis kinetics of organic pollutants. The few examples of enhanced hydrolysis of organic chemicals in the presence of DOM suggests that catalysis will be limited to a few classes of chemicals. For example, hydrolysis of the chloro-1,3,5-triazines, which are subject to acid-catalyzed hydrolysis, appear to be catalyzed by DOM (Khan, 1978; Li and Felbeck, 1972). Khan (1978) measured a rate enhancement of a factor of 10 for the hydrolysis of atrazine in the presence of fulvic acid. Choundry (1984) proposed that the observed rate enhancement was due to the interaction of acidic functional groups of the fulvic acid with the ring N-atom adjacent to the C–Cl bond, resulting in the weakening of the C–Cl bond and a lowering of the activation energy for hydrolysis (Figure 2.18).

In contrast, Perdue and Wolfe (1982) observed that DOM retarded the basic hydrolysis of the octyl ester of 2,4-D. They found that the base-catalyzed rate constant was reduced in proportion to the fraction of the hydrophobic ester that was associated with the DOM. Perdue (1983) has proposed a micelle-type model to rationalize these results and those presented for atrazine. Perdue compares the physical characteristics of DOM, which is negatively charged at environmental pHs, to those of anionic surfactants. Anionic surfactants have been demonstrated to increase hydrolysis rates for acid-catalyzed processes and decrease rates for base-catalyzed processes (Fendler and Fendler, 1975). Rate enhancements for acid-catalyzed hydrolysis reactions are attributed to stabilization of the positive charge that is developed in the transition state, whereas base-catalyzed hydrolysis reactions are impeded due to destabilization of the negatively charged transition state. Although this is an attractive model, it remains largely untested.

b. Soil and Sediment-Associated Organic Matter

Because many organic chemicals are nonionic and have low water solubilities, they will exist primarily in the sorbed state in soil- and sediment-water systems. The sorption of nonionic chemical occurs through hydrophobic sorption or partitioning to the organic matter associated with the soil or sediment (Karickhoff, 1980; Chiou et al., 1983). Furthermore, because desorption kinetics may be slow relative to hydrolysis kinetics, to accurately predict the fate of hydrolyzable chemicals in soil- and sediment-water systems an understanding of hydrolysis kinetics in the sorbed

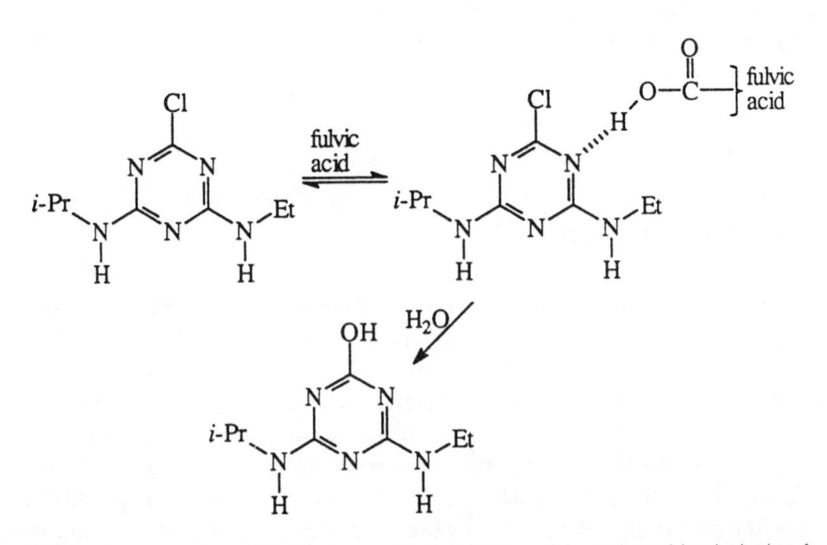

Figure 2.18. Proposed reaction pathway for the fulvic acid-catalyzed hydrolysis of atrazine.

state is necessary (Karickhoff, 1980 and 1984). Generally, it has been assumed that only the fraction of compound in the aqueous phase would hydrolyze (Wolfe et al., 1977). Detailed studies of the hydrolysis kinetics of chemicals in the sorbed state, however, have demonstrated this assumption is not always valid.

Numerous studies have demonstrated that the effect of sorption on hydrolysis rates is dependent on the hydrolysis pathway (i.e., neutral, acid, or base hydrolysis). For example, sediment organic matter does not appear to significantly alter neutral hydrolysis rates for a variety of organic chemicals including a number of organophosphorothioates (Macalady and Wolfe, 1984 and 1985), halogenated aliphatics (Haag and Mill, 1988a; Deeley et al., 1991) and epoxides (Haag and Mill, 1988a; Metwally and Wolfe, 1990). On the other hand, base-catalyzed hydrolysis of the octyl ester of 2,4-D (Macalady et al., 1989) and chlorpyrifos (Macalady and Wolfe, 1984) in the presence of sediments with high organic carbon content were retarded.

Inhibition of base-catalyzed processes in sediment-water systems has been rationalized in terms of the negative charge associated with sediment particles at pHs common in natural waters. The concentration of negative ions, such as OH^-, are expected to be lower near the sediment surface, thus resulting in a decrease in the effective pH at the sediment surface. Additionally, stabilization of the negatively charged transition state formed during base-catalysis may occur. Based on this simplistic model, acid-catalyzed processes would be expected to be accelerated in sediment-water systems at pH values above the zero point of charge (zpc) or isoelectric point of the sediments (above the zpc, the sediment surface will have a negative potential and therefore, a higher concentration of H^+ counter ions in the diffuse layer). For example, the methyl ester of 2,4-D (2,4-DME) was observed to hydrolyze faster in sediment-water systems at pH values above the expected zpc of sediment organic matter (about 2) (Coleman, 1988). At pH values near the expected zpc of sediment organic matter (2 to 2.5), accelerated hydrolysis rates for 2,4-DME were not observed. Similarly, styrene oxide, which exhibits acid catalysis below pH 7, was found to hydrolyze approximately 4 times faster in saturated subsurface sediment (0.02% organic carbon) than in distilled water at the same pH (Haag and Mill, 1988a). By contrast, however, the hydrolysis of 4-chlorostilbene oxide (CSO) in acidic sediment-water systems, well above the expected zpc's of sediment organic matter, was slower than in distilled water (Metwally and Wolfe, 1990). Similarly, hydrolysis studies of aziridine derivatives in sediment-water systems were not consistent with the proposed model (Macalady et al., 1989). These conflicting results suggest that our understanding of the effect of sediment surfaces on acid-catalyzed hydrolysis is incomplete and will require further investigation.

Changes in hydrolysis pathways also have been observed in the presence of sediment materials. For example, Deeley et al. (1991) found that the favored reaction pathway for 1,2-dibromo-3-chloropropane (DBCP) in phosphate buffer solution was dehydrohalogenation, which is in agreement with the finding of Burlinson et al. (1982) that was discussed earlier (see Figure 2.4). In contrast, however, these investigators found that nucleophilic substitution, resulting in the formation of glycerol, was the dominant transformation process in aquifer solids/groundwater systems (2.117).

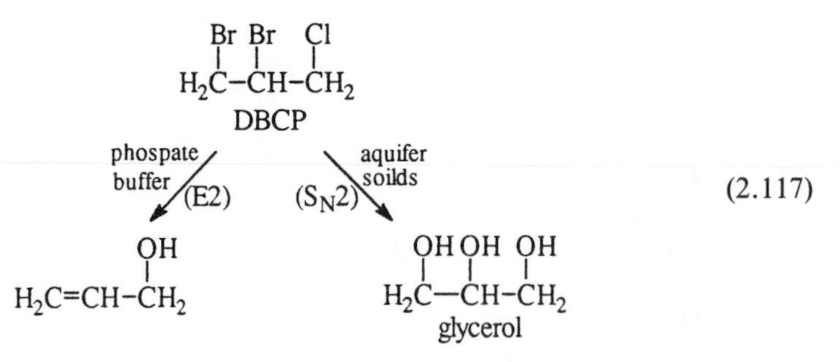

(2.117)

This type of observation illustrates the problems that can be encountered when extrapolating data between reaction systems.

REFERENCES

Bailey, G. W., J. L. White, and T. Rothberg. 1968. Adsorption of organic herbicides by montmorillonite: Role of pH and chemical character of adsorbate, *Soil Sci. Soc. Am. Proc.*, 32:222–234.

Bank, S., and R. J. Tyrrell. 1984. Kinetics and mechanisms of alkaline and acid hydrolysis of aldicarb, *J. Agric. Food Chem.*, 32:1223–1232.

Bank, S., and R. J. Tyrrell. 1985. Copper(II)-promoted aqueous decomposition of aldicarb, *J. Org. Chem.*, 50:4938–4943.

Barbash, J. E., and M. Reinhard. 1989a. Abiotic dehalogenation of 1,2-dichloroethane and 1,2-dibromoethane in aqueous solution containing hydrogen sulfide, *Environ. Sci. Technol.* 23(11):1349–1358.

Barbash, J. E., and M. Reinhard. 1989b. Reactivity of sulfur nucleophiles toward halogenated organic compounds in natural waters, in *Biogenic Sulfur in the Environment*, Saltzman, E. and W. Cooper, Eds. (Washington, D.C.: American Chemical Society), pp. 1011–1139.

Bartlett, P. D., and C. G. Swain. 1949. Kinetics of hydrolysis and displacement reactions of β,β′-dichlorodiethyl sulfide (mustard gas) and of β-chloro-β′-hydroxydiethyl sulfide (mustard chlorohydrin), *J. Am. Chem. Soc.*, 71:1406–1415.

Becker, A. R., J. M. Janusz, and T. C. Bruice. 1979. Solution chemistry of the *syn*- and *anti*-tetrahydrodiol epoxides, the *syn*- and *anti*-tetrahydrodimethoxy epoxides, and the 1,2- and 1,4-tetrahydro epoxides of naphthalene, *J. Am. Chem. Soc.*, 101:5679–5687.

Bender, M. L. 1960. Mechanisms of catalysis of nucleophilic reactions of carboxylic acid derivatives, *Chem. Rev.*, 60:53–113.

Bender, M. L., and L. J. Brubacher. 1973. *Catalysts and Enzyme Action*. (New York, NY: McGraw-Hill).

Bender, M. L., and R. J. Thomas. 1961. The concurrent alkaline hydrolysis and isotopic oxygen exchange of a series of *p*-substituted acetanilides, *J. Am. Chem. Soc.* 83:4183–4196.

Bergon, M., N. J. Hamida, and J. Calmon. 1985. Isocyanate formation in the decomposition of phenmedipham in aqueous media, *J. Agric. Food Chem.* 33:577–583.

Beyer, E. M., M. J. Duffy, J. V. Hay, and D. D. Schlueter. 1987. Sulfonylureas, in *Herbicides: Chemistry, Degradation and Mode of Action*, Vol. 3, P. C. Kearney, and D. D. Kaufman, Eds. (New York, NY: Dekker), pp. 117–189.

Blanchet, P.-F., and A. St. George. 1982. Kinetics of chemical degradation of organophosphorus pesticides; hydrolysis of chlorpyrifos and chlorpyrifos-methyl in the presence of copper(II), *Pesticide Sci.*, 13:85–91.

Bolt, H. M., R. J. Laib, and J. G. Filser. 1982. Reactive metabolites and carcinogenicity of halogenated ethylenes, *Biochem. Pharmacol.*, 31:1–4.

Brown, H. M. 1990. Mode of action, crop selectivity, and soil relations of the sulfonylurea herbicides, *Pestic. Sci.*, 29:263–281.

Buckingham, D. A. 1977. Metal-OH and its ability to hydrolyze (or hydrate) substrates of biological interest, in *Biological Aspects of Inorganic Chemistry*, A. W. Addison, W. R. Cullen, D. Dolphin, and B. R. James, Eds. (New York, NY: Wiley).

Buckingham, D. A., and C. R. Clark. 1981. Reactions of acyl phosphates. Base hydrolysis of $[Co(NH_3)_5OPO_3COCH_3]^+$ and the metal hydroxide promoted hydrolysis of acetyl phenyl phosphate, *Aust. J. Chem.*, 34:1769–1773.

Buckingham, D. A., and C. R. Clark. 1982. Metal-hydroxide promoted hydrolysis of activated esters. Hydrolysis of 2,4-dinitrophenyl acetate and 4-nitrophenyl acetate, *Aust. J. Chem.*, 35:431–436.

Burlinson, N. E., L. A. Lee, and D. H. Rosenblatt. 1982. Kinetics and products of hydrolysis of 1,2-dibromo-3-chloropropane, *Environ. Sci. Technol.*, 16(9):627–632.

Camilleri, P. 1984. Alkaline hydrolysis of some pyrethroid insecticides, *J. Agric. Food Chem.*, 32:1122–1124.

Capon, B., and S. P. McManus. 1976. *Neighboring Group Participation*, New York, NY: Plenum.

Chiou, C. T., P. E. Porter, and D. W. Schmedding. 1983. Partition equilibria of nonionic organic compounds between soil organic matter and water, *Environ. Sci. Technol.*, 17:227–231.

Choundry, G. C. 1984. In *Humic Substances*, Vol. 7, Current Topics in Environmental and Toxicological Chemistry Series. (New York, NY: Gordon and Breach), pp. 143–169.

Coleman, K. D. 1988. "Acid Catalyzed Abiotic Hydrolysis of Sorbed Pesticides," M.S. Thesis, Colorado School of Mines.

Deeley, G. M., M. Reinhard, and S. M. Stearns. 1991. Transformation and sorption of 1,2-dibromo-3-chloropropane in subsurface samples collected at Fresno, California, *J. Environ. Qual.*, 20:547–556.

Duboc, C. 1978. The correlation analysis of nucleophilicity, in *Correlation Analysis in Chemistry: Recent Advances*, N.B. Chapman, and J. Shorter, Eds., (New York, NY: Plenum Press), pp. 313–315.

Edwards, J. O. 1954. Correlation of relative rates and equilibria with a double basicity scale, *J. Am. Chem. Soc.*, 76:1540–1547.

Edwards, J. O. 1956. Polarizability, basicity and nucleophilic character, *J. Am. Chem. Soc.*, 78:1819–1820.

El-Amamy, M. M., and T. Mill. 1984. Hydrolysis kinetics of organic chemicals on montmorillonite and kaolinte surfaces as related to moisture content, *Clays & Clay Minerals*, 32(1):67–73.

Emerson, S. 1976. Early diagenesis in anaerobic lake sediments: Chemical equilibriums in in interstitial waters, *Geochim. Cosmochim. Acta,* 40:925–934.

Fendler, J. H., and E. J. Fendler. 1975. *Catalysis in Micellar and Macromolecular Systems* (New York, NY: Academic Press).

Fest, C., and K. J. Schmidt. 1983. Organophosphorus insecticides in *Chemistry of Pesticides*, K. H. Buchel, Ed. (New York, NY: John Wiley & Sons), pp. 49–125.

Frenkel, M. 1974. Surface acidity of montmorillonites, *Clays and Clay Minerals*, 22:435–441.

Gleave, J. L., E. D. Hughes, and C. K. Ingold. 1935. Mechanism of substitution at a saturated carbon atom, *J. Chem. Soc.* p. 236.

Gonzalez, A. C., and M. R. Hoffman. 1989. The kinetics and mechanism of the catalytic hydrolysis of nitrophenyl acetates by metal oxide surfaces, *J. Phys. Chem.*

Haag, W. R., and T. Mill. 1988a. Effect of a subsurface sediment on hydrolysis of haloalkanes and epoxides, *Environ. Sci. Technol.*, 22(6):658–663.

Haag, W. R., and T. Mill. 1988b. Some reactions of naturally occurring nucleophiles with haloalkanes in water, *Environ. Toxicol. Chem.* 7:917–924.

Harris, J. C. 1981. Rate of hydrolysis, in *Handbook of Chemical Property Estimation Methods. Environmental Behavior of Organic Compounds.* W. J. Lyman, W. F. Reehl, and D. H. Rosenblatt, Eds. (New York, NY: McGraw-Hill), pp. 7–1 to 7–48.

Hay, J. V. 1990. Chemistry of sulfonylurea herbicides, *Pestic. Sci.*, 29:247–261.

Hay, R. W. 1987. Lewis acid catalysis and the reactions of coordinated ligands, in *Comprehensive Coordination Chemistry, Vol. 6 Applications*, G. Wilkinson, J. A. Gillard, and J. A. McCleverty, Eds. (Oxford, England: Pergamon).

Hay, R. W., and P. J. Morris. 1976. Metal ion-promoted hydrolysis of amino acid esters and peptides, in *Metal Ions in Biological Systems*, Vol. 5, Sigel, H., Ed., (New York, NY: Marcel Dekker), pp. 173–243.

Hegarty, A. F., and L. N. Frost. 1973. Elimination-addition mechanism for the hydrolysis of carbamates. Trapping of an isocyanate intermediate by an o-amino group, *J. Chem. Soc., Perkin Trans. 2*, 1719.

Hill IV, J., H. P. Kollig, D. F. Paris, N. L. Wolfe, and R. G. Zepp. 1976. Dynamic behavior of vinyl chloride in aquatic ecosystems, U.S. Environmental Protection Agency. EPA-600/3-76-001.

Hine, J., A. M. Dowell, Jr., and J. E. Singley, Jr. 1956. Carbon dihalides as intermediates in the basic hydrolysis of haloforms. IV. Relative reactivities of haloforms, *J. Am. Chem. Soc.*, 78:479–482.

Hine, J., and P. B. Langford. 1956. The effect of halogen atoms on the reactivity of other halogen atoms in the same molecule. VII. The reaction of β-haloethyl bromides with sodium hydroxide, *J. Am. Chem. Soc.*, 78:5002–5004.

Hoffmann, M. R., 1990. Catalysis in aquatic environments, in *Aquatic Chemical Kinetics*, W. Stumm, Ed. (New York, NY: John Wiley & Sons), pp. 71–112.

Houghton, R. P. 1979. *Metal Complexes in Organic Chemistry*, (Cambridge, England: Cambridge University Press).

Hudson, R. F. 1965. *Structure and Mechanism in Organophosphorus Chemistry*, (New York, NY: Academic Press).

Ingold, C. K. 1969. *Structure and Mechanism in Organic Chemistry*, 2nd ed., (Ithaca, NY: Cornell University Press), p. 1131.

IUPAC. 1981. Manual of symbols and terminology for physicochemical quantities and units, Appendix V, Symbolism and terminology in chemical kinetics, *Pure Appl. Chem.*, 53:753.

Jackson, R. E., R. J. Patterson, B. W. Graham, J. Bahr, D. Belanger, J. Lockweed, and M. Priddle. 1985. Series No. 141, NHRI Paper No. 23: National Hydrology Research Institute Inland Water Directorate: Ottawa, Canada.

Jeffers, P. M., L. M. Ward, L. M. Woytowitch and N. L. Wolfe. 1989. Homogeneous hydrolysis rate constants for selected chlorinated methanes, ethanes, ethenes, and propanes, *Environ. Sci. Technol.*, 23(8):965–969.

Jencks, W. P. 1987. Covalent catalysis, in *Catalysis in Chemistry and Enzymology*, (New York, NY: Dover), pp. 78–107.

Jensen, W. B. 1978. The Lewis acid-base definitions: A status report, *Chemical Reviews*, 78(1):1–22.

Jerina, D. M., H. Yagi, R. E. Lehr, D. R. Thakker, M. Schaeffer-Ridder, J. M. Karle, W. Levin, A. W. Wood, R. L. Chang, and A. H. Conney. 1978. The bay-region theory of carcinogenesis by polycyclic aromatic hydrocarbons, in *Polycyclic Hydrocarbons and Cancer*, Vol 1., H. V. Gelboin and P. O. Ts'o, Eds., (New York, NY: Academic Press), pp. 173–188.

Junglaus, G. A., and S. Z. Cohen. 1986. Hydrolysis of ethylene dibromide, *Abstr. Div. Env. Chem. Am. Chem. Soc.*, 26(1):12–16.

Karickhoff, S. W. 1980. Sorption kinetics of hydrophobic pollutants in natural sediments, in *Contaminants and Sediments*, R. A. Baker, Ed. (Ann Arbor, MI: Ann Arbor Science), pp. 193–205.

Karickhoff, S. W. 1984. Organic pollutant sorption in aquatic systems, *J. Hydraulic Engineering*, 110(6):707–735.

Karickhoff, S. W., and G. W. Bailey. 1976. Protonation of organic bases in clay-water systems, *Clays and Clay Minerals*, 24:170–176.

Keith, L. H., and W. A. Telliard. 1979. Priority pollutants. I—A perspective view, *Environ. Sci. Technol.*, 13:416–423.

Ketalaar, J. A., H. R. Gersmann, and M. M. Beck. 1956. Metal-catalyzed hydrolysis of thiophosphoric esters, *Nature*, 177:392–393.

Khan, S. U. 1978. Kinetics of hydrolysis of atrazine in aqueous fulvic acid solution, *Pestic. Sci.* 9:39–43.

Kirby, A. J., and S. G. Warren. 1967. *The Organic Chemistry of Phosphorus*, (New York, NY: Elsevier Publishing Company).

Li, G. C., and G. T. Felbeck, Jr. 1972. Atrazine hydrolysis as catalyzed by humic acids, *Soil Sci.* 114:201–209.

Livingston, D. A. 1963. Chemical composition of rivers and lakes, *Geol. Surv. Prof. Paper. 440-G.*

Long, F. A., and J. G. Pritchard. 1956. Hydrolysis of substituted ethylene oxides in H_2O^{18} solutions, *J. Am. Chem. Soc.*, 78(12):2663-2667.

Mabey, W., and T. Mill. 1978. Critical review of hydrolysis of organic compounds in water under environmental conditions, *J. Phys. Chem. Ref. Data*, 7(2):383–415.

Mabey, W. R., A. Baraze, and T. Mill. 1978. In *Environmental Pathways of Environmental Pathways of Selected Chemicals, Part I*, EPA Final Report, EPA 600/7–77–113.

Mabey, W. R., H. Drossman, A. M. Liu, and T. Mill. 1984. Toxic Substances Process Data Generation, EPA Final Report, EPA Contract 68–03–2921, Task 18.

Macalady, D. L., and N. L. Wolfe. 1984. Abiotic hydrolysis of sorbed pesticides, in *ACS Symposium Series, No. 259 Treatment & Disposal of Pesticide Wastes*, Krueger, R. F., and J. N. Seiber, Eds., (Washington, DC: American Chemical Society), pp. 221–244.

Macalady, D. L., and N. L. Wolfe. 1985. Effects of sediment sorption and abiotic hydrolyses. 1. Organophophorothioate esters, *J. Agric. Food Chem.*, 33(2):167–173.

Macalady, D. L., P. G. Tratnyek, and N. L. Wolfe. 1989. Influences of natural organic matter on the abiotic hydrolysis of organic contaminants in aqueous systems, in *Aquatic Humic Substances*, Suffet, I.H., and P. MacCarthy, Eds., (Washington, DC: American Chemical Society), pp. 323–332.

March, J. 1985. *Advanced Organic Chemistry*, 3rd ed. (New York, NY: John Wiley & Sons).

Mason, R. E., D. D. McFadden, V. G. Iannacchione, and D. S. McGrath. 1981. Survey of DBCP Distribution in Groundwater Supplies and Surfacewater Ponds, U.S. Environmental Protection Agency, EPA Report No. 68–015848.

Metwally, M. E., and N. L. Wolfe. 1990. Hydrolysis of chlorostilbene oxide II. Modeling of hydrolysis in aquifer samples and in sediment-water systems, *Environ. Toxicol. Chem.*, 9:963–973.

Mill, T., and W. Mabey. 1988. Hydrolysis of organic chemicals, in *The Handbook of Environmental Chemistry, Volume 2D: Reactions and Processes*, Hutzinger, O. Ed., (New York, NY: Springer-Verlag), pp. 71–111.

Mingelgrin, U., S. Saltzman and B. Yaron. 1977. A possible model for the surface-induced hydrolysis of organophosphorus pesticides on kaolinite clays, *Soil Sci. Soc. Am. J.*, 41:519–523.

Mortland, M. M. 1970. Clay-organics complexes and interactions, *Advan. Agron.*, 22:75–117.

Muir, D. C.G. 1988. Phosphate esters, in *The Handbook of Evironmental Chemistry*, Volume 3, Part C: Anthropogenic Compounds, O. Hutzinger, Ed., (New York, NY: Springer-Verlag), pp. 41–66.

Nelson, S. J., M. Iskander, M. Volz, S. Khalifa, and R. Haberman. 1981. Studies of DBCP in subsoils, *The Science of the Total Environment*, 21:35–40.

Okamoto, T., K. Shuddo, N. Miyata, Y. Kitahara, and S. Nagata. 1978. Reactions of K-region oxides of carcinogenic and noncarcinogenic aromatic hydrocarbons. Comparative studies on reactions with nucleophiles and acid-catalyzed reactions, *Chem. Pharm. Bull.*, 26(7):2014–2026.

Page, G. W. 1981. Comparison of groundwater and surface water for patterns and levels of contamination by toxic substances, *Environ. Sci. Technol.*, 15:1475–1481.

Parker, R. E., and N. S. Isaacs. 1959. Mechanisms of epoxide reactions, *Chem. Rev.*, 59:737–799.

Pearson, R. G. 1963. Hard and soft acids and bases, *J. Am. Chem. Soc.*, 85:3533–3539.

Pearson, R. G. 1973. *Hard and Soft Bases,* (Stroudsburg, PA: Dowden, Hutchinson, and Ross).

Pearson, R. G. 1983. Absolute hardness: companion parameter to absolute electronegativity, *J. Am. Chem. Soc.,* 105:7512–7516.

Perdue, E. M. 1983. Association of organic pollutants with humic substances: Partitioning equilibria and hydrolysis kinetics, in *Aquatic and Terrestrial Humic Materials*, R. F. Christman and E. T. Gjessing, Eds. (Ann Arbor, MI: Ann Arbor Science), pp. 441–460.

Perdue, E. M., and N. L. Wolfe. 1982. Modification of pollutant hydrolysis kinetics in the presence of humic substances, *Environ. Sci. Technol.* 16:847–852.

Perdue, E. M., and N. L. Wolfe. 1983. Prediction of buffer catalysis in field and laboratory studies of pollutant hydrolysis reactions, *Environ. Sci. Technol.*, 17:635–642.

Pignatello, J. J. 1986. Ethylene dibromide mineralization in soils under aerobic conditions, *Appl. Environ. Microbiol.*, 51(3):588–592.

Plastourgou, M., and M. R. Hoffmann. 1984. Transformation and fate of organic esters in layered-flow systems: The role of trace metal catalysis, *Environ. Sci. Technol.*, 18:756–764.

Pritchard, J. G., and F. A. Long. 1956. Hydrolysis of substituted ethylene oxides in H_2O^{18} solutions, *J. Am. Chem. Soc.*, 78:2663–2667.

Pusino, A., C. Gessa and H. Kozlowski. 1988. Catalytic hydrolysis of quinalphos on homoionic clays, *Pestic. Sci.*, 24:1–8.

Roberts, A. L., P. N. Sanborn, and P. M. Gschwend. 1992. Nucleophilic substitution reactions of dihalomethanes with hydrogen sulfide, *Environ. Sci. Technol.*, 26:2263–2274.

Ross, A. M., T. M. Pohl, K. Pizza, M. Thomas, B. Fox and D. L. Whalen. 1982. Vinyl epoxide hydrolysis reactions, *J. Am. Chem. Soc.*, 104:1658–1665.

Saltzman, S., B. Yaron and U. Mingelgrin. 1976. The surface catalyzed hydrolysis of parathion on kaolinite, *Soil Sci. Soc. Amer. Proc.*, 38:231–234.

Saltzman, S., U. Mingelgrin, and B. Yaron. 1976. Role of water in the hydrolysis of parathion and methylparathion on kaolinite, *J. Agric. Food Chem.*, 24(4):739–743.

Samuel, D., and B. L. Silver. 1965. Oxygen isotope exchange reactions of organic compounds, *Advan. Phys. Org. Chem.*, 3:123–186.

Santodonato, J., S. S. Lande, P. H. Howard, D. Orzel, and D. Bogy. 1980. Investigation of selected potential environmental contaminants: Epichlorohydrin and epibromohydrin, U.S. Environmental Protection Agency. EPA-560/11–80-006.

Schmidt, K. J. 1975. Chemical aspects of organophosphate pesticides in view of the environment, in *Environmental Quality and Safety, Vol. 4: Global Aspects of Chemistry, Toxicology and Technology as Applied to the Environment*, F. Coulston, and F. Korte, Eds. (New York, NY: Academic Press), pp. 96–108.

Schmidt, K. J., and C. Fest. 1982. *The Chemistry of Organophosphorus Pesticides*, (New York, NY: Springer-Verlag),

Schwarzenbach, R. P., W. Giger, C. Schaffner, and O. Wanner. 1985. Groundwater contamination by volatile halogenated alkanes: Abiotic formation of volatile sulfur compounds under anaerobic conditions, *Environ. Sci. Technol.* 19(4):322–327.

Stone, A. T. 1988. The effect of Dismal Swamp dissolved organic matter on the adsorption and surface-enhanced hydrolysis of monophenyl terephthalate in aluminum oxide suspension, *J. Colloid and Interface Sci.*, 132(1):81–87.

Stone, A. T. 1989. Enhanced rates of monophenyl terephthalate hydrolysis in aluminum oxide suspensions, *J. Colloid Interface Sci.*, 127(2):429–441.

Streitwieser, A. 1962. *Solvolytic Displacement Reactions*, (New York, NY: McGraw-Hill).

Swain, C. G., and C. B. Scott. 1953. Quantitative correlation of relative rates. Comparison of hydroxide ion with other nucleophilic reagents toward alkyl halides, esters epoxides and acyl halides, *J. Am. Chem. Soc.*, 75:141–147.

Sykes, P. 1986. *A Guidebook to Mechanism in Organic Chemistry*, (New York, NY: John Wiley), pp. 95–96.

Takahashi, N., N. Mikami, T. Matsuda and J. Miyamoto. 1985a. Hydrolysis of the pyrethroid insecticide cypermethrin in aqueous media, *J. Pesticide Sci.*, 10:643–648.

Takahashi, N., N. Mikami, H. Yamada and J. Miyamoto. 1985. Hydrolysis of the pyrethroid insecticide fenpropathrin in aqueous media, *Pestic. Sci.*, 16:113–118.

Tinsley, I. J. 1979. Modifications of chemicals in the environment, in *Chemical Concepts in Pollutant Behavior*, (New York, NY: John Wiley & Sons), p.113.

Torrents, A., and A. T. Stone. 1991. Hydrolysis of phenyl picolinate at the mineral/water interface, *Environ. Sci. Technol.*, 25:143–149.

Turekian, K. K. 1969. The oceans, streams, and atmosphere, in *Handbook of Geochemistry,* Vol. 1, K. H. Wedepohl, Ed. (New York, NY: Springer-Verlag) pp. 297–323.

Ukachukwu, V. C., J. J. Blumenstein, and D. L. Whalen. 1986. Evidence for reversible formation of an intermediate in the spontaneous hydrolysis reaction of *p*-methoxystyrene oxide, *J. Am. Chem. Soc.*, 108(16):5039-5040.

Van Duuren, B. L., C. Katz, B. M. Goldschmidt, K. Frenkel, and A. Sivak. 1972. Carcinogenicity of halo-ethers. II. Structure-activity relationships of analogs of bis(chloromethyl) ether. *J. Natl. Cancer Inst.* 48(5): 1431–1439.

Vogel, T. M., and M. Reinhard. 1986. Reaction products and rates of disappearance of simple bromoalkanes, 1,2-dibromopropane and 1,2-dibromoethane in water, *Environ. Sci. Technol.*, 20(10):992-997.

Voudrias, E. A., and M. Reinhard. 1986. Abiotic organic reactions at mineral surfaces: A review, in *Geochemical Processes at Mineral Surfaces*, J. A. Davis, and K. F. Hayes, Eds., ACS Symp. Series. pp. 642–486.

Wanner, O., T. Egll, T. Fleischmann, K. Lanz, P. Reichert, and R. P. Schwarzenbach. 1989. Behavior of the insecticides disulfoton and thiometon in the Rhine River: A chemodynamic study, *Environ. Sci. Technol.*, 23:1232–1242.

Weintraub, R. A., G. W. Jex, and H. A. Moye. 1986. Chemical and microbial degradation of 1,2-dibromoethane (EDB) in Florida groundwater, soil and sludge, in *Evaluation of Pesticides in Ground Water*, W. Y. Garner, R. C. Honeycutt and H. N. Nigy, Eds., (Washington, DC: American Chemical Society), pp. 294–310.

Weintraub, R. A., and H. A. Moye. 1987. Ethylene dibromide (EDB) transformations in abiotic-reducing aqueous solutions in the presence of hydrogen sulfide, *Abstr. Amer. Chem. Soc. Div. Env. Chem.*, 27(2):236–240.

Williams, A. J. 1972. Alkaline hydrolysis of substituted phenyl N-phenylcarbamates. Structure-reactivity relations consistent with an E1cB mechanism, *J. Chem. Soc., Perkin Trans. 2*, 6:808–812.

Wolfe, N. L., R. G. Zepp, D. F. Paris, G. L. Baughman and R. C. Hollis. 1977. Methoxychlor and DDT degradation in water: Rates and products, *Environ. Sci. Technol.* 11(12):1077–1081.

Wolfe, N. L., R. G. Zepp, and D. F. Paris. 1978. Use of hydrolytic persistence of carbamate pesticides, *Water Res.* 12:561–563.

Wolfe, N. L., W. C. Steen, and L. A. Burns. 1980. Phthalate ester hydrolysis: Linear free energy relationships, *Chemosphere*, 9:403–408.

Wolfe, N. L., U. Mingelgrin, and G. C. Miller. 1990. Abiotic transformations in water, sediments, and soil, in *Pesticides in the Soil Environment*, H. H. Cheng, Ed., (Madison, WI: Soil Science Society of America) pp. 103–168.

CHAPTER 3

REDUCTION

A. INTRODUCTION

Reducing environments abound in nature (subsurface waters and soils, aquatic sediments, sewage sludge, waterlogged peat soils, hypolimnia of stratified lakes, and oxygen-free segments of eutrophic rivers); until recently, however, very few investigations of organic reactions characteristic of these systems have been undertaken. The realization that pollutants, which are normally considered persistent "aboveground" (or in aerobic environments), may not be nearly as persistent in a reducing environment has generated a great deal of interest in the behavior of organic chemicals in anaerobic environments (Macalady et al., 1986). This interest has been intensified because of the recognition that the reduction of certain classes of chemicals can result in the formation of reaction products that potentially are of more concern than the parent compound. Additionally, the elucidation of reductive transformation pathways may lead to the development of remediation technologies for the removal of organic pollutants from contaminated ecosystems.

Reduction, which is defined as a gain of electrons, occurs when there is a transfer of electrons from an electron donor "or reductant" to an electron acceptor "or oxidant" (March, 1985). The oxidant in this case is the organic chemical or pollutant of interest. Reductive (and oxidative) transformations can be distinguished from other processes (e.g., hydrolysis) by determining if a change in the oxidation state of the atoms involved in the reaction process has occurred. For example, the transformation of 1,1,2,2-tetrachloroethane can result in the formation of two reaction products, 1,1,2-trichloroethylene and 1,2-dichloroethylene (Equation 3.1).

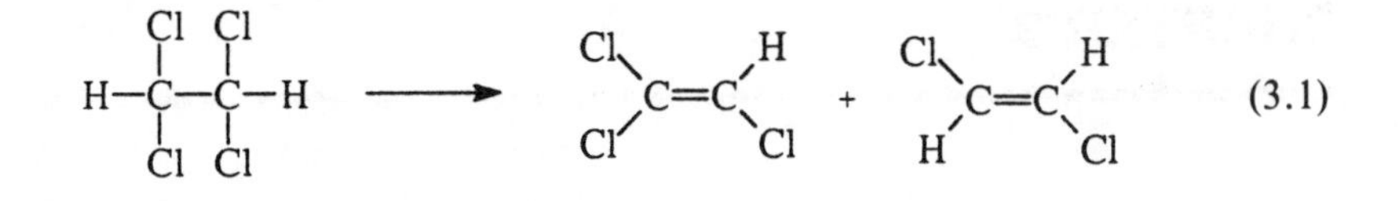

(3.1)

We can determine if either of the reaction products result from reduction of the parent compound by determining if a change in the oxidation state of the chloride-bearing carbons has occurred. Oxidation states of carbon are determined in the following manner: (1) a value of +1 is assigned to the carbon atom of interest for each substituent that is more electronegative than carbon (e.g., Cl), (2) a value of 0 is assigned to the carbon atom for other carbon substituents because their electronegativity is identical, and (3) a value of −1 is assigned to the carbon atom for each substituent that is less electronegative than carbon (e.g., H). Once a value for each substituent on carbon is determined, the assigned values are summed to provide the oxidation state of the carbon atom. This process is performed for both the parent compound and reaction product(s). As shown below, each carbon in 1,1,2,2-tetrachloroethane has an oxidation state of +1, which gives a total of +2. Repeating the same procedure for each of the reaction products, we find that the overall oxidation state of the carbon atoms in 1,1,2-trichloroethene remained unchanged (i.e., there has been no net transfer of electrons).

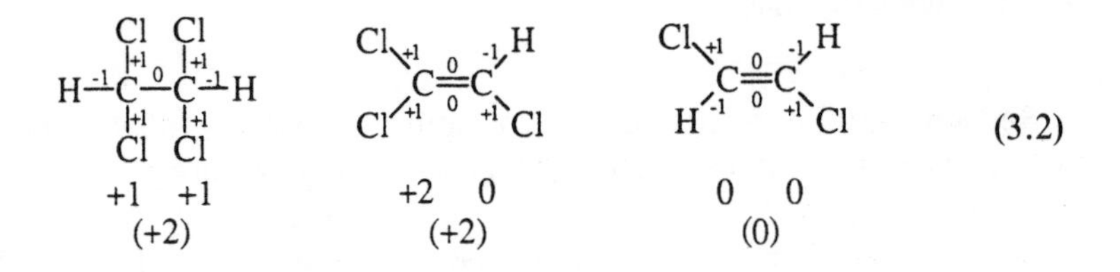

(3.2)

We recognize from previous discussions in Chapter 2 that 1,1,2-trichloroethane results from the base-catalyzed dehydrochlorination of 1,1,2,2-dichloroethane. From this analysis, however, we observe that there has been a change in the overall oxidation state of the carbon atoms in 1,2-dichloroethylene (0) compared to that of the parent compound (+2). We conclude that the formation of 1,2-dichloroethylene results from a process involving the transfer of two electrons. For larger, more complex molecules, we must only consider the atoms directly involved in the reaction process to determine if a change in oxidation state has occurred.

Although our understanding of reductive transformations in the environment has progressed to the point that we can identify the types of functional groups that will be susceptible to reduction in the environmental systems, our limited understanding of reaction mechanisms for such transformations currently is a barrier to the prediction of absolute reduction rates, and how reaction rates will vary from one environmental system to the next. The obvious question that arises from this discussion is, "What is the source of electrons in natural reducing environments?" The cumulative knowledge in this area suggests that naturally occurring reductants are not limited to two or three reactive species, as is the case for hydrolysis, but that a complex array of species is involved, ranging from chemical or "abiotic" reagents such as sulfide minerals, reduced metals, and natural organic matter, through extracellular biochemical reducing agents such as iron porphyrins, corrinoids, and bacterial transition-metal coenzymes, to biological systems such as microbial populations. Furthermore, the relationship between these various reductants in natural systems is most likely quite complex. For example, chemical species such as reduced metals and sulfide ion may be the direct result of microbial metabolism. As a result, though the main focus of this book is to address chemical reactions in the environment, because of the difficulty in distinguishing between abiotic and biological processes, examples of reductive transformations will be presented that probably occur to a significant extent through direct microbial metabolism.

B. REDUCTIVE TRANSFORMATION PATHWAYS

Reductive transformations are most conveniently categorized according to the type of functional group that is reduced. General schemes illustrating the reductive transformations that are known to occur in natural reducing environments are summarized in Table 3.1.

For many organic chemicals containing these functional groups, reduction will be their dominant transformation pathway in reducing environments. A significant portion of this chapter will provide in-depth discussion and examples of the reaction pathways illustrated in Table 3.1. For environmental assessment purposes, just as important is knowledge concerning reducible organic functional groups whose reductive transformations are not thought to occur in environmental systems. Some of the most common organic functional groups found in organic pollutants that are resistant to reduction in the environment are illustrated in Table 3.2.

Although the reduction of these organic functional groups is performed routinely by synthetic chemists in the laboratory using strong reductants (i.e., metal hydrides and dissolving metals), the abiotic reduction of these functional groups in natural aquatic ecosystems is not likely to occur (Hudlicky, 1984). The lack of reactivity observed for these classes of chemicals in natural reducing systems is probably due to a combination of unfavorable kinetic and thermodynamic factors.

1. Reductive Dehalogenation

The reductive transformation of halogenated aliphatic and aromatic compounds appears to be a general phenomenon that occurs in reducing environments. The reduction of these chemicals has been observed in a variety of environmental and laboratory systems, including anaerobic sediments (Jafvert and Wolfe, 1987; Peijnenburg et al., 1992), soils (Glass, 1972), anaerobic sewage sludge (Mikesell and Boyd, 1985; Fathepure et al., 1988), groundwaters (Criddle et al., 1986), aquifer materials (Curtis and Reinhard, 1989), reduced iron porphyrin systems (Castro, 1964; Klecka and Gonsior, 1984), bacterial transition-metal coenzymes (Gantzer and

Table 3.1. Reductive Transformations Known to Occur in Natural Reducing Environments

1. Reductive Dehalogenation

Hydrogenolysis

$$R{-}X + 2e^- + H^+ \longrightarrow R{-}H + X^-$$

Vicinal Dehalogenation

2. Nitroaromatic Reducion

$$Ar{-}NO_2 + 6e^- + 6H^+ \longrightarrow Ar{-}NH_2 + 2H_2O$$

3. Aromatic Azo Reduction

$$Ar{-}N{=}N{-}Ar' + 4e^- + 4H^+ \longrightarrow ArNH_2 + H_2NAr'$$

4. Sulfoxide Reduction

$$R_1{-}S(=O){-}R_2 + 2e^- + 2H^+ \rightleftharpoons R_1{-}S{-}R_2 + H_2O$$

5. N-Nitrosoamine Reduction

$$R_1{-}N(N{=}O){-}R_2 + 2e^- + 2H^+ \longrightarrow R_1{-}N(H){-}R_2 + HNO$$

6. Quinone Reduction

$$O{=}C_6H_4{=}O + 2e^- + 2H^+ \rightleftharpoons HO{-}C_6H_4{-}OH$$

7. Reductive Dealkylation

$$R_1{-}X{-}R_2 + 2e^- + 2H^+ \longrightarrow R_1{-}XH + R_2H$$

X = NH, O, or S

Table 3.2. Examples of Functional Groups That Are Resistant to Reduction in Naturally Reducing Environments

1. Aldehyde to Alcohol

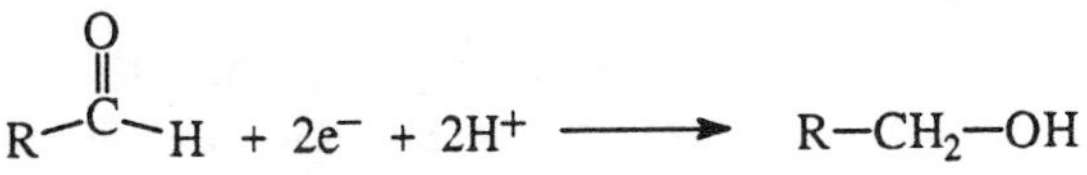

$$R-C(=O)-H + 2e^- + 2H^+ \longrightarrow R-CH_2-OH$$

2. Ketone to Alcohol

$$R_1-C(=O)-R_2 + 2e^- + 2H^+ \longrightarrow R_1-CH(OH)-R_2$$

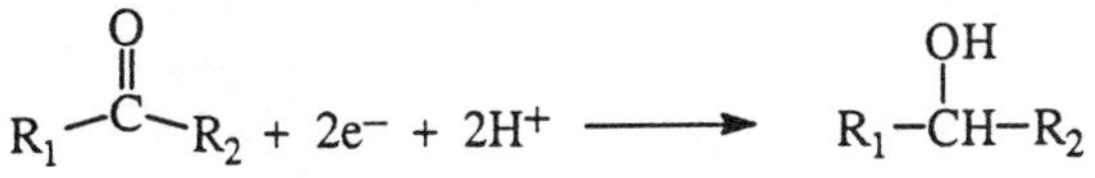

3. Carboxylic Acid to Alcohol

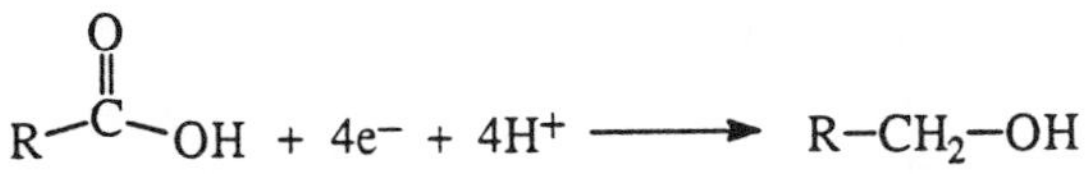

$$R-C(=O)-OH + 4e^- + 4H^+ \longrightarrow R-CH_2-OH$$

4. Carboxylic Acid Ester to Alcohol

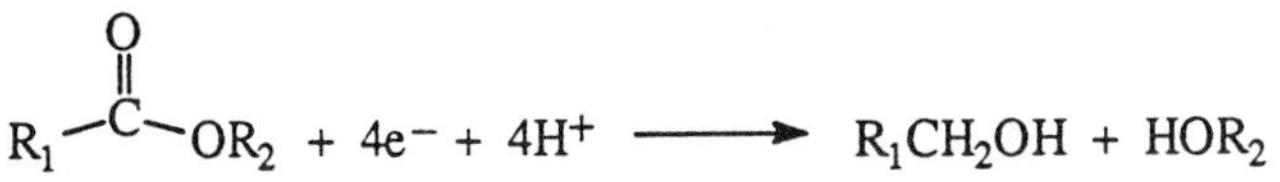

$$R_1-C(=O)-OR_2 + 4e^- + 4H^+ \longrightarrow R_1CH_2OH + HOR_2$$

5. Amide to Amine

$$R_1-C(=O)-NHR_2 + 4e^- + 4H^+ \longrightarrow R_1CH_2OH + H_2NR_2$$

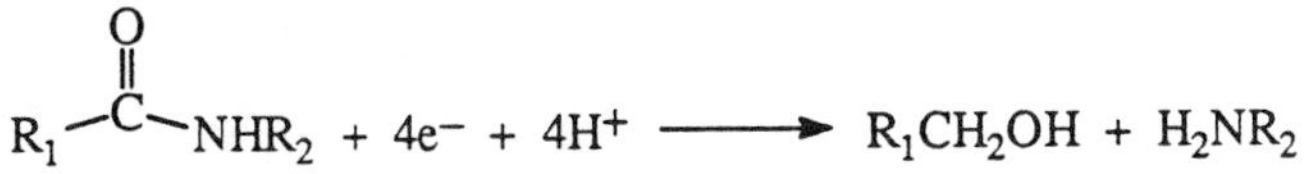

6. Alkene to Alkane

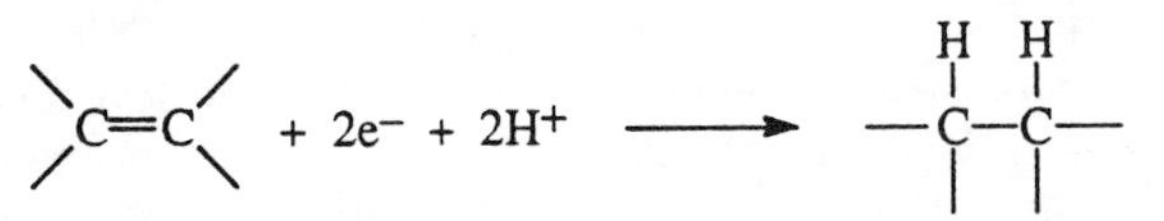

7. Aromatic Hydrocarbon to Saturated Hydrocarbon

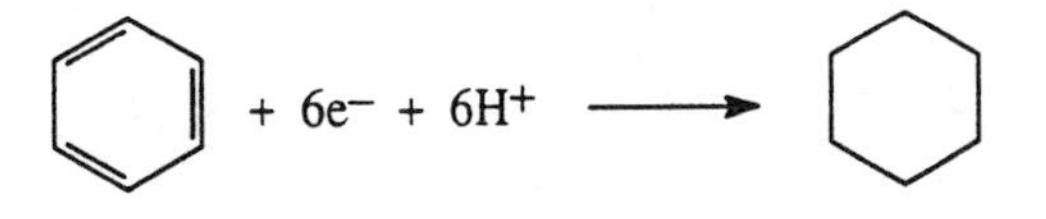

Wackett, 1991), and mineral sulfides (Kriegman-King and Reinhard, 1991 and 1992; Yu and Bailey, 1992) and various chemical reducing agents (Castro and Kray, 1963; Kray and Castro, 1964). The impetus to understand the reaction pathways of halogenated chemicals in these types of systems results from their widespread contamination of sediments, soils, and groundwaters due to their frequent use as agrochemicals, solvents, and synthetic intermediates.

Numerous remediation schemes have been developed around the premise that reductive dehalogenation is a detoxification process. Reductive dehalogenation processes, however, can result in the formation of lower-halogenated products that may present a greater health hazard than the parent compound (Bolt et al., 1982). For example, reductive dehalogenation of hexachloroethane forms tetrachloroethylene (Equation 3.3), which can be sequentially dehalogenated to give trichloroethylene (Equation 3.4), cis-1,2-dichloroethylene (Equation 3.5), and finally vinyl chloride (Equation 3.6), a suspected carcinogen (Suflita et al., 1982).

$$Cl_3C\text{-}CCl_3 \rightarrow Cl_2C=CCl_2 \tag{3.3}$$

$$Cl_2C=CCl_2 \rightarrow Cl_2C=CHCl \tag{3.4}$$

$$Cl_2C=CHCl \rightarrow ClCH=CHCl \tag{3.5}$$

$$ClCH=CHCl \rightarrow ClCH=CH_2 \tag{3.6}$$

Halogenated Aliphatics

Reductive dehalogenation of halogenated aliphatics can occur through several mechanisms, each of which begins by a electron transfer that results in the formation of a carbon centered radical (Equation 3.7) (Vogel et al., 1987). Subsequent reactions of this radical intermediate include (1) abstraction of a H-atom from a suitable donor (*hydrogenolysis*) (Equation 3.8), (2) loss of a proton from an adjacent carbon to form a C–C double bond (*dehydrodehalogenation*) (Equation 3.9), *radical coupling* (or dimerization) (Equation 3.10), and (4) elimination of vicinal halides (*vicinal dehalogenation*), resulting in the formation of a C–C double bond (Equation 3.11).

$$-\overset{|}{\underset{|}{C}}-\overset{|}{\underset{|}{C}}-X + e^- \longrightarrow -\overset{|}{\underset{|}{C}}-\overset{|}{\underset{|}{C}}\cdot + X^- \tag{3.7}$$

Hydrogenolysis

$$-\overset{|}{\underset{|}{C}}-\overset{|}{\underset{|}{C}}-X \xrightarrow{e^-} -\overset{|}{\underset{|}{C}}-\overset{|}{\underset{|}{C}}\cdot \xrightarrow{H\cdot} -\overset{|}{\underset{|}{C}}-\overset{|}{\underset{|}{C}}-H \qquad (3.8)$$

Dehydrodehalogenation

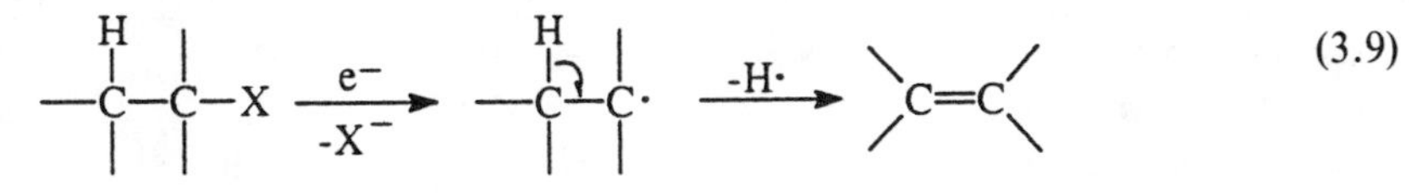

(3.9)

Radical coupling

$$2\,-\overset{|}{\underset{|}{C}}-\overset{|}{\underset{|}{C}}-X \xrightarrow[-2X^-]{2e^-} 2\,-\overset{|}{\underset{|}{C}}-\overset{|}{\underset{|}{C}}\cdot \longrightarrow -\overset{|}{\underset{|}{C}}-\overset{|}{\underset{|}{C}}-\overset{|}{\underset{|}{C}}-\overset{|}{\underset{|}{C}}- \qquad (3.10)$$

Vicinal Dehalogenation

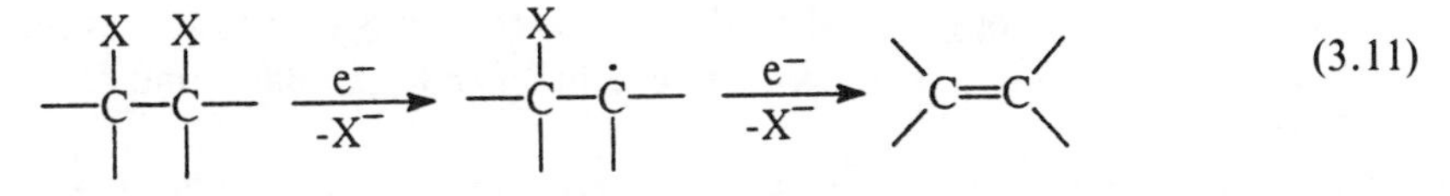

(3.11)

Because radical coupling is a bimolecular reaction involving two reactive chemical species, these processes are generally not thought to be significant in environmental systems due to the lower concentration of these species.

As with the hydrolytic transformation of halogenated hydrocarbons, the reactivity of these chemicals toward reductive transformation varies considerably. Chemical-physical properties of halogenated hydrocarbons that are expected to affect reactivity in reducing environments include the bond strength of the carbon-halogen bond, the electron affinity of the carbon-halogen bond (or the ability of the halogen to accept electrons), and the stability of the carbon-radical species resulting from initial electron transfer. The dependence of reactivity on structure is illustrated in the data set summarized in Table 3.3 for the reductive dehalogenation of a series of halogenated ethanes in an anaerobic sediment-water system (Jafvert and Wolfe, 1987). Analysis of reaction products indicated that vicinal dehalogenation resulting in the formation of alkenes was the primary transformation pathway of these chemicals in sediment slurries. The degradation half-lives ranged from minutes to greater than 735 days. The order of reactivity for the chlorinated ethanes ($Cl_3C\text{-}CCl_3 > Cl_2HC\text{-}CHCl_2 > ClH_2C\text{-}CH_2Cl$) was found to correlate with the C-X bond dissociation energy (a measure of bond strength), given in Table 3.3. The reactivity order for the dihalogenated ethenes ($IH_2C\text{-}CH_2I > BrH_2C\text{-}CH_2Br > ClH_2C\text{-}CH_2Cl$) also correlates with the bond dissociation energies, and to lesser extent with the electronegativity of the leaving group. Based on transition state theory, a decrease in the dissociation energy of the covalent bond to be broken, or an increase in the electro-

Table 3.3. Relative Rates for the Vicinal Dehalogenation of Halogenated Ethanes in a Reducing Sediment-Water Slurry[a,b]

Compound	k_{obs} (min^{-1})	$t_{1/2}$	Bond Dissociation Energy (C-X; kcal/mol)	Electron Affinity of Associated Halogen (kcal/mol)
Hexachloroethane	1.9×10^{-2}	36 min	72	83(Cl)
1,2-Diiodoethane	2.9×10^{-2}	24 min	74.3	83(Cl)
1,2-Dibromoethane	2.1×10^{-4}	55 h	77.9	83(Cl)
1,1,2,2-Tetrachloroethane	7.3×10^{-5}	6.6 d	69.8	77.5(Br)
1,2-Dichloroethane	$< 1.4 \times 10^{-5}$	735 d	<47	70.5(Cl)

[a]Eh of the sediment-water slurry was −140 mV; pH = 6.5; sediment-to-water ratio = 0.075.
[b]Data taken from Jafvert and Wolfe (1987).

negativity of the leaving group should stabilize the transition state, resulting in an increase in the reaction rate for reductive dehalogenation (Bank and Juckett, 1975 and 1976).

It is interesting to note the stark contrast in the reactivity of hexachloroethane toward reductive dehalogenation and hydrolysis. The half-life for hexachloroethane in the reducing sediment-water slurry was 36 min (Table 3.3), compared to its hydrolysis half-life of 1.8×10^9 years at pH 7 at 25°C. Clearly, the extent to which hexachloroethane will persist in a given environmental system will be dependent on the reducing characteristics of the reaction system.

As was observed for the chlorinated ethanes in Table 3.3, vicinal dehalogenation is typically the dominant pathway for the reductive dehalogenation of polyhalogenated aliphatics having vicinal halides. The predominance of vicinal dehalogenation for these polyhalogenated aliphatics is attributed to the formation of an intermediate in which stabilization of the radical occurs through bridging of the neighboring halogen adjacent to the carbon from which the initial halogen was removed (Equation 3.12).

$$\text{X}\text{–C–C–X} + e^- \longrightarrow \text{C}\overset{\dot{\text{X}}}{\triangle}\text{C} + X^- \qquad (3.12)$$

Examples of halogenated organics that undergo vicinal dehalogenation are γ-BHC, commonly known as lindane, and DTE, an analog of DDT. The vicinal dehalogenation of lindane has been observed in flooded rice soils (Tsukano and Kobayashi, 1972) and anaerobic sewage sludge (Beland et al., 1976).

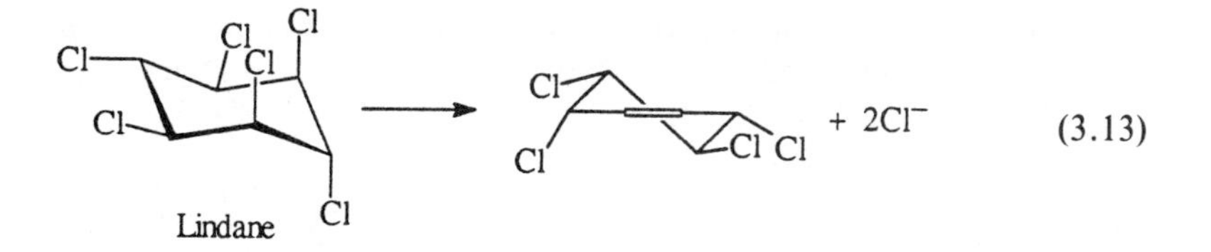

The electrochemical reduction of DTE results in the formation of DDE (Beland et al., 1974).

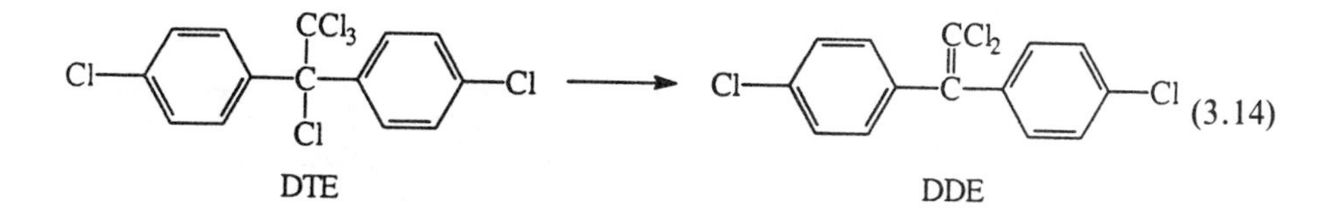

Examples of environmentally significant halogenated aliphatics that are reductively dehalogenated through the hydrogenolysis mechanism include toxaphene and mirex. Both of these chemicals are highly persistent in aerobic environments. Toxaphene components are reduced at the geminal-dichloro group to give hexachlorobornanes (Khalifa et al., 1976; Saleh et al., 1977).

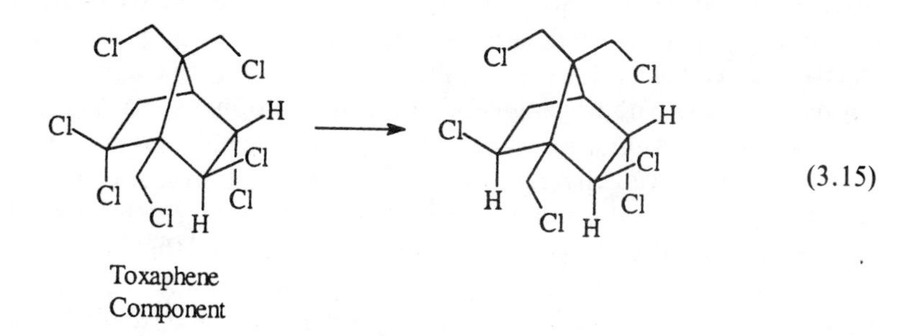

Likewise, reduction of mirex in an iron(II)porphyrin model system occurs to give mono-, di-, tri, and tetrahydro derivatives as well as other more polar reaction products (Holmstead, 1976).

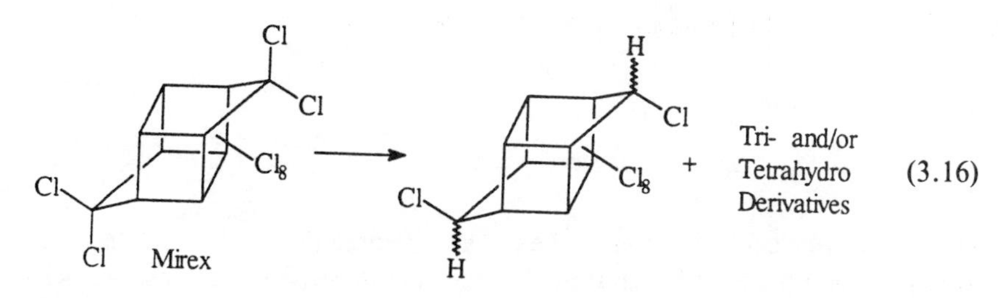

The iron-mediated reductive dechlorination of DDT has been observed to occur by both hydrogenolysis and dehydrodehalogenation to give DDD and DDE, respectively (Glass, 1972; Zoro et al., 1974).

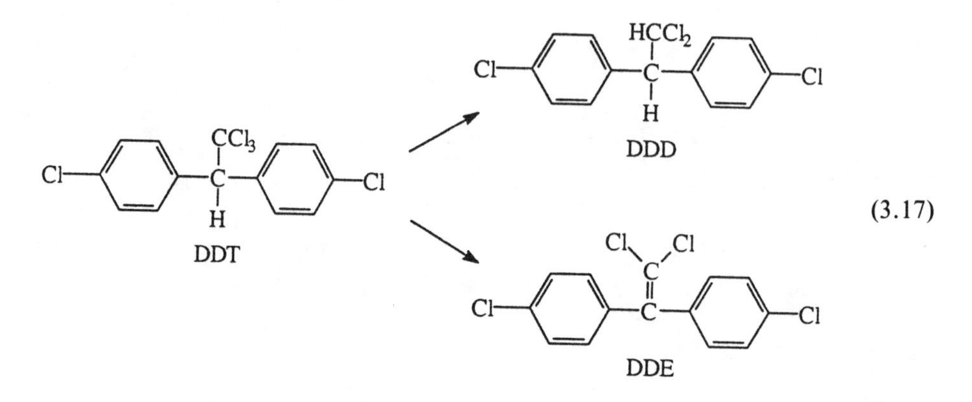

(3.17)

Halogenated Aromatics

The reduction of halogenated aromatics occurs through hydrogenolysis (i.e., replacement of halide with hydrogen). The reactivity of halogenated aromatic hydrocarbons in reducing sediments has been correlated with a number of molecular descriptors that include carbon-halogen bond strength, the summation of the Hammett σ constants (used to describe inductive and resonance electronic effects on the reactive center for additional substituents), the inductive constant, $\sigma_{I,}$ (describes intrinsic electron withdrawing capability) of additional substituents, and the summation of steric factors, Es, for additional substituents (Peijnenburg et al., 1992b). Generally, reductive dehalogenation of halogenated aromatics occurs at a slower rate than their aliphatic counterparts, and usually requires direct microbial mediation. Several studies, however, have provided evidence that reductive dehalogenation of halogenated aromatics can occur through abiotic reaction pathways. For example, pentachlorophenol has been reported to undergo reductive dechlorination at the *meta* and *para* positions in sterilized flooded soils (Kuwatsuka, 1977).

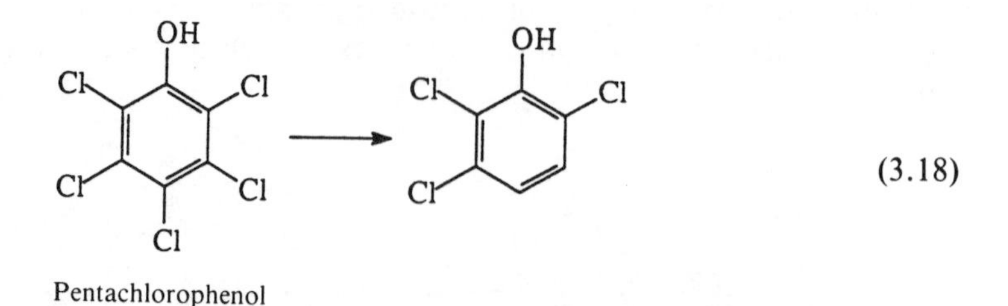

(3.18)

The reductive dechlorination of hexa- and pentachlorobenzene has been demonstrated to be mediated by vitamin B_{12} and hematin (Gantzer and Wackett, 1991).

In other studies, the reductive transformation of halogenated benzenes in sediment-water slurries has been proposed to occur by both an abiotic and a biotic pathway (Peijnenburg et al.,1992a). For example, in nonsterile systems, the initial slow degradation process for 1-fluoro-2-iodo-benzene was attributed to an abiotic process (e.g., electron transfer by biologically reduced iron), whereas the increase in the removal of 1-fluoro-2-iodobenzene between days 10 and 15 was attributed to biodegradation (Figure 3.1). In contrast to the nonsterile system, no increase in the degradation rate of 1-fluoro-2-iodobenzene in the sterilized system was observed within 50 days of incubation.

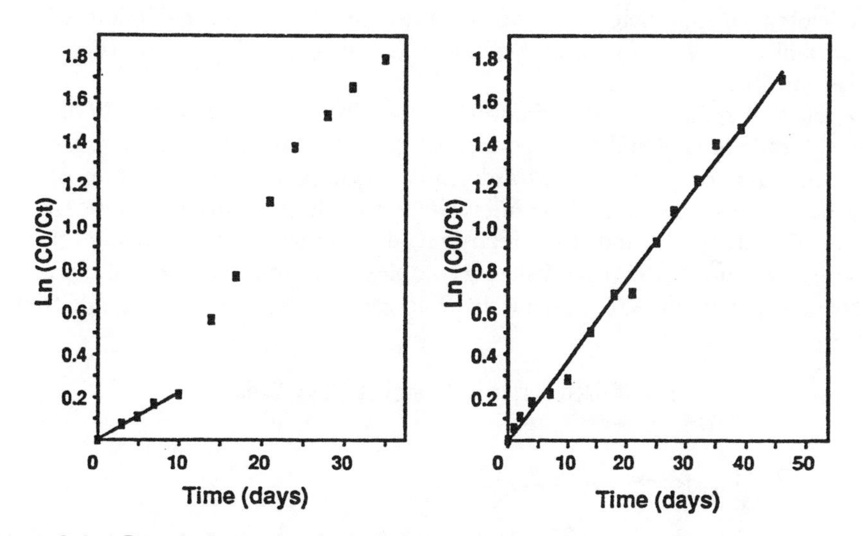

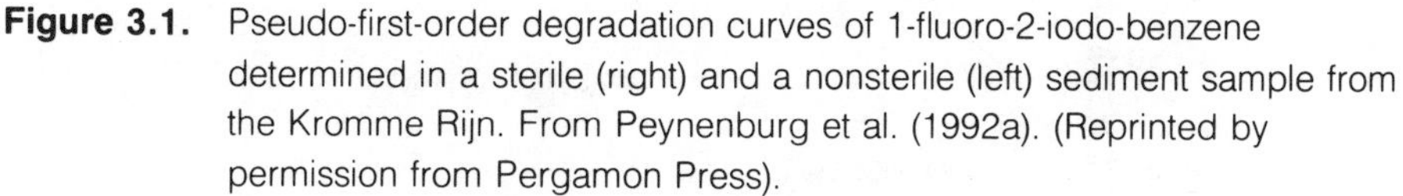

Figure 3.1. Pseudo-first-order degradation curves of 1-fluoro-2-iodo-benzene determined in a sterile (right) and a nonsterile (left) sediment sample from the Kromme Rijn. From Peynenburg et al. (1992a). (Reprinted by permission from Pergamon Press).

Biologically mediated reduction of halogenated aromatics has focused primarily on the chlorinated phenols, anilines, and biphenyls because of their widespread contamination and concerns over ecotoxicity. Reductive dechlorination of these chemicals may be the rate-limiting step in their complete mineralization. Reductive dehalogenation for chlorinated phenols and anilines in sediment- and aquifer-water slurries appears to be regioselective (i.e., reduction at one position on the ring is favored over that of another). For example, reductive dechlorination of 2,3,4,5-tetrachloroaniline in a methanogenic aquifer resulted in the formation of 2,3,5-trichloroaniline, and eventually, 3,5-dichloroaniline (Kuhn and Suflita, 1989b) (Equation 3.19).

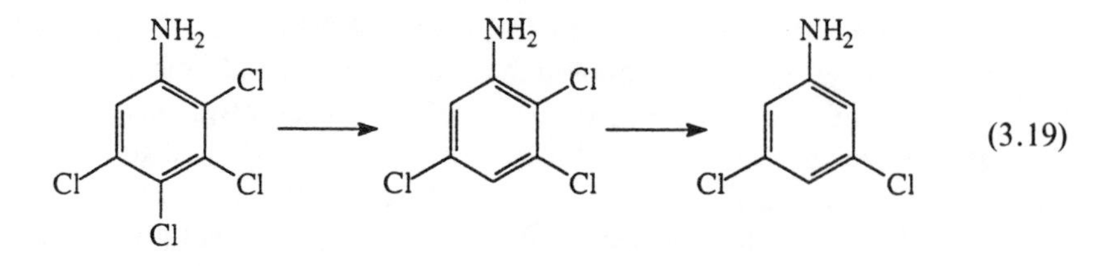

(3.19)

Similar selectivity has been observed for the reduction of chlorinated phenols in anoxic sediments (Hale et al., 1991, Bryant et al., 1991). Studies in acclimated pond sediments suggest that the same microbial enzyme systems are involved in the reduction of chlorinated phenols and anilines. For example, preexposure of a sediment-water slurry to 3,4-dichlorophenol resulted in the facile reduction of 3,4-dichloroaniline to 4-chloroaniline without a lag or adaptation period (Figure 3.2) (Struijs and Rogers, 1989).

Reports that reductive dechlorination of polychlorinated biphenyls (PCBs) may occur naturally in lake and river sediments has spurred efforts to develop remediation techniques for contaminated sediments. Widespread contamination of sediments has occurred because of the intensive uses of these chemicals as heat transfer fluids, hydraulic fluids, and flame retardants due to their excellent stability properties (Hutzinger et al., 1974). PCBs are a complex mixture of chlorinated biphenyls. Theoretically, there are 209 possible PCB congeners; because of steric hindrance,

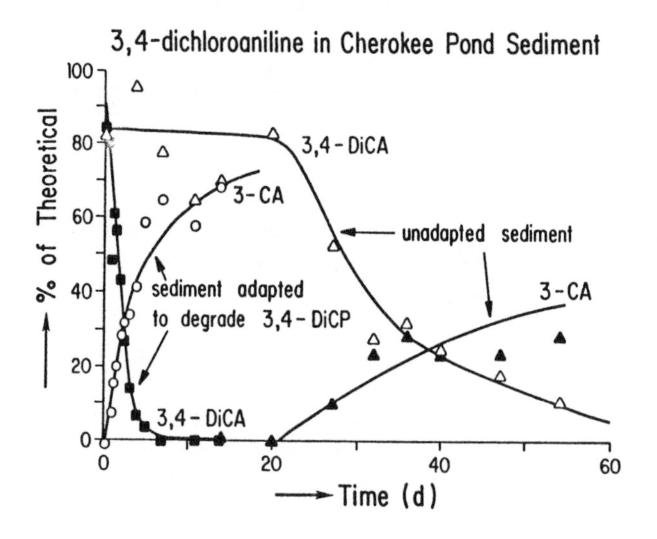

Figure 3.2. Reductive dechlorination of 3,4-dichloroaniline (3,4-DiCA) in unacclimated sediment and sediment acclimated to reductively dechlorinate 3,4-dichlorophenol (3,4-DiCP). Shown are the loss in 3,4-DiCA in acclimated and unacclimated sediments and the formation of the 3-CA in acclimated and unacclimated sediments (d, Days). (Reprinted by permission from the American Society for Microbiology).

however, only about 50% these congeners are synthesized. Studies of the reduction of PCBs in anaerobic sediments and by microorganisms isolated from contaminated sediments suggest a general relationship between reduction potential, number of chlorine substituents, chlorine substituent pattern, and dechlorination rates (Brown, et al., 1987; Nies and Vogel, 1990; Quensen et al., 1988; Quensen et al., 1990). Rates for reductive dechlorination are found to increase with increasing chlorine substitution. This observation has been attributed to the fact that the reduction potential, or "the willingness to except electrons," increases with increasing chlorine substitution (Farwell et al., 1975; Rusling and Miaw, 1989). Reductive dechlorination also is found to occur preferentially at the *para*- and *meta*-positions, resulting in the formation of lower-chlorinated predominantly *ortho*-substituted biphenyls (Equation 3.20).

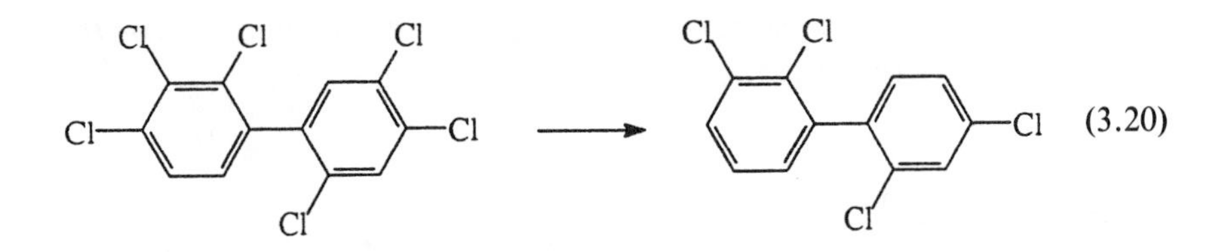

(3.20)

2. Nitroaromatic Reduction

The reduction of nitroaromatics in anaerobic systems has received considerable attention because of their importance as agrochemicals, munitions, textile dyes, and dye intermediates. The reaction products resulting from the reduction of nitroaromatic compounds are aromatic amines. The formation of aromatic amines occurs through a series of electron transfer reactions with nitroso compounds and hydroxylamines as even-electron intermediates (Equation 3.21). Typically, these intermediates are more reactive than the parent nitroaromatic compound, and do not persist to any great extent in reducing environments.

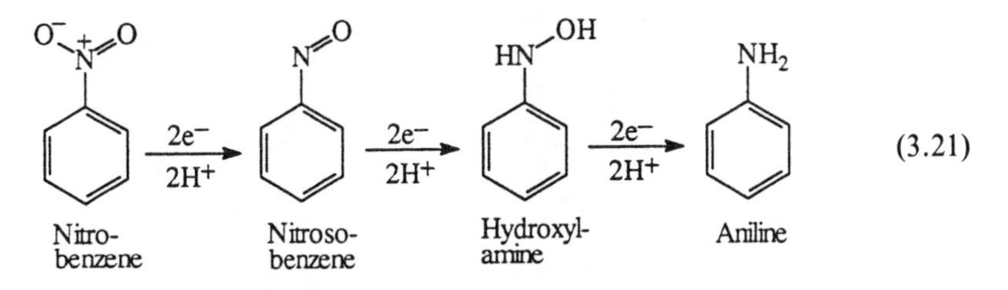

(3.21)

Nitroaromatic reduction has been observed in natural systems and in well defined laboratory model systems. Nitro reductions are known to be mediated by anaerobic soils and sediments (Wahid and Sethunathan, 1979; Wahid et al., 1980; Adhya et al., 1981a and 1982b; Gambrell et al., 1984; Wolfe et al., 1986; Sanders and Wolfe, 1985; Weber, 1988), sulfide minerals (Yu and Bailey, 1992), sewage sludge (Geer,

1978), model quinone-hydroquinone redox couples (Tratnyek and Macalady, 1989), iron porphyrins (Ong and Castro, 1977; Schwarzenbach et al., 1990) and dissolved organic carbon (Dunnivant et al., 1992). The facile reductions observed in these systems (half-lives are often measured on the order of minutes to hours), as well as the lack of induction periods in natural systems and reactivity observed in sterile systems suggests that the reduction of nitroaromatics can occur through abiotic pathways.

Because the reduction of aromatic nitro groups is such a facile process, reductive transformation of chemicals containing this moiety is often the dominant pathway for their transformation in the environment. For example, the reduction of methyl parathion, which is representative of the nitro-containing organophosphorus insecticides, to amino methyl parathion has been observed in anaerobic sediments (Wolfe et al., 1986) and flooded soils (Wahid and Sethunathan, 1979; Wahid et al., 1980; Adhya et al., 1981a and 1981b; Gambrell et al., 1984), with half-lives on the order of minutes to hours (Equation 3.22).

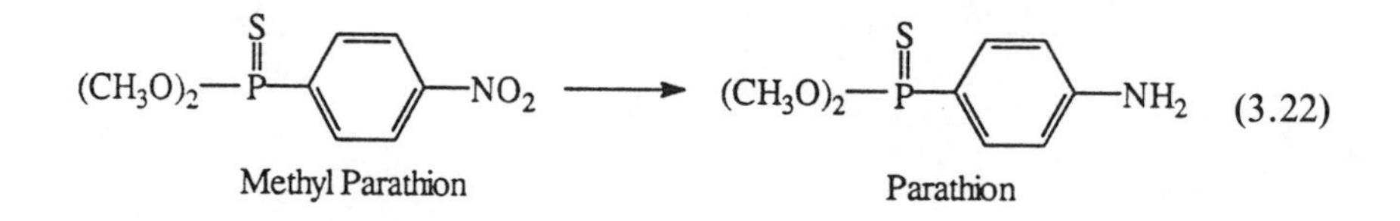

(3.22)

These half-lives are significantly shorter than the measured hydrolysis half-life of 20 weeks at pH 7 at 25 °C (Smith et al., 1978).

Pentachloronitrobenzene (PCNB) is another example of a nitroaromatic agrochemical that is known to undergo facile reduction in anaerobic systems. PCNB has been identified as a pollutant in river water and groundwater (Fushiwaki et al., 1990). Reduction of PCNB results in the formation of pentachloroaniline, which is fairly resistant to further transformation pathways (Wang and Broadbent, 1973; Kuhn and Suflita, 1989a) (Equation 3.23).

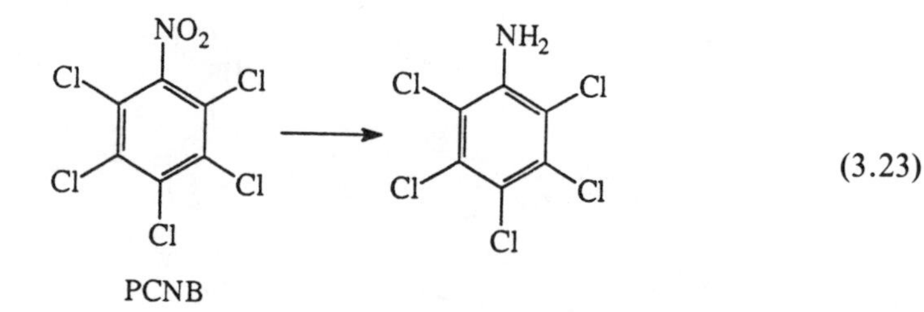

(3.23)

Polynitro Aromatics

A number of environmentally significant nitroaromatics, such as the dinitro herbicides, nitroaromatic munitions, and textile dye intermediates, contain two or more nitro groups. Because of the presence of more than one nitro group, the question of regioselectivity (i.e., the preferential reaction at one site versus another site in the same chemical) becomes important. As a result, the reaction pathways for the reductive transformation of these chemicals can be somewhat complex. For example, dinitro herbicides, which represent an important class of agrochemicals, are known to be susceptible to reduction in anaerobic systems. The reduction of trifluralin has been observed in sediments and flooded soils (Willis et al., 1974; Helling, 1976; Southwick and Willis, 1974). Although reduction of the first nitro group is a facile process, the second nitro group is much more resistant to reduction.

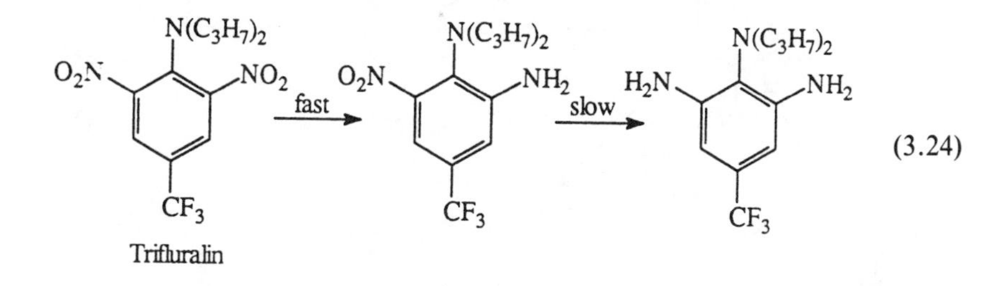

(3.24)

The diamino compound has been observed in only trace amounts in an anaerobic soil suspension (Southwick and Willis, 1974). This result, which is often observed for the reduction of polynitroaromatics, can be explained in the following manner. Nitro groups have strong electron-withdrawing properties. As a result, their reduction potentials are more positive, suggesting that the ease with which they receive electrons will be greater. When the first nitro group of a polynitroaromatic is reduced to an amino group, which is an electron-donating group by resonance effects, the net result is that the reduction potential of the reaction product will be lower than that of the parent compound. Consequently, subsequent reduction of remaining nitro groups is expected to occur at slower rates.

There has been considerable interest in the reductive transformation of the polynitroaromatics, 2,4-dinitrotoluene (2,4-DNT), and 2,4,6-trinitrotoluene (TNT) in anaerobic environments. These chemicals are high volume munitions, which because of their improper disposal and storage, have contaminated soils and aquifers at numerous sites (Kaplan and Kaplan, 1982). Enhancing the reduction of these contaminants in situ using abiotic and biotic processes has been proposed as a remediation technique for contaminated soils and sediments (Kaplan, 1990; Ou et al., 1992). The reduction of TNT and DNT, as well as other polynitroaromatics, has been studied in some detail in anaerobic microbial enzyme systems (McCormick et al., 1976). The reactivity of the nitro groups of these chemicals was found to be dependent on the type and position of other substituents on the benzene ring. The initial reactions in TNT reduction are stepwise reductions of the TNT nitro groups to yield amino and hydroxylamino compounds, which may undergo coupling reactions to yield azoxy compounds (Figure 3.3). The number of nitro groups reduced was found to be dependent on the reducing capacity of the enzyme system. The initial site for reduction is thought to be the nitro group in the 4-position. Facile reduction of 2,4-

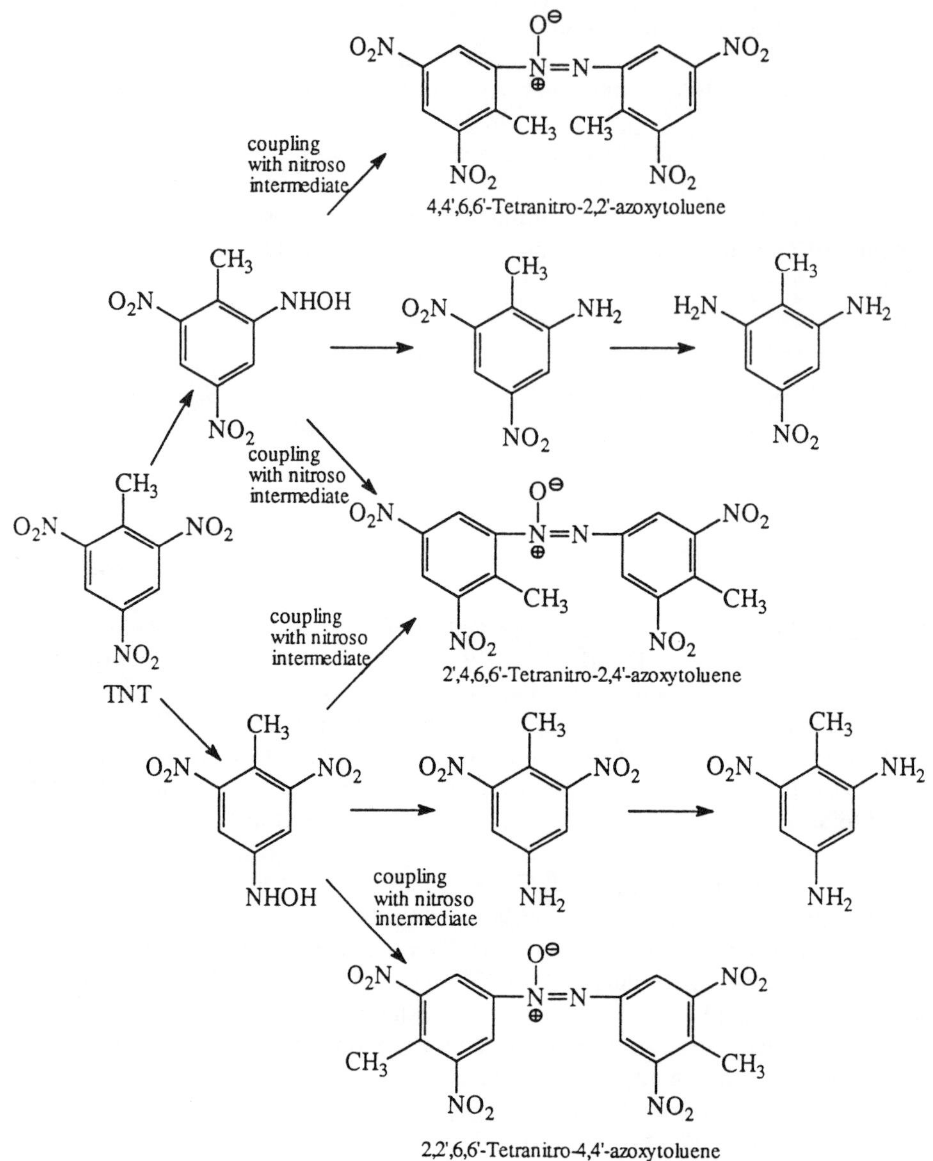

Figure 3.3. Reaction pathways proposed for the reduction of TNT in an anaerobic microbial enzyme system (McCormick et al., 1976).

DNT was also observed in the enzyme system. The reduction of 2,4-DNT was very selective; reduction of the 2-nitro group did not take place until reduction of the 4-nitro group was complete.

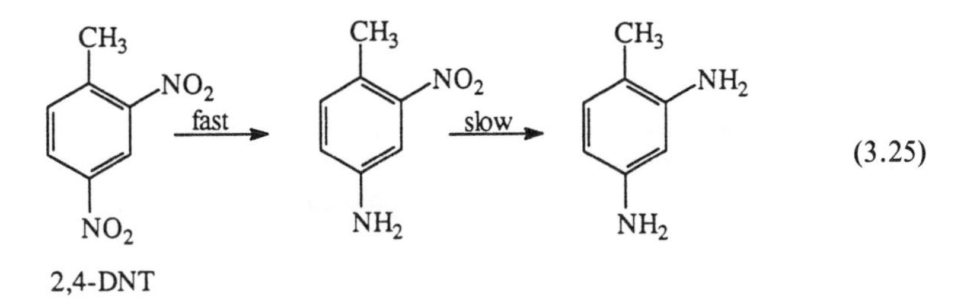

(3.25)

In contrast, reduction of 2,4-dinitroaniline ($X = NH_2$) and 2,4-dinitrophenol ($X = OH$) was selective for the nitro groups in the 2-position.

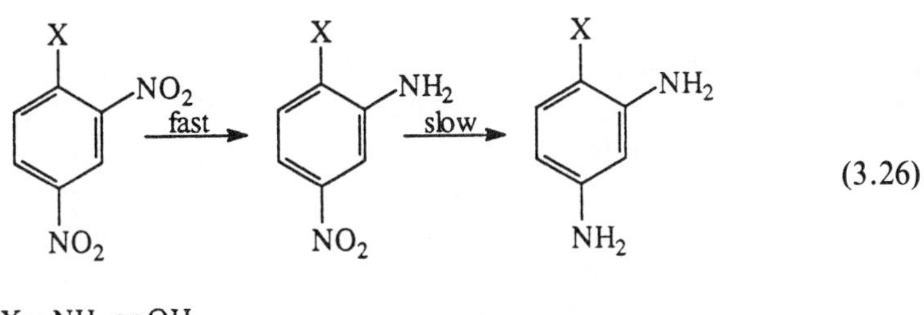

(3.26)

Furthermore, reduction of the second nitro group occurred at a much slower rate than that of the first nitro group.

The reduction of 2-bromo-4,6-dinitroaniline (BDNA), an important intermediate in the preparation of textile dyes that has been detected in river waters (Maguire and Tkacz, 1991), in anaerobic sediment-water systems occurred in an analogous manner to that of 2,4-dinitroaniline and 2,4-dinitrophenol (Equation 3.27).

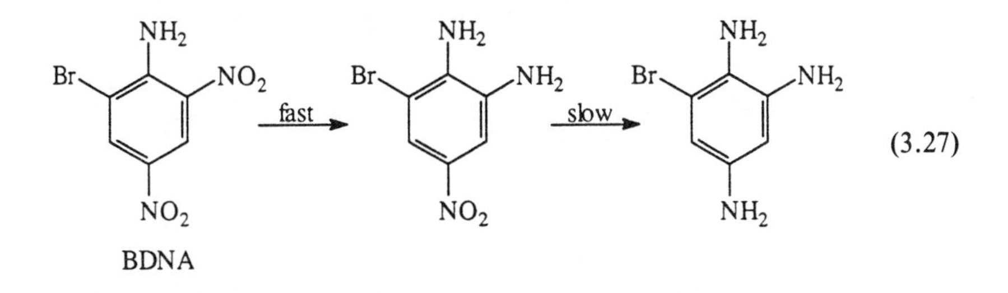

(3.27)

The reduction of BDNA was very selective for the nitro group in the 2-position (Weber, 1988). Also, as was the case for the reduction of trifluralin, 2,4-dinitroaniline, and 2,4-dinitrophenol, the resulting diamino benzene was much more persistent than the parent compound. Reduction of the second nitro group to give the triamino benzene did occur, but very slowly. Only trace amounts of this compound were observed in the sediment slurries.

Regioselectivity

For environmental assessment needs, the ability to predict the initial site of reduction is important because reaction products that are regioisomers (i.e., isomers that differ in the position of substituent groups) can have very different physical and toxicity characteristics. The reduction of the more sterically-hindered nitro group in the *ortho*-position of BDNA (Equation 3.27), 2,4-dinitroaniline, and 2,4-dinitrophenol (Equation 3.26) appears to be a general phenomenon for the reduction of 2,4-dinitrobenzenes substituted with resonance electron-donating groups (e.g., $-NH_2$ or –OH) in the 1-position (Hudlicky, 1984). Both chemical (Hartman and Silloway, 1955; Terpko and Heck, 1980) and enzymatic reduction (McCormick et al., 1976) of these derivatives result in the preferential reduction of the *ortho*-nitro group. By contrast, the less sterically hindered nitro group in the *para*-position of TNT (Figure 3.3) and 2,4-DNT (Equation 3.25) is preferentially reduced by chemical (Hartman and Silloway, 1955) and enzymatic (McCormick et al., 1976) reduction.

Comparison of the resonance structures of 2,4-dinitrobenzenes substituted with electron-donating resonance groups which illustrate the shift of electron density onto the *para*- and *ortho*-nitro groups by through resonance, provides insight into the observed regioselectivity of these aromatic nitro reductions (Equation 3.28). The resonance structure resulting from the shift of electron density onto the *para*-nitro group has maximum like-charge separation resulting in a more favorable electrostatic interaction. In contrast, the shift of electron density onto the *ortho*-nitro group results in a resonance structure with an unfavorable electrostatic interaction because of the close proximity of the like-charges. As a result, the *para*-nitro group is deactivated toward reduction to a greater extent by through-resonance than the *ortho*-nitro group.

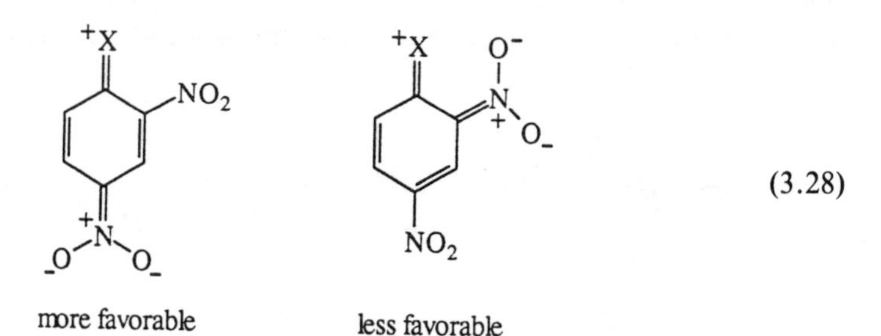

(3.28)

Furthermore, deactivation of the *ortho*-nitro group toward reduction by through-resonance is minimized because the 1-substituent (e.g., $-NH_2$ or CH_3) forces the *ortho*-nitro group out of the plane of the benzene ring, which reduces π-orbital overlap. This phenomenon is referred to as *steric inhibition of resonance* (Wheland, 1944). The result is that resonance electron-donating groups deactivate *para*-nitro groups toward reduction to a greater extent than *ortho*- nitro groups. In the case of TNT and 2,4-DNT (i.e., the 1-substituent is $-CH_3$), steric interactions appear to be the dominant factor controlling the regioselectivity of nitro reduction.

3. Aromatic Azo Reduction

The reduction of aromatic azo compounds is a four-electron process that proceeds through the hydrazobenzene (Florence, 1965).

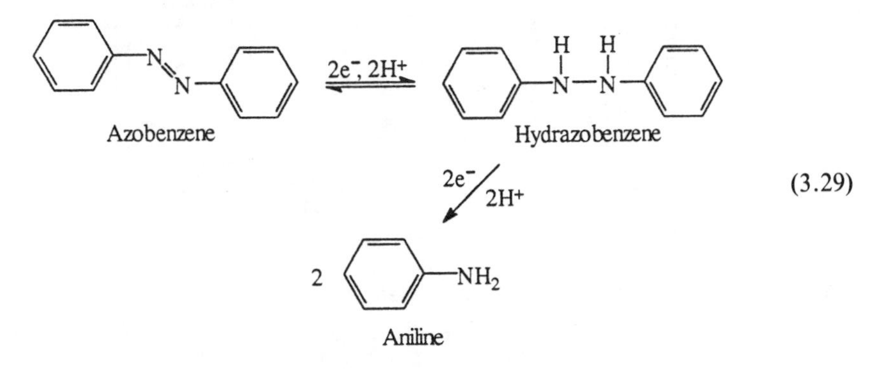

(3.29)

Because the hydrazobenzene is more labile than the parent azo compound, its formation is generally not detected and complete reductive cleavage of the azo linkage occurs, resulting in the formation of aromatic amines.

Concern over the environmental fate of aromatic azo compounds arises primarily from their importance in the textile dye industry. Of the dyes available on the market today, approximately 50% are azo compounds (Kulkarni et al., 1985). It is estimated that approximately 12% of the synthetic textile dyes used each year are lost to waste streams during manufacturing and processing operations, and 20% of these losses will enter the environment through effluents from wastewater treatment plants (Clarke and Anliker, 1980). The reductive cleavage of the azo linkage of the dyestuffs in aquatic ecosystems represents a possible pathway for the entry of aromatic amines into these systems. The aromatic amines are considered a hazardous class of compounds because of their mutagenic and carcinogenic properties.

Although the reductive cleavage of the azo linkage of aromatic azo compounds in anaerobic biological systems is well documented (Walker, 1970; Chung et al., 1978), relatively few studies on the fate of these compounds in natural aquatic ecosystems have been reported. Studies in this area have focused primarily on the reduction of aromatic azo compounds in anaerobic sediment-water slurries. Typically, the reduction kinetics of aromatic azo compounds in these systems are fast, with half-lives ranging from several minutes to several days. A plot of concentration versus time for reduction of azobenzene and formation of aniline in a pond sediment is illustrated in Figure 3.4 (Weber and Wolfe, 1987).

As predicted, the initial rate of formation of aniline was twice the disappearance rate of azobenzene (2 molecules of aniline are formed for every molecule of azobenzene that is reduced). However, the apparent rate of formation of aniline falls off with time, due to reactions of aniline with the sediment-associated organic matter. In what appears to be a general phenomenon for reductive transformations, no reduction of azobenzene was observed in the filtrate of the sediment-water system, sug-

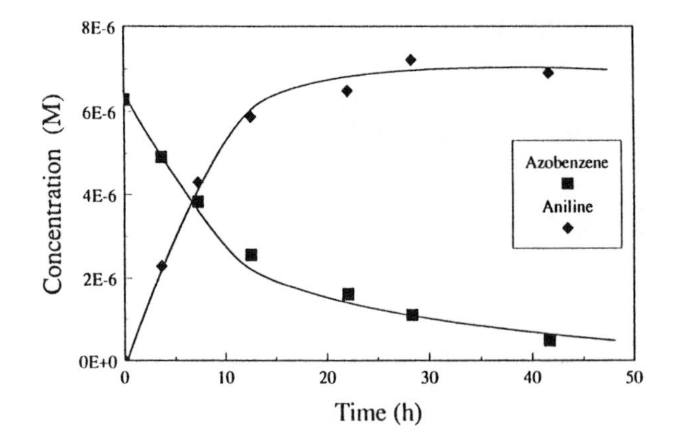

Figure 3.4. Rate of disappearance of azobenzene and the rate of formation from aniline in an anaerobic sediment-water system (pH 6.5, 5% solids). (Reprinted by permission from Pergamon Journals Ltd.).

gesting that the reductants are in some way associated with the sediment. Either the reductant(s) are bound to the sediment surface or the sediment acts as a source; that is, the sediment releases short lived reductants into the aqueous phase, where reduction takes place. Furthermore, reactivity observed in chemically sterilized sediment-water systems (i.e., the microbial activity has been inhibited) suggests that the reduction of aromatic azo compounds has a significant abiotic component.

More recently, the study of the reduction of aromatic azo compounds in anaerobic sediment-water systems has been extended to the azo dyes, particularly those for which there is concern that reductive cleavage of the azo linkage will result in the formation of mutagenic and carcinogenic aromatic amines. For example, reductive cleavage of the bisazo linkages of Direct Red 28 in anoxic sediments would result in the release of benzidine into these systems.

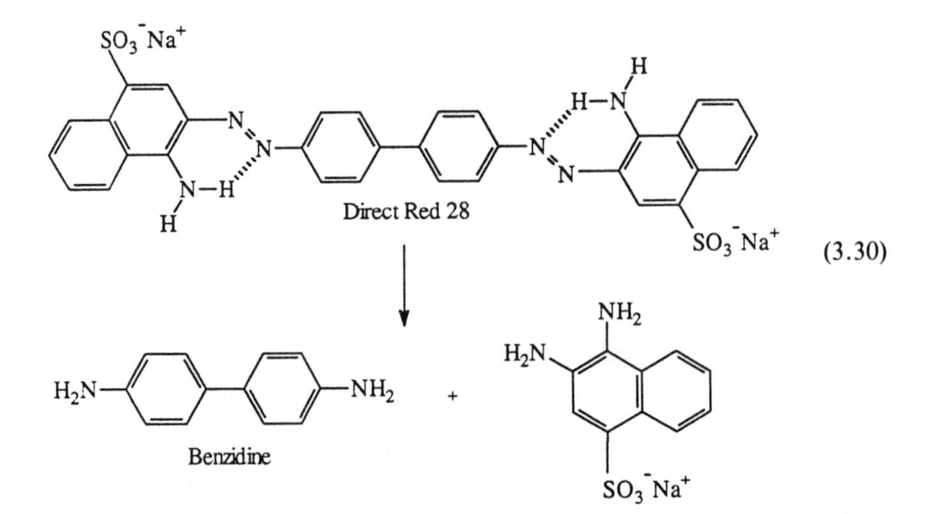

(3.30)

The carcinogenicity and mutagenicity of benzidine is well documented (Ames et al., 1975). Although reduction of Direct Red 28 has been observed in anoxic sediment-water systems, the accumulation of benzidine did not occur because secondary reactions result in its irreversible binding to the sediment (Weber, 1991). It is interesting to note that though Direct Red 28 has considerable water solubility, it is sorbed significantly to the sediment. The enhancement of sorption observed with decreasing pH and the addition of inorganic salts is consistent with an anion-adsorption mechanism. Studies of pH-amended sediment-water systems demonstrated that degradation was inhibited when the dyes were strongly bound to the sediment.

The reduction of several water-insoluble disperse azo dyes in anaerobic sediment-water systems also has been reported. For example, reduction of Disperse Blue 79, the largest volume textile dye on the market today, was reduced in anoxic sediment slurries with half-lives on the order of 1 to 5 hours (Weber, 1988). Reduction appears to occur initially across the azo linkage and results in formation of 2-bromo-4,6-dinitroaniline (BDNA) and the substituted 1,4-phenylenediamine, (A) (Figure 3.5).

BDNA is formed as a transient intermediate which, as previously described, undergoes facile reduction to the diamino compound (B), which is reduced at a much slower rate to the triaminobenzene (C). The substituted 1,4-phenylenediamine (A) undergoes a cyclization through the intramolecular attack of the 4-amino group on the acetamide to form the unstable intermediate (D), which loses H_2O to give the benzimidazole (E). All of these reaction products have considerably greater water solubility than the parent dye. The potential for the release of these reduction products into the water column, where they may have greater stability, does exist.

Product studies of Disperse Red 1, another large volume textile dye, in anoxic sediment slurries demonstrates that reduction of nitro group substituents can occur prior to the reduction of the azo linkage (Equation 3.31) (Yen et al., 1991).

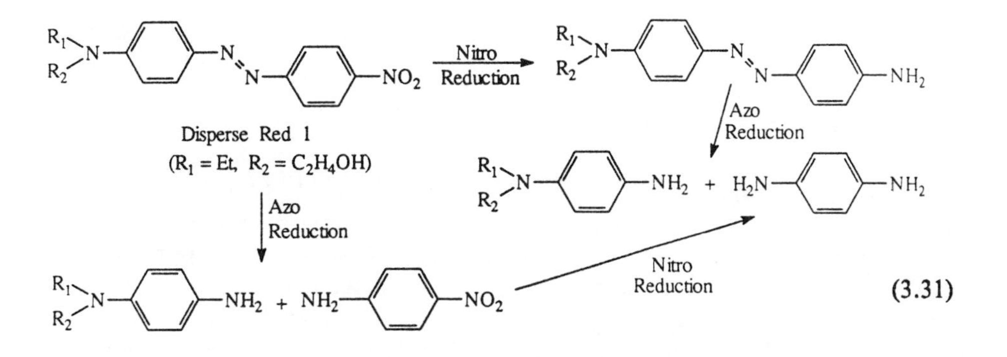

(3.31)

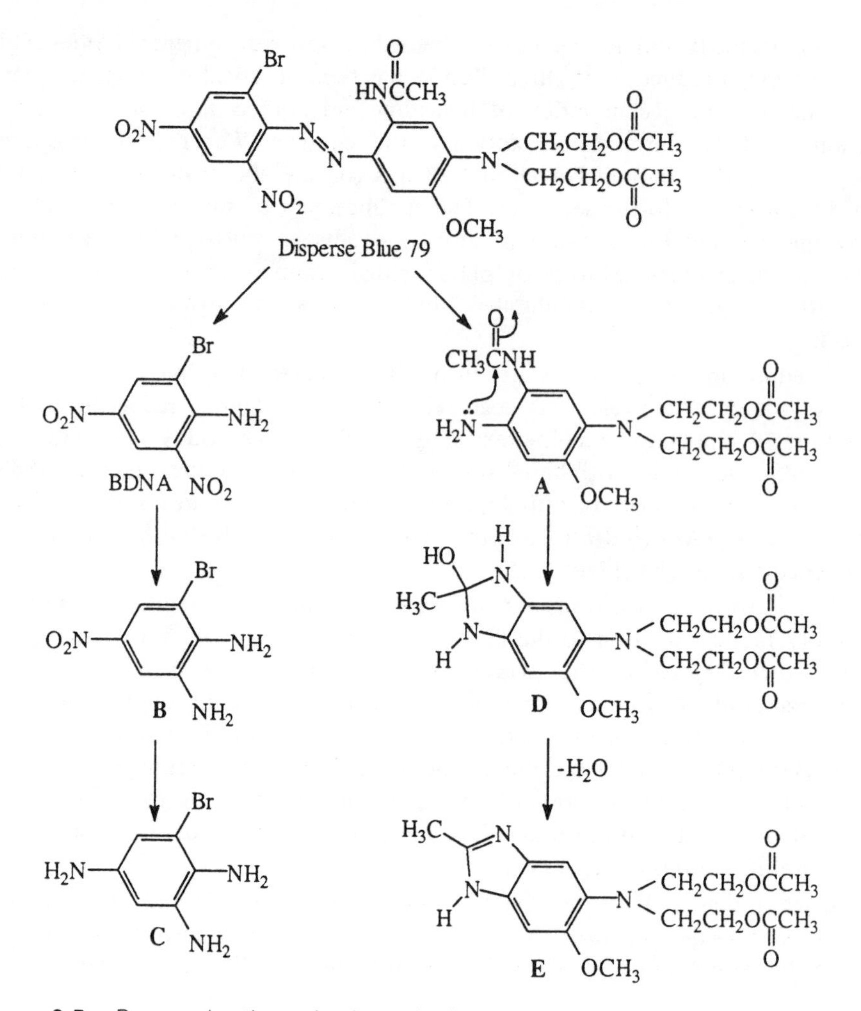

Figure 3.5. Proposed pathway for the reductive transformation of Disperse Blue 79 in anaerobic sediments.

4. N-Nitrosoamine Reduction

The N-nitrosoamines are an important class of chemicals whose chemistry has been studied extensively due to their mutagenic and carcinogenic properties (Preussmann and Stewart, 1984). Much of this work has focused on their formation and occurrence in foodstuffs. A number of studies have suggested that the formation of N-nitrosoamines may occur in crop soils that have been treated with agrochemicals containing secondary nitrogens (e.g, atrazine) and heavy applications of nitrogen fertilizers (Ayanaba et al., 1973; Kearney et al., 1977; Oliver and Kontson, 1978). Furthermore, N-nitrosoamines have been detected as components of large volume agrochemicals that were presumably formed during the manufacturing process. Several N-nitrosoamines also have important uses in industrial processes. For example, N-nitrosodiphenylamine is widely utilized in the rubber industry to retard vulcanization.

Reduction of N-nitrosoamines can occur across either the N-N bond (Equation 3.32) or the N–O bond (Equation 3.33).

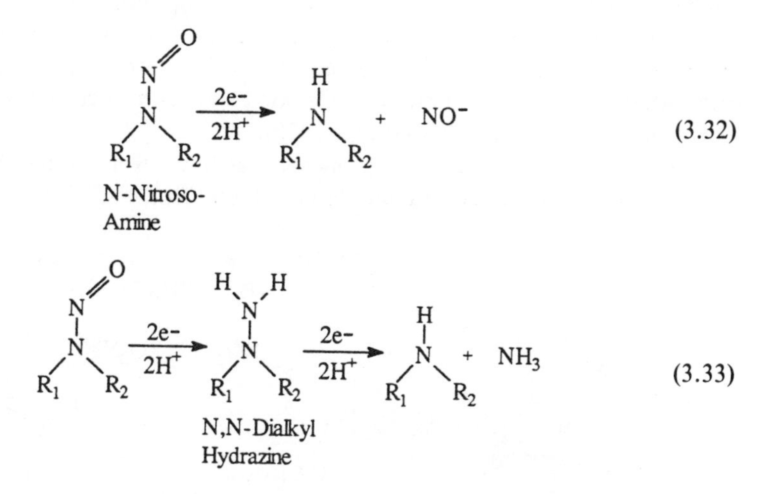

Reduction of the N–N bond affords the parent amine and nitrous oxide. Reduction of the N–O bond gives the unsymmetrical hydrazine, which can be further reduced to give the parent amine and ammonia. Because both reaction pathways ultimately result in the formation of the parent amine, it can be difficult to discern the reaction mechanism for reduction. Reduction of N-nitrosoamines to the parent amine is generally considered a detoxification pathway for nitrosoamines; however, formation of the hydrazine has been suggested to be a possible pathway for bioactivation (Tatsumi et al., 1983).

The existing database concerning the behavior of N-nitrosoamines in reducing environments is somewhat limited. Chemical reduction or reduction using electrochemical means results in mixtures of the hydrazine and the parent amine (Entwistle et al., 1982). Presumably, reduction of the more polar N–O bond occurs initially in these systems, giving the hydrazine, followed by reduction to the parent amine (Equation 3.33). The reductive metabolism of N-nitrosoamines also has been reported. N-Nitrosodiphenylamine was reduced by guinea pig liver supernatant to give 1,1-diphenylhydrazine (Tatsumi et al., 1983). The formation of diphenylamine also was observed; it was concluded from control studies, however, that its formation was due to an abiotic process.

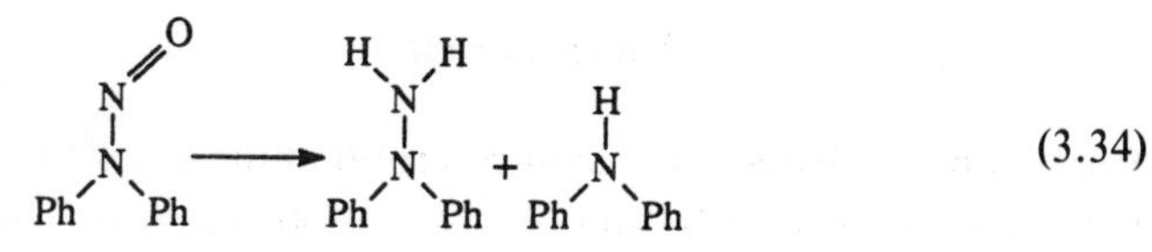

In contrast to this result, the degradation of several nitrosoamines in bacteria gave the parent secondary amine and nitrite ion (Rowland and Grasso, 1975).

The existing data for the reduction of N-nitrosoamines in anaerobic sediments and soils suggests that the aromatic substrates will be much more reactive. For example, N-nitrosodiethylamine and N-nitrosodisopropylamine were found to be very persistent in sediment slurries. In contrast, the reduction of N-nitrosodiphenylamine was quite fast ($t_{1/2}$ = 1.5 days), resulting in the formation of diphenylamine (Weber, 1988). Formation of the intermediate, 1,1-diphenylhydrazine, was not observed. Several N-nitrosodialkylamines also have been found to be very persistent in flooded soils and microbial enrichments from bog sediments (Tate and Alexander, 1975).

Examples of nitrosated agrochemicals that have been studied in environmental systems include N-nitrosoatrazine, N-nitrosobutralin, and N-nitrosopendimethalin.

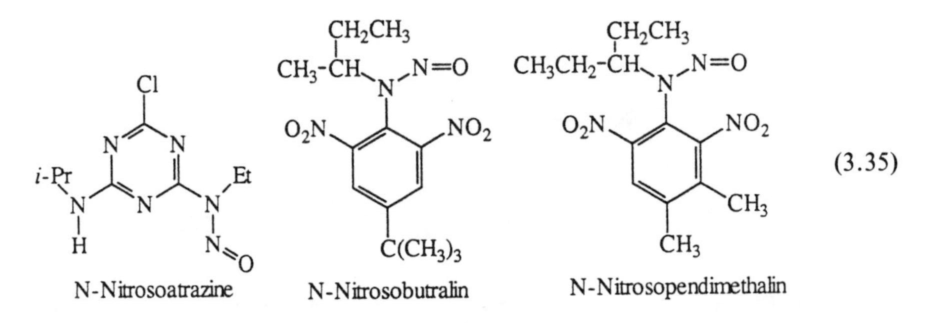

(3.35)

Reduction of N-nitrosoatrazine to the parent compound, atrazine, appears to occur readily in soils; thus, buildup of this compound is not likely to occur (Kearney et al., 1977). Similar studies, however, suggest that N-nitrosobutalin will be quite persistent in soils (Oliver and Konston, 1978). Reduction of N-nitrosopendimethalin has been observed in anaerobic soils (Smith et al., 1979). Product studies indicate, however, that the nitroso moiety remains intact and reduction of the aromatic nitro group in the 6-position occurs to give the aromatic amine (Equation 3.36).

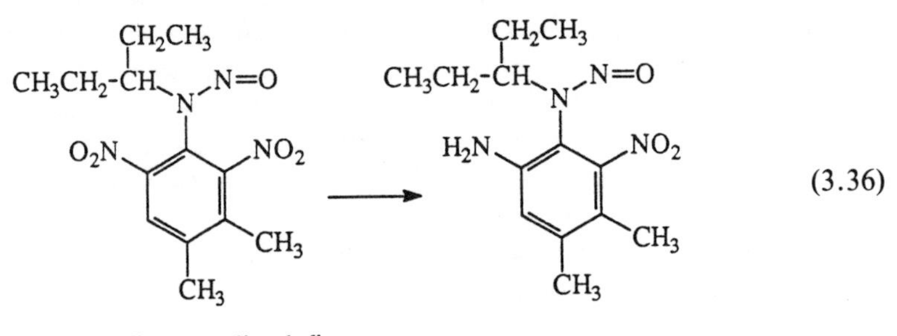

(3.36)

Presumably, reduction of the nitro group in the 6-position is favored because of the bulky methyl group adjacent to the nitro group in the 2-position.

5. Sulfoxide Reduction

The reduction of sulfoxides is a two-electron transfer process that results in the formation of thioethers.

$$R_1-\overset{\overset{O}{\|}}{S}-R_2 + 2e^- + 2H^+ \rightleftharpoons R_1-S-R_2 + H_2O \quad (3.37)$$

Because the reduction of sulfoxides to thioethers is reversible, reoxidation of the thioether to the sulfoxide may occur in aerobic environments. Although oxidation of sulfoxides to sulfones (RSO_2R) has been observed in aerobic systems, no evidence exists that suggests reduction of the sulfone functional group to the sulfoxide moiety will occur in anaerobic systems. The chemical reduction of sulfones in the laboratory also is known to be quite difficult. Reduction of organic sulfoxides has been observed in anaerobic soils (Walter-Echols and Lichtenstein, 1977) and sediments (Tratnyek and Wolfe, 1988), flooded paddy soil (Tomizawa, 1975), groundwaters (Lightfoot et al., 1987; Miles and Delfino, 1985) and microbial systems (Timms and MacRae, 1983). The existing data suggest that both abiotic and microbial processes contribute to the reduction of these compounds in natural environmental systems.

The environmental significance of organic sulfoxides results primarily from their importance as agrochemicals; specifically, the organophosphorus and carbamate insecticides containing sulfoxide moieties. Studies suggest that reduction of these chemicals can be the primary pathway for their transformation in anoxic sediments. For example, reduction of phorate sulfoxide to phorate in anoxic pond sediments occurred with half-lives ranging from 2 to 41 days (Equation 3.38) (Tratnyek and Wolfe, 1988).

$$\underset{\text{Phorate Sulfoxide}}{(EtO)_2\overset{\overset{S}{\|}}{P}SCH_2\overset{\overset{O}{\|}}{S}Et} \longrightarrow \underset{\text{Phorate}}{(EtO)_2\overset{\overset{S}{\|}}{P}SCH_2SEt} \quad (3.38)$$

Oxidation of phorate back to phorate sulfoxide was not observed in these systems. Because only a small portion of phorate sulfoxide was sorbed to the sediment, sorption did not have a significant effect on the reduction kinetics. As is often the case, removal of the sediment phase by filtration inhibited reduction, suggesting that the reducing agent was sediment associated. Studies in heat-sterilized and amended sediments suggested that the reduction of phorate sulfoxide was due to extracellular soil enzymes or microbial cometabolism.

The reduction of aldicarb sulfoxide, which is the major metabolite in aerobic soils, has been observed in groundwater systems (Miles and Delfino, 1985; Lightfoot et al., 1987).

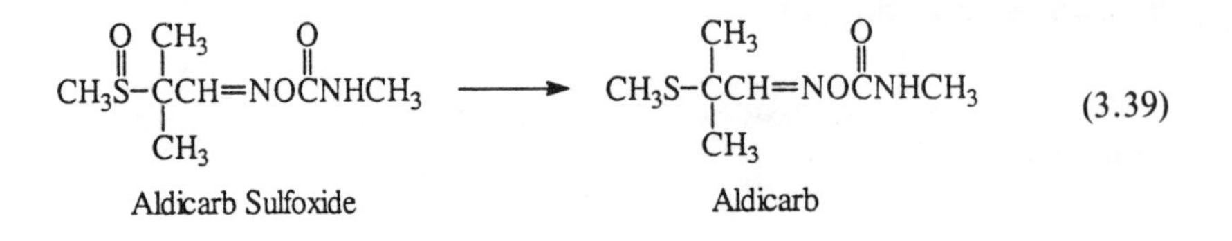

(3.39)

An understanding of the factors that control the redox equilibria for aldicarb and aldicarb-sulfone is important because hydrolysis of aldicarb-sulfoxide will occur at a much faster rate than hydrolysis of aldicarb.

6. Quinone Reduction

Quinones are reduced by a one-electron transfer reaction to the semiquinone radical:

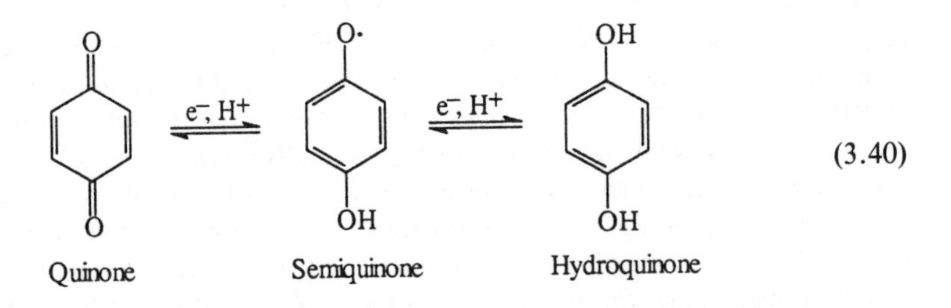

(3.40)

Transfer of the second electron results in formation of the hydroquinone. Because this reaction is readily reversible, the equilibrium between the quinone, semiquinone, and hydroquinone will be dependent on the reduction potential of the system of interest. Few studies of quinone-containing pollutants in natural reducing systems exist. Studies of quinones have focused primarily on elucidating their role as electron "mediators" in the reduction of other pollutants of interest (see Section 3.C.2 for further discussion of electron-mediated reductions) (Schwarzenbach, et al., 1990; Tratnyek and Macalady, 1989). Quinoid-type compounds are thought to be constituents of natural organic matter (Thurman, 1985; see Chapter 1.B.3c). It has been hypothesized that some free radicals in humic substances are quinone-hydroquinone redox couples (Tollin et al., 1963; Steelink and Tollin, 1967).

The use of quinone and quinone-like chemicals as redox indicators has provided insight into the reactivity of these compounds in anoxic sediment slurries (Tratnyek and Wolfe, 1990; ZoBell, 1946). For example, indophenols, such as 2,6-dichloroindophenol, were found to undergo rapid reduction (half-lives measured in minutes) in each of a number of anoxic sediment slurries (Equation 3.41).

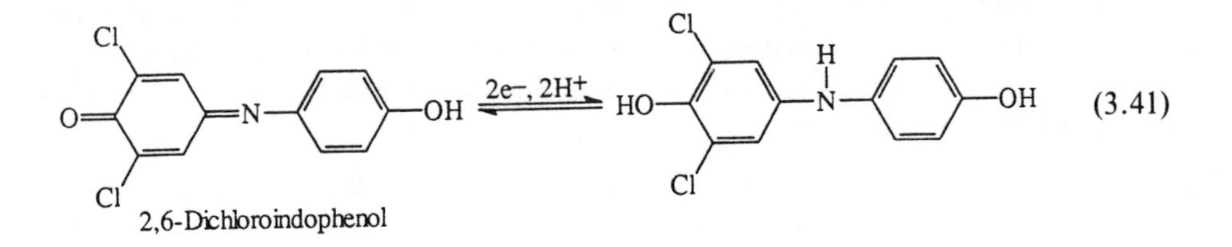

By contrast, in the same sediment slurries, sulfonated indigoes, such as indigo-5,5′-disulfonate, were found to be much less reactive, with half-lives of one or more days (Equation 3.42). For a series of such redox indicators, it was found that structure type was the primary factor influencing reaction rates, and that reduction potentials of the indicators would not be useful for predicting reactivity in reducing sediment-water systems.

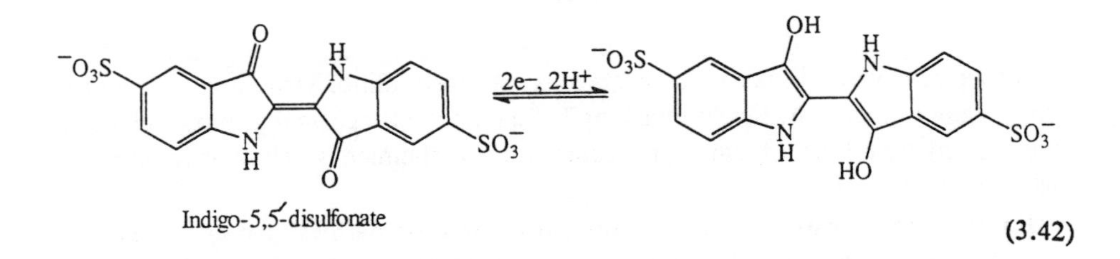

Regarding the significance of quinones as environmental pollutants, the anthraquinone dyes are of greatest concern because of their heavy use in the textile industry for dyeing fibers. A number of anthraquinone dyes have been detected in rivers receiving wastewater of textile dying operations (Tincher, 1978). Studies of several disperse anthraquinone dyes indicate that their reductive transformation will occur in anoxic sediment slurries (Baughman et al., 1992). For example, half-lives for the reduction of Disperse Red 9 in pond sediment ranged from 1 to 10 days. Although the possibility exists that these dyes exist in their hydroquinone form in reducing environments, product studies of Disperse Red 9 demonstrate that irreversible reduction of the quinone moiety to the thermodynamically stable anthrones can occur.

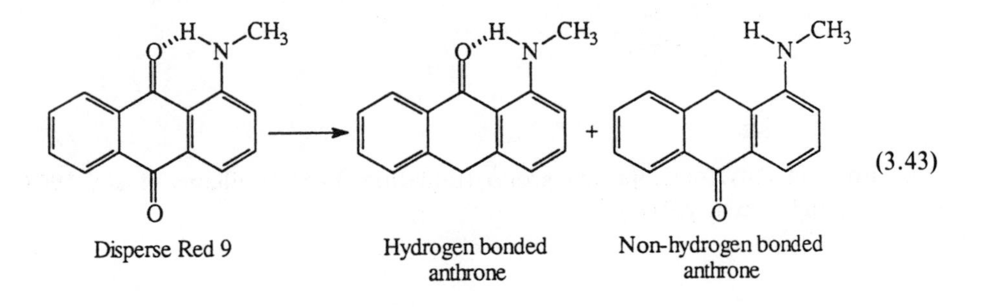

The intramolecularly, hydrogen-bonded anthrone isomer, 1-methylamino-(9,10H)anthraceneone, is formed more slowly, and its stability in the sediment-water system was much greater than the other nonhydrogen bonded isomer, 4-methylamino(9,10H)anthraquinone.

7. Reductive Dealkylation

Reductive dealkylation is the replacement of an alkyl group on a heteroatom with hydrogen.

$$R_1\text{-}X\text{-}R_2 \rightarrow R_1\text{-}XH + R_2H \tag{3.44}$$
$$X = NH, O, \text{ or } S$$

Although dealkylation may be considered a relatively minor structural change for many organic chemicals, the "unmasking" of a reactive functional group, such as an amino or hydroxyl group, can significantly alter a chemical's behavior in environmental systems.

Dealkylation appears to be a common transformation pathway for agrochemicals with the requisite functionality; however, the reaction mechanisms for this process have not been thoroughly investigated. The lack of reactivity often observed in sterile systems has led to the general conclusion that reductive dealkylation requires the presence of viable microorganisms. Examples of N-dealkylation include the dealkylation of carbofuran (Equation 3.45) (Hassall, 1982) and atrazine, which may

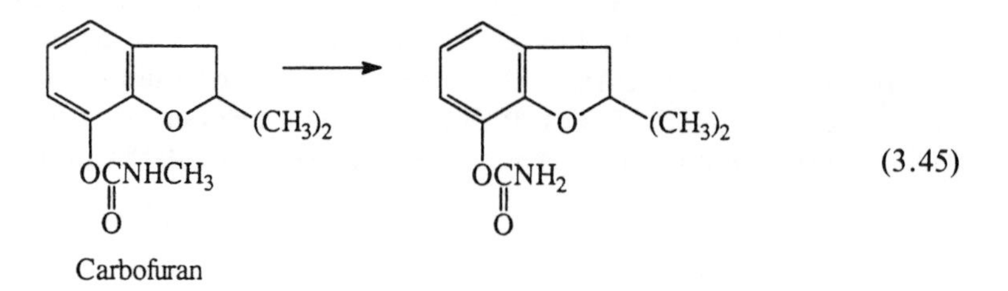

(3.45)

lose either an ethyl or isopropyl group (Equation 3.46) (Williams et al., 1968; Goswami and Green, 1971).

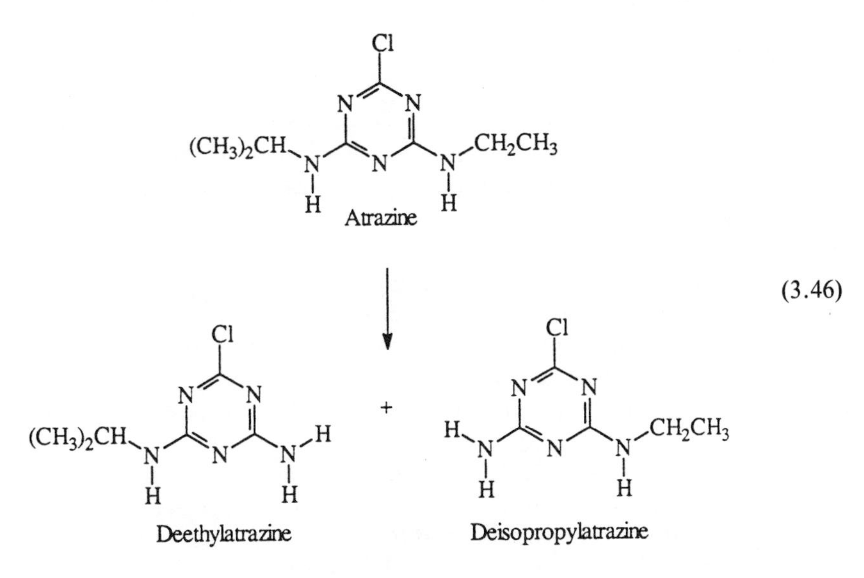

The dealkylation of methoxychlor, which is an example of O-dealkylation, has also been observed in anaerobic soils (Equation 3.47)

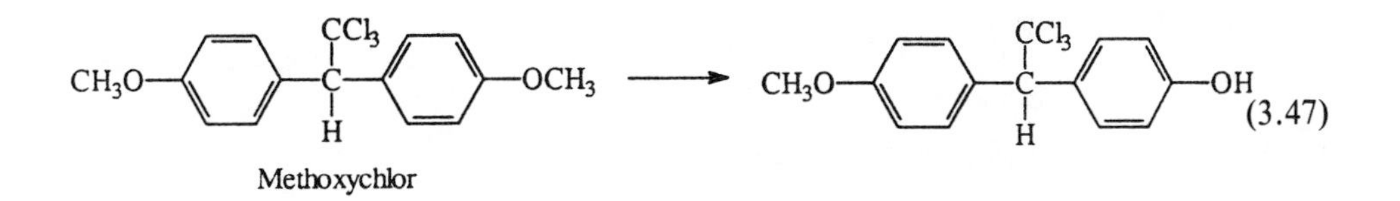

More recently, Peijnenburg et al. (1992a) reported that O-demethylation appears to be the initial transformation pathway for iodoanisoles in anaerobic sediments (Equation 3.48).

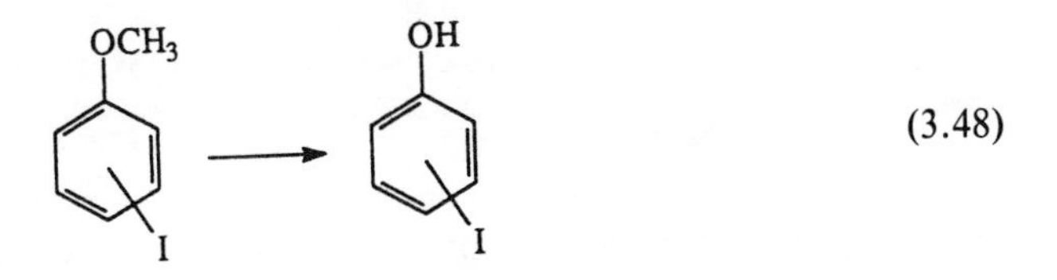

The fast rates of degradation (disappearance half-lives of 1 to 3 days) and the absence of a lag phase observed for loss of the iodoanisoles suggests that this may be an example of an O-demethylation that is occurring through an abiotic process.

C. REDUCTION KINETICS

Although our understanding of the classes of organic chemicals that are subject to reduction in anaerobic environments has been significantly advanced in recent years, our ability to predict reaction rates in these complex environments is very limited. The complexity of these environmental systems and the inability to identify the reductants may provide insurmountable barriers to the development of predictive models. Although a variety of reducing agents have been tested in model experiments designed to mirror environmental processes, none has been adequately identified as an important agent of reducing activity in natural systems. Reduced transition metal ions (particularly iron[II] in various complexes), iron-sulfur proteins, sulfides, humic materials, and polyphenols have all been suggested as potential reductants.

The development of kinetic models to describe the behavior of organic pollutants in reducing environments is further complicated by the fact that reduction probably occurs by a combination of abiotic and biotic processes, depending on the reaction system of interest. To some extent, a kinetic distinction can be made between biological processes, which usually display a lag period as the growth of microorganisms capable of degrading a substrate builds up; abiotic chemical reductions usually occur at a more constant rate. Furthermore, biodegradation usually shows an optimum temperature, whereas chemical reactions normally proceed more rapidly as temperature increases.

1. One-Electron Transfer Scheme

The reductive transformation of organic functional groups is often described as occurring through the transfer of electron pairs. As a result, the chemistry of radical ions, products of one-electron transformations of neutral organic compounds, is often ignored. On the contrary, the accepted theory is that electrons are transferred one by one (Semenov, 1958). The general mechanism for one-electron transfer, as described by Schwarzenbach and Gschwend (1990), is:

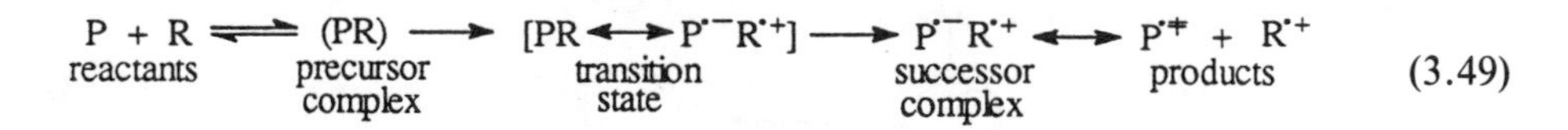

(3.49)

where P is an organic pollutant (the electron acceptor) and R is a reductant (the electron donor). The precursor complex, (PR), describes the electron coupling prior to the one-electron transfer, which occurs in the transition state. After electron transfer, a successor complex is proposed that subsequently dissociates to provide the radical ions. There are two well-established general mechanisms for electron transfer processes; the *outer-sphere mechanism* and the *inner-sphere mechanism* (Eberson, 1987). These mechanisms are differentiated by the degree to which the oxidant and reductant are associated in the transition state for electron transfer. In the inner-sphere mechanism there is intimate contact between the oxidant and reductant in the transition state. An atom or ligand is bonded by both the reductant and oxidant, which acts as a bridge for electron transfer. In contrast, for outer sphere electron transfer there is no sharing of an atom or ligand between the oxidant and reductant in the transition state. As a result, reactivity for electron transfer occurring through the outer-sphere mechanism is more likely to reflect the ease of electron gain or loss. For inner-sphere mechanisms, reactivity is expected to be dependent on chelating groups or atoms. Because the terms "outer-sphere" and "inner-sphere" mechanisms are not directly applicable to electron transfer reactions between organic chemicals, it has been proposed to replace these terms with "non-bonded" and "bonded" (Littler, 1970). With respect to the reduction of organic chemicals, electron transfer by the outer-sphere or "non-bonded" mechanism is probably of greater significance.

The concept of the one-electron transfer scheme provides a common feature to the seemingly unrelated reductive transformation pathways presented in Section 3.B. Each of these processes occurs initially by transfer of a single electron in the rate-determining step. In each case, a radical anion is formed that is more susceptible to reduction than the parent compound.

2. Structure Reactivity Relationships for Reductive Transformations

Although it is not possible to predict the absolute rates for the reduction of organic chemicals in environmental systems, recent laboratory studies have established relationships between substrate properties and relative reactivities (Tratnyek et al., 1991). For example, Schwarzenbach et al. (1990) studied the quinone and iron porphyrin mediated reduction of a series of nitrobenzenes and nitrophenols (see Section 3.C.2 for further discussion of electron mediated reductions). From the experimentally determined rate law, it was determined that the transfer of the first electron from the monophenolate species of the hydroquinone, lawsone ($HLAW^-$), to the nitroaromatic was rate determining (Equation 3.50).

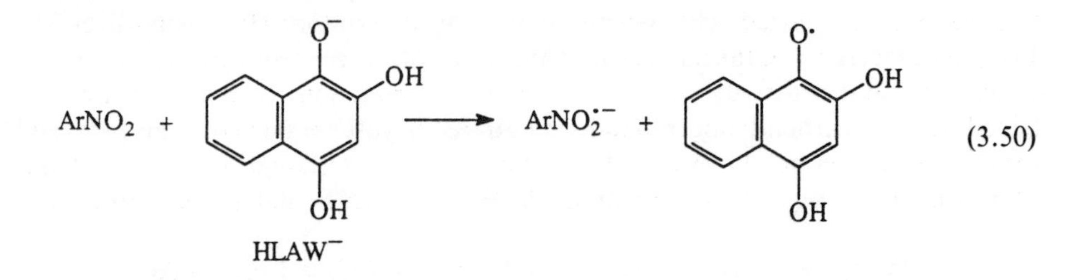

(3.50)

The one-electron reduction potentials for the nitroaromatics, which are related to the "willingness" of the nitroaromatic to accept an electron, correlated with the second-order rate constants for the reduction by $HLAW^-$ (Figure 3.6). These investi-

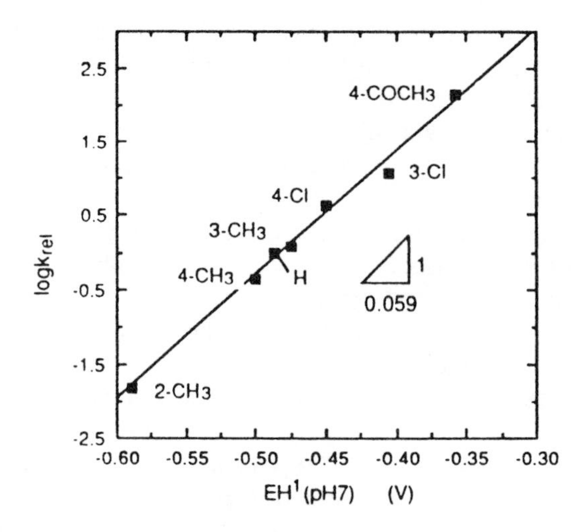

Figure 3.6. Relationship between relative reaction rates and one-electron potentials for the reaction of a series of substituted nitrobenzenes with HLAW. From Schwarzenbach and Gschwend (1990). Reprinted by permission of John Wiley and Sons.

gators concluded that one-electron reduction potentials could be used to predict relative reaction rates in well-defined reducing systems.

Hammett σ constants also have been successfully correlated with reduction rate constants. For example, the reduction rate constants of a series of 4-substituted nitrobenzenes in anaerobic sediment-water systems and the Hammett σ constants exhibited a positive correlation (Delgado and Wolfe, 1992). The slopes of the linear-free energy plots for a river sediment, pond sediment, and aquifer material were similar, suggesting that reduction was occurring through the same mechanism in each of these systems.

The disappearance rate constants for a large number of halogenated hydrocarbons, which include halogenated alkanes and alkenes, in anoxic sediment-water systems were correlated with several molecular descriptors (Peijnenburg et al., 1991). The reactivity of the halogenated hydrocarbons in the sediment-water systems varied over four orders of magnitude. The best correlation for reactivity was obtained with the carbon-halogen bond strength (BS), Taft's sigma constant (σ^*), and the carbon-carbon bond energy (BE) (Equation 3.51). Although reductive dehalogenation was occurring through both reductive dehalogenation and

$$\log k_{cal} = -0.142 \times BS + 0.483 \, x\sigma^* + 0.039 \times BE + 3.03 \qquad (3.51)$$

elimination, the successful correlation with the molecular descriptors is consistent with the idea that the reduction processes involve an initial rate-determining electron transfer step. Figure 3.7 illustrates the good agreement between k_{cal} (calculated from

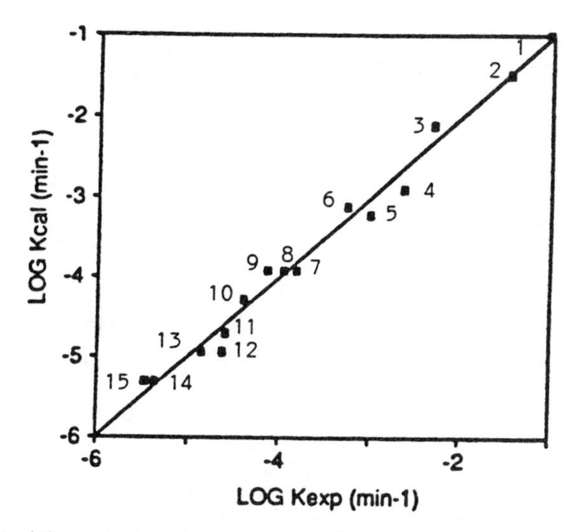

Figure 3.7. Plot of the calculated rate constant (k_{cal}) using Equation 3.51 versus the measured rate constant (k_{exp} for the reductive dehalogenation of a number of halogenated aliphatics: (1) I_2CHCHI_2, (2) ICH_2CH_2I, (3) BrClCHCHBrCl, (4) Cl_3CCCl_3, (5) $CH_3CHBrCHBrCH_3$, (6) $BrCH_2CH_2Br$, (7) γ-Hexachlorocyclohexane, (8) CCl_4, (9) Cl_2CHCH_2Cl, (10) $Cl_2C=CCl_2$, (11) α-Hexachlorocyclohexane, (12) $ClCH_2CH_2Cl$, (13) $ClCH=CH_2$, (14) *trans*-ClCH = CHCl, (15) *cis*-ClCH = CHCl. (Reprinted with permission from Elsevier Science Publishers).

Equation 3.51) and experimentally determined rate constants (k_{exp}) for the series of halogenated alkanes and alkenes.

3. Electron-Mediated Reductions

Numerous studies have demonstrated that the facile reduction of organic pollutants such as nitroaromatics, azo compounds, and polyhalogenated hydrocarbons can occur in anaerobic environments with exceedingly fast reaction rates. The most abundant electron donors in these systems include reduced iron and sulfur species. Laboratory studies, however, indicate that the rates of reduction by these bulk reductants are much slower than those observed in natural reducing systems. The addition of electron carriers or mediators to the laboratory systems is found to greatly accelerate the measured reduction rates. An electron shuttle system has been postulated to account for the enhanced reactivity observed in these laboratory studies and presumably in reducing natural systems (Dunnivant et al., 1992).

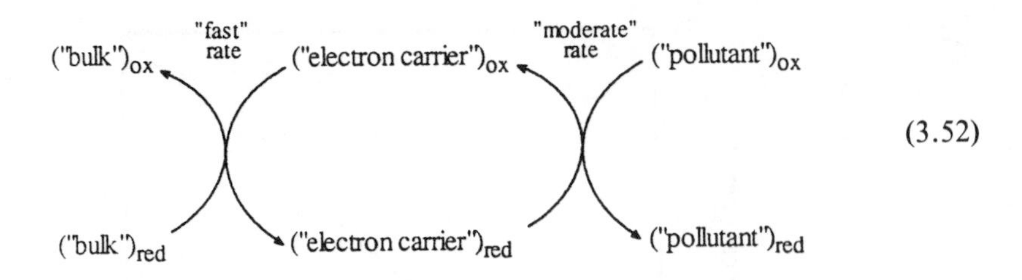

(3.52)

In this scenario, the bulk electron donor or reductant rapidly reduces an electron carrier or mediator, which in turn transfers electrons to the pollutant of interest. The oxidized electron mediator is then rapidly reduced again by the bulk reductant, which enables the redox cycle to continue.

Several chemical substances have been proposed as electron mediators including natural organic matter (NOM) (Dunnivant et al., 1992), iron porphyrins (Baxter, 1990; Holmstead, 1976; Khalifa et al., 1976; Quirke et al., 1979; Klecka and Gonsior, 1984; Wade and Castro, 1973), corrinoids (Krone et al., 1989) and bacterial transition-metal coenzymes such as vitamin B_{12} and hematin (Gantzer and Wackett, 1991). Chemical structures of several of the proposed electron-mediators are illustrated in Figure 3.8.

a. Natural Organic Matter

The role that NOM may play as an electron mediator in the reduction of organic pollutants is probably of greatest significance because of its ubiquitous nature in the environment. The electron-mediating properties of NOM are believed to be associated with the iron complexes and quinoid-type moieties known to occur in NOM (Thurman, 1985; Buffle and Altmann, 1987). Model studies have demonstrated that quinones can rapidly reduce nitro aromatic agrochemicals (Tratnyek and Macalady, 1989). Although limited data sets currently exist, the available information suggests that the ability of NOM to mediate reductions is independent of its source. For example, NOM from a variety of sources, including anaerobic groundwaters, streams, landfill leachates, and extracts from tree materials, has been shown to mediate the reduction of aromatic nitro compounds in aqueous solution containing H_2S as the bulk electron donor (Dunnivant et al., 1992). The carbon-normalized second-order rate constants for k_{NOM} (h-mg C/L)$^{-1}$ for the reduction of 3-chloronitrobenzene in a number of different NOM aqueous solutions are summarized in Figure 3.9. The enhanced reactivity of the NOM isolated from walnut trees was attributed to its high quinone content.

b. Mineral Systems

Recent laboratory studies suggest that electron-mediated reductions may not be limited to environmental systems containing NOM. Mineral systems found in groundwaters, such as iron-bearing phyllosilicates, iron sulfides, and sulfide miner-

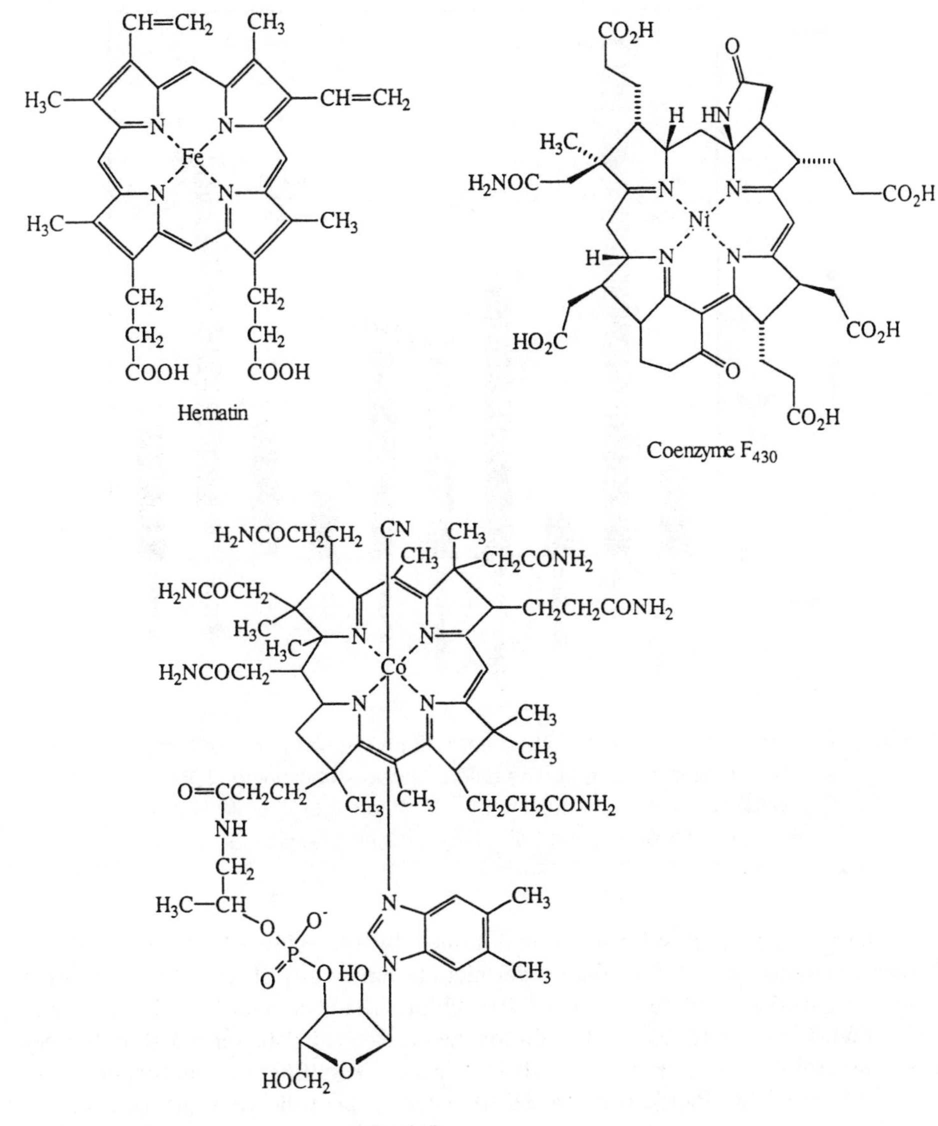

Figure 3.8. Chemical structures of biomolecules that behave as electron-mediators.

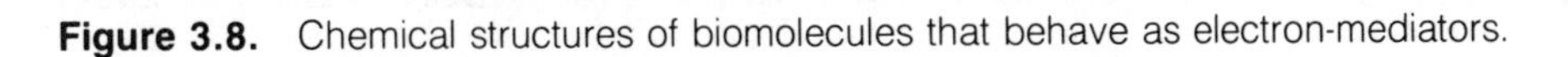

als also may participate in electron-mediated reactions. The heterogeneous oxidation of ferrous iron present in sheet silicates is described by:

$$[Fe^{+2},nM^{+}]_{mineral} \rightarrow [Fe^{+3},(n-1)M^{+}]_{mineral} + M^{+}_{aq} + e^{-} \qquad (3.53)$$

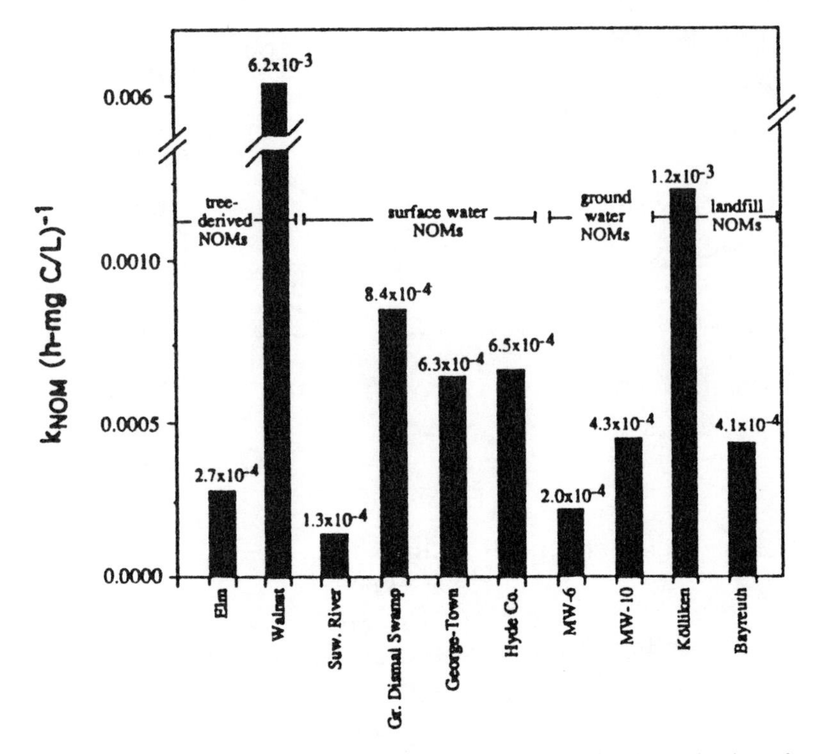

Figure 3.9. Carbon-normalized second-order rate constants for the reduction of 3-chloronitrobenzene in nine different waters containing NOM. Experimental conditions were: 1:1 dilution of NOM waters, pH 7, 5mM H_2S, and 25°C. Reprinted by permission from the American Chemical Society.

The charge balance in solution is maintained by the release of a cation (M^+). Mineral systems such as biotite and vermiculite have been observed in laboratory studies to mediate the reduction of hexachloroethane and carbon tetrachloride (Kriegman-King, 1991, 1992). The heterogeneous electron transfer reaction between hexachloroethane (HCA) and the sheet silicates, resulting in the formation of tetrachloroethylene (PCE), is proposed to occur by the following mechanism:

$$2[Fe^{+3},nM^+]_{mineral} + HCA \rightarrow 2[Fe^{+3},(n-1)M^+] + PCE + 2Cl^- + 2M^+ \qquad (3.54)$$

Although the reduction of halogenated aliphatics by mineral systems is generally not fast enough to be considered environmentally significant, the addition of HS^- to the mineral systems greatly accelerated the reductive dehalogenation of hexachloroethane and carbon tetrachloride. Furthermore, the reaction of HS^- with the halogenated aliphatics in homogeneous solution was quite slow, suggesting that reaction with HS^- in the sulfide-mineral systems is heterogeneous. One possible mechanism to account for the influence of HS^- on reaction rates involves the regeneration of

Fe^{+2} sites on the mineral surface by HS^- according to the following electron-transfer reaction:

$$2[Fe^{3+},(n\text{-}1)M^+]_{mineral} + 2HS^- + 2M^+ \rightarrow HSSH + 2[Fe^{2+},nM^+]_{mineral} \quad (3.55)$$

Another possible mechanism involves the reaction of HS^- with ferrous iron that has dissolved from the minerals to form an iron sulfide precipitate, which can then act as an electron donor.

Nitroaromatic reduction has also been demonstrated to occur in mineral systems (Yu and Bailey, 1992). Laboratory studies indicated that the order of reactivity for the reduction of nitrobenzene by several sulfide minerals was $Na_2S > MnS > ZnS > MoS_2$. This order of reactivity paralleled the solubility of the mineral sulfides, which suggests that reduction rates were dependent on sulfide ion concentrations in solution. The sulfide ions and other sulfur species were hypothesized to play a "bridging role" for the transfer of electrons from the mineral to nitrobenzene.

c. *Microbial-Mediated Reductions*

Electron shuttle systems as described in Equation 3.52 are known to play an important role in the intracellular reduction of organic chemicals in biological systems. Electron shuttle systems also may be involved in the extracellular reduction processes. For example, the electron shuttle scheme has been used to provide a rationale for the reaction kinetics observed for the reduction of a series of azo food dyes in anaerobic cultures of *Proteus vulgaris* (Dubin and Wright, 1975). The reduction kinetics for the dyes were found to be zero-order with respect to dye concentration. Furthermore, reduction rates correlated well with the redox potentials of the dyes. This correlation would not be expected if transport of the dye across the cell membrane was rate-determining. To account for this correlation and the observed zero-order kinetics, a reduction mechanism was proposed that involved an extracellular reducing agent, B/BH_2 (e.g., soluble flavin) that acts as an electron shuttle between the dye and cellular-reducing enzymes, E/EH_2.

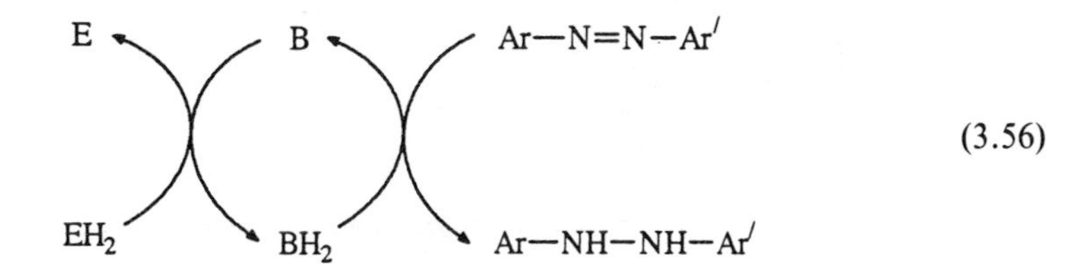

(3.56)

4. Effects of Sorption on Reduction Kinetics

Although molecular descriptors that measure the ability of chemicals to accept electrons (e.g., one-electron reduction potentials) have been used successfully to predict relative reaction rates in well-defined homogeneous systems, their use for predicting rates in more complex heterogeneous systems may be limited. For example, based on the LFER for the reduction of aliphatic halogenated aliphatics illustrated in Figure 3.7, the half-life for the reduction of DDT in an anoxic sediment-water system would be approximately 8 days. In laboratory studies, however, no reduction of DDT was observed in an anoxic sediment-water system over a period of 90 days (Macalady and Wolfe, 1992). Based on DDT's partition coefficient (K_p), it was estimated that >99.99% of DDT was sorbed to the sediment. Sediment sorption also appears to affect the reaction kinetics of substituted nitro aromatics and azobenzenes. The pseudo-first-order rate constants for the reduction of a series of alkylsubstituted nitrobenzenes in a sediment-water system, which should react at similar rates based on LFER considerations, were found to decrease with increasing sorption to the sediment phase (Figure 3.10) (Sanders and Wolfe, 1985). Similarly, rates of reduction for a series of substituted azobenzenes were found to be primarily controlled by the amount of partitioning onto the sediment, with increased partitioning inhibiting the reduction process. After correction for sorption to the sediment, the first-order rate constants for the substituted azobenzenes, with the exception of azobenzene, were found to be quite similar (Figure 3.11: Weber and Wolfe, 1987).

Although these observations suggest that sorption inhibits reduction, removal of the sediment by filtration generally results in loss of reactivity, suggesting that the reducing agent or agents is associated with the solid phase (i.e., the reductants are either part of the sediment or derived from the sediment). To explain this apparent

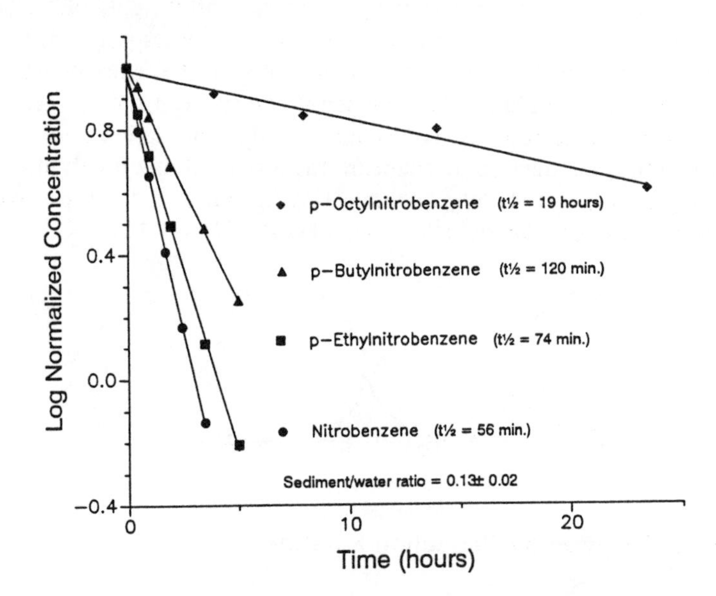

Figure 3.10. Reduction kinetics of a series of 4-alkyl substituted nitrobenzenes in an anaerobic sediment-water system. From Sanders and Wolfe (1985). Reprinted by permission of the American Chemical Society.

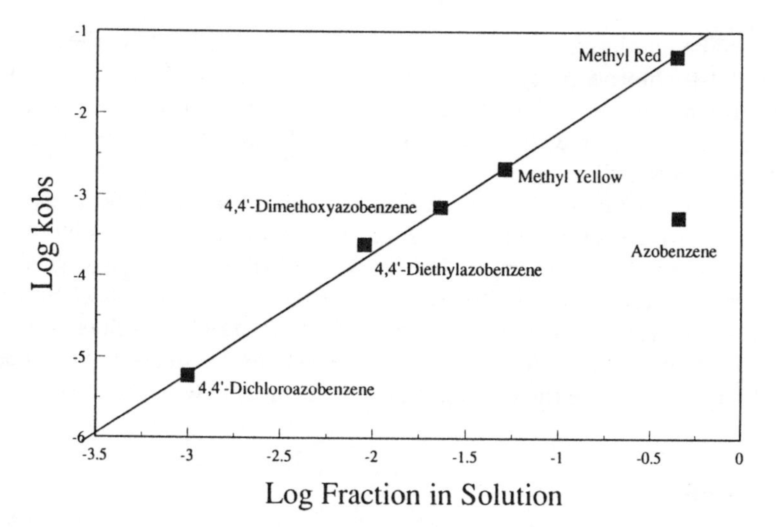

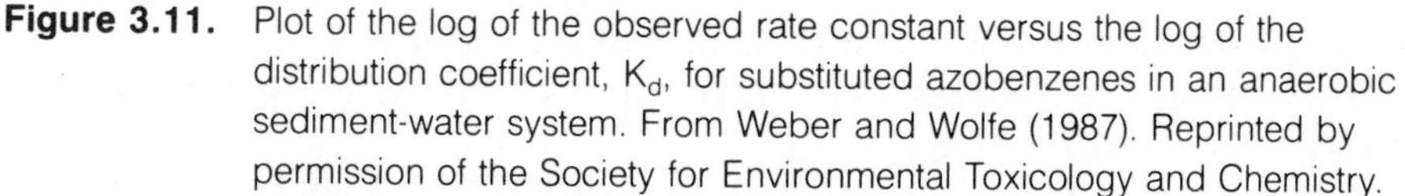

Figure 3.11. Plot of the log of the observed rate constant versus the log of the distribution coefficient, K_d, for substituted azobenzenes in an anaerobic sediment-water system. From Weber and Wolfe (1987). Reprinted by permission of the Society for Environmental Toxicology and Chemistry.

dichotomy, a kinetic model describing the reduction process has been proposed that incorporates a nonreactive sorptive sink and a reactive sink, both of which are associated with the sediment:

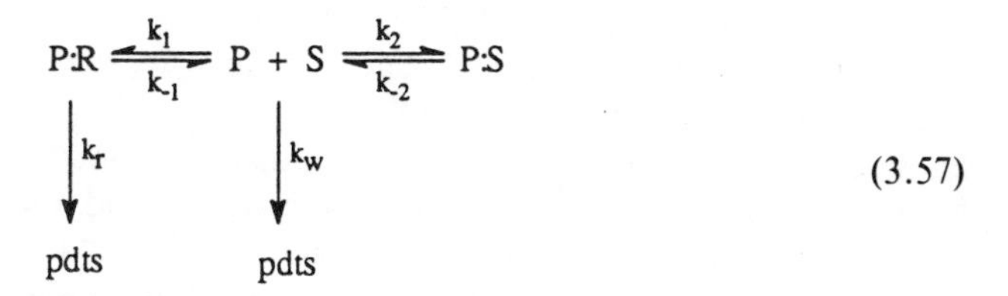

(3.57)

where P is the pollutant in water; S is the sediment; P:S is the sediment-sorbed pollutant (nonreactive sink); k_r is the first-order rate constant for reduction at the reactive site; k_w is the second-order rate constant for reduction in the bulk aqueous phase; k_1 and k_{-1} are the respective sorption-desorption rate constants for partitioning to the nonreactive sink; k_2 and k_{-2} are the respective sorption-desorption rate constants for partitioning to a reactive site; and pdts are products (Weber and Wolfe, 1987). This model describes the scenario in which the pollutant is partitioned between the unreactive and the reactive site, and that partitioning to the unreactive site retards the rate of reduction.

Returning to our initial example of DDT, based on DDT's extremely large sediment-water partition coefficient, this model suggests that DDT is strongly sorbed to the nonreactive sink ($k_2 >> k_{-2}$) and that transport to reactive sites on the sediment will be rate-limiting. By contrast, this model would predict that a chemical such as 1,1,1-trichloroethane, which has the same reactive functional group (R-CCl_3) as DDT but whose sediment-water distribution coefficient is several orders of magnitude smaller, would be much more susceptible to reduction in anoxic sediment-water systems.

Our accumulated knowledge concerning reductive transformations in environmental systems suggests that natural surfaces (e.g., sediments, soils, mineral oxides) play crucial roles in these processes. This is in stark contrast to hydrolysis reactions for which surface effects are limited to a relatively small number of chemicals that have unique functionality. Clearly, the ability to understand, and thus predict reaction rates for reductive transformations in environmental systems will be dependent on our ability to delineate the role that natural surfaces play in the overall process.

REFERENCES

Adhya, T. K., Sudhakar-Bakar and N. Sethunathan. 1981a. Fate of fenitrothion, methyl parathion, and parathion in anoxic sulfur-containing soil systems, *Pestic. Biochem. Phys.* 16:14–20.

Adhya, T. K., Sudhakar-Barik and N. Sethunathan. 1981b. Stability of commercial formulations of fenitrothion, methyl parathion, and parathion in anaerobic soils, *J. Agric. Food Chem.*, 29:90–93.

Ames, B. N., J. McCann and E. Yamasaki. 1975. Methods for detecting carcinogens and mutagens with the *salmonella*-mammalian microsome mutagenicity, *Mutat. Res.*, 31:347-364.

Ayanaba, A., W. Verstraete and M. Alexander. 1973. Formation of dimethylnitrosamine, a carcinogen and mutagen, in soils treated with nitrogen compounds, *Proc. Am. Soil Sci.*, 37:565–568.

Bank, S. and D. A. Juckett. 1975. Reactions of aromatic radical anions. 11: kinetic studies of the reaction of sodium naphthalene and anthracene with *n*-hexyl bromide and chloride, *J. Am. Chem. Soc.*, 97:567–573.

Bank, S. and D. A. Juckett. 1976. Reactions of aromatic radical anions. 12: kinetic studies of the reaction of radical anions of varying reduction potential with *n*-hexyl bromides, iodides, and chlorides, *J. Am. Chem. Soc.*, 98:7742–7746.

Baughman, G. L., E. J. Weber and M. S. Brewer. 1992. Sediment reduction of anthraquinone dyes and related compounds: anthrone formation, Presented at the American Chemical Symposium on the Oxidation-Reduction Transformations of Inorganic and Organic Species in the Environment, San Francisco, CA, April 5–10, 1992.

Baxter, R.M. 1990. Reductive dechlorination of certain chlorinated organic compounds by reduced hematin compared with their behavior in the environment, *Chemosphere*, 21(4 and 5):451–458.

Beland, F. A., S. O. Farwell, and R. D. Geer. 1974. Anaerobic degradation of 1,1,1,2-tetrachloro-2,2-bis(p-chlorophenyl)ethane (DTE), *J. Agric. Food Chem.*, 22:1148–1149.

Beland, F. A., S. O. Farwell, A. E. Robocker, and R. D. Geer. 1976. Electrochemical reduction and anaerobic degradation of lindane, *J. Agric. Food Chem.*, 24(4):753–756.

Bolt, H. M., R. J. Laib and J. G. Filser. 1982. Reactive metabolites and carcinogenicity of halogenated ethylenes, *Biochem. Pharmac.*, 31(1):1–4.

Bryant, F. O., D. D. Hale and J. E. Rogers. 1991. Regiospecific dechlorination of pentachlorophenol by dichlorophenol-adapted microorganisms in freshwater, anaerobic sediment slurries, *Appl. Environ. Microbiol.*, 57(8):2293–2301.

Buffle, J. and R. S. Altmann. 1987. Interpretation of metal complexes by heterogeneous complexants, in *Aquatic Surface Chemistry*, Stumm, W. ed. (New York, NY: Wiley).

Castro, C. E., and W. C. Kray, Jr. 1963. The cleavage of bonds by low valent transition metal ions. The homogeneous reduction of alkyl halides by chromous sulfate, *J. Am. Chem. Soc.* 85:2768–2773.

Castro, C. E. 1964. The rapid oxidation of iron(II)porphyrins by alkyl halides. A possible mode of intoxification or organisms by alkylhalides, *J. Am. Chem. Soc.* 86:2310–2311.

Chung, K., B. E. Fulk and M. Egan. 1978. Reduction of azo dyes by intestinal anaerobes, *Appl. Environ. Microbiol.*, 35:558–562.

Clarke, E. A. and R. Anliker, 1980. Organic dyes and pigments, in *Handbook of Environmental Chemistry: Vol. 3, Part A, Anthropogenic Compounds*, O. Hutzinger, Ed. (Berlin: Springer), pp. 181–215.

Criddle, C. S., P. L. McCarty, M. C. Elliott and J. F. Barker. 1986. Reduction of hexachloroethane to tetrachloroethylene in groundwater, *J. Contam. Hydrol.* 1:133–142.

Curtis, G. P. and M. R. Reinhard. 1989. Reductive dehalogenation of hexachloroethane by aquifer materials, paper presented at the 197th meeting of the American Chemical Society, Dallas, TX, April 9–14.

Delgado, M. C. and N. L. Wolfe. 1992. Structure-activity relationships for the reduction of *p*-substituted nitrobenzenes in anaerobic sediments, Presented at the American Chemical Symposium on the Oxidation-Reduction Transformations of Inorganic and Organic Species in the Environment, San Francisco, CA, April 5–10, 1992.

Dubin, P. and K. L. Wright. 1975. Reduction of azo food dyes in cultures of *proteus vulgaris*, *Xenobiotica*, 5:563–571.

Dunnivant, F. M., R. P. Schwarzenbach and D. L. Macalady. 1992. Reduction of substituted nitrobenzenes in aqueous solutions containing natural organic matter, *Environ. Sci. Technol.*, 26(11):2133–2141.

Eberson, L. 1987. *Electron Transfer Reactions in Organic Chemistry*, (Springer-Verlag: Berlin, Germany)

Entwistle, I. D., R. A. W. Johnstone and A. H. Wilby. 1982. Rapid reduction of N-nitrosoamines to N,N-disubstituted hydrazines; the utility of some low-valent titanium reagents, *Tetrahedron*, 38(3):419–423.

Farwell, S. O., F. A. Beland and R. D. Geer. 1975. Interrupted-sweep voltammetry for the identification of polychlorinated biphenyls and naphthalenes, *Anal. Chem.*, 47:895–903.

Fathepure, B. Z., J. M. Tiedge and S. A. Boyd. 1988. Reductive dechlorination of hexachlorobenzene to tri- and dichlorobenzenes in anaerobic sewage sludge, *Appl. Environ. Microbiol.*, 54(2):327–330.

Florence, T. M. 1965. Polarography of aromatic azo compounds, *Aust. J. Chem.*, 18:609–618.

Fushiwaki, Y., N. Tase, A. Saeki and K. Urano. 1990. Pollution by the fungicide pentachloronitrobenzene in an intensive farming area in Japan, *Sci. Tot. Environ.*, 92:55–67.

Gambrell, R. P., B. A. Taylor, K. S. Reddy, and W. H., Jr., Patrick. 1984. Fate of toxic compounds under controlled redox potential and pH conditions in soil and sediment-water systems, EPA-600/3-83-018, U.S. Environmental Protection Agency, Springfield, VA, NTIS PB81-213266.

Gantzer, C. J. and L. P. Wackett. 1991. Reductive dechlorination catalyzed by bacterial transition-metal coenzymes, *Envrion. Sci. Technol.* 25:715–722.

Geer, R. D. 1978. Predicting the anaerobic degradation of organic chemical pollutants in waste water treatment plants from their electrochemical behavior. Montana University Joint Water Resources Research Center Rep. No. 95, Bozeman, Montana, 61 pp.

Glass, B. L. 1972. Relation between the degradation of DDT and the iron redox system in soils, *J. Agr. Food Chem.* 20(2):324–327.

Goswami, K. P. and R. E. Green. 1971. Microbial degradation of the herbicide atrazine and its 2-hydroxy analog in submerged soils, *Environ. Sci. Technol.*, 5:426–430.

Hale, D. D., J. E. Rogers and J. Wiegel. 1991. Environmental factors correlated to dichlorophenol dechlorination in anoxic freshwater sediments, *Environ. Toxicol. Chem.* 10:1255–1265.

Hartman, W. W. and H. L. Silloway. 1955. *Org. Syn. Coll. Vol 3*, (New York, NY: John Wiley), p. 82.

Hassall, K. A. 1982. *The Chemistry of Pesticides*, (New York, NY: John Wiley), 373 pp.

Helling, C. S. 1976. Dinitroaniline herbicides in soils, *J. Environ. Qual.*, 5:1–15.

Holmstead, R. L. 1976. Studies of the degradation of mirex with an iron(II)porphyrin model system, *J. Agric. Food Chem.*, 24(3):620–624.

Hudlicky, M. 1984. *Reductions in Organic Chemisty*, (New York, NY: Halsted Press), p. 74.

Hutzinger, O., S. Safe and V. Zitko. 1974. *The Chemistry of PCB's*, (Cleveland, OH: CRC Press).

Jafvert, C. T. and N. L. Wolfe. 1987. Degradation of selected halogenated ethanes in anoxic sediment-water systems, *Environ. Toxicol. Chem.*, 6:827–837.

Kaplan, D. L. and A. M. Kaplan. 1982. 2,4,6-Trinitrotoluene-surfactant complexes: decomposition, mutagenicity, and soil leaching studies. *Environ. Sci. Technol.* 16:566–571.

Kaplan, D. L. 1990. Biotransformation pathways of hazardous energetic organo-nitro compounds, in *Advances in Applied Biotechnology, 4: Biotechnology and Biodegradation*, Kamely, D., A. Chakrabarty and G. S. Omen, eds., (The Woodlands, TX: Portfolio Publishing), pp. 155–182.

Kearney, P. C., J. E. Oliver, C. S. Helling, A. R. Isensee and A. Konston. 1977. Distribution, movement, persistence, and metabolism of N-nitrosoatrazine in soils and a model aquatic ecosystem, *J. Agric. Food Chem.*, 25(5):1177–1181.

Khalifa, S., R. L. Holmstead and J. E. Casida. 1976. Toxaphene degradation by iron(II)protoporphyrin systems, *J. Agric. Food Chem.*, 24:277–282.

Klecka, G. M. and S. J. Gonsior. 1984. Reductive dechlorination of chlorinated methanes and ethanes by reduced iron(II)porphyrins, *Chemosphere*, 13:391–402.

Kray Jr., W. C. and C. E. Castro. 1964. The cleavage of bonds by low-valent transition metal ions. The homogeneous dehalogenation of vicinal dihalides by chromous sulfate, *J. Am. Chem. Soc.*, 86:4603–4608.

Kriegman-King, M. R. and M. Reinhard. 1991. Reduction of hexachloroethane and carbon tetrachloride at surfaces of biotite, vermiculite, pyrite, and marcasite, in *Organic Substances and Sediments in Water, Vol 2.* Baker, R. A., ed., (Chelsea, MI: Lewis Publisher). pp. 349–364.

Kriegman-King, M. R. and M. Reinhard. 1992. Transformation of carbon tetrachloride in the presence of sulfide, biotite, and vermiculite, *Environ. Sci. Technol.* 26(11):2198–2206.

Krone, U. E., R. K. Thauer and H. P. C. Hogenkamp. 1989. Reductive dehalogenation of chlorinated C_1-hydrocarbons mediated by corrinoids, *Biochemistry*, 28(11):4908–4914.

Kuhn, E. P. and J. M. Suflita. 1989a. Dehalogenation of pesticides by anaerobic microorganisms in soils and groundwater—a review, in *Reactions and Movement of Organic Chemicals in Soils*, Sawhney, B. L. and K. Brown, Eds. SSSA special publication No.22, Madison, WI.

Kuhn, E. P. and J. M. Suflita. 1989b. Sequential reductive dehalogenation of chloroanilines by microorganisms from a methanogenic aquifer, *Environ. Sci. Technol.*, 23:848–852.

Kulkarni, S. V., C. D. Blackwell, A. L. Blackard, C. W. Stackhouse. 1985. Textile dyes and dyeing equipment: classification, properties, and environmental aspects, U.S. Environmental Protection Agency, Resource Triangle Park: NC, EPA-600/2-85/010.

Kuwatsuka, S. 1977. Studies of the fate and behavior of herbicides in soil and plants, *Nippon Noyaku Gakkaishi*, 2:201–213.

Lightfoot, E. N., P. S. Thorne, R. L. Jones, J. L. Hansen and R. R. Romine. 1987. Laboratory studies on mechanisms for the degradation of aldicarb, aldicarb sulfoxide and aldicarb sulfone, *Environ. Toxicol. Chem.*, 6:377–394.

Littler, J. S. 1970. Homolytic oxidation and reduction of organic compounds by metallic ions, In *Essays on Free-Radical Chemistry. Special Publication No. 24*, (London, Chemical Society), pp. 383–408.

Macalady, D. L. and N. L. Wolfe. 1992. Hydrophobic effects in the reduction of anthropogenic organic chemicals by natural organic matter, paper presented at the American Chemical Symposium on Environmental Chemistry. San Francisco, CA., April 5-10.

Macalady, D. L., P. G. Tratnyek and T. J. Grundl. 1986. Abiotic reactions of anthropogenic organic chemicals in anaerobic systems: a critical review, *J. Contam. Hydrol.* 1:1-28.

Maguire, R. J. and R. J. Tkacz. 1991. Occurrence of dyes in the Yamaska River, Quebec, *Water Poll. Res. J. Canada*, 26(2):145-161.

March, J. 1985. *Advanced Organic Chemistry*, 3rd Ed., (New York, NY: Wiley).

McCormick, N. G., F. E. Feeherry, and H. S. Levinson. 1976. Microbial transformation of 2,4,6-trinitrotoluene and other nitroaromatic compounds, *Appl. Environ. Microbiol.* 31(6):949-958.

Mikesell, M. D. and S. A. Boyd. 1985. Reductive dechlorination of the pesticides 2,4-D, 2,4,5-T, and pentachlorophenol in anaerobic sludges, *J. Environ. Qual.*, 14:337-340.

Miles, C. J. and J. J. Delfino. 1985. Fate of aldicarb, aldicarb sulfoxide, and aldicarb sulfone in Floridian groundwater, *J. Agric. Food Chem.*, 33:455-460.

Nies, L. and T. M. Vogel. 1990. Effects of organic substrates on dechlorination of aroclor 1242 in anaerobic sediments, *Appl. Environ. Microbol.*, 56(9):2612-2617.

Oliver, J. E. and A. Kontson. 1978. Formation and persistence of N-nitrosobutralin in soil, *Bull. Env. Contam. Tox.*, 20:170-173.

Ong, J. M. and C. E. Castro. 1977. Oxidation of iron(II)porphyrins and hemoproteins by nitro aromatics, *J. Am. Chem. Soc.*, 99:6740-6745.

Ou, T., L. H. Carreira, R. W. Schottman and N. L. Wolfe. 1992. Nitroreduction of 2,4,6-trinitrotoluene (TNT) in contaminated soils, presented at the 203rd American Chemical Society Symposium on Oxidation-Reduction Transformations of Inorganic and Organic Species in the Environment. San Francisco, CA, April 5-10. 32(1):513-514.

Peijnenburg, W. J. G. M., M. J. 't Hart, H. A. den Hollander, D. van de Meent, H. H. Verboom and N. L. Wolfe. 1991. QSARs for predicting biotic and abiotic reductive transformation rate constants of halogenated hydrocarbons in anoxic sediment systems, *Sci. Total Environ.* 109/110: 283-300.

Peijnenburg, W. J. G. M., M. J. 't Hart, H. A. den Hollander, D. van de Meent, H. H. Verboom and N. L. Wolfe. 1992a. Reductive transformation of halogenated aromatic hydrocarbons in anaerobic water-sediment systems: kinetics, mechanisms, and products, *Environ. Toxicol. Chem.* 11:289-300.

Peijnenburg, W. J. G. M., M. J. 't Hart, H. A. den Hollander, D. van de Meent, H. H. Verboom and N. L. Wolfe. 1992b. QSARs for predicting reductive transformation rate constants of halogenated aromatic hydrocarbons in anoxic water-sediment systems, *Environ. Toxicol. Chem.* 11:301-314.

Preussmann, R. and B. W. Stewart. 1984. N-nitroso carcinogens, in C. E. Searle (Ed.), Chemical Carcinogens, Vol. 2, ACS Monograph 182, American Chemical Society, Washington, pp. 643-828.

Quensen III, J. F., J. M. Tiedje and S. A. Boyd. 1988. Reductive dechlorination of polychlorinated biphenyls by anaerobic microorganisms from sediments, *Science*, 242:752–754.

Quensen III, J. F., S. A. Boyd and J. M. Tiedje. 1990. Dechlorination of four commercial polychlorinated biphenyl mixtures (aroclors) by anaerobic microorganisms from sediments, *Appl. Environ. Microbiol.*, 56(8):2360–2369.

Quirke, J. M. E., A. S. M. Marei and G. Eglinton. 1979. The degradation of DDT and its degradative products by reduced iron(II)porphyrins and ammonia, *Chemosphere*, 3:151–155.

Rowland, I. R. and P. Grasso. 1975. Degradation of N-nitrosamines by intestinal bacteria, *Appl. Microbiol.*, 29(1):7–12.

Rusling, J. F. and C. L. Miaw. 1989. Kinetic estimation of standard reduction potentials of polyhalogenated biphenyls, *Environ. Sci. Technol.*, 23:476–479.

Saleh, M. A., W. V. Turner and J. E. Casida. 1977. Polychlorobornane components of toxaphene: structure-toxicity relations and metabolic reductive dechlorination, *Science*, 198:1256–1258.

Sanders, P. and N. L. Wolfe. 1985. Reduction of nitroaromatic compounds in anaerobic, sterile sediments, presented at the 190th National Meeting of the American Chemical Society, Chicago, IL, September.

Schwarzenbach, R. P. and P. M. Gschwend. 1990. Chemical transformations of organic pollutants in the aquatic environment, in *Aquatic Chemical Kinetics*, Stumm, W. Ed., (New York, NY: John Wiley), pp. 199–233.

Schwarzenbach, R. P., R. Stierli, K. Lanz, and J. Zeyer. 1990. Quinone and iron porphyrin mediated reduction of nitroaromatic compounds in homogeneous aqueous solution, *Environ. Sci. Technol.* 24:1566–1574.

Semenov, N. N. 1958. *Some Problems in Chemical Kinetics and Reactivity*, (Princeton, NJ: Princeton University Press).

Smith, R. H., J. E. Oliver and W. R. Lusby. 1979. Degradation of pendimethalin and its N-nitroso and N-nitro derivatives in anaerobic soils, *Chemosphere*, 11/12:855–861.

Southwick, L. M. and G. H. Willis. 1974. Interaction of trifluralin with soil components under flooded anaerobic conditions, Presented in the Division of Pesticide Chemistry at the 176th National American Chemical Society Meeting, Washington, D.C.

Steelinck, C. and G. Tollin. 1967. Free radicals in soil, in *Soil Biochemistry*, McLaren, A. D. and G. H. Peterson, eds., (New York, NY: Marcel Dekker), pp. 147–169.

Struijs, J. and J. E. Rogers. 1989. Reductive dehalogenation of dichloroanilines by anaerobic microorganisms in fresh and dichlorophenol-acclimated pond sediment, *Appl. Environ. Microbiol.*, 55(10):2527–2531.

Suflita, J. M., A. Horowitz, D. R. Shelton, and J. M. Tiedje. 1982. Dehalogenation: a novel pathway for the anaerobic biodegradation of haloaromatic compounds, *Science*, 218:1115–1117.

Tate III, R. L. and M. Alexander. 1975. Stability of nitrosamines in samples of lake water, soil, and sewage, *J. Natl. Cancer Inst.*, 54:327–330.

Tatsumi, K., H. Yamada and S. Kitamura. 1983. Reductive metabolism of N-nitrosodiphenylamine to the corresponding hydrazine derivative, *Archives of Biochemistry and Biophysics*, 226(1):174–1881.

Terpko, M. O. and R. F. Heck. 1980. Palladium-catalyzed triethylammonium formate reductions. 3. Selective reduction of dinitroaromatic compounds, *J. Org. Chem.*, 45:4992.

Thurman, E. M. 1985. *Organic Geochemistry of Natural Waters*, (Boston, MA: Nijhoff)

Timms, P. and I. C. MacRae. 1983. Reduction of fensulfonthion and accumulation of the product, fensulfothion sulfide, by selected microbes, *Bull. Environ. Contam. Toxicol.*, 31:112–115

Tincher, W. 1978. Survey of the Coosa Basin for organic contaminants from carpet processing, Final Report, Project E-27-630. Textile Engineering Department, Georgia Institute of Technology, Atlanta, GA.

Tollin, G., T. Reid and C. Steelink. 1963. Structure of humic acid. IV. Electron paramagnetic resonance studies, *Biochimica et Biophysica*, 66(2):444–447.

Tomizawa, C. 1975. Degradation of organophophorus pesticides in soils with special reference to anaerobic soil conditions, *Environ. Qual. Saf.*, 4:117–127.

Tratnyek, P. G. and N. L. Wolfe. 1988. Reduction of phorate sulfoxide in anaerobic sediment slurries, Paper presented at the 195th meeting of the American Chemical Society, Dallas, TX, April 9–14.

Tratnyek, P. G. and N. L. Wolfe. 1990. Characterization of the reducing properties of anaerobic sediment slurries using redox indicators, *Environ. Toxicol. Chem.*, 9:289–295.

Tratnyek, P. G. and D. L. Macalady. 1989. Abiotic reduction of nitro aromatic pesticides in anaerobic laboratory systems, *J. Agric. Food Chem.* 37:248–254.

Tratnyek, P. G., J. Hoigne, J. Zeyer and R. P. Schwarzenbach 1991. QSAR analyses of oxidation and reduction rates of environmental organic pollutants in model systems, *Sci. Total Environ.* 109/110:327–341.

Tsukano, Y., and A. Kobayashi. 1972. Formation of γ-BTC in flooded rice field soils treated with γ-BHC. *Agric. Biol. Chem.*, 36:116–117.

Vogel, T. M.; C. S. Criddle and P. L. McCarty. 1987. Transformations of halogenated aliphatic compounds, *Environ. Sci. Technol.*, 21(8):722–736.

Wade, R. S. and C. E. Castro. 1973. Oxidation of iron(II) porphyrins by alkyl halides, *J. Am. Chem. Soc.*, 95(1):226–230.

Wahid, P. A. and A. Sethunathan. 1979. Involvement of hydrogen sulfide in the degradation of parathion in flooded acid sulphate soil, *Nature (London)*, 282:401–402.

Wahid, P. A., C. Ramakrishna and N. Sethunathan. 1980. Instantaneous degradation of parathion in anaerobic soils, *J. Eviron. Qual.* 9:127–130.

Walker, R. 1970. The metabolism of azo compounds: a review of the literature, *Food Cosmet. Toxicol.*, 8:659–676.

Walter-Echols, G. and E. P. Lichtenstein. 1977. Microbial reduction of phorate sulfoxide to phorate in a soil-lake mud-water microcosm, *J. Econ. Entomol.*, 70:505–509.

Wang, C. H. and F. E. Broadbent. 1973. Effect of soil treatments on losses of two chloronitrobenzene fungicides, *J. Environ. Qual.*, 2:511–515.

Weber, E. J. and N. L. Wolfe. 1987. Kinetic studies of the reduction of aromatic azo compounds in anaerobic sediment/water systems, *Environ. Toxicol. and Chem.* 6:911–919.

Weber, E J. Environmental fate studies of N-nitosoamines, paper presented at the Ninth Annual Meeting of the Society of Environmental Toxicology and Chemistry, Arlington, VI (1988), p. 73.

Weber, E.J. 1991. Studies of benzidine-base dyes in sediment-water systems, *Environ. Sci. Chem.* 10:609–618.

Weber, E. J. 1988. Reduction of disperse blue 79 in anaerobic sediment-water systems, in Preprints of Papers Presented at the 196th ACS National Meeting, Los Angeles, CA, September, 25–30. pp. 177–180.

Wheland, G. W. 1944. *The Theory of Resonance*, (New York, NY: Wiley), p. 195.

Williams, P. P., K. L. Davidson and E. J. Thacker. 1968. In vitro and in vivo rumen microbiological studies with 2-chloro-4,6-bis(isopropylamino)-s-triazine (propazine), *J. Anim. Sci.*, 27:1472–1476.

Willis, G. H., R. C. Wander and L. M. Southwick. 1974. Degradation of trifluralin in soil suspensions as related to redox potential, *J. Environ. Quality*, 3(3):262–265.

Wolfe, N. L., B. E. Kitchens, D. L. Macalady and T. J. Grundl. 1986. Physical and chemical factors that influence the anaerobic degradation of methyl parathion in sediment systems, *Environ. Toxicol. Chem.* 5:1019–1026.

Yen, C-P. C., T. A. Perenich and G. L. Baughman. 1991. Fate of commercial disperse dyes in sediment, *Environ. Toxicol. Chem.* 10:1009–1017.

Yu, Y. S. and G. W. Bailey. 1992. Reduction of nitrobenzene by four sulfide minerals: kinetics, products, and solubility, *J. Environ. Qual.* 21:86–94.

ZoBell, C. E. 1946. Studies on redox potential of marine sediments, *Bull. Am. Assoc. Petrol. Geol.*, 30:477–513.

Zoro, J. A., J. M. Hunter, G. Eglinton and G. C. Ware. 1974. Degradation of *p,p'*-DDT in reducing environments, *Nature*, 247:235–236.

CHAPTER 4

ENVIRONMENTAL OXIDATIONS

Oxidation is defined as a loss of electrons. Oxidizing **agents** gain electrons and are by definition electrophiles. In organic chemistry, oxidation can either be associated with the introduction of oxygen into a molecule or the conversion of a molecule to a higher oxidation state. For example, in Equation 4.1

$$RCH_2\text{-}H \rightarrow R\text{-}CH_2\text{-}OH \rightarrow R\text{-}CHO \tag{4.1}$$

the first step in the sequence is the incorporation of oxygen by a formal "oxygen atom donor," whereas the second is formally a dehydrogenation, or oxidation of the carbon atom to a higher formal oxidation state.

Almost all oxidations are kinetically second-order reactions in which the rate is proportional to the concentrations of both the oxidizing agent, Ox, and the substrate, A (Equation 4.2):

$$-dA/dt = k\,[Ox]\,[A] \tag{4.2}$$

For reactions in water, there are few values for the rate constant, k, and also few data on the concentrations of potential oxidants.

In environmental chemistry, there are many sources of oxidants whose importance is highly variable due to changes in concentration or reactivity as one moves from one compartment or region of the environment to another. A potent oxidant such as the hydroxyl radical, $\cdot OH$, that is of critical importance for organic transformations in the gas phase or during combustion, may not be important at all in another environmental compartment such as the soil. This could either be due to inadequate production rates for the species or to rapid side reactions that diminish its steady-state concentration. These considerations will be addressed as each oxidizing species is discussed in turn for the remainder of the chapter.

A. MOLECULAR OXYGEN

Oxygen, O_2, is the most abundant oxidizing agent in both the troposphere and the water column. In the lower troposphere it accounts for nearly 21% of the gas mixture, falling off at higher elevations not only because of reduced total numbers of atmospheric gas molecules, but also because of the reactivity of oxygen toward ultraviolet wavelengths of less than about 240 nm. At altitudes of greater than 110 km, above the stratosphere, molecular oxygen concentrations become smaller and smaller; it does not exist above 400 km. The absorption spectrum of O_2 shows several bands between 130 and 250 nm; uptake of light energy in this region leads to homolytic dissociation of O_2. Above 20 km, around the middle of the stratosphere, this photolysis of O_2 to O atoms (Equation 4.3) begins to become important; the subsequent "dark" reaction of O with molecular oxygen is the source of the so-called "ozone layer," which is actually a broad zone of enhanced ozone concentration centered at about 20–25 km (Figure 4.1).

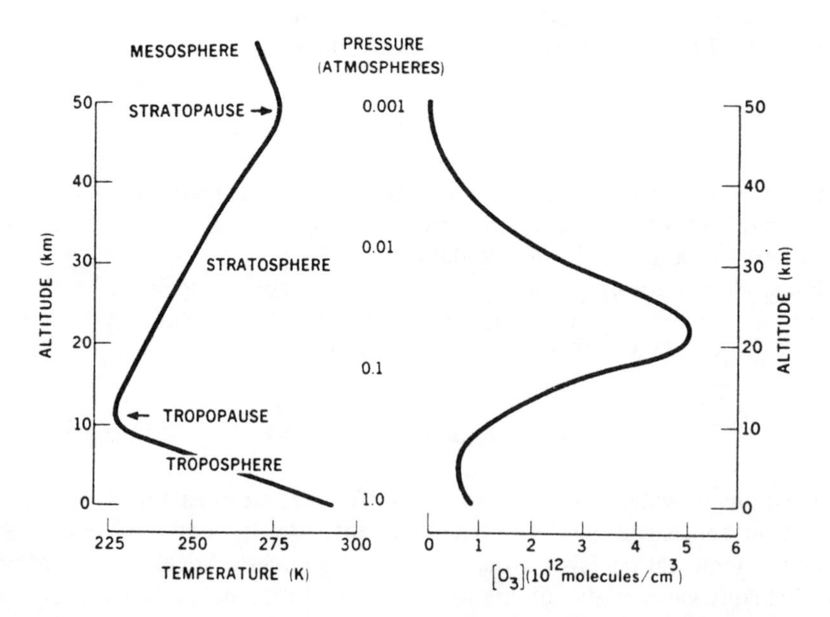

Figure 4.1. The stratospheric ozone "layer"; variation in ozone concentration with altitude, showing a maximum at 20–25 km, just above the temperature minimum for the atmosphere. From Howard (1981). Reprinted by permission of Academic Press.

$$O_2 \xrightarrow{h\nu} O \xrightarrow{O_2} O_3 \tag{4.3}$$

At lower elevations O atoms take part in less important reactions with water, ammonia, oxides of sulfur and nitrogen, and organic compounds.

In water, oxygen is present near the surface of water bodies in amounts that depend on temperature, ionic strength, and chemical and biological redox activity. The solubility of oxygen in liquid water falls off with increasing temperature. At 0°C its solubility is 14.4 ppm, corresponding to 4.5×10^{-4} M, falling to 8.3 ppm (2.7×10^{-4} M) at 25°. The presence of other solutes, such as salts, diminishes the solubility of oxygen, so that in marine waters its solubility drops by about 30%. Natural waters, of course, are not always at saturation due to metabolic action of microorganisms; in well-lighted, transparent waters that are high in photosynthetic activity, it may be supersaturated, whereas in water containing actively respiring bacteria and fungi, its concentration can drop to zero.

The oxygen concentration in natural waters is closely related to their redox potential as measured by the parameter E_h (see also Chapter 1). This quantity is the potential difference, in mV, between a water sample and the hydrogen electrode. The upper limit (at pH 7.0) for the value of E_h is about 800 mV; above this value, water becomes oxidized to oxygen gas. The lower limit for E_h is -400 mV at pH 7; at lower voltages, water is reduced to hydrogen. Natural water in equilibrium with air has an E_h of about 750 mv. Figure 4.2 shows an E_h-pH diagram indicating typical values for various types of commonly encountered water bodies.

In waters with high values of E_h, many more biotic and abiotic oxidation reactions become thermodynamically probable (though not necessarily feasible from a kinetic perspective). Similarly, a low oxygen concentration or E_h makes reductive processes more likely.

In aqueous surface microlayers and in surface films caused by discharges of hydrophobic liquids, it is likely that higher concentrations of oxygen are present. Because of its nonpolar nature, oxygen is several times more soluble in organic media than in water. In liquid hydrocarbon solvents, its solubility is about eightfold higher than in water, and in a dispersion of lipids in water, its solubility was increased by about a factor of 5 (Smotkin et al., 1991).

Ground-state molecular oxygen has the structure ·O–O·, in which the two electrons have parallel spins and predominantly exist in a "triplet" configuration having one electron localized on each of the two oxygen atoms. This means that oxygen is fundamentally a diradical species. A quantum-mechanical spin barrier exists for interactions of oxygen and most organic molecules, whose structures do not incorporate unpaired electrons. Accordingly, molecular oxygen's rapid reactions with organic species are restricted (at low temperatures) to free radicals. Molecular oxygen is, however, the principal initiator of oxidation in high-temperature combustion processes (see Section 4.D.1). With inorganic species that possess redox states that vary by one unit of charge, oxygen can act as an oxidant and become reduced itself:

$$Fe^{2+} + O_2 \rightarrow Fe^{3+} + O_2^{-\cdot} \tag{4.4}$$

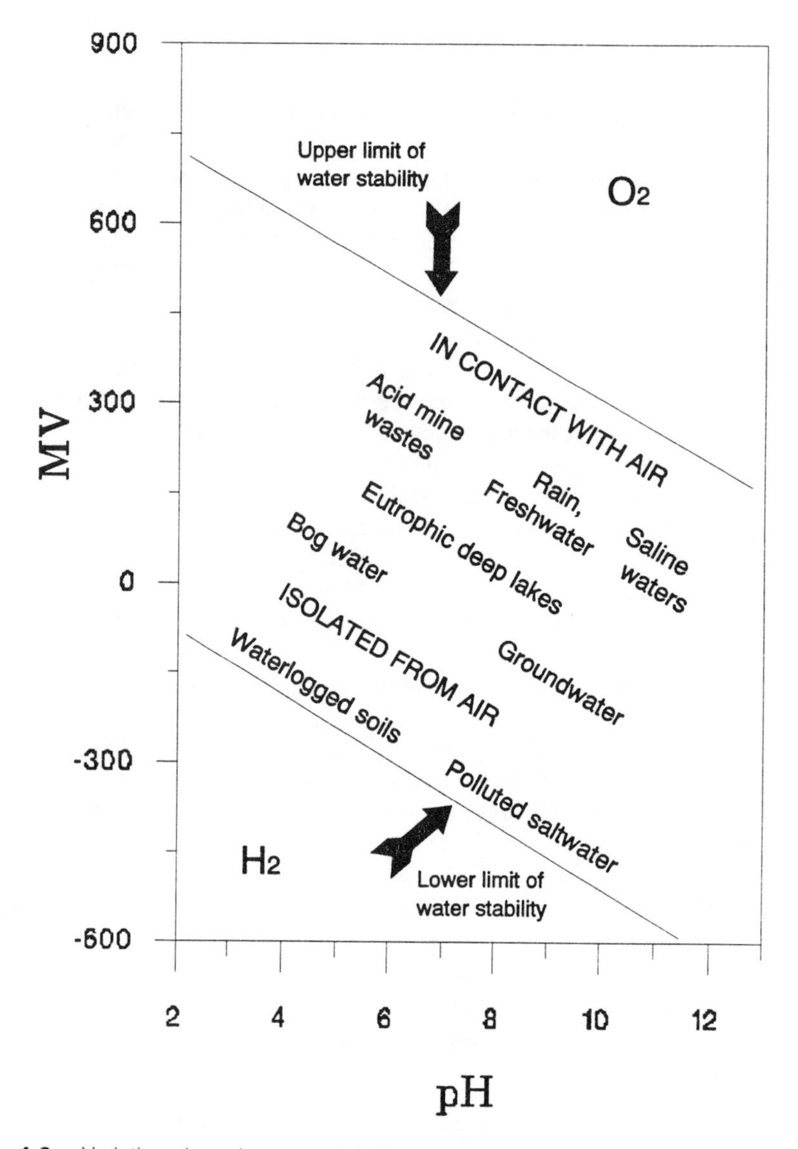

Figure 4.2. Variations in redox potential (E_h) for natural waters.

The above reaction has a half-life of about 30 minutes at a pH of 6.9 and 20°C. The reduction product, superoxide radical anion, has redox chemistry of its own which is discussed in detail in Section 4.A.2.

The overall, four-electron reduction of oxygen is strongly favored by thermodynamics ($E^\circ = +0.82$ v referenced to the normal hydrogen

$$O_2 + 4\,H^+ + 4\,e^- \rightarrow 2\,H_2O \tag{4.5}$$

electrode), but in practice it is a one-electron step that predominates. The formation of $\cdot O_2^-$ in Equation 4.4 or a related reaction is normally followed (in water) by a rapid self-redox reaction to O_2 and H_2O_2, which represents a two-electron reduction. Further conversion of H_2O_2 to $\cdot OH$ requires a third electron, and a fourth one-electron transfer provides water as the ultimate reduction product. The thermodynamics for the reduction of oxygen in solutions of varying pH are summarized in Figure 4.3.

Oxygen is known to form metastable complexes with olefins and some aromatic hydrocarbons which are capable of absorbing sunlight UV wavelengths, unlike the parent hydrocarbons (Khalil and Kasha, 1978). In some cases, the existence of these complexes may lead to reactions of the substrate hydrocarbons or formation of singlet oxygen (see Section 4.A.3).

1. Autooxidation

Autooxidation may be defined as the oxidative degradation of materials exposed to the air. An often-used synonym is **weathering**, though geochemists also use this term to refer to the dissolution of minerals. "Things fall apart," said the poet

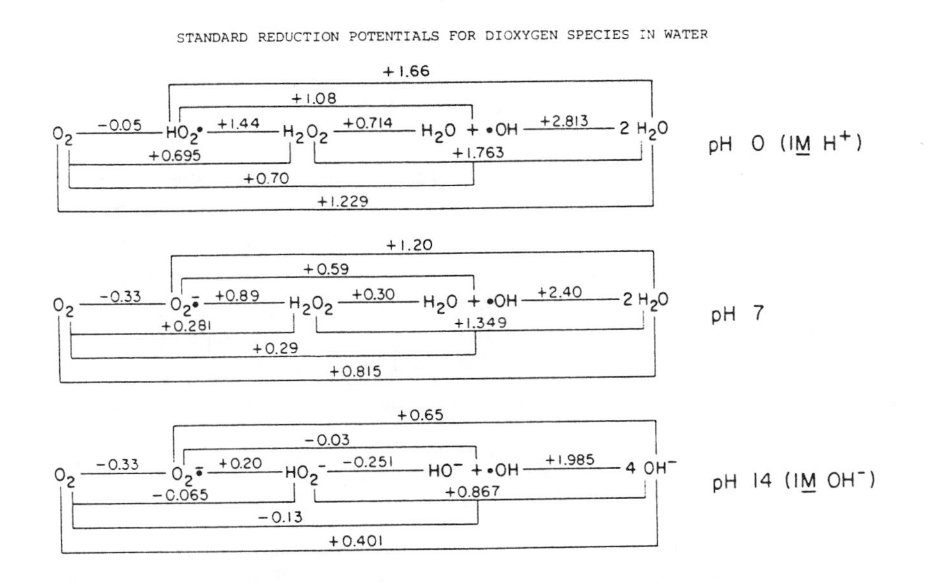

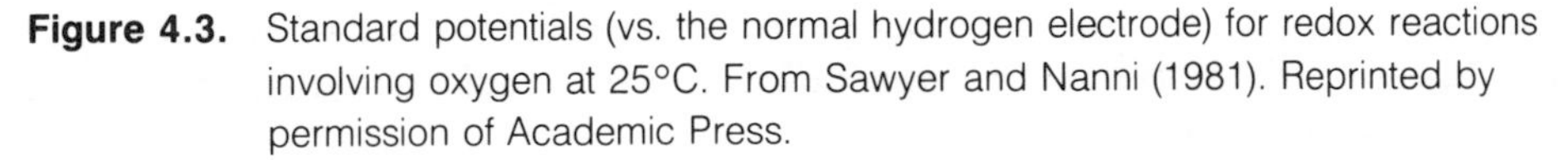

Figure 4.3. Standard potentials (vs. the normal hydrogen electrode) for redox reactions involving oxygen at 25°C. From Sawyer and Nanni (1981). Reprinted by permission of Academic Press.

William Butler Yeats (although admittedly in a different context), zeroing in on the destiny of virtually all organic materials (whether they be rubber tires, tree stumps, or human bodies) as they move through time.

The mechanism of autooxidation was something of a mystery for many years, but in the 1930s and 1940s it was discovered that in the chemistry of weathering, free radicals were key intermediates. Currently, the theory of Bolland (1949) accounts for most of the observed phenomena of autooxidation. This theory postulates that autooxidation is actually a low-temperature form of combustion (see Section 4.D.1); autooxidation is sometimes called "flameless oxidation."

The first step of the accepted mechanism of autooxidation is **initiation** (Equation 4.6), in which a substrate molecule AB is converted to one or more free radicals by some agent which may be chemical (transition metal ions, ozone, hydroxyl or other pre-existing radicals, etc.) or physical (ultraviolet light, heat, gamma radiation, ultrasonic energy, or mechanical disintegration).

$$\text{A–B} \rightarrow \text{A}\cdot + \text{other products} \quad (4.6)$$

In most cases, radical A· will react very rapidly (to begin the **propagation** phase of autoxidation) with molecular oxygen, to form a new radical AOO·, commonly referred to as a peroxy radical. In the case of carbon-centered radicals, this step is extremely fast (usually approximating diffusion control).

Although peroxy radicals have many possible routes of reaction (see Section 4.B.2 for a brief summary of some of these), in autooxidation the most significant propagation step is a hydrogen abstraction,

$$\text{AOO}\cdot + \text{RH} \rightarrow \text{AOOH} + \text{R}\cdot \quad (4.7)$$

In this usually slow reaction, RH may represent another molecule of AB, the solvent, or some other constituent with a relatively active hydrogen atom, such as a compound containing allylic hydrogen. The formation of new radicals in this step provides other species able to continue to react with O_2 and thus to continue the autooxidation by a so-called chain reaction.

Hydroperoxides are normally unstable to thermal or photochemical decomposition and often decompose to provide additional radical species. One such example is ultraviolet-induced homolysis of the O–O bond:

$$\text{AOOH} \xrightarrow{h\nu} \text{AO}\cdot + \cdot\text{OH} \quad (4.8)$$

Although the absorption of sunlight UV by hydroperoxides is usually quite small, the quantum yield of homolysis for the photons that are absorbed is usually quite high. Reactions of this type are unusually favorable for the occurrence of chain processes since they increase the number of reactive free radicals that are capable of initiating chains of their own.

Termination of autooxidation normally occurs (at least under laboratory conditions) by self-destruction of peroxy radicals,

$$2\ AOO\cdot \rightarrow \text{products (see Section 4.B.2 for details)} \qquad (4.9)$$

although sometimes (especially in the environment) a reduced metal ion or an easily reduced organic compound (antioxidant) may be present that is able to trap or scavenge them. An example of an antioxidant whose mechanism of action is well known is 2,6-di-*tert*-butyl-4-methyl phenol (also known as "butylated hydroxytoluene" or BHT). One mole of this compound is able to destroy two moles of peroxy radicals by the following mechanism:

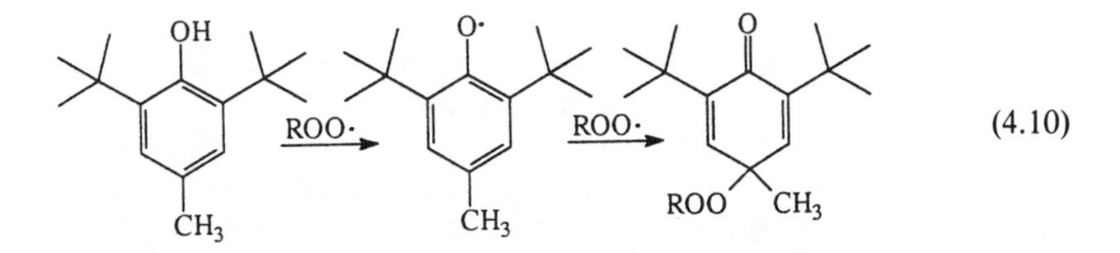

(4.10)

Recent studies in biological systems have revealed the presence of a variety of compounds of many structural types that appear to function as antioxidants. The plant kingdom is especially rich in these materials (Larson, 1988). They include flavonoids and other phenolic compounds, alkaloids and other amines, reduced sulfur compounds, uric acid, ascorbic acid, Vitamin E, carotenoids, and many other substances. Often, their principal mechanism of action appears to be the quenching of peroxy radicals, removing them from the autooxidation chain. Vitamin E, for example, reacts with these radicals in a manner entirely analogous to that of BHT:

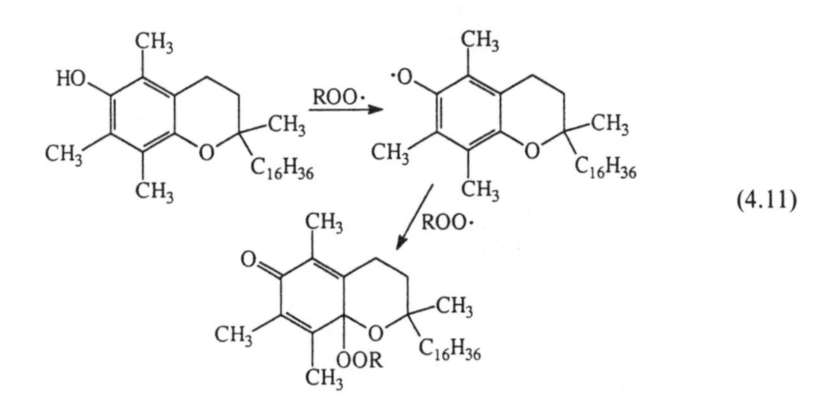

(4.11)

Ascorbic acid (Vitamin C, cf. Figure 4.4) exists in cells, often in rather high (millimolar) concentrations, as a negatively charged lactone that is very susceptible to one-electron oxidation (Swartz and Dodd, 1981), giving a free radical that is rather stable due to resonance stabilization. Further one-electron oxidation affords

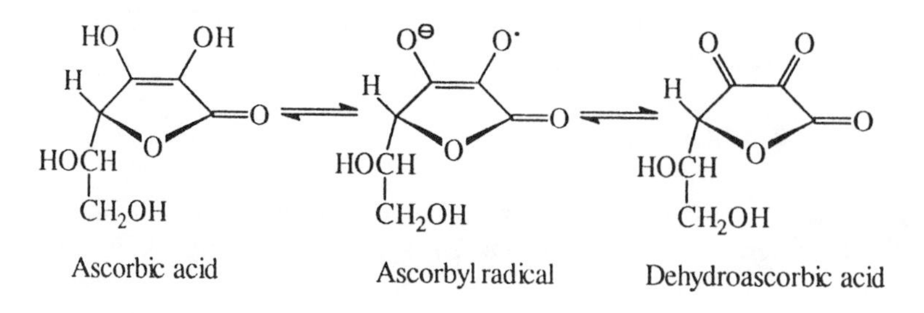

Figure 4.4. Redox reactions involving ascorbic acid. From Swartz and Dodd (1981). Reprinted by permission of Academic Press.

dehydroascorbic acid, which can also be formed by disproportionation from the radical (along with another molecule of ascorbate).

The kinetics of autooxidation by radical chain mechanisms are characteristic (Figure 4.5), and include a more or less extended **induction period**, during which the initiation phases are predominant and there is a slow uptake of O_2. During this period peroxide concentrations are very low. In the second or **autocatalytic stage**, peroxide concentrations increase rapidly as does oxygen uptake; during this period new chain sequences are being initiated faster than the radicals can be destroyed. During the final, **termination** phase, oxygen or oxidizable substrates become exhausted and radical recombination reactions become dominant.

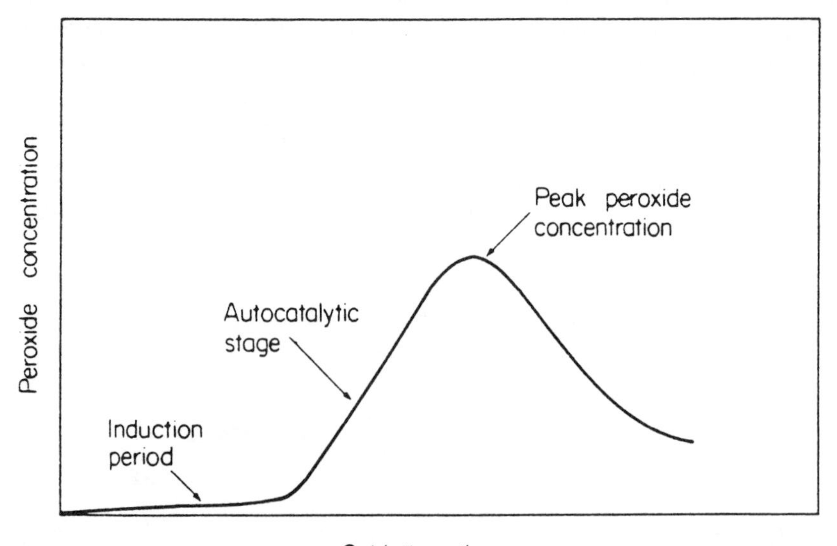

Figure 4.5. Idealized kinetics for a typical autooxidation reaction. From Rånby and Rabek (1975). Reprinted by permission of John Wiley and Sons.

The end result of the operations of weathering in a material is the enrichment of the material with oxidized compounds at the expense of a substantial portion of reduced matter.

a. Polymers

In the autooxidation of a polymer such as polystyrene, the initiation step consists of the formation of a benzylic free radical via homolysis of the Ph–C–H bond. Rapid reaction with molecular oxygen normally follows, to give a polymeric benzylperoxy radical:

$$\left[CH_2-CH(Ph)\right]_n \longrightarrow \left[CH_2-\dot{C}(Ph)\right]_n \longrightarrow \left[CH_2-C(OO\cdot)(Ph)\right]_n \tag{4.12}$$

The polymer benzylperoxy radical can react with nearby C-H bonds to form hydroperoxides, or undergo a variety of other scission and crosslinking reactions (for a summary, see Rånby and Rabek [1975]). Because of proximity effects, chain propagation in polymers can be highly efficient. The overall result of prolonged exposure of polystyrene to air and light is yellowing of the residual polymer due to the formation of carbonyl groups and other light-absorbing chromophores, and also the emission from the polymer of small, volatile fragments that result from radical elimination or abstraction processes. Some of these latter compounds include hydrogen, water, benzene, styrene, and benzophenone (Rånby and Rabek, 1975). The remaining polymer becomes extremely brittle due to chain fragmentation processes that greatly diminish the average molecular weight of the polymer molecules. An example of such chain-fracturing reactions is a β-scission that commonly occurs in polymer alkoxy radicals:

$$-CH_2-C(R)(O\cdot)-CH_2-CH(R)- \longrightarrow -CH_2-C(R)(=O) + \cdot CH_2-CH(R)- \tag{4.13}$$

Termination processes in weathering polymers include a variety of potential reactions of free radicals with one another. When the oxygen concentration is high, the predominant termination reaction (Equation 4.14) is usually the formation of unstable tetroxide intermediates (for mechanisms of decomposition of tetroxides, see Section 4.B.2):

$$2\ ROO\cdot \rightarrow ROO\text{-}OOR \rightarrow \text{nonradical products} \tag{4.14}$$

Polymers vary greatly in their susceptibility to autooxidation. Polystyrenes, because they have benzylic carbon-hydrogen bonds that afford highly stabilized free radicals on hydrogen atom abstraction, are among the most readily oxidized polymers under environmental conditions. As would be expected on these grounds, polymers that contain alkene groups, such as rubber (**1**), are readily autooxidized. The

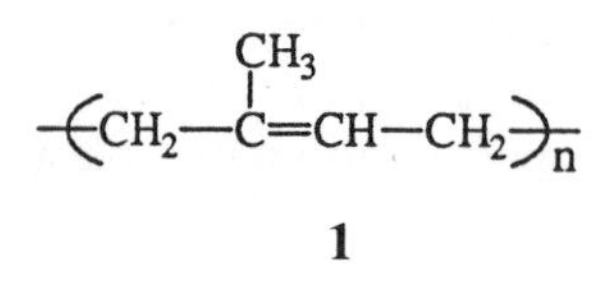

allylic methylene groups of this structure are the weakest points in the chain as far as susceptibility to radical formation are concerned. The double bonds of this polymer are also very reactive toward ozone, which is always present in trace quantities in the air, resulting in the formation of carbonyl compounds and possibly other oxidized groupings (Amberlang et al., 1963). To forestall these destructive processes, rubber articles usually contain high levels of added antioxidants or radical scavengers such as carbon black (also a potent absorber of light), or are "vulcanized" by heating with elemental sulfur, which also has antioxidant properties.

Polyethylene, $(-CH_2-)_n$, on the other hand, is an example of a polymer that really has no good sites for attack. The secondary free radicals that would form are not stabilized in any significant manner; as a result, polyethylene objects can be expected to remain in a more or less undamaged condition for many years. Polyethylene might be almost everlasting if it were not for scattered impurity oxidized sites (-CH=CH-, hydroxyl, hydroperoxyl, and carbonyl groups) that are introduced during processing of the material. These few functional groups provide sites where free-radical and other chemical reactions can be inaugurated.

b. Petroleum

Petroleum-derived fuels can be autooxidized in a similar fashion. Since some refined fractions, such as No. 2 fuel oil, are often rich in two- and three-ring condensed aromatic compounds that strongly absorb light, the possibility of sensitized processes involving singlet oxygen or electron transfer intermediates is quite likely in these materials (Larson et al. 1977, 1979: Lichtenthaler et al., 1989). In addition, readily oxidized compounds such as tetrahydronaphthalenes and indans (**2**) are often abundant. When a No. 2 fuel oil was exposed to sunlight, a variety of physical and chemical changes took place; the oil darkened in color and became cloudy due to the formation of insoluble, oxidized, hydrocarbon-derived compounds. Included were alkylated phenols, naphthols, ketones, alcohols, and perhaps most interesting, a group of hydroperoxides of the general formula **3** (Larson et al., 1979). The concentrations of these compounds built up over a period of days and have been shown to reach as much as 70 mM. Hydroperoxides are interesting environmental pollutants because of their high toxicity and (sometimes) mutagenicity, and because their water solubilities are much greater than those of their parent hydrocarbons (Larson et al., 1979; Callen and Larson, 1978).

2. Superoxide

One-electron reduction of molecular oxygen occurs in a variety of systems including autooxidation of ferrous iron (Equation 4.4) or cuprous copper, photoionization reactions of many compounds such as humic materials, and also in biologically mediated reactions such as enzymatic oxidations. In the atmosphere, gas phase reactions of hydroxyl radicals with carbon monoxide or of oxygen molecules with formyl and alkoxy radicals also result in hydrogen atom transfers to oxygen. The reduction product, the hydroperoxyl radical, has the structure **4**. It is the conjugate acid (pKa = 4.8) of the superoxide radical anion, $\cdot O_2^-$, which is also a reduction product of molecular oxygen via reactions with an electron. The salts of the anion, for example potassium superoxide, KO_2, are well known and stable reagents.

$$H{:}O{-}O\cdot$$

4

In the atmosphere, estimates of HOO· concentrations appear to be on the order of 1 part in 10^{10}–10^{11} (Anderson et al., 1981; Logan et al., 1981). In seawater, an estimated upper limit for $\cdot O_2^-$ concentration of about 10^{-8} M has been arrived at by considering the concentration and rates of formation and loss for H_2O_2 (Zafiriou et al., 1984).

Elimination of HOO· from peroxy radicals having α-hydrogens is another common pathway for the formation of superoxide. For example, ethanol reacts with ·OH by hydrogen atom transfer to give a mixture of several radicals dominated by **5**, which rapidly reacts with oxygen to form a peroxy radical that eliminates ·OOH (with the concomitant production of acetaldehyde). This reaction is not as efficient with the peroxy radical of *t*-butanol, which has no α-hydrogens (von Sonntag, 1987).

$$CH_3CH_2OH \xrightarrow{\cdot OH} \underset{\mathbf{5}}{CH_3\dot{C}HOH} \xrightarrow{O_2} CH_3CH(O{-}O\cdot)OH \longrightarrow CH_3CH(=O) + HOO\cdot$$

It is often stated that superoxide is an oxidizing agent or that it has "both reducing and oxidizing properties." Superoxide itself reacts mostly as a reductant or nucleophilic reactant (see Section 2.B.1c), but it is converted into potent oxidants when it undergoes self-disproportionation, or **dismutation**. This bimolecular reaction produces hydrogen peroxide and oxygen. The reaction can be visualized as an electron transfer from a superoxide anion to a hydroperoxyl radical to yield molecular oxygen and the anion of H_2O_2, which is readily protonated:

$$HOO\cdot + \cdot O_2^- \xrightarrow{H^+} H_2O_2 + O_2 \quad (4.16)$$

The reaction as written is more than a hundred times faster than the reaction between two HOO· radicals, and the reaction between two $\cdot O_2^-$ radicals is almost immeasurably slow (Bielski, 1978). Therefore, the decay of superoxide is strongly pH-dependent (Figure 4.6). At the usual pHs of the environment, dismutation will probably be fast enough (in water) to outcompete most other potential reactions. In seawater, however, Millero (1987) has shown that $\cdot O_2^-$ is stabilized by its interactions with divalent cations so that its half-life should be on the order of 2–9 minutes. In aprotic solvents, dismutation of superoxide is likely to be much slower, and other

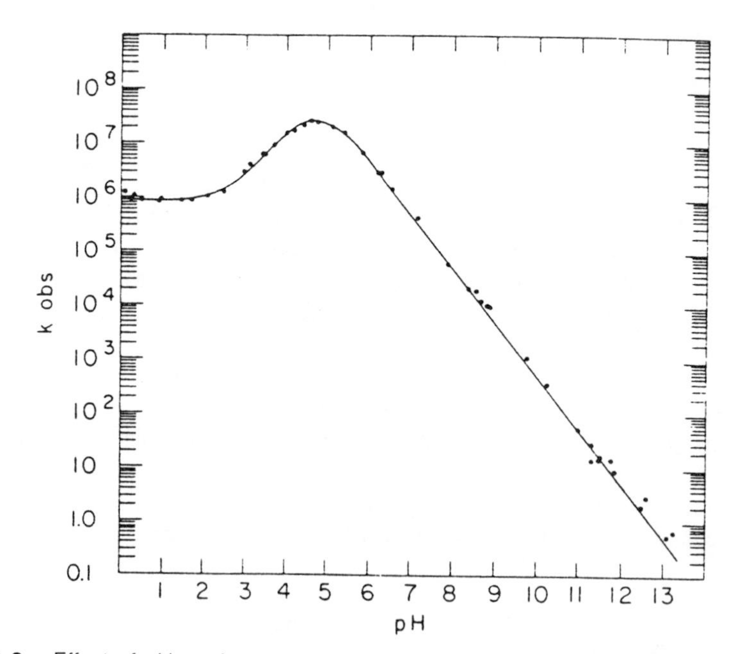

Figure 4.6. Effect of pH on the second-order rate constant for the dismutation of superoxide (Equation 4.16). From Bielski (1978). Reprinted by permission of American Society for Photobiology.

types of reactions may be better able to compete kinetically with its self-reaction. Niehaus (1978) has demonstrated that $\cdot O_2^-$ is stable for about 20 min in 85:15 DMSO:deionized water.

Further reactions of H_2O_2, the product of superoxide dismutation, include the Fenton reaction (Walling, 1975; Section 4.B.1), which produces the potent oxidizing species ·OH. A related mechanism is a proton abstraction by $\cdot O_2^-$ to give the anion of the substrate and the dismutation products HOO– and O_2 which oxidize the anion (Sawyer and Nanni, 1981). Therefore, although superoxide itself has powerful reducing properties, it can be involved in reactions that generate powerful oxidizing agents. This is apparently the reason why living organisms who live in contact with oxygen all contain a potent enzymatic catalyst for Equation 4.16, **superoxide dismutase**; the cell has to be defended from damaging attacks of oxidizing agents such as HO· on biologically important targets such as DNA or enzymes (Fridovich, 1974).

The rate constants for reactions of $\cdot O_2^-$ and HOO· have been collected by Bielski (1983) and Bielski et al. (1985). It has been observed that the reactivities of the two forms are often quite different for many compounds (Table 4.1). For example, superoxide anion reacted with tetranitromethane at an almost diffusion-controlled rate of about 2×10^9 l/mol sec, whereas HOO· was more than five orders of magnitude less reactive (Rabani et al., 1965).

The differences apparently reflect the fact that HOO· is a fairly reactive abstractor of hydrogen atoms, whereas the chemistry of $\cdot O_2^-$ is dominated by its greater nucleophilicity and electron-conferring properties (von Sonntag, 1987). Bielski (1983) and Bielski and Shiue (1979) point out that many compounds of biological interest such as some phenols, amino acids, and unsaturated fatty acids were several orders of magnitude more reactive with HOO· than with $\cdot O_2^-$.

With formaldehyde, at least in the vapor phase, HOO· exhibits an interestingly fast reaction involving addition to the C=O bond, giving an intermediate alkoxy

Table 4.1. Rate Constants for Reactions of O_2^- and HOO· in Various Compounds

	Rate Constant ($l\ mol^{-1}\ sec^{-1}$)	
Substrate	**With HOO·**	**With $\cdot O_2^-$**
Cu^+	1×10^8	$\sim 10^{10}$
Cu^{2+}	$>10^9$	8×10^9
Tetranitromethane	$<10^4$	1.9×10^9
Adrenaline	1.9×10^4	2.3×10^7
Cysteine	600	<15
Glycine	<49	<0.42
Linoleic acid	1.2×10^3	N.D.
Catechol	4.7×10^4	2.7×10^5
H_2O_2	0.50	0.13
α-Tocopherol	2.0×10^5	<6

N.D. = not detectable

radical that rapidly goes through a formal 1,4-hydrogen atom shift; this peroxy radical, for reasons that are not entirely clear, appears to be rather stable (Niki et al., 1980).

$$HOO\cdot + HCHO \rightarrow HOO\text{-}CH_2O\cdot \rightarrow \cdot OOCH_2OH \quad (4.17)$$

Unlike most free radicals, $\cdot O_2^-$ has little tendency to add to carbon-carbon double bonds: it has been shown to be inert with most simple olefins as well as many aromatic hydrocarbons. It reacts with electron-poor aromatic compounds such as quinones and nitro derivatives to produce radical anions (Wilshire and Sawyer, 1979). Certain metalloproteins such as ferricytochrome c are rapidly reduced by $\cdot O_2^-$ (Rao and Hayon, 1975) and other transition metal complexes also undergo reduction.

3. Singlet Oxygen

An interesting and reactive, if selective, oxidant for organic molecules is singlet oxygen ($^1O_2$ or, more correctly, O_2 $[^1\Delta_g]$). This form of molecular oxygen is an electronically excited state which is no longer diradical-like (as is ground state, triplet oxygen, ·O-O·), but instead is electronically similar (O=O) to spin-paired, "singlet" molecules with pi bonds. For the photochemical details of the sensitized conversion of ordinary O_2 to $^1O_2$, see Section 6.C.2. Another pathway for $^1O_2$ production that does not require light is the reaction of chlorine (or hypochlorite) with H_2O_2 (Arnold et al., 1964):

$$Cl_2 + H_2O_2 \rightarrow 2\ HCl + {}^1O_2 \quad (4.18)$$

Although singlet oxygen was postulated as the reactive agent in sensitized photooxidations over 60 years ago (Kautsky and de Bruin, 1931), its importance was not recognized until it was rediscovered by Foote and Wexler (1964) more than 30 years later. An explosion of interest in the species followed, which is still continuing today.

In water, $^1O_2$ is rapidly deactivated back to the ground state by collision with solvent molecules; the lifetime has been calculated to be approximately 4 microseconds (Rodgers and Snowden, 1982). In organic solvents, particularly those with halogen atoms, the rate of quenching by solvent decreases greatly; in CCl_4, for example, its lifetime is 700 microseconds (Merkel and Kearns, 1972). In the vapor phase it can have lifetimes of several seconds.

Singlet oxygen appears to be generated by exposure of natural waters to light (Zepp et al., 1977; Baxter and Carey, 1982; Momzikoff et al., 1983; Haag et al., 1984; Frimmel et al., 1987), as shown by the disappearance of particularly reactive acceptors in illuminated samples of such waters. Dienes such as 2,5-dimethylfuran (**6**), histidine (**7**), and furfuryl alcohol (**8**) undergo reactions to ring-opened products as described below. Although the estimated quantum yield for $^1O_2$ is about 1-3% based on total light absorbed, its rapid quenching by water is by far the predominant pathway for its disappearance. Haag and Hoigné (1986) calculated a mean steady-state concentration of $^1O_2$ of no more than 4×10^{-14} M for 18 different natural waters exposed to midday summer sunlight. Its role as an active oxidizing agent in aqueous solution appears to be limited, therefore, to only the most reactive acceptor molecules (such as tryptophan, histidine, phenolates, and perhaps a few sulfur compounds and reduced transition metal ions), due to its rapid deactivation (Braun et al., 1986; Scully and Hoigné, 1987). It is conceivable that $^1O_2$ might be of more importance in hydrophobic environments, such as surface microlayers or petroleum films, where its lifetime would be longer and ground-state oxygen and sensitizers would be present in higher concentration (Larson et al., 1979). Lichtenthaler et al. (1989) calculated that $[^1O_2]$ could be as high as 6×10^{-9}M at the surface of a crude oil film.

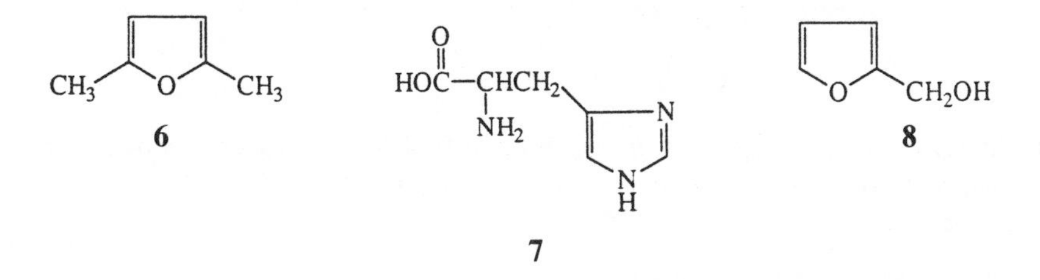

Singlet oxygen has been reported to be generated at soil surfaces (Gohre and Miller, 1983; Gohre et al., 1986), based on analyses of the products formed when 2,5-dimethylfuran, tetramethylethylene, and other acceptors were adsorbed to soil and exposed to sunlight or fluorescent lamps. Free-radical reactions were not completely ruled out, and some disappearance of the acceptors in a nitrogen atmosphere was observed. The authors suggested that the organic matter of the soil was the sensitizing agent, although they pointed out that some metal oxides could have played roles; and in fact, in a later paper (Gohre and Miller, 1985) they presented evidence for $^1O_2$ formation on the surfaces of Al_2O_3, SiO_2, and MgO (but not TiO_2). It is conceivable that $^1O_2$ or other oxidizing species formed at soil surfaces could be important in the destruction of pesticides or other organic molecules or in their chemical binding to soil constituents.

Weathering of untreated wood exposed to sunlight is a major problem. The formation of $^1O_2$ at wood surfaces was reported by Hon et al. (1982). Studies of illuminated wood samples (pine sawdust) by ESR (electron spin resonance) showed that free radicals were also formed under the influence of light. The nature of the possible sensitizers and radical sources were not discussed, although it should be noted that some experiments were performed in acetone, a potential source of free radicals (see Chapter 6). Berenbaum and Larson (1988) demonstrated that $^1O_2$ was formed at the surfaces of some leaves exposed to light; it was able to migrate from the leaf and photodecompose a $^1O_2$ acceptor, furfuryl alcohol, in solutions a few mm away. The authors proposed that the reaction might play a role in the defense of plants against leaf-eating insects.

Singlet oxygen interacts with organic molecules either by collisional (physical) quenching or by chemical reaction. (For a review of the kinetics of $^1O_2$ reactions, see Wilkinson and Brummer, 1981.) The first type of quenching is exemplified by aliphatic amines, which very effectively deactivate $^1O_2$ without undergoing chemical change themselves. Only a few classes of compounds react readily to form new chemical substances, usually by ene and diene reactions, with $^1O_2$:

(a) Olefins, with the most electron-rich (usually tetrasubstituted) varieties reacting fastest. The principal products of these reactions are hydroperoxides, the result of an ene reaction in which the oxygen molecule has formally been added to one end of the original double bond, which then shifts into an allylic position:

$$(R_1)(R_2)C{=}CH{-}CH_2{-}R_3 + {}^1O_2 \longrightarrow R_1{-}C(OOH)(R_2){-}CH{=}C(H)(R_3) \qquad (4.19)$$

(b) Polycyclic aromatic hydrocarbons, exemplified by anthracene, which forms a cyclic endoperoxide (**9**) by addition of $^1O_2$ in Diels-Alder (4 + 2 cycloaddition) fashion:

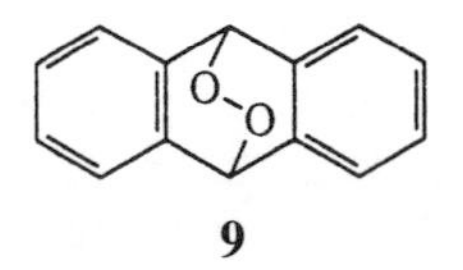

9

(c) Cyclic dienes, including heterocyclic dienes such as furans and pyrroles as well as carbocyclic dienes. These compounds also usually react initially by the Diels-Alder route, producing endoperoxides that are usually unstable to further thermal reac-

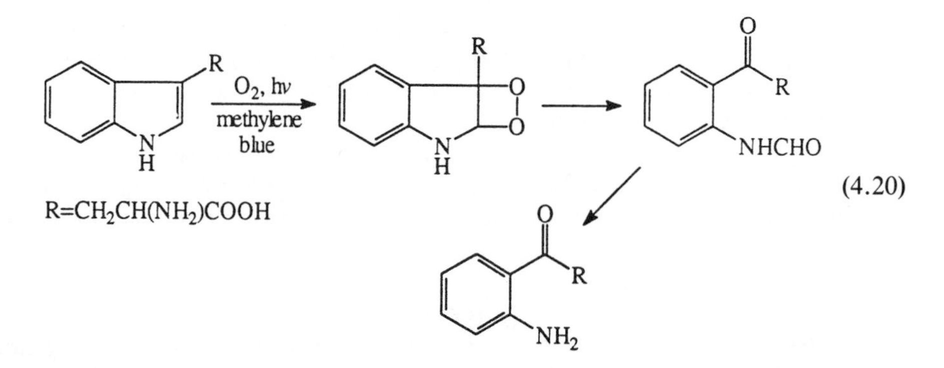

tions. Tryptophan, however, appears to add $^1O_2$ via a 2 + 2 cycloaddition to its indole double bond (Saito et al., 1977): ring-opening of this intermediate gives the observed product, N-formyl kynurenine. Tryptophan, one of the most reactive amino acids toward $^1O_2$, may possibly disappear from marine waters by a sensitized pathway (Momzikoff et al., 1983), although tryptophan does absorb some solar UV and could be destroyed directly.

The natural peroxide, ascaridole (**10**) could be biosynthesized by a $^1O_2$ reaction from its diene precursor, α-terpinene (**11**). This reaction has been accomplished in vivo (Gollnick and Schenk, 1967).

(d) Sulfur compounds undergo rapid reactions with sensitizing dyes that have not been fully characterized mechanistically. Reactions with the triplet state of the sensitizer as well as $^1O_2$ reactions are likely. The products include sulfoxides and disulfides. The pesticide disulfoton was photooxidized by a sensitized reaction due to aquatic humic materials to the corresponding sulfoxide (Draper and Crosby, 1976).

$$\underset{\text{Disulfoton}}{\mathrm{EtO{-}\overset{\overset{S}{\|}}{\underset{\underset{OEt}{|}}{P}}SCH_2CH_2SCH_2CH_3}} \longrightarrow \underset{\text{Disulfoton Sulfoxide}}{\mathrm{EtO{-}\overset{\overset{S}{\|}}{\underset{\underset{OEt}{|}}{P}}SCH_2CH_2\overset{\overset{O}{\|}}{S}CH_2CH_3}} \quad (4.21)$$

(e) Phenols, especially those with multiple alkyl substitution or those that are otherwise electron-rich, form cyclohexadienone hydroperoxides with $^1O_2$. Again, the mechanism appears to be a preliminary Diels-Alder addition followed by rearrangement (Thomas and Foote, 1978). An example, using the antioxidant BHT, is shown in Equation 4.22. In cases where the hydroperoxide can eliminate water, quinones

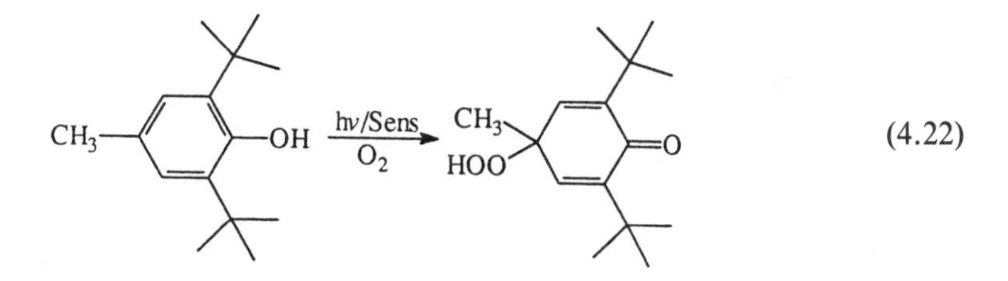

were the end-products (Pfoertner and Böse, 1970). Other extremely electron-rich aromatic compounds, such as polyalkoxylated benzenes (Saito et al., 1972) and polyalkylated naphthalenes (Sakugari et al., 1982) were also reactive toward $^1O_2$.

The rate constants for the reactions of $^1O_2$ with most phenols are probably too slow for these compounds to be significantly degraded by sunlight-sensitized reactions in most natural waters (Tratnyek and Hoigné, 1991). However, phenolate anions are much more electron-rich than the parent phenols due to the increased electron-donating ability of the phenoxide substituent, and at relatively high pHs the contribution of their faster reactions with $^1O_2$ may become important (Scully and Hoigné, 1987; Gsponer et al., 1987).

An unusual reaction of $^1O_2$ with 1,5-dihydroxynaphthalene (**12**) has been described by Duchstein and Wurm (1984). The product, juglone (**13**) is formed in high yield in aprotic solvents such as acetonitrile. The authors have proposed the reaction as a specific probe for the presence of $^1O_2$.

Many reactions with sensitizing agents such as methylene blue, riboflavin, or rose bengal have been reported and the products have been attributed to $^1O_2$ reactions; however, it should be kept in mind that other oxidation mechanisms exist for these compounds, many involving direct reactions between the sensitizer and the substrate (Bonneau et al., 1975; Foote, 1976). These mechanisms have been designated "Type I" reactions (Foote, 1976, 1991) to distinguish them from the presumed "Type II" chemistry typical of true $^1O_2$ reactions. (Electron transfer from sensitizer to oxygen, to give superoxide and the radical cation of the sensitizer, are also formally Type II reactions, although $^1O_2$ is not involved.) It is often quite difficult to distinguish the two mechanisms, since very few oxidation processes that are specific to $^1O_2$ have been reported. For example, many well-known antioxidants react rapidly both with free radicals and $^1O_2$ (Bellus, 1979; Chou and Khan, 1983; Burton and Ingold, 1984; Mukai et al., 1991).

4. Ozone and Related Compounds: Photochemical Smog

Stratospheric ozone has been much in the minds of the public recently because of concern over the decrease in the UV-absorbing "ozone layer." Ozone is always present, as well, in low concentrations in the troposphere, due to lightning as well as photochemically driven oxidations in polluted atmospheres (see Figure 4.1).

Ozone is formed in the stratosphere by the short-wavelength (<240-nm) homolysis of molecular oxygen (O_2) to two oxygen atoms (O), followed by the subsequent collision of an oxygen atom with an oxygen molecule to produce O_3. The absorption spectrum of ozone (λ_{max} = 255 nm) is such that virtually all the potentially damaging ultraviolet wavelengths between 200 and 300 nm are screened out before they reach the earth's surface. The uptake of UV energy in this region dissociates the ozone molecule into oxygen and an O atom.

Ozone disappears in the atmosphere by reactions with both inorganic and organic species. In cloud droplets, either $O_2 \cdot^-$ or bisulfite (HSO_3^-) are likely to be the major reactants with ozone (Hoigné, 1984). Ozone reacts at a significant rate with only very few organic molecules, particularly olefins, to which it adds rapidly (Atkinson and Carter, 1984) to form ozonides (for details of the reaction, see Section 5.C.2). These intermediates break down to produce a variety of aldehydes and carboxylic acids. Under typical daytime tropospheric conditions, the fraction of olefinic materials that react with ozone are roughly equal to that which react with ·OH (Atkinson et al., 1979; Finlayson-Pitts and Pitts, 1986: see Section 4.B.2). Because ·OH is principally a daytime species, reactions of olefins with ozone become much more significant at night.

Ozone reacts rapidly with some PAHs sorbed on surfaces. In a study of nine PAHs on cellulose, Katz et al. (1979) showed that benzanthracenes and benzo[a]pyrene disappeared rapidly when exposed to ozone ($t_{1/2}$ = 0–4 min), whereas pyrene, benzo[e]pyrene and benzofluoranthenes were more resistant ($t_{1/2}$ = 7–53 min). In the presence of light and ozone, the rates of disappearance for most of the PAHs increased, but a few rates were even slower than in the absence of light. Glass fiber sorption of PAHs seems to lead to a somewhat different order of reactivity with pyrene becoming one of the more reactive compounds toward ozonolysis (Pitts et al., 1986). Fly ash particles are also catalysts for PAH oxidation and photooxidation, but large differences exist in their catalytic activities (Fox and Olive, 1979; Korfmacher et al., 1980; Dlugi and Güsten, 1983; Yokley et al., 1986; Behymer and Hites, 1988). A complex mixture of reaction products is observed in these reactions, with quinones, phenols, epoxides, and ring-cleavage products having been identified (Van Cauwenberghe et al., 1979; Pitts et al., 1980). Quinones exposed to simulated sunlight on glass surfaces undergo ozonolysis at a much greater rate than in darkness (Cope and Kalkwarf, 1987). Little is known about the mechanisms of these surface reactions.

Ozone (as well as other oxidants) has long been known to be formed in the tropospheric oxidation of organic compounds. The complete oxidation of methane to CO_2 will be briefly considered in Section 4.B.2b. The key initiation step for this process is the interaction of CH_4 with ·OH to produce a methyl free radical. In the stratosphere, the usual fate of this radical and other radicals is combination with ground-state molecular O_2. In the troposphere, however, and especially the polluted troposphere, other species capable of intercepting free radicals may be present in significant concentrations. Changes in the reaction mechanisms for environmental transformations of hydrocarbons and other organic molecules in the atmosphere may occur, and the products formed may be important to human or ecosystem health. Studies of these reaction mechanisms gained great emphasis after the recognition of a new form of air pollution in the Los Angeles basin at around the end of the second World War. The phenomenon, termed "smog," was characterized by atmospheric hazes ranging from nearly colorless to distinctly red-brown, resulting in irritation to the human eyes and mucous membranes. It was quickly established that the hazes were oxidizing in character and that they developed most noticeably on bright, dry days, with peak concentrations occurring at or somewhat after noon. Haagen-Smit and his colleagues showed that smog could be simulated by irradiating rather simple hydrocarbons in the presence of automobile exhaust or NO_2 (Haagen-Smit, 1952). Large flasks or plastic bags used for this and subsequent studies have been given the name "smog chambers." It was soon realized that nitrogen oxides initiate a complex series of atmospheric reactions, leading to a net production of "photochemical oxidants"—ozone and nitrate esters having peroxide functionality.

The chemical course of a typical smoggy southern California day is summarized in Figure 4.7. Concentrations of nitric oxide (NO) begin to rise coincident with early morning traffic and reach a maximum at around 7 A.M. The high temperatures attained in the spark chamber of an internal-combustion engine make oxidations of N_2 by O_2 possible; almost all of the oxidized N in automobile exhaust is NO. A sharp decline in NO concentration then follows as nitrogen dioxide (NO_2) increases. "Oxidant" concentration then rises to a maximum at midday; finally, NO_2 and oxidant decline until low levels are once again reached after sunset.

The initiation step for the smog-producing reaction sequence is the photodissocia-

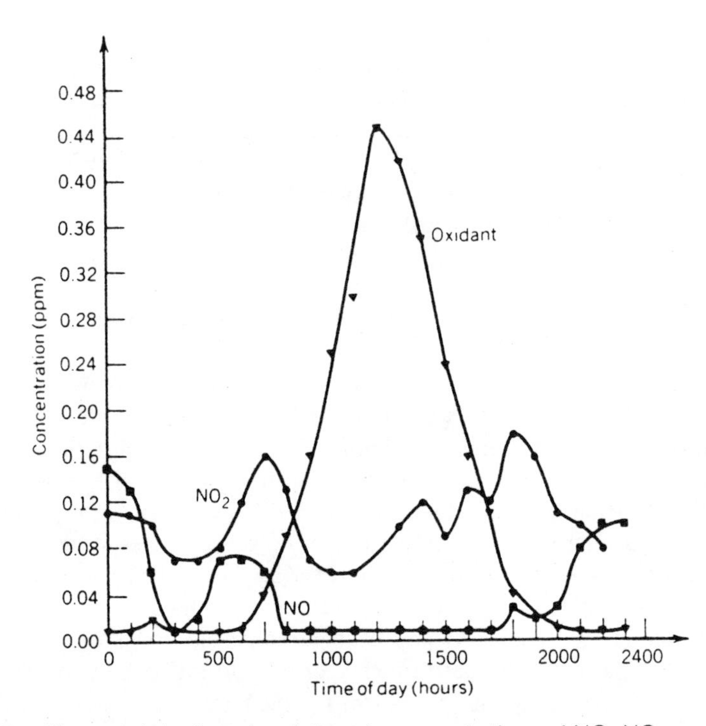

Figure 4.7. Characteristic diurnal variations in concentrations of NO, NO_2, and "total oxidant" during a photochemical smog episode in Pasadena, CA (25 July 1973). From Finlayson-Pitts and Pitts (1986). Reprinted by permission of John Wiley and Sons.

tion of NO_2, whose absorption extends broadly into the visible region (Figure 4.8). Light of less than 395 nm wavelength is effective; at longer wavelengths, excited states of NO_2 are produced, which fluoresce or have photosensitizing properties, but do not dissociate. A variable quantum yield for NO_2 dissociation is noted at wavelengths greater than 395 nm; the reaction has a strong temperature dependence and is presumably due to dissociation of a fraction of thermally excited NO_2 molecules.

In air, the O atoms produced photochemically combine with O_2 in the presence of a third body to yield O_3 by the reaction already discussed (Section 1.A.4). Ozone itself, however, is readily photodissociated, and also reacts with NO to reform O_2 and NO_2. The rates of production and destruction of O_3 in an air-nitrogen oxide system are such that the concentrations of NO and O_3 should be roughly equal. In practice, however, the combination of NO and hydrocarbons plus solar ultraviolet leads to a net production of O_3 and other oxidants.

The products of the photochemical reactions of aliphatic hydrocarbons in the presence of nitrogen oxides are reasonably well understood. For example, *n*-butane is converted to CH_2O, CH_3CHO, and methyl ethyl ketone, together with the nitrogen-containing species methyl, ethyl and isopropyl nitrate (**14–16**) and "peroxyacetyl nitrate" ("PAN," **17**; Altshuller et al., 1969). Radical (RO·, RC=O and ROO·) recombination reactions with NO_2 probably account for the formation of **14–17**; **17** is of special interest because of its phytotoxic, oxidizing, and irritant properties. PAN is also a nearly universal constituent of the troposphere, occurring even in extremely clean mid-ocean samples (Temple and Taylor, 1983; Singh and Salas, 1983).

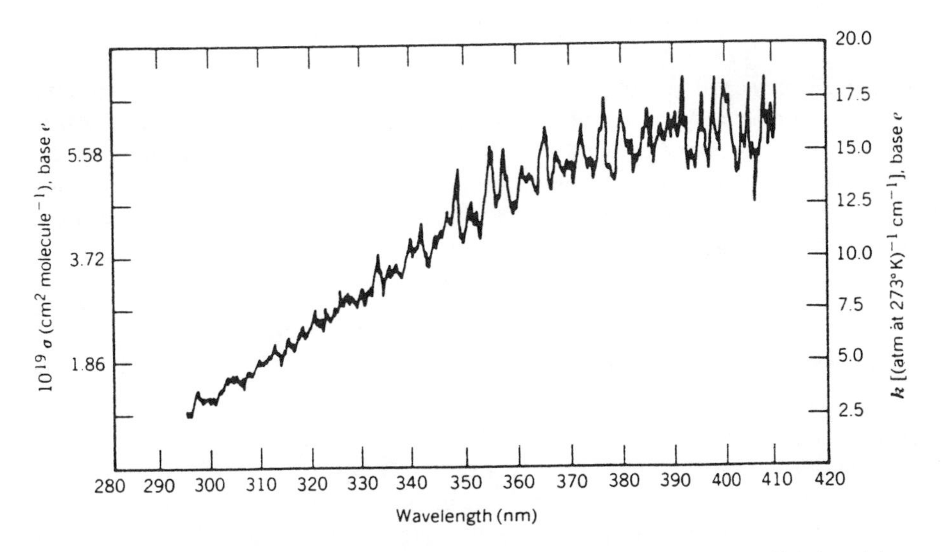

Figure 4.8. Absorption spectrum of NO_2 in the solar ultraviolet region. Reprinted from A. M. Bass et al., "Extinction coefficients of NO_2 and N_2O_4," *J. Res. U.S. Natl. Bur. Stand.* A80, 143 (1976).

Other peroxynitrate esters, including the benzoyl derivative (suspected to be a potent eye irritant), also occur in polluted atmospheres; although often somewhat thermally unstable relative to reversion to RC(O)OO· and NO_2, they are relatively stable to sunlight photodecomposition and can build up to constitute an appreciable fraction of the total oxidant. PAN and related compounds therefore act as sinks for storage of atmospheric NO_2.

The reactions of PAN and other pernitrate esters with organic compounds have not been studied very much. Simple alkenes can be epoxidized by PAN in laboratory experiments (Darnall and Pitts, 1970), but whether this reaction occurs in the environment is unknown.

The many atmospheric reactions which convert NO to NO_2, and therefore increase its photolysis to O, effectively lead to a net production of the oxidant O_3 during periods when light is intense. It might appear paradoxical that oxidizing species should be formed during the oxidation of hydrocarbons to aldehydes and CO_2. The overall sum of the reactions leading to the oxidation of CH_4 in the presence of nitrogen oxides is:

$$CH_4 + 8\,O_2 \rightarrow CO_2 + 2H_2O + 4\,O_3 \qquad (4.23)$$

NO and ·OH play catalytic roles; it can be seen that O_2 is both oxidized and reduced.

Ozone also reacts with low concentrations of NO_2 to form (Equation 4.24) an oxidant that is important in nighttime atmospheric chemistry, the nitrate radical, $\cdot NO_3$. This species, which is unstable in the presence of light, reacts fairly rapidly with many compounds such as alkenes (including terpenes), phenols and other aromatic compounds,

$$O_3 + \cdot NO_2 \rightarrow O_2 + \cdot NO_3 \qquad (4.24)$$

$$RH + \cdot NO_3 \rightarrow R\cdot + HNO_3 \qquad (4.25)$$

aldehydes, and sulfur compounds (Morris and Niki, 1974: Atkinson et al., 1984a, 1984b; Atkinson, 1991), usually by hydrogen atom abstraction (Equation 4.25) to generate organic free radicals and nitric acid. Subsequent reaction of the organic free radicals with oxygen can produce hydroxy radicals (Platt et al., 1990). In the case of alkenes, spectroscopic and kinetic evidence suggests that addition of the radical to the double bond also occurs (Morris and Niki, 1974; Atkinson, 1991). These radicals, too, can (depending on their structures) be precursors of organoperoxy free radicals that also contain nitrogen, or eliminate NO_2 to form epoxides or aldehydes (Equation 4.26).

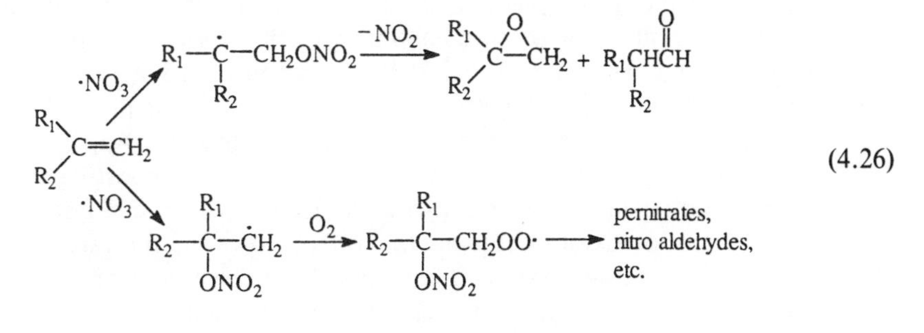

(4.26)

Hydrogen atom abstraction reactions of $\cdot NO_3$ with phenol, cresols, and methoxybenzenes are particularly rapid, indicating that this free radical has electrophilic properties. Nitrophenols are among the products of this reaction. Aromatic hydrocarbons, however, do not appear to add $\cdot NO_3$ unless they contain two or more fused rings (Atkinson et al., 1987; Atkinson, 1991).

B. HYDROGEN PEROXIDE AND ITS DECAY PRODUCTS

1. H_2O_2

The formation of hydrogen peroxide by photolysis of natural waters is discussed in Chapter 6. It is also formed by illumination of some sands and semiconductor oxides (Kormann et al., 1988; see also Section 6.E.3). Other sources of H_2O_2 include formation in the gas phase of the troposphere by the self-termination (dismutation) reaction of $\cdot OOH$ and the autooxidation of reduced transition metals such as iron (Equation 4.4). The formation and fate of H_2O_2 in the atmosphere has been reviewed (Gunz and Hoffmann, 1990; Sakugawa et al., 1990).

Several dozen measurements of atmospheric H_2O_2 levels in air, precipitation, and cloudwater have been summarized by Gunz and Hoffmann (1990). A strong tendency for H_2O_2 to partition from the vapor phase to the liquid phase is reflected by its high Henry's law coefficient of about 10^5 M/atm (Yoshizumi et al., 1984; Martin and Damschen, 1981); there are also several mechanisms for direct *in situ* formation in the aqueous phase.

Concentrations of hydrogen peroxide in atmospheric water samples such as rainwater are usually considerably higher than those determined in surface water samples. For example, Zika et al. (1982) found rainwater collected in southern Florida and the Bahamas to contain levels ranging from $1\text{–}7 \times 10^{-5}$ M. Levels of up to 3×10^{-5} M were detected in southern California and Tokyo rainwater (Kok, 1980; Yoshizumi et al., 1984), although the usual hourly mean concentration was about an order of magnitude lower. An extreme value of more than 2×10^{-4} M was reported in cloudwater from a mountaintop location in Virginia (Olszyna et al., 1988).

H_2O_2 derived from atmospheric sources seems to be able to contribute significantly to the concentration of surface waters during rainstorms (Cooper et al., 1987). Increases of more than tenfold have been demonstrated in a marine system. A more important mechanism for natural waters, however, is the photochemical generation of H_2O_2 from the excited states of humic materials (Draper and Crosby, 1983; Cooper and Zika, 1983; Cooper et al., 1988). It has been postulated that these compounds photoionize, releasing a hydrated electron which can attack dissolved oxygen, forming superoxide (Zepp et al., 1987a); however, the rate of release of electrons appears too low to account for the rate of formation of H_2O_2 in waters exposed to light. Other electron-transfer mechanisms, such as charge-transfer processes between oxygen and humic chromophores, may be responsible for superoxide and H_2O_2 formation in water.

In the atmosphere, H_2O_2 is important as an oxidant for inorganic compounds such as SO_2 (Freiberg and Schwartz, 1981; Calvert et al., 1985). Its reactions with organic compounds are not as well studied and are probably not as important as those reactions with transition metals such as iron that generate the hydroxyl radical (Faust and Hoigné, 1990). Hydrogen peroxide in water, in the absence of light or metal ion catalysts, is a rather mild and ineffective oxidant. Very electron-rich organic compounds such as anilines are stable in pure water for long periods of time in the dark, even in the presence of approximately 1000x molar excess of H_2O_2 (Larson and Zepp, 1988). There have been a few studies of the oxidation of organic sulfur compounds by H_2O_2 under atmospheric conditions; for example, dimethyl sulfide is converted by H_2O_2 to dimethylsulfoxide (Equation 4.27) in acidic droplets (Adewuyi and Carmichael, 1986).

$$(CH_3)_2S + H_2O_2 \rightarrow (CH_3)_2S{=}O + H_2O \tag{4.27}$$

The disappearance of H_2O_2 from natural waters is probably largely due to reactions involving particles, possibly including living microorganisms such as bacteria and algae (Zepp et al., 1987c). It was much more stable in filtered freshwater samples than in unfiltered water (Cooper et al., 1988; Cooper and Lean, 1989). However, according to calculations by Moffett and Zika (1987), H_2O_2 reacts significantly with reduced forms of iron and copper; these Fenton reactions (see Section 4.B.2a) may contribute importantly to the formation of hydroxyl radicals in the ocean.

2. Hydroxyl Radical

a. Formation

The hydroxyl radical, •OH, is formed in the environment by a number of important routes:

(a) photolysis of ozone in the upper troposphere:

$$O_3 + h\nu\ (<310\ nm) \rightarrow O_2 + O \xrightarrow{H_2O} 2\ HO\cdot \qquad (4.28)$$

(b) decomposition of H_2O_2 by reduced transition metal ions (the Fenton reaction:)

$$M^{n+} + H_2O_2 \rightarrow M^{n+1} + \cdot OH + HO^- \qquad (4.29)$$

(c) photolysis of iron(III) aequo complexes, especially $Fe(OH)^{2+}$, which predominates in slightly acidic (pH 3–5) waters such as rain droplets (Weschler et al., 1986; Faust and Hoigné, 1990):

$$Fe(OH)^{2+} + h\nu \rightarrow Fe^{3+} + \cdot OH \qquad (4.30)$$

(d) reaction of NO with the hydroperoxyl radical:

$$HOO\cdot + NO\cdot \rightarrow HO\cdot + NO_2\cdot \qquad (4.31)$$

(e) photolysis of water vapor, important in the upper stratosphere above 70 km, or of hydrogen peroxide, important in the lower stratosphere and troposphere:

$$H_2O + h\nu\ (<125\ nm) \rightarrow HO\cdot + H\cdot \qquad (4.32)$$

$$H_2O_2 + h\nu\ (<300\ nm) \rightarrow 2\ HO\cdot \qquad (4.33)$$

(f) reaction of water with oxygen atoms from ozone photolysis:

$$H_2O + O\cdot \rightarrow 2\ HO\cdot \qquad (4.34)$$

(g) photolysis of nitrate and nitrite anions:

$$NO_3^- + H_2O + h\nu \rightarrow \cdot NO_2 + HO\cdot + OH^- \qquad (4.35)$$

$$NO_2^- + H_2O + h\nu \rightarrow \cdot NO + HO\cdot + OH^- \qquad (4.36)$$

(h) reaction of superoxide with ozone (see Section 5.C.1), an important reaction in atmospheric water droplets and also in water treatment:

$$\cdot O_2^- + O_3 \xrightarrow{H^+} 2\ O_2 + HO\cdot \qquad (4.37)$$

Direct and indirect measurements of [HO·] in the atmosphere have indicated a range of concentrations around 10^6 molecules/cm^3 (Prinn et al., 1987; McKeen et al., 1990; Hofzumahaus et al., 1991). In natural surface freshwaters steady-state [HO·] has been estimated to be within a factor of 10 of 10^{-16} M (Mill et al., 1980: Zepp et al., 1987b; Zhou and Mopper, 1990), but in seawater its steady-state concentration is 1–2 orders of magnitude lower due to lower concentrations of nitrate and dissolved organic carbon, its likely precursors in this milieu, and high concentrations of bromide, a potent quencher (Zepp et al., 1987b; Zhou and Mopper, 1990).

The hydroxyl radical is an extraordinarily potent and unselective oxidizing agent. In recent years, it has been recognized as a fundamentally important intermediate for reactions taking place not only in the atmosphere, where its importance was first appreciated, but also in the water column and in some surface reactions.

b. Reactions with Organic Compounds

For most organic compounds that enter the troposphere, reactions with HO· govern their disappearance (Carter and Atkinson, 1985). For example, methane, which is by far the most abundant tropospheric hydrocarbon, is virtually inert to atmospheric reactions except for its reaction with ·OH (4.38), which initiates a series of free-radical reactions that lead to its conversion to oxidized forms:

$$CH_4 + \cdot OH \rightarrow \cdot CH_3 + H_2O \tag{4.38}$$

$$\cdot CH_3 + O_2 \rightarrow CH_3OO\cdot \tag{4.39}$$

$$CH_3OO\cdot + RH \rightarrow CH_3OOH \tag{4.40}$$

$$\text{(or) } CH_3OO\cdot + NO \rightarrow CH_3O\cdot + NO_2 \tag{4.41}$$

$$CH_3OOH + h\nu \rightarrow CH_3O\cdot + \cdot OH \tag{4.42}$$

$$CH_3O\cdot + O_2 \rightarrow CH_2O + HOO\cdot \tag{4.43}$$

$$CH_2O + HO\cdot \rightarrow \cdot CHO + H_2O \tag{4.44}$$

$$\cdot CHO + O_2 \rightarrow CO + HOO\cdot \tag{4.45}$$

$$CO + HO\cdot \rightarrow CO_2 + H\cdot \tag{4.46}$$

The methylperoxy radical (and probably other peroxy radicals) reacts very rapidly with NO (4.41) relative to hydrogen abstraction (4.40), so only in very clean environments will the latter reaction occur to a significant extent (McFarland et al., 1979; Hanst and Gay, 1983). In polluted atmospheres, nitrogen oxides interfere with several other of the above reactions, leading to the formation of nitrate and peroxynitrate esters (see Section 4.A.4).

Methane is probably the least reactive of all atmospheric hydrocarbons (Vaghijiani and Ravishankara, 1991). Its average atmospheric turnover time is probably about four years. Longer-chain and branched aliphatic compounds offer the possibility of more reactive secondary and tertiary alkyl radical formation that have many other subsequent possibilities for rearrangement and H-abstraction reactions that are lacking in the case of methane. The fact that higher *n*-alkanes are not produced at the earth's surface in such high amounts, combined with their greater rates of reaction with ·OH (Atkinson, 1986), explain their very small atmospheric concentrations.

Olefins react extremely rapidly with HO·, with most rate constants within an order of magnitude of the diffusion-controlled limit (Atkinson, 1986). Probably the most reactive hydrocarbon yet tested is the monoterpene myrcene (**18**), with two vinyl groups and a trisubstituted double bond. It reacts with ·OH about three times as fast as isoprene and about thirty times as fast as ethylene (Altshuller, 1983). Aromatic compounds usually react somewhat faster than alkanes and somewhat slower than alkenes.

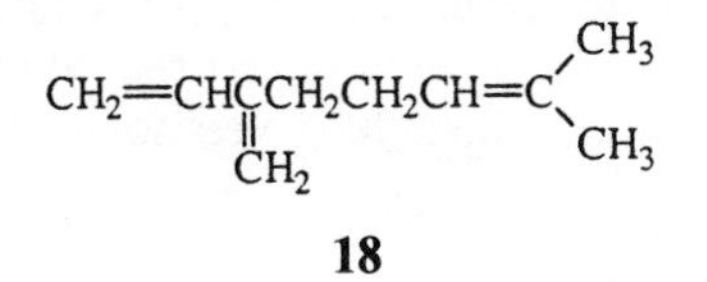

18

Aldehydes are oxidized by ·OH to acyl radicals, which may also combine with O_2 and NO but also have other modes of decomposition:

$$RCHO + \cdot OH \rightarrow RCO\cdot + H_2O \tag{4.47}$$

$$RCO\cdot + O_2 \rightarrow RCO_3\cdot \rightarrow RO\cdot + CO_2 \tag{4.48}$$

$$RCO_3\cdot + NO \rightarrow RCO_2\cdot + NO_2 \tag{4.49}$$

Aldehydes may be partly protected from further oxidation either by reaction with water to form hydrates (*gem*-diols), a reaction that is highly favored for formaldehyde but less so for higher aldehydes:

$$RCHO + H_2O \rightarrow RCH(OH)_2 \tag{4.50}$$

or by reaction with sulfur oxides or bisulfite to give acidic adducts:

$$RCHO + HSO_3^- \rightarrow RCHOHSO_3 \tag{4.51}$$

Evidence suggests that aerosol droplets containing sulfur oxides are environments where aldehydes accumulate (Waldman et al., 1982; Munger et al., 1984).

The reaction types of the hydroxyl radical with organic compounds include only two of significant importance: electron (or hydrogen-atom) abstraction, as shown above for methane, and addition to double bonds. An example of the latter reaction is the addition of HO· to benzene:

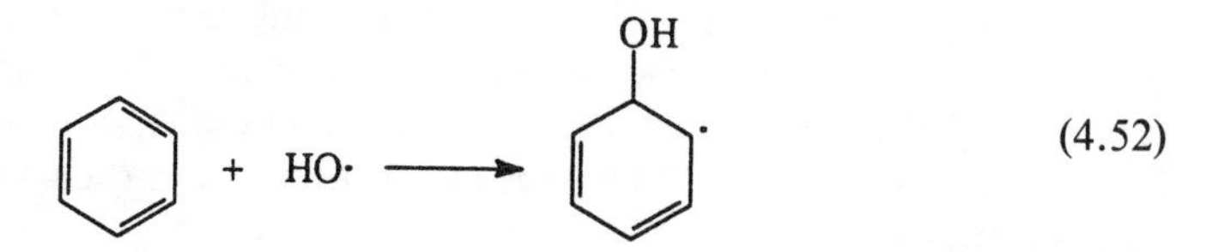

(4.52)

Although ·OH radicals are very unselective, in practice there is some discrimination among potential sites of attack. For H-abstraction, tertiary hydrogens are often somewhat more reactive than secondary or primary types, but this simplifying assumption does not hold in all cases (Atkinson, 1986); neighboring substituents on the carbon from which the hydrogen is being abstracted often influence the rate of abstraction significantly. For addition of the radical to an unsymmetrical double bond, the formation of the more stable product radical would be expected, and in most cases this does appear to occur. For example, an analysis of the products derived from attack of HO· on propylene suggest that about two-thirds of the initial adduct was **19** and the remaining third **20** (Cvetanovic, 1976).

19 **20**

For aromatic compounds, again there is some, but not much, discrimination based on electron density of the ring. The electron-rich compound anisole (methoxybenzene), for example, has a rate constant for reaction with HO· only 2.6 times larger than that of nitrobenzene, a much more electron-poor aromatic compound (Zepp et al., 1987b). Addition to the aromatic ring (to form the intermediate cyclohexadienyl radicals **21**) often predominates by a factor of approximately 10 over

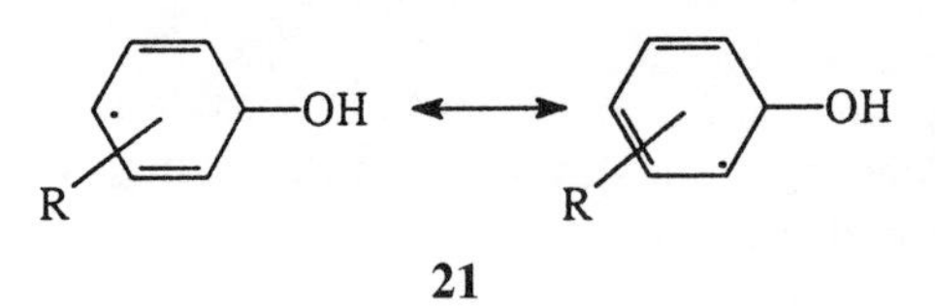

21

hydrogen atom abstraction from a side chain or functional group (Dorfman and Adams, 1973). The addition of HO· to phenol gave a mixture of *o*- and *p*-dihydroxybenzenes, with *o* predominating by nearly 2:1, suggesting an electrophilic substitution without much steric discrimination (Matsuura and Omura, 1966); halophenols, after addition of HO·, lost a halogen in the form of F^-, Cl^-, or Br^- (Ye and Schuler, 1990) and afforded the corresponding dihydroxybenzene. Benzoic acid reacts with HO· to produce, as major products, hydroxybenzoic acids (Equation 4.53a), along with varying amounts of phenol (Equation 4.53b) and benzene (Equation 4.53c), and products of cleavage of the aromatic ring (Equation 4.53d) (Ogata et al., 1980).

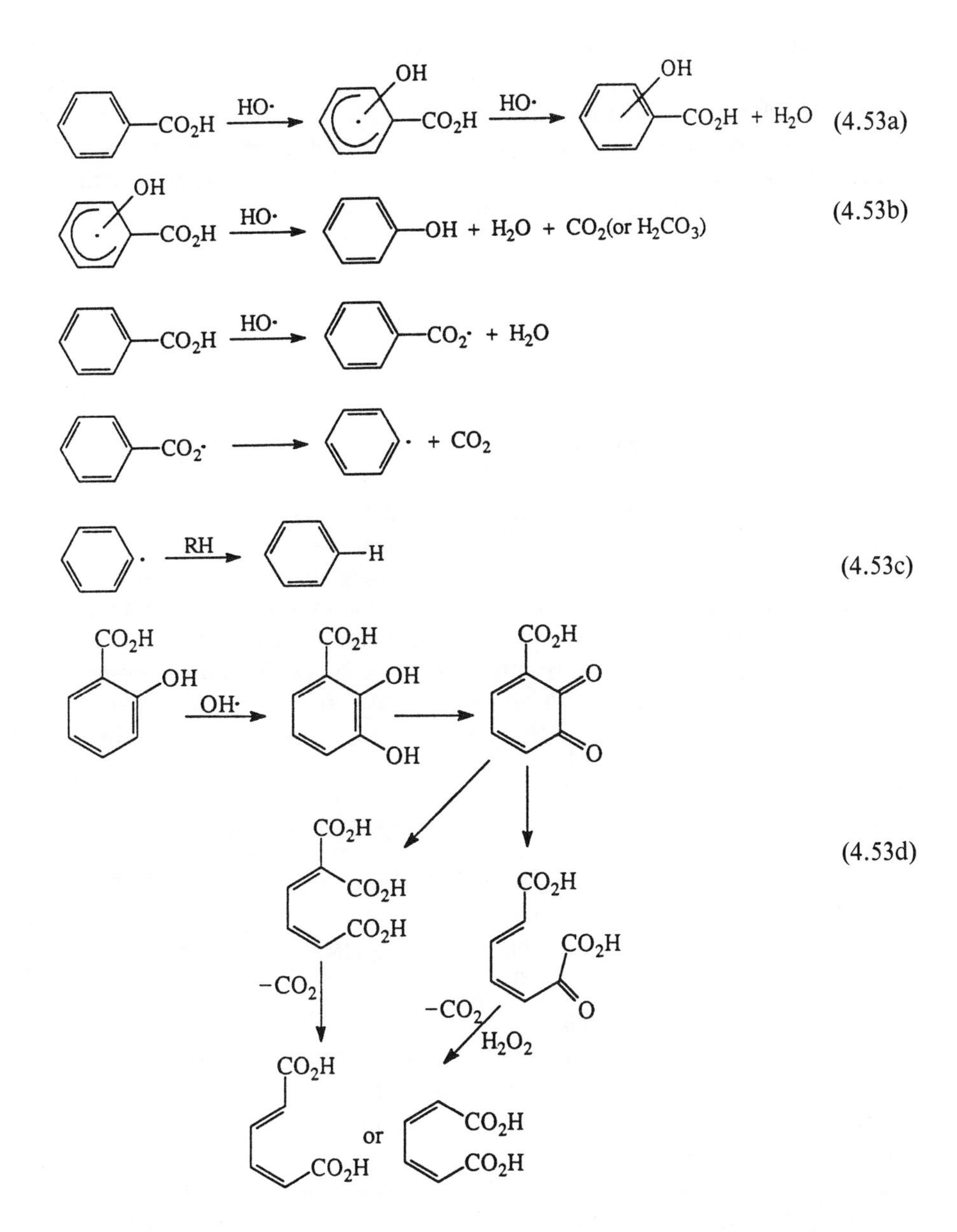

Molecules that contain both unsaturation and abstractable hydrogen atoms, such as alkylated olefins or alkylbenzenes, almost always give products derived from both pathways, although, as mentioned, addition is usually much more important. Thus, for example, in a study in which ·OH radicals were generated by photolysis of HONO, benzene was converted to phenol; toluene, mostly to cresols with some benzaldehyde; and ethylbenzene to ethylphenols, acetophenone, and benzaldehyde (Hoshino et al., 1978). The cresols were dominated by *o*-cresol, which made up about 80% of the total; the reason for this predominance is not clear. Aniline, however, appears to be an exception in that the formation of the anilino radical (**22**) is favored, whether the ·OH radical attacks at either the amino group (about 36% of all encounters) or the *o*-position (about 54%; Solar et al., 1986). Products of dimerization of the anilino radical, such as azobenzenes, azoxybenzenes, and aminobiphenyls, predominate in HO· reactions as well as many other one-electron oxidations (Chan and Larson, 1991).

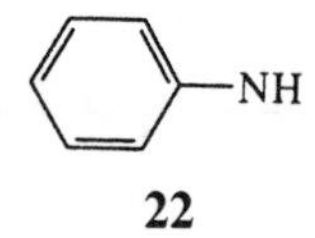

22

c. Daughter Radicals: Bromide, Carbonate, etc.

In aqueous solution, HO· has the potential for reacting with many dissolved substrates, both organic and inorganic. The principal paths for reaction will depend on the concentrations of potentially reactive solutes in the particular water under consideration. Hoigné (1983) has analyzed the situation kinetically for a typical lake water and for seawater. As Table 4.2 indicates, in seawater the principal fate of ·OH is to react with bromide by electron transfer, producing HO^- and a bromide atom, ·Br, which readily combines with another Br^- to form the bromine radical anion, $\cdot Br_2^-$. Despite the large excess of Cl^- in seawater, no evidence exists for a Br–Cl

Table 4.2. Kinetic Analysis of Typical Lake Water and Seawater

		Seawater		Lakewater	
Species	k_{OH} ($M^{-1}s^{-1}$)	Conc. (M)	Rate k_{OH}·[M] (s^{-1})	Conc. (M)	Rate k_{OH}·[M] (s^{-1})
Cl^-	<0.002	0.6	$\ll 60 \times 10^3$	—	
Br^-	$\sim 10^9$	0.8×10^{-3}	800×10^3	—	
I^- +	1×10^{10}	$\sim 10^{-7}$	1×10^3	—	
NO_2^-	10^{10}	$\sim 10^{-7}$	2×10^3	—	
HCO_3^- free	1×10^7			$\sim 1.2 \times 10^{-3}$	$\sim 12 \times 10^3$
HCO_3^- tot	10^7 (?)	2×10^{-3}	2×10^3 (?)		
CO_3^{2-}, free	3×10^8			$\sim 7 \times 10^{-6}$	$\sim 2 \times 10^3$
CO_3^{2-}, tot	3×10^8 (?)	10^{-4} (?)	30×10^3 (?)	—	—
DOM	1×10^9	0.5×10^{-6}	2.5×10^3	2.5×10^{-5}	$25 \times \ 0^3$
Xorg	5×10^9	$<10^{-8}$	<50	$<10^{-8}$	<50
$(CH_3)_2S$	5×10^9	2×10^{-9}	10		
H_2O_2	2×10^7				

Source: Hoigné (1983)

radical species (Zafiriou et al., 1987). Only trace amounts of ·OH attack other targets, principally carbonate and dissolved organic matter. The fates of the bromine radical anion in seawater are not adequately understood, although it appears to react with various components of the carbonate-bicarbonate system (including ion-pair complexes such as $MgCO_3$) to give unknown products (True and Zafiriou, 1987).

In freshwaters, where the Br^- concentrations are typically far lower, ·OH may react either with bicarbonate, carbonate, or dissolved organic matter, depending on the relative concentrations of these species. Reactions with bicarbonate and carbonate produce either the bicarbonate radical, $\cdot HCO_3$, or the carbonate radical anion, $\cdot CO_3^-$; the pKa of the acid form is about 7.6 (Eriksen et al., 1985).

In the polluted atmosphere, the principal reactions of HO· with inorganic species are its termination reactions with nitrogen oxides to form nitrous and nitric acids:

$$HO\cdot + NO\cdot \rightarrow HNO_2 \tag{4.54}$$

$$HO\cdot + NO_2\cdot \rightarrow HNO_3 \tag{4.55}$$

3. Peroxy Radicals

Some of the reactions of peroxy radicals were discussed in Section A.1 of this chapter during our analysis of autooxidation reactions. These reactive intermediates also undergo many other processes of great importance in natural systems and also in treatment systems where high concentrations of free radicals are generated.

There are few direct measurements of peroxy radical concentrations in the atmosphere (Mihelcic et al., 1978; Cantrell et al., 1984). Like many other atmospheric radical species, their numbers seem to be greatest when solar UV intensity is strongest. For 28 samples collected from flights over West Germany, values ranging from 0.05 to 0.3 ppb were obtained (Mihelcic et al., 1982). In the atmosphere, oxidizing radicals such as $RCH_2OO\cdot$ are excellent scavengers for NO, converting it to NO_2 :

$$RCH_2OO\cdot + NO \rightarrow RCH_2O\cdot + NO_2 \tag{4.56}$$

Further reaction of the alkoxy radicals, formed in this step, with O_2 affords aldehydes and hydroperoxy radicals (also capable of oxidizing NO with regeneration of ·OH):

$$RCH_2O\cdot + O_2 \rightarrow RCHO + \cdot OOH \tag{4.57}$$

$$\cdot NO + \cdot OOH \rightarrow \cdot NO_2 + \cdot OH \tag{4.58}$$

Peroxy radicals can also react directly with NO_2 to form pernitrate esters:

$$ROO\cdot + \dot{N}O_2 \rightarrow ROONO_2 \tag{4.59}$$

Another possible reaction of peroxy radicals in the atmospheric environment, particularly when nitrogen oxide concentrations are low (as they are in very clean air), is a hydrogen abstraction from HOO· to produce a hydroperoxide, ROOH. Since all the above reactions of ROO· have comparable rates, the relative concentrations of HOO· and NO_x species will determine the extent of formation of hydroperoxides, alkoxy radicals, and pernitrate esters (McFarland et al., 1979; Hanst and Gay, 1983).

Mill et al. (1980) were the first to attempt an estimation of the concentrations of peroxy radicals in natural waters. Their technique was based on an analysis of the reaction products observed when cumene and pyridine were exposed to light in natural waters, together with the known kinetics and products (Figure 4.9) of ROO· and HO· with these substrates. The calculated steady-state concentration for ROO· in a typical sunlit freshwater was about 10^{-10} M. Faust and Hoigné (1987) also proposed on kinetic grounds that transient peroxy radicals were formed when natural waters were irradiated with sunlight. They proposed that reactions with such species were the dominant sink for alkylated phenolic compounds in illuminated water bodies, and that the contributions of HO·, $^1O_2$, and other potential oxidizing species were minimal.

The fate of peroxy radicals at sediment or soil surfaces has been considered by Pohlman and Mill (1983). They examined the ability of common soil constituents to quench the reaction of an alkylperoxy radical with a reactive probe molecule, *p*-isopropylphenol. They found that reactions with organic matter were dominant over possible reactions with the mineral constituents, except for possibly Cu^{2+}; but natural humic materials appeared to be a rather poor scavenger, presumably because the bulk of their structure consisted of unreactive moieties such as polysaccharides and aliphatic chains. They also concluded that phenols and other reactive xenobiotics might be partially susceptible to removal through reaction with these radicals.

Intramolecular decay reactions of some peroxy radicals, particularly α-hydroxy types, are known (Bothe et al., 1978) in which superoxide is released.

$$RO_2\cdot + CuH \longrightarrow RO_2H + Cu\cdot$$
$$Cu\cdot + O_2 \longrightarrow CuO_2\cdot$$
$$CuO_2\cdot + CuH \longrightarrow CuO_2H + Cu\cdot$$
$$CuO_2\cdot + RO_2\cdot \longrightarrow CuO\cdot + RO\cdot + O_2$$
$$CuO\cdot \longrightarrow C_6H_5COCH_3 + CH_3\cdot$$
$$CH_3\cdot + O_2 + CuO_2\cdot \longrightarrow CH_2O + CuOH + O_2 \text{ (2 steps)}$$
$$HO\cdot + CuH \longrightarrow H_2O + Cu\cdot$$
$$HO\cdot + O_2 + CuH \longrightarrow HOC_6H_5\text{-}i\text{-}C_3H_7 + HO_2\cdot \text{ (2 steps)}$$

Figure 4.9. Reactions of cumene (CuH) with hydroxyl and peroxyl radicals. From Mill et al. (1980). Reprinted by permission of the American Association for the Advancement of Science.

$$R_2\underset{\displaystyle OO\cdot}{\underset{|}{C}}-OH \longrightarrow R_2C{=}O + HOO\cdot \tag{4.60}$$

Many of these reactions are extremely fast, with rate constants within a factor of 2–3 of pure diffusion control. The elimination need not take place in an aliphatic structure, as implied above; peroxy radicals derived from subsequent HO· and O_2 addition to aromatic rings undergo analogous eliminations (Reaction 4.52).

In addition, bimolecular decay of peroxy radicals is commonly observed. The intermediate in this type of reaction is generally accepted to be a tetroxide (**23**), that is, a species with four successive oxygen atoms (Russell, 1957). These intermediates break down by several routes to afford oxidized products and O_2 or H_2O_2 (Figure 4.10). It has even been suggested that such reactions could occur in the unpolluted troposphere (in the absence of NO), although it appears likely that in most environments the reaction with HOO· to form a hydroperoxide would prevail. This appears to be true at least for methylperoxy radicals (Cox and Tyndall, 1980).

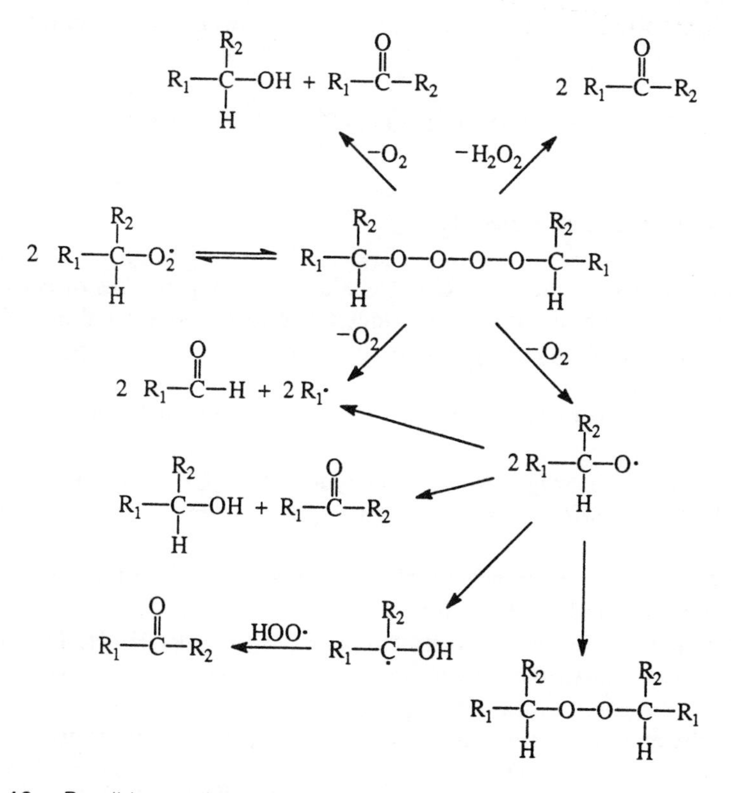

Figure 4.10. Possible reactions of secondary alkylperoxy radicals. From von Sonntag (1987). Reprinted by permission of Taylor and Francis.

ROO-OOR

23

The rates of decay of peroxy radicals are widely variable; some degrade at almost diffusion-controlled rates and others are far more stable (Howard and Scaiano, 1984). Obviously the radicals most likely to take part in reactions with other organic species will mostly be those whose self-termination rates are lower.

In the presence of oxides of nitrogen (especially NO, present in polluted atmospheres), peroxy radicals form pernitrate esters in a two-step process:

$$ROO\cdot + NO\cdot \rightarrow RO\cdot + NO_2\cdot \tag{4.61}$$

$$ROO\cdot + NO_2\cdot \rightarrow ROONO_2 \tag{4.62}$$

Addition reactions of peroxy radicals with olefins (Equation 4.63) have often been described (Mayo, 1958; Hamberg and Gotthammar, 1973). Among the stable products are epoxides, possibly formed by elimination of alkoxy radicals. The structural constraints on epoxide formation are quite stringent and the overall rate constants for their formation can vary by three or more orders of magnitude.

$$ROO\cdot + >C{=}C< \rightarrow ROO\text{-}C\text{-}C\cdot \rightarrow RO\cdot + >\overset{O}{C - C}< \tag{4.63}$$

4. Alkoxy and Phenoxy Radicals

The sources of alkoxy radicals, RO·, include the decay processes of peroxy radicals such as breakdown of Russell tetroxides, the direct or sensitized photolysis (or metal-catalyzed heterolysis) of hydroperoxides, and the reaction of peroxy radicals with NO_x or olefins.

Alkoxy radicals undergo a variety of different reactions including:

(a) reactions with oxygen to give aldehydes: $RCH_2O\cdot + O_2 \rightarrow RCHO + \cdot OOH$. This is the principal source of atmospheric formaldehyde when R = H (Lorenz et al., 1985).

(b) radical-radical reactions with NO or NO_2 in the polluted troposphere to give alkyl nitrites and nitrates, RONO and $RONO_2$, respectively.

(c) H-atom shifts or hydride abstractions: $RCH_2O\cdot \rightarrow RCHOH$. The hydrogen can come from further down the chain (intramolecular abstraction) if a suitably-sized cyclic transition state can occur (Carter et al., 1976). A 1,5-shift is by far the most favorable, since a 6-membered, unstrained transition state develops.

(d) Cleavage at the α position to give an alkyl radical and a carbonyl compound: $RR_1R_2C\text{-}O\cdot \rightarrow R\cdot + R_1R_2C{=}O$ (Al Akeel and Waddington, 1984;

Drew et al., 1985). In these decomposition reactions, the most stable alkyl radical is typically formed.

Phenoxy radicals are produced by one-electron oxidation of phenols (Equation 4.64) and also by the addition of HO· to phenols followed by water elimination (Equation 4.65):

$$C_6H_5\text{-}OH \xrightarrow{-e\cdot} [C_6H_5\text{-}O\cdot \leftrightarrows \cdot C_6H_4\text{-}OH] \qquad (4.64)$$

$$\cdot C_6H_5\text{-}OH + HO\cdot \rightarrow [HO\text{-}C_6H_5\text{-}OH] \rightarrow C_6H_5\text{-}O\cdot + H_2O \qquad (4.65)$$

The principal reaction of these compounds is dimerization ("phenol coupling") to produce new C–C or C–O bonded hydroxylated biphenyls or hydroxydiphenyl ethers. These reactions have been observed in a wide variety of environmental and biological media (Taylor and Battersby, 1967).

$$2\ PhO\cdot \rightarrow Ph\text{-}O\text{-}PhOH \text{ or } HOPh\text{-}PhOH \qquad (4.66)$$

In most instances, phenoxy radicals do not engage in propagation steps in chain reactions, consistent with their use as free radical inhibitors. They are, however, reactive enough to oxidize ascorbic acid or to reduce some quinones (Neta and Steenken, 1981).

C. SURFACE REACTIONS

Heterogeneous oxidation reactions involving solid surfaces are important contributors to the interconversion of organic substances both in the natural environment and in industrial processes, including wastewater treatment. A tremendous variety of heterogeneous-phase reactions has been studied, but very few of the data are applicable to environmental conditions. In most instances, surface reactions have been found valuable because of their preparative utility, and studies of their theoretical or mechanistic aspects have lagged. A complicating factor is that solid oxidants such as metal oxides seldom occur in nature as pure solids; they are very often part of heterogeneous mineral aggregations, such as films on other mineral surfaces. Furthermore, oxides occur in a range of particle sizes, from large rock outcrops down to barely visible clay-sized specks, with correspondingly immense fluctuations of surface areas and porosities. Oxidation reactions at oxide surfaces, like other surface reactions, nearly always depend on the sorption of the substrate on the surface (a process that can be slowed by electronic and steric incompatibilities, as well as limitations on the number of active sorption sites in the crystal), and subsequent oxidation steps reveal a variety of mechanisms, such as charge- or electron-transfer and the formation of intermediate oxidizing species such as free radicals (Voudrias and Reinhard, 1986).

Most laboratory studies of solids that have oxidizing activity have been performed using metal oxides such as those of silver, manganese, chromium, mercury, and lead (Pickering, 1966). In the environment, however, from the limited number of studies that have been performed, it appears that iron and manganese oxides are probably the most important naturally occurring catalysts.

1. Clays

Although clays and related aluminosilicates have been used for decades to catalyze a tremendous variety of reactions of industrial significance, such as the isomerization of petroleum hydrocarbons, less attention has been given to their roles as environmental reactants. Most (possibly all) of the reactions of clays as oxidation catalysts appear to depend on redox-active metal ions such as Cu, Fe, and other trace elements. Soil- and clay-catalyzed oxidations have been reviewed by Theng (1974), Pinnavaia (1983), and Dragun and Helling (1985). The latter authors classified 93 compounds that had been reported to undergo such oxidations into four structural groups: (1), aromatic compounds with electron-withdrawing and weak electron-donating fragments; (2), aromatic compounds with electron-withdrawing and strong electron-donating groups; (3), aromatic compounds with only electron-donating substituents; (4), aromatic compounds containing extensive conjugation. In addition, there is one report (Mortland and Boyd, 1989) of a polymerization reaction of chlorinated ethylenes with Cu(II) smectite in refluxing, nitrogen-purged CCl_4, though the environmental significance of this observation remains to be established. Aliphatic compounds are apparently unlikely, in general, to be oxidized by soils or clays. A useful table of redox potentials in Dragun and Helling's paper indicates a likelihood that many compounds having redox potentials of less than 0.8 V (the redox potential of some highly oxidized soils) would be predicted to undergo oxidation.

For a long time it has been known that aromatic amines sorbed to certain clay minerals undergo color-forming reactions (Hauser, 1955; Solomon et al., 1968). In the last decade or so, some of the mechanisms of these transformations are beginning to be worked out; generally, they appear to involve electron transfer from the amine to a transition metal salt bound to the clay lattice. Iron appears to coordinate with the π-electrons, whereas Cu^{2+} tends to coordinate with the lone pair on nitrogen (Moreale et al., 1985). A radical cation of the amine and a reduced metal ion are the preliminary products. Air-dried smectite and montmorillonite, for example, sorbed aniline and *p*-chloroaniline, forming charge-transfer complexes as shown by infrared and electron spin resonance spectroscopic evidence (Cloos et al., 1979, 1981). Formation of the complexes was facilitated by the presence of Fe(III) on the clay surface. The oxidation products were not characterized, but other investigators have shown that dimers, trimers, and higher polymers of anilines and partial oxidation products such as azobenzene (Ph–N=N–Ph) are formed under similar conditions, presumably by coupling of intermediate free radicals such as PhNH· and subsequent oxidation of the coupling products (see,

e. g., Pillai et al., 1982). In some cases, oxygen was shown not to be necessary for aniline oxidation to proceed if the clay contained Fe^{3+} or Cu^{2+}, but when these ions were almost completely replaced with alkali or alkaline earth metals, O_2 was required (Furukawa and Brindley, 1973).

Similar mechanisms appear to be predominant in the oxidative oligomerization or polymerization of other types of aromatic compounds (hydrocarbons, ethers, phenols, etc.) on smectites containing Fe(III) or Cu(II) (Mortland and Halloran, 1976; Soma and Soma, 1983; Sawnhey et al., 1984). Electron transfer to Cu(II) smectite by 4-chloroanisole (**24**) results in a radical cation intermediate that undergoes dechlorination to chloride ion and the coupling product, 4,4′-dimethoxybiphenyl (Govindaraj et al., 1987). Even dibenzodioxin (unchlorinated) forms a radical cation on Cu(II)-smectite and polymerizes (Boyd and Mortland, 1985).

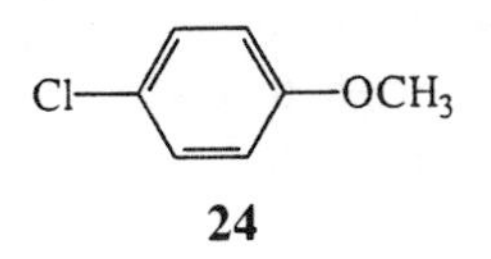

Clays, sediments, and soil dusts promote the oxidation of various reactive organic species such as catechol (**25**: Larson and Hufnal, 1980), pyrogallol (**26**: Wang et al., 1983) and parathion (**27**: Spencer et al., 1980). In the case of parathion, the reaction product was the highly toxic derivative, paraoxon (**28**). Kaolinite clays were more reactive than montmorillonite in this reaction, which may also have involved atmospheric ozone.

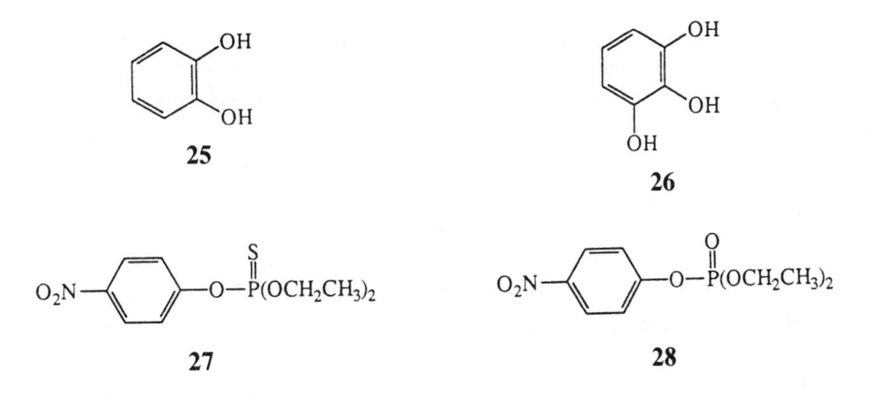

2. Silicon Oxides

Dechlorination of chlorofluorocarbons adsorbed to dry desert sand has been reported by Bahadir et al. (1978). As much as 71% of adsorbed CCl_2F_2 was observed to react in this system. No input of external heat or light energy was required. Sunlight has also been observed to promote oxidation of organic matter in the presence of sand or silica gel (see Section 6.E.3a). The mechanisms of these reactions are poorly understood. They may involve trace amounts of redox-active cations such as iron, managanese, or copper (Schofield et al., 1964).

3. Aluminum Oxides

As is shown by its usefulness in chromatography, alumina is a powerful sorbent for many classes of organic compounds. It has been long known as a reagent that can bring about changes in organic molecules such as isomerizations, hydrolyses, addition reactions, etc. (Posner, 1983), but its use as a redox catalyst has been sporadic. Cannizzaro-like disproportionation reactions of aldehydes to the corresponding alcohols and carboxylic acids in the presence of neutral alumina at room temperature have been reported (Lamb et al., 1974). Both soluble aluminum ($AlCl_3$) and aluminum oxides were reported to catalyze the oxidative polymerization of catechol in aqueous acidic solution (pH <4.4) to green, melanin-like products (McBride et al., 1988). The reactions observed were slow, on the order of weeks, and required concentrations of the order of 10 mM. The authors proposed that the complexed aluminum, by displacing protons, permitted greater electron density to be localized into the aromatic ring, facilitating oxidation.

4. Iron Oxides

As one of the most geologically common elements, iron forms minerals such as oxides, hydrous oxides, and sulfides that are practically ubiquitous constituents of soils and sediments. In aerobic environments at ordinary pHs, the thermodynamically stable form of iron is the ferric ($3+$) state, which is a reactive enough oxidant to be reduced under a variety of conditions. In the presence of light, for example, dissolved or suspended iron oxyhydroxides react with natural humic materials, becoming reduced to the ferrous (Fe^{2+}) state (see Section 6.F.2). Several authors have shown that various iron oxides have the ability to oxidize polyphenols, often to polymeric species (Larson and Hufnal, 1980; McBride, 1987). Goethite and hematite both promoted oxygen uptake in solutions of catechol or hydroquinone (McBride, 1987). Hydroquinone took part in redox reactions on the surface of a goethite-ferrihydrite mixture to give, among other products, *p*-benzoquinone (Kung and McBride, 1988b). In the process, an initial fast electron transfer that converted the Fe(III) on the oxide surface to Fe(II) was followed by a partial back-reaction between the quinone and Fe(II) to re-form hydroquinone. Over a long period of time (>50 hr), polymeric material was produced on the oxide surface and in solution.

Iron(III) oxides, especially hematite, Fe_2O_3, were found to catalyze the oxidation of the common soil phenolic acids, sinapic acid (**29**), and (to a lesser extent) ferulic acid (**30**; Lehmann et al., 1987). Salicylic acid (**31**), while strongly bound by iron oxide, was resistant to oxidation (McBride, 1987).

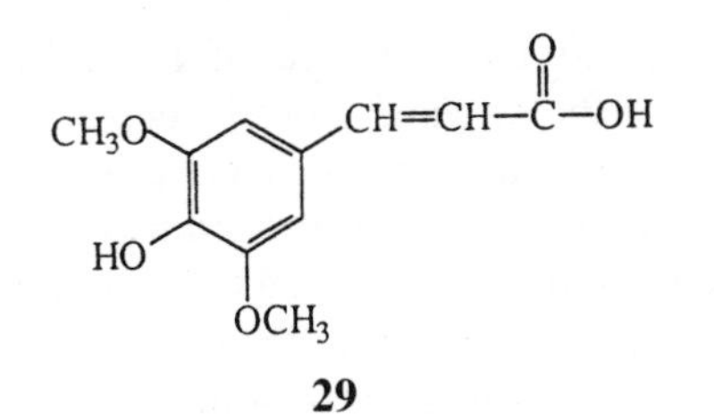

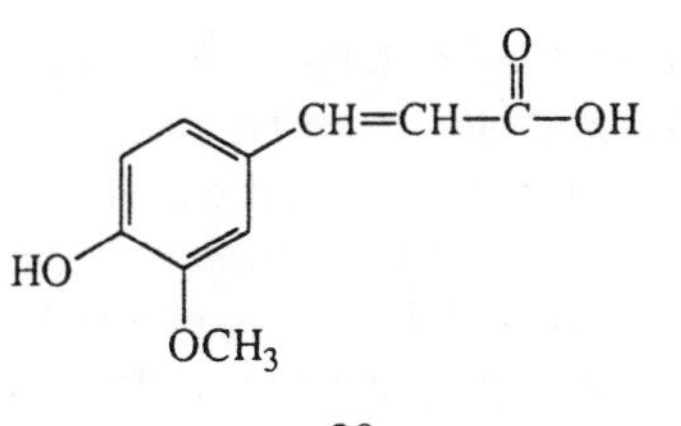

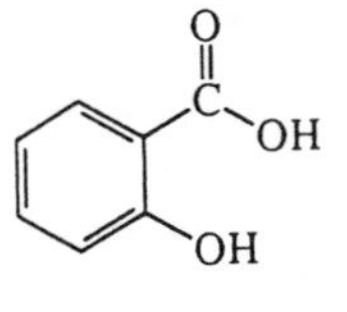

31

5. Manganese Oxides

Manganese dioxide is a well-known synthetic oxidant in preparative organic chemistry, being especially useful as a mild oxidant for certain alcohols to carbonyl compounds (Pickering, 1966). It is also a common mineral phase, often forming nodules in seawater and freshwater, as well as occurring as films on other minerals. Manganese oxides have been shown to be photoreduced by the natural organic matter of seawater or freshwater, converting them to the soluble Mn(II) form which is important in algal nutrition (Sunda et al., 1983; Waite et al., 1988).

A number of investigations have indicated that manganese oxides are potent catalysts for oxidation, especially of phenols (Larson and Hufnal, 1980; Stone and Morgan, 1984; McBride, 1987; Stone, 1987; Ulrich and Stone, 1989). Phenols with electron-donating substituents such as hydroxyl, alkyl, alkoxyl, etc., are usually oxidized much more rapidly than those possessing electron-withdrawing groups. One suggested mechanism for oxidation on these surfaces is electron transfer from the phenol to the oxide, followed by release of Mn^{2+} into solution, and further reaction of the phenol-derived radicals; in high enough phenol concentrations, the preferred reaction is probably oxidative coupling to form polymers. Colored products were observed in activated MnO_2-promoted reactions of catechol (Larson and Hufnal, 1980). Catechol and hydroquinone were oxidized by birnessite, a form of manganese oxide found in soils. No products were identified except for semiquinone and hydroxylated semiquinone radicals (McBride, 1989). Hydroquinone was also oxidized in an aqueous suspension of hausmannite (Mn_3O_4) by a process that led to the dissolution of the oxide, the release of Mn^{2+}, and the formation of phenolic polymers. Electron spin resonance data revealed the presence of the *p*-benzosemiquinone radical anion, a likely precursor of the polymeric products (Kung and McBride, 1988). The manganese oxide was shown to be a much better oxidation catalyst for this phenol than was the mixed iron oxide reagent also studied by these authors (*vide supra*). Manganese silicates such as tephroite and other Mn-containing minerals also polymerized hydroquinone (Shindo and Huang, 1985).

Manganese dioxide oxidized five common soil phenolic acids—vanillic (**32**), syringic (**33**), *p*-coumaric (**34**), ferulic (**30**), and sinapic (**29**) acids—although *p*-hydroxybenzoic acid (**35**) was inert (Lehmann et al., 1987). Salicylic acid (**31**), although strongly held by manganese oxide, was resistant to oxidation (McBride, 1987), as were several aliphatic acids such as citric, lactic, pyruvic and tartaric acids (Stone and Morgan, 1984).

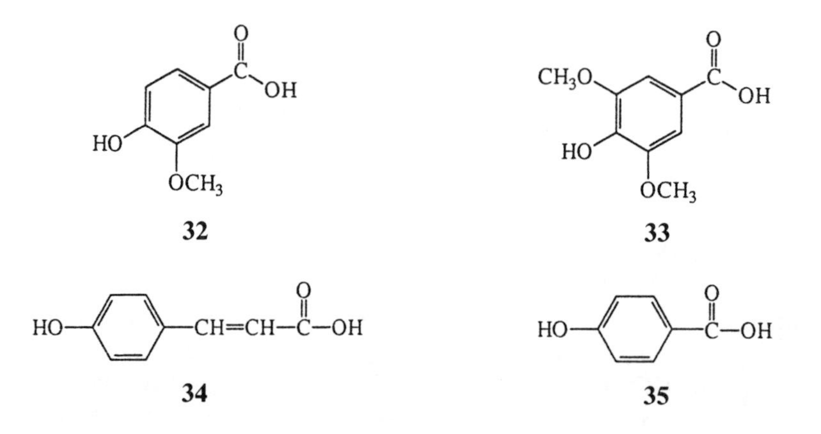

A citrate extract from soil, previously thought to contain oxidative enzymes, was shown to contain manganese in a form capable of oxidative coupling activity toward 2,6-dimethoxyphenol and other reactive aromatic substrates. A solution of MnO_2 in citrate buffer exhibited many of the same chemical properties as the soil extract (Loll and Bollag, 1985).

Manganese dioxide was also reactive toward other electron-rich aromatic compounds such as anilines (Laha and Luthy, 1990). The order of reactivity of substituted anilines toward MnO_2 (alkoxy > alkyl > Cl > -COOH > NO_2) was that expected for other oxidizing agents. The reaction products were mainly azobenzenes (PhN-=NPh) and azoxybenzenes

$$\begin{array}{c} O^- \\ | \\ (PhN = N_+ Ph) \end{array}$$

suggesting a radical-mediated oxidative coupling mechanism.

D. THERMAL OXIDATIONS

1. Combustion and Incineration

Molecular oxygen is the principal oxidant in the ordinary combustion of fuels. Burning is essentially a high-temperature analog of autooxidation, in which the spin barrier to the initiation step (Equation 4.67) is overcome by use of high temperatures.

$$RH + \cdot O\text{-}O\cdot \rightarrow R\cdot + HOO\cdot \tag{4.67}$$

Oxygen atoms and hydroxyl and hydroperoxyl radicals are also important oxidants in flames (Gardiner, 1984; Miller and Fisk, 1987).

A great variety of organic combustion products have been identified (Schumacher et al., 1977; Junk and Ford, 1980; Hawthorne et al., 1988). The mechanisms of the reaction of oxygen with combustible species are, however, very poorly understood. Even such a simple fuel molecule as methane has very complex behavior during combustion. The methyl free radical formed during the initiation step has a number of possible fates including recombination to form ethane (Equation 4.68), common in fuel-rich flames:

$$2\ \cdot CH_3 \rightarrow H_3C\text{-}CH_3 \tag{4.68}$$

Further oxidation of $\cdot CH_3$ proceeds by intermediate steps that are not entirely certain, involving formaldehyde, $\cdot CHO$, and CO, which is finally converted to the ultimate combustion product, CO_2, by the reaction

$$CO + \cdot OH \rightarrow CO_2 + \cdot H \tag{4.69}$$

This reaction is very exothermic and is responsible for much of the heat generation during the combustion of fuels (Miller et al., 1990). The hydrogen atoms formed in the reaction are very important chain-carriers because they produce highly reactive oxidizing free radicals when they combine with molecular oxygen:

$$\cdot H + O_2 \rightarrow \cdot OH + \cdot O\cdot \tag{4.70}$$

Polycyclic aromatic hydrocarbons (PAHs) are produced in virtually all combustion processes due to chemical recombination of organic radical intermediates produced by "cracking" of larger organic molecules. The center of a flame has a reducing atmosphere; under these conditions small C_2–C_4 unsaturated fragments combine in radical chain reactions with other hydrocarbon fragments before the radical intermediates can be intercepted by oxygen. The formation of the "first ring" having six carbon atoms, either molecular benzene or the phenyl radical, is a key target of current research in this area (Miller et al., 1990). Badger and associates have shown that PAHs form by successive addition of C_2 building units onto the "first ring" $^-C_6$-C_2, C_6-C_4, and alkylated tetrahydronaphthalene intermediates (Figure 4.11, Badger and Spotswood, 1960; Badger et al., 1966). The intermediate radicals are strongly resonance-stabilized.

Preformed aromatic hydrocarbons appear to serve as "templates" for the formation of PAHs; thus, burning benzene leads to a far greater yield of PAHs than burning aliphatic hydrocarbons. Unsubstituted PAHs are typical of high-temperature combustion processes, whereas lower-temperature flames afford a large number of PAHs substituted with alkyl (primarily methyl) groups (Hites, 1976).

The presence of chlorinated organic matter in a fuel will result in the formation of objectionable substances like HCl, halogenated acids, and other organochlorine compounds in the incinerator exit gases and particles (Eiceman et al., 1979: Mowrer and Nordin, 1987). During the combustion of chlorine-containing organic materials such as polyvinyl chloride and polyvinylidene chloride (**36** and **37**, respectively), new

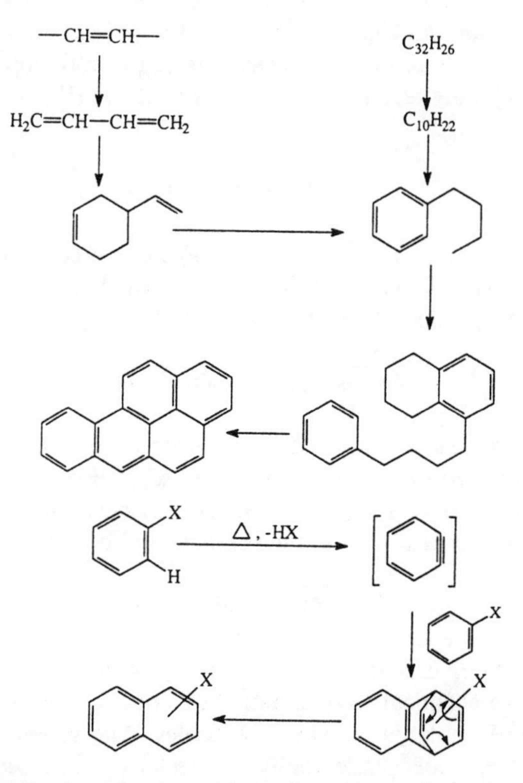

Figure 4.11. Generalized mechanism of formation of aromatic hydrocarbons in reducing flames. From J. M. Neff, *Polycyclic Hydrocarbons in the Aquatic Environment* (1979). Reprinted by permission of Applied Science Publishers.

organic compounds can be produced, including chlorinated benzenes, styrenes, naphthalenes, and biphenyls (Ahling et al., 1978; Yasuhara and Morita, 1988). Chlorinated hydrocarbons are also known to produce the highly toxic chlorinated dibenzodioxins (CDDs: **38**) and chlorinated dibenzofurans (CDFs: **39**) upon partial oxidation (Choudhry et al., 1982: Shaub and Tsang, 1983). Model calculations using rate constants for gas-phase reactions whose temperature variances are known suggest that CDDs and CDFs will form in the vapor phase in appreciable amounts from precursors such as chlorophenols and chlorinated diphenyl ethers at temperatures from about 600° to 900°C (Shaub and Tsang, 1983).

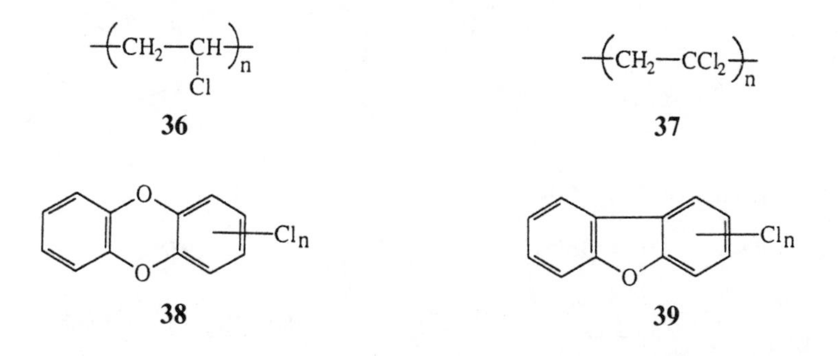

Several researchers have also reported the formation of chlorinated organic compounds during the combustion of organic compounds that do not contain chlorine in the presence of an inorganic chlorine source. Eklund et al. demonstrated that phenol and HCl, when heated to 550°C for 5 min, produced more than 50 identifiable chlorinated organic products (1986). Ferric chloride and benzene have been shown to produce CDDs and CDFs upon heating (Nestrick et al., 1987).

Surface reactions, such as free-radical combustion reactions that take place in the presence of fly ash (a mixture of silica and metal oxides) also appear to be of great potential importance in producing organochlorine compounds. For example, Hoffman et al. (1990) demonstrated that the interaction of gaseous HCl with fly ash produced a reagent that was capable of chlorinating a wide range of aromatic compounds at the rather low temperature of 80°C. These authors interpreted their data to mean that iron(III) was taking part in the chlorination reactions, but the roles of various other transition metal ions such as Cu and Mn species also need to be examined. Furthermore, the roles of HCl and the chlorine atom, Cl·, are still uncertain.

2. Wet Oxidation

Wet air oxidation is a combustion process, with atmospheric oxygen as the oxidant, that takes place in the presence of water. Very high pressures (2–15 MPa) and rather elevated temperatures (125–325°C) are normally required to accomplish significant destruction of dissolved organic pollutants (Laughlin et al., 1983). Com-

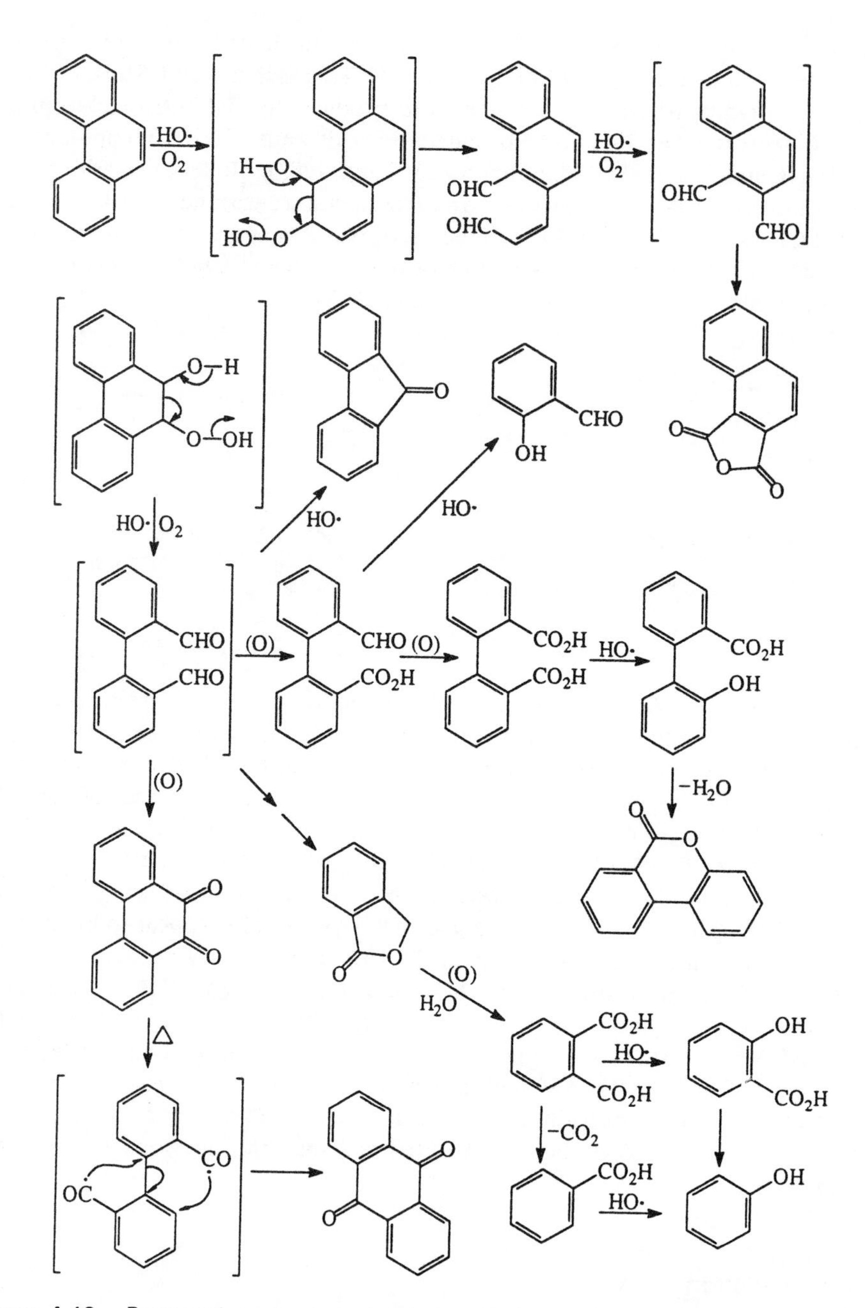

Figure 4.12. Proposed mechanism of wet air oxidation of phenanthrene. Compounds in brackets are postulated intermediates. Reprinted with permission from *Water Research*, Vol. 22, Larson et al. Some intermediates in the wet air oxidation of phenanthrene adsorbed on powdered activated carbon. Copyright 1988 Pergamon Press PLC.

mercial reactors are available that allow continuous processing of wastewater, but most laboratory studies have been performed in pressurized batch reactors.

Many studies of wet air oxidation have claimed highly efficient conversion of DOC to low molecular weight polar species such as short-chain organic acids and diacids. Not much is known about the mechanisms of the reactions taking place in wet oxidation. Hydroxyl radicals have been suggested as intermediates, but very little information of a mechanistic nature is available to prove or disprove this hypothesis. In a study of wet air oxidation of phenanthrene (Larson et al., 1988), the structures of the partial oxidation products identified suggested a radical mechanism that could have been initiated by ·OH (Figure 4.12).

REFERENCES

Adewuyi, Y. G., and G. R. Carmichael. 1986. Kinetics of oxidation of dimethyl sulfide by hydrogen peroxide in acidic and alkaline medium. *Environ. Sci. Technol.* 20: 1017–1022.

Ahling, B., A. Bjørseth, and G. Lunde. 1978. Formation of chlorinated hydrocarbons during combustion of poly(vinyl chloride). *Chemosphere* 7: 799–806.

Al Akeel, N. Y., and D. J. Waddington. 1984. Reactions of oxygenated radicals in the gas phase. Part 16. Decomposition of isopropoxyl radicals. *J. Chem. Soc. Perkin Trans. II*: 1575–1579.

Altshuller, A. P. 1983. Natural volatile organic substances and their effect on air quality in the United States. *Atmos. Environ.* 17: 2131–2165.

Altshuller, A. P., S. L. Kopczynski, D. Wilson, W. A. Lonneman, and F. D. Sutterfield. 1969. Photochemical reactivities of n-butane and other paraffinic hydrocarbons. *J. Air Pollut. Control Assoc.* 19: 787–796.

Amberlang, J. C., R. H. Kline, O. M. Lorenz, C. R. Parks, C. Wadelin, and J. R. Shelton. 1963. Antioxidants and antiozonants for general purpose elastomers. *Rubber Chem. Technol.* 36: 1497–1541.

Anderson, J. G., H. J. Grassl, R. E. Shetter, and J. J. Margitan. 1981. HO_2 in the stratosphere: three *in situ* observations. *Geophys. Res. Lett.* 8: 289–292.

Arnold, S. J., E. A. Ogryzlo, and H. Witzke. 1964. Some new emission bands of molecular oxygen. *J. Chem. Phys.* 40: 1769–1770.

Atkinson, R. 1986. Kinetics and mechanisms of the gas phase reactions of the hydroxyl radical with organic compounds under atmospheric conditions. *Chem. Rev.* 86: 69–201.

Atkinson, R. 1991. Kinetics and mechanism of the gas-phase reaction of the NO_3 radical with organic compounds. *J. Phys. Chem. Ref. Data* 20: 459–507.

Atkinson, R. and W. P. L. Carter. 1984. Kinetics and mechanism of the gas phase reactions of ozone with organic compounds under atmospheric conditions. *Chem. Rev.* 84: 437–470.

Atkinson, R., K. R. Darnall, A. C. Lloyd, A. M. Winer, and J. N. Pitts, Jr. 1979. Kinetics and mechanisms of the reaction of the hydroxyl radical with organic compounds in the gas phase. *Advan. Photochem.* 11: 375–488.

Atkinson, R., C. N. Plum, W. P. L. Carter, A. M. Winer, and J. N. Pitts, Jr. 1984a. Rate constants for the gas phase reactions of NO_3 radicals with a series of organics in air at $298 \pm 1°K$. *J. Phys. Chem.* 88: 1210–1215.

Atkinson, R., S. M. Aschmann, A. M. Winer, and J. N. Pitts, Jr. 1984b. Kinetics of the gas phase reactions of NO_3 radicals with a series of dialkenes, trialkenes, cycloalkenes, and monoterpenes at $295 \pm 1°$ K. *Environ. Sci. Technol.* 18: 370–375.

Atkinson, R., S. M. Aschmann, and A. M. Winer. 1987. Kinetics of the reactions of NO_3 radicals with a series of aromatic compounds. *Environ. Sci. Technol.* 21: 1123–1126.

Badger, G. M. and T. M. Spotswood. 1960. The formation of aromatic hydrocarbons at high temperatures. Part IX. The pyrolysis of toluene, ethylbenzene, propylbenzene, and butylbenzene. *J. Chem. Soc.* 4420–4444.

Badger, G. M., J. K. Donnelly, and T. M. Spotswood. 1966. The formation of aromatic hydrocarbons at high temperatures. Part XXVII. The pyrolysis of isoprene. *Austr. J. Chem.* 19: 1023–1044.

Bahadir, M., S. Gäb, J. Schmitzer, and F. Korte. 1978. Degradation of CCl_2F_2; formation of CO_2 upon adsorption on Mecca sand. *Chemosphere* 941–942.

Baxter, R. M. and J. H. Carey. 1982. Reactions of singlet oxygen in humic waters. *Freshwat. Biol.* 12: 285–292.

Behymer, T. D. and R. A. Hites. 1988. Photolysis of polycyclic aromatic hydrocarbons adsorbed on fly ash. *Environ. Sci. Technol.* 22: 1311–1319.

Bellus, D. 1979. Physical quenchers of molecular oxygen. *Advan. Photochem.* 11: 105–205.

Berenbaum, M. R. and R. A. Larson. 1988. Flux of singlet oxygen from leaves of phototoxic plants. *Experientia* 44: 1030 -1032.

Bielski, B. H. J. 1978. Reevaluation of the spectral and kinetic properties of $HO_2\cdot$ and of $\cdot O_2^-$ free radicals. *Photochem. Photobiol.* 28: 645–649.

Bielski, B. H. J. 1983. Evaluation of the reactivities of HO_2/O_2^- with compounds of biological interest. In Cohen, G. and R. A. Greenwald, Eds. *Oxy Radicals and Their Scavenger Systems*. Vol. 1, Molecular aspects. (Elsevier, Amsterdam), pp. 1–7.

Bielski, B. H. J. and Shiue, G. G. 1979. Reaction rates of superoxide radicals with the essential amino acids. In *Oxygen Free Radicals and Tissue Damage*, CIBA foundation symposium 65. *Excerpta Medica*, Amsterdam, pp. 43–56.

Bielski, B. H. J., D. E. Cabelli, R. L. Arudi, and A. B. Ross. 1985. Reactivity of HO_2/O_2^- radicals in aqueous solution. *J. Chem. Phys. Ref. Data* 14: 1041–1100.

Bolland, J. L. 1949. Kinetics of olefin oxidation. *Quart. Rev.* 3: 1–21.

Bonneau, R., R. Pottier, O. Bagno, and J. Joussot-Dubien. 1975. pH dependence of singlet oxygen production in aqueous solutions using thiazine dyes as photosensitizers. *Photochem. Photobiol.* 21: 159–163.

Bothe, E., Schuchmann, M. N., D. Schulte-Frohlinde, and C. von Sonntag. 1978. $HOO\cdot$ elimination from α-hydroxyalkylperoxyl radicals in aqueous solution. *Photochem. Photobiol.* 28: 639–644.

Boyd, S. A. and M. M. Mortland. 1985. Dioxin radical formation and polymerization on Cu(II)-smectite. *Nature* 316: 532–535.

Braun, A. M., F. H. Frimmel, and J. Hoigné. 1986. Singlet oxygen analysis in irradiated surface waters. *Internat. J. Environ. Anal. Chem.* 27: 137–149.

Burton, G. W. and K. U. Ingold. 1984. β-Carotene: an unusual type of lipid antioxidant. *Science* 224: 569–573.

Callen, D. F. and R. A. Larson. 1978. Toxic and genetic effects of fuel oil hydroperoxides in *Saccharomyces cerevisiae*. *J. Toxicol. Environ. Health* 4: 913–917.

Calvert, J. G., A. Lazrus, G. L. Kok, B. G. Heikes, J. G. Walega, J. Lind, and C. A. Cantrell. 1985. Chemical mechanisms of acid generation in the troposphere. *Nature* 317: 27–35.

Cantrell, C. A., D. H. Stedman, and G. J. Wendell. 1984. Measurement of atmospheric peroxy radicals by chemical amplification. *Anal. Chem.* 56: 1496–1502.

Carter, W. P. L. and R. Atkinson. 1985. Atmospheric chemistry of alkanes. *J. Atmos. Chem.* 3: 377–405.

Carter, W. P. L., K. R. Darnall, A. C. Lloyd, A. M. Winer, and J. N. Pitts, Jr. 1976. Evidence for alkoxy radical isomerization in photooxidations of C_4-C_6 alkanes under simulated atmospheric conditions. *Chem. Phys. Lett.* 42: 22–27.

Chan, W. F. and R. A. Larson. 1991. Formation of mutagens from the aqueous reactions of ozone and anilines. *Water Res.* 25: 1529–1538.

Chou, P.-T. and A. U. Khan. 1983. L-ascorbic acid quenching of singlet delta molecular oxygen in aqueous media: generalized antioxidant property of vitamin C. *Biochem. Biophys. Res. Commun.* 115: 932–937.

Choudhry, G. G., K. Olie, and O. Hutzinger. 1982. In *Chlorinated dioxins and related compounds*. O. Hutzinger, R. W. Frei, E. Merian, and F. Pocchiari, Eds. (Pergamon Press, New York).

Cloos, P., A. Moreale, C. Broers, and C. Badot. 1979. Adsorption and oxidation of aniline and p-chloroaniline by montmorillonite. *Clay Min.* 14: 307–321.

Cloos, P., C. Badot, and A. Herbillon. 1981. Interlayer formation of humin in smectites. *Nature* 289: 391–393.

Cooper, W. J. and D. R. S. Lean. 1989. Hydrogen peroxide concentration in a northern lake: photochemical formation and diel variability. *Environ. Sci. Technol.* 23: 1425–1428.

Cooper, W. J. and R. G. Zika. 1983. Photochemical formation of hydrogen peroxide in surface and ground waters exposed to sunlight. *Science* 222: 711–712.

Cooper, W. J., E. S. Saltzman, and R. G. Zika. 1987. The contribution of rainwater to variability in surface ocean hydrogen peroxide. *J. Geophys. Res.* (*Oceanogr.*) 92: 2970–2980.

Cooper, W. J., R. G. Zika, R. G. Petasne, and J. M. Plane. 1988. Photochemical formation of H_2O_2 in natural waters exposed to sunlight. *Environ. Sci. Technol.* 22: 1156–1160.

Cope, V. W. and D. R. Kalkwarf. 1987. Photooxidation of selected polycyclic aromatic hydrocarbons and pyrenequinones coated on glass surfaces. *Environ. Sci. Technol.* 21: 643–648.

Cox, R. A. and G. S. Tyndall. 1980. Rate constants for the reactions of CH_3O_2 with

HO_2, NO and NO_2 using molecular modulation spectrometry. *J. Chem. Soc. Faraday Trans. II* 76: 153–163.

Cvetanovic, R. J. 1976. Chemical kinetic studies of atmospheric interest. Paper presented at 12th Sympos. on Free Radicals, Laguna Beach, CA.

Darnall, K. R. and J. N. Pitts, Jr. 1970. Peroxyacetyl nitrate, a novel reagent for oxidation of organic compounds. *J. Chem. Soc. Chem. Commun.* 1305–1306.

Dlugi, R. and H. Güsten. 1983. The catalytic and photocatalytic activity of coal fly ashes. *Atmos. Environ.* 17, 1765–1771.

Dorfman, L. M. and G. E. Adams. 1973. Reactivity of the hydroxyl radical in aqueous solutions. Report NSRDS-NBS 46. National Bureau of Standards, Washington, DC.

Dragun, J. and C. S. Helling. 1985. Physicochemical and structural relationships of organic chemicals undergoing soil- and clay-catalyzed free-radical oxidation. *Soil Sci.* 139: 100–111.

Draper, W. M. and D. G. Crosby. 1976. Measurement of photochemical oxidants in agricultural field water. Paper presented at 172nd Natl. Mtg., Amer. Chem. Soc., San Francisco, CA.

Draper, W. M. and D. G. Crosby. 1983. The photochemical generation of hydrogen peroxide in natural waters. *Arch. Environ. Contam. Toxicol.* 12: 121–126.

Drew, R. M., J. A. Kerr, and J. Olive. 1985. Relative rate constants of the gas phase decomposition of the *s*-butoxy radical. *Internat. J. Chem. Kinet.* 17: 167–176.

Duchstein, H.-J. and G. Wurm. 1984. Juglonbildung, eine Indikatorreaktion für molekularen aktivieren Sauerstoff in Singlettzustand. *Arch. Pharm.* 317: 809–812.

Eiceman, G. A., R. E. Clement, and F. W. Karasek. 1979. Analysis of fly ash from municipal incinerators for trace organic compounds. *Anal. Chem.* 51: 2343–2350.

Eklund, G., J. R. Pedersen, and B. Stromberg. 1986. Phenol and HCl at 550°C yield a large variety of chlorinated toxic compounds. *Nature* 320: 155–156.

Eriksen, T. E., J. Lind, and G. Merényi. 1985. On the acid-base equilibrium of the carbonate radical. *Radiat. Phys. Chem.* 26: 197–199.

Faust, B. C. and J. Hoigné. 1987. Sensitized photooxidation of phenols by fulvic acid in natural waters. *Environ. Sci. Technol.* 21: 957–964.

Faust, B. C. and J. Hoigné. 1990. Photolysis of Fe(III)-hydroxy complexes as sources of OH radicals in clouds, fog, and rain. *Atmos. Environ.* 23: 235–240.

Finlayson-Pitts, B. J. and J. N. Pitts, Jr. 1986. Atmospheric chemistry: fundamentals and experimental techniques. (New York, NY: Wiley-Interscience).

Foote, C. S. 1976. Photosensitized oxidation and singlet oxygen: consequences in biological systems. In *Free Radicals in Biology*, Vol. 2, W. A. Pryor, Ed., (New York, NY: Academic Press), pp. 85–133.

Foote, C. S. 1991. Definition of type I and type II photosensitized oxidation. *Photochem. Photobiol.* 54: 659.

Foote, C. S. and S. Wexler. 1964. Olefin oxidations with excited singlet molecular oxygen. *J. Amer. Chem. Soc.* 86: 3879–3880.

Fox, M. A. and S. Olive. 1979. Photooxidation of anthracene on atmospheric particulate matter. *Science* 205: 582–583.

Freiberg, J. E. and S. E. Schwartz. 1981. Oxidation of SO_2 in aqueous droplets: mass-transport limitation in laboratory studies and in the ambient atmosphere. *Atmos. Environ.* 15: 1145–1154.

Fridovich, I. 1974. Superoxide dismutases. *Advan. Enzymol.* 41: 35–97.

Frimmel, F. H., H. Bauer, J. Putzien, P. Murasecco, and A. M. Braun. 1987. Laser flash photolysis of dissolved aquatic humic material and the sensitized production of singlet oxygen. *Environ. Sci. Technol.* 21: 541–545.

Furukawa, T. and G. W. Brindley. 1973. Adsorption and oxidation of benzidine and aniline by montmorillonite and hectorite. *Clays Clay Min.* 21: 279–288.

Gardiner, W. C., Jr. 1984. *Combustion Chemistry.* (New York, NY: Springer Verlag).

Gohre, K. and G. C. Miller. 1983. Singlet oxygen generation on soil surfaces. *J. Agric. Food Chem.* 31: 1104–1108.

Gohre, K. and G. C. Miller. 1985. Photochemical generation of singlet oxygen on non-transition-metal oxide surfaces. *J. Chem. Soc. Faraday Trans. 1*, 81: 793–800.

Gohre, K., R. Scholl, and G. C. Miller. 1986. Singlet oxygen reactions on irradiated soil surfaces. *Environ. Sci. Technol.* 20: 934–938.

Gollnick, K. and G. O. Schenk. 1967. In Hamer, J., Ed. *1,4-Cycloaddition Reactions: Diels-Alder Reaction in Heterocyclic Synthesis.* (New York, NY: Academic Press), pp. 255–300.

Govindaraj, N., M. M. Mortland, and S. A. Boyd. 1987. Single electron transfer mechanism of oxidative dechlorination of 4-chloroanisole on copper(II)-smectite. *Environ. Sci. Technol.* 21: 1119–1123.

Gsponer, H. E., C. M. Previtali, and N. A. Garcia. 1987. Kinetics of the photosensitized oxidation of polychlorophenols in alkaline aqueous solution. *Toxicol. Environ. Chem.* 16: 23–37.

Gunz, D. and M. R. Hoffmann. 1990. Atmospheric chemistry of peroxides: a review. *Atmos. Environ.* 24A: 1601–1633.

Haag, W. and J. Hoigné. 1986. Singlet oxygen in surface waters. 3. Photochemical formation and steady-state concentration in various types of waters. *Environ. Sci. Technol.* 20: 341–348.

Haag, W., J. Hoigné, E. Gassmann and A. M. Braun. 1984. Singlet oxygen in surface waters. Part 1. Furfuryl alcohol as a trapping agent. *Chemosphere* 13: 631–640.

Haagen-Smit, A. J. 1952. Chemistry and physiology of Los Angeles smog. *Industr. Eng. Chem.* 44: 1342–1346.

Hamberg, M. and B. Gotthammar. 1973. A new reaction of unsaturated fatty acid hydroperoxides: formation of 11-hydroxy-12,13-epoxy-9-octadecenoic acid from 13-hydroperoxy-9,11-octadecadienoic acid. *Lipids* 8: 737–744.

Hanst, P. L. and B. W. Gay, Jr. 1983. Atmospheric oxidation of hydrocarbons: formation of hydroperoxides and peroxyacids. *Atmos. Environ.* 17: 2259–2265.

Hauser, E. A. 1955. *Silicic Science.* (Princeton, NJ: Van Nostrand).

Hawthorne, S. B., D. J. Miller, R. M. Barkley, and M. S. Krieger. 1988. Identification of methoxylated phenols as candidate tracers for atmospheric wood smoke pollution. *Environ. Sci. Technol.* 22: 1191–1196.

Hites, R. A. 1976. Sources of polycyclic aromatic hydrocarbons in the aquatic environment. In F. T. Weiss, Ed. *Sources, Effects, and Sinks of Hydrocarbons in the Aquatic Environment.* (Washington, DC: Amer. Inst. Biol. Sci.), pp. 325–332.

Hoffman, R. V., G. A. Eiceman, Y.-T. Long, M. C. Collins, and M.-Q. Liu. 1990. Mechanism of chlorination of aromatic compounds adsorbed on the surface of fly ash from municipal incinerators. *Environ. Sci. Technol.* 24: 1635–1641.

Hofzumahaus, A., H.-P. Dorn, J. Callies, U. Platt, and D. H. Ehhalt. 1991. Tropospheric hydroxyl radical concentration measurements by laser long-path absorption spectroscopy. *Atmos. Environ.* 25A: 2017–2022.

Hoigné, J. 1983. Reactions of ozone and inorganic radicals in natural waters. Presented at NATO-ARI workshop, Photochemistry of Natural Waters, Woods Hole, MA. Abstracts, p. 77.

Hoigné, J. 1984. Kinetics of reactions of aqueous ozone and of its decomposition products. Paper presented at Conference on Gas-Liquid Chemistry of Natural Waters, Brookhaven, NY. Abstracts, pp. 13–1 to 13–6.

Hon, D. N.-S., S.-T. Chang, and W. C. Feist. 1982. Participation of singlet oxygen in the photodegradation of wood surfaces. *Wood Sci. Technol.* 16: 193–201.

Hoshino, M., H. Akimoto, and M. Okuda. 1978. Photochemical oxidation of benzene, toluene, and ethylbenzene initiated by OH radicals in the gas phase. *Bull. Chem. Soc. Japan* 51: 718–724.

Howard, J. A. and J. C. Scaiano. 1984. Oxyl-, Peroxyl und verwandte Radikale. In *Landoldt-Bornstein, Neue Serie II*, H. Fischer, Ed. (Berlin: Springer), pp. 5–431.

Junk, G. A. and C. S. Ford. 1980. A review of organic emissions from selected combustion processes. *Chemosphere* 9: 187–230.

Katz, M., C. Chan, H. Tosine, and T. Sakuma. 1979. Relative rates of photochemical and biological oxidation (in vitro) of polycyclic aromatic hydrocarbons. In P. W. Jones and P. Leber, Eds. *Polynuclear Aromatic Hydrocarbons.* (Ann Arbor, MI: Ann Arbor Sci. Pub.), pp. 171–189.

Kautsky, H. and H. de Bruin. 1931. Die Aufklärung der Photoluminescenztilgung fluorescerender Systeme durch Sauerstoff: die Bildung aktiver, diffussionsfähiger Sauerstoffmoleküle durch Sensibilisierung. *Naturwissenschaften* 19: 1043–1058.

Khalil, G.-E. and M. Kasha. 1978. Oxygen-interaction luminescence spectroscopy. *Photochem. Photobiol.* 28: 435–443.

Kok, G. L. 1980. Measurements of hydrogen peroxide in rainwater. *Atmos. Environ.* 14: 653–656.

Korfmacher, W. A., D. F. S. Natusch, D. R. Taylor, G. Mamantov, and E. L. Wehry. 1980. Oxidative transformation of polycyclic aromatic hydrocarbons adsorbed on coal fly ash. *Science* 207: 763–765.

Kormann, C., D. W. Bahnemann, and M. R. Hoffmann. 1988. Photocatalytic production of H_2O_2 and organic peroxides in aqueous suspensions of TiO_2, ZnO, and desert sand. *Environ. Sci. Technol.* 22: 798–806.

Kung, K.-H. and M. B. McBride. 1988a. Electron transfer processes between hausmannite (Mn_3O_4) and hydroquinone. *Clays Clay Min.* 36: 297–302.

Kung, K.-H. and M. B. McBride. 1988b. Electron transfer processes between hydroquinone and iron oxides. *Clays Clay Min.* 36: 306–309.

Laha, S., and R. G. Luthy. 1990. Oxidation of aniline and other primary aromatic amines by manganese dioxide. *Environ. Sci. Technol.* 24:363–373.

Lamb, F. A., P. N. Cote, B. Slutsky, and B. M. Vittimberga. 1974. Oxidation-reduction of 9-(*p*-methoxyphenyl)-9-fluorenylacetaldehyde on activated alumina. *J. Org. Chem.* 39: 2796–2797.

Larson, R. A. 1988. The antioxidants of higher plants. *Phytochemistry* 27: 969–978.

Larson, R. A. and J. M. Hufnal, Jr. 1980. Oxidative polymerization of dissolved phenols by soluble and insoluble inorganic species. *Limnol. Oceanogr.* 25: 505–512.

Larson, R. A. and R. G. Zepp. 1988. Reactivity of the carbonate radical with aniline derivatives. *Environ. Toxicol. Chem.* 7: 265–274.

Larson, R. A., L. L. Hunt, and D. W. Blankenship. 1977. Formation of toxic products from a #2 fuel oil by photooxidation. *Environ. Sci. Technol.* 11: 492–496.

Larson, R. A., T. L. Bott, L. L. Hunt, and K. Rogenmuser. 1979. Photooxidation products of a fuel oil and their antimicrobial activity. *Environ. Sci. Technol.* 13: 965–969.

Larson, R. A., H.-L. Ju, V. L. Snoeyink, M. A. Recktenwalt, and P. A. Dowd. 1988. Some intermediates in the wet air oxidation of phenanthrene adsorbed on powdered activated carbon. *Water Res.* 22: 337–342.

Laughlin, R. G. W., T. Gallo, and H. Robey. 1983. Wet air oxidation for hazardous waste control. *J. Hazard. Mater.* 8: 1–9.

Lehmann, R. G., H. H. Cheng, and J. B. Harsh. 1987. Oxidation of phenolic acids by soil iron and manganese oxides. *Soil Sci. Soc. Amer. J.* 51: 352–356.

Lichtenthaler, R. G., W. R. Haag, and T. Mill. 1989. Photooxidation of probe compounds sensitized by crude oils in toluene and as an oil film on water. *Environ. Sci. Technol.* 23: 39–45.

Logan, J. A., M. J. Prather, S. C. Wofsy, and M. B. McElroy. 1981. Tropospheric chemistry: a global perspective. *J. Geophys. Res.* 86: 7210–7254.

Loll, M. J. and J. M. Bollag. 1985. Characterization of a citrate-buffer soil extract with oxidative coupling activity. *Soil Biol. Biochem.* 17: 115–117.

Lorenz, K., D. Rhäsa, R. Zellner, and B. Fritz. 1985. Laser photolysis-LIF kinetic studies of the reactions of CH_3O and CH_2CHO with O_2 between 300 and 500K. *Ber. Bunsenges. Phys. Chem.* 89: 341–342.

Martin, L. R. and D. E. Damschen. 1981. Aqueous oxidation of sulfur dioxide by hydrogen peroxide at low pH. *Atmos. Environ.* 15: 1615–1621.

Matsuura, T. and K. Omura. 1966. The hydroxylation of phenols by the photolysis of hydrogen peroxide in aqueous media. *J. Chem. Soc. Chem. Commun.* 127–128.

Mayo, F. R. 1958. The oxidation of unsaturated compounds. 7. The oxidation of aliphatic unsaturated compounds. *J. Amer. Chem. Soc.* 80: 2497–2500.

McBride, M. B. 1987. Adsorption and oxidation of phenolic compounds by iron and manganese oxides. *Soil Sci. Soc. Amer. J.* 51: 1466–1472.

McBride, M. B. 1989. Oxidation of dihydroxybenzenes in aerated aqueous suspensions of birnessite. *Clays Clay Min.* 37: 341–347.

McBride, M. B., F. J. Sikora, and L. G. Wesselink. 1988. Complexation and catalyzed oxidative polymerization of catechol by aluminum in acidic solution. *Soil Sci. Soc. Amer. J.* 52: 985–992.

McFarland, D., D. Kley, J. W. Drummond, A. L. Schmellekopf, and R. H. Winkler. 1979. Nitric oxide measurements in the equatorial Pacific region. *Geophys. Res. Lett.* 6: 605–608.

McKeen, S. A., M. Trainer, E. Y. Hsie, R. K. Tallamraju, and S. C. Liu. 1990. On the indirect determination of atmospheric hydroxyl radical concentrations from reactive hydrocarbon measurements. *J. Geophys. Res.* 95: 7493–7500.

Merkel, P. B. and D. R. Kearns. 1972. Radiationless decay of singlet molecular oxygen in solution. An environmental and theoretical study of electronic-to-vibrational energy transfer. *J. Amer. Chem. Soc.* 94: 7244–7253.

Mihelcic, D., M. Helten, H. Fark, P. Müsgen, H. W. Pätz, M. Trainer, D. Kempa, and D. H. Ehhalt. 1982. Tropospheric airborne measurements of NO_2 and RO_2 using the technique of matrix isolation and electron spin resonance. 2nd Sympos. on Composition of the Nonurban Troposphere, Williamsburg, VA. Abstracts, p. 327–329.

Mihelcic, D., D. H. Ehhalt, G. Kulessa, J. Klomfass, M. Trainer, A. Schmidt, and H. Rohrs. 1978. Measurements of free radicals in the atmosphere by matrix isolation and electron paramagnetic resonance. *Pure Appl. Geophys.* 116: 530–536.

Mill, T., D. Hendry, and D. Richardson. 1980. Free-radical oxidants in natural waters. *Science* 207: 886–887.

Miller, J. A. and G. A. Fisk. 1987. Combustion chemistry. *Chem. Eng. News.* (Aug. 31). pp. 22–46.

Miller, J. A., R. J. Kee, and C. K. Westbrook. 1990. Chemical kinetics and combustion modeling. *Ann. Rev. Phys. Chem.* 41: 345–387.

Millero, F. J. 1987. Estimate of the life time of superoxide in water. *Geochim. Cosmochim. Acta* 51: 351–353.

Moffett, J. W. and R. G. Zika. 1987. Reaction kinetics of hydrogen peroxide with copper and iron in seawater. *Environ. Sci. Technol.* 21: 804–810.

Momzikoff, A., R. Santus, and M. Giraud. 1983. A study of the photosensitizing properties of seawater. *Mar. Chem.* 11: 1–14.

Mopper, K. and X. Zhou. 1990. Hydroxyl radical photoproduction in the sea and its potential impact on marine processes. *Science* 250: 661–664.

Moreale, A., P. Cloos, and C. Badot. 1985. Differential behavior of Fe(III) and Cu(II) montmorillonite with aniline. I. Suspensions with constant solid:liquid ratio. *Clay Min.* 20: 29–37.

Morris, E. D., Jr. and H. Niki. 1974. Reaction of the nitrate radical with acetaldehyde and propylene. *J. Phys. Chem.* 78: 1337–1338.

Mortland, M. M. and S. A. Boyd. 1989. Polymerization and dechlorination of chloroethenes on Cu(II)-smectite via radical-cation intermediates. *Environ. Sci. Technol.* 23: 223–227.

Mortland, M. M. and L. J. Halloran. 1976. Polymerization of aromatic molecules on smectite. *Soil Sci. Soc. Amer. J.* 40: 367–370.

Mowrer, J. and J. Nordin. 1987. Characterization of halogenated organic acids in flue gases from municipal incinerators. *Chemosphere* 16: 1181–1192.

Mukai, K., K. Daifuku, K. Ozabe, T. Tanigaki, and K. Inoue. 1991. Structure-activity relationship in the quenching of singlet oxygen by tocopherol (vitamin E) derivatives and related phenols. Finding of linear correlation between the rates of quenching of singlet oxygen and scavenging of peroxyl and phenoxy radicals in solution. *J. Org. Chem.* 56: 4188–4192.

Munger, J. W., D. J. Jacob, and M. R. Hoffmann. 1984. The occurrence of bisulfite-aldehyde addition products in fog- and cloudwater. *J. Atmos. Chem.* 1: 335–350.

Nestrick, T. J., L. L. Lamparski, and W. B. Crummett. 1987. Thermolytic surface-reaction of benzene and iron(III) chloride to form chlorinated dibenzo-p-dioxins and dibenzofurns. *Chemosphere* 16: 777–790.

Neta, P. and S. Steenken. 1981. Phenoxyl radicals: formation, detection, and redox properties in aqueous solutions. In M. A. J. Rodgers and E. L. Powers, Eds. *Oxygen and Oxy-Radicals in Chemistry and Biology*. (New York, NY: Academic Press), pp. 83–88.

Niehaus, W. G. Jr. 1978. A proposed role of superoxide anion as a biological nucleophile in the deesterification of phospholipids. *Bioorg. Chem.* 7: 77–84.

Niki, H., P. D. Maker, C. M. Savage, and L. P. Breitenbach. 1980. FTIR studies of the Cl atom initiated oxidation of formaldehyde: detection of a new metastable species in the presence of NO_2. *Chem. Phys. Lett.* 72: 71–73.

Ogata, Y., K. Tomizawa, and Y. Yamashita. 1980. Photoinduced oxidation of benzoic acid with aqueous hydrogen peroxide. *J. Chem. Soc. Perkin Trans.* II: 616–619.

Olszyna, K. J., J. F. Meagher, and E. M. Bailey. 1988. Gas-phase, cloud and rainwater measurements of hydrogen peroxide at a high-elevation site. *Atmos. Environ.* 22: 1699–1706.

Pfoertner, K. and D. Böse. 1970. Die photosensibilisierte Oxydation einwertiger Phenole zu Chinonen. *Helv. Chim. Acta* 53: 1555–1566.

Pickering, W. F. 1966. Heterogeneous oxidation reactions. *Rev. Pure Appl. Chem.* 16: 185–208.

Pillai, P., C. S. Helling, and J. Dragun. 1982. Soil-catalyzed oxidation of aniline. *Chemosphere* 11: 299–317.

Pinnavaia, T. J. 1983. Intercalated clay catalysts. *Science* 220: 365–371.

Pitts, J. N., Jr., D. M. Lokensgard, P. S. Ripley, K. A. van Cauwenberghe, L. van Vaeck, S. D. Shaffer, A. J. Thill, and W. L. Belser, Jr. 1980. "Atmospheric" epoxidation of benzo[a]pyrene by ozone: formation of the metabolite benzo[a]pyrene-4,5-oxide. *Science* 210: 147–149.

Pitts, J. N., Jr., H. R. Paur, B. Zielinska, J. A. Sweetman, A. M. Winer, T. Ramdahl, and V. Mejia. 1986. Factors influencing the reactivity of poycyclic aromatic hydrocarbons adsorbed on model substrates and in ambient POM with ambient levels of ozone. *Chemosphere* 15: 675–685.

Platt, U., G. LeBras, G. Poulet, J. P. Burrows, and G. Moortgat. 1990. Peroxy radicals from night-time reaction of NO_3 with organic compounds. *Nature* 348: 147–149.

Pohlman, A. A. and T. Mill. 1983. Peroxy radical interaction with soil constituents. *Soil Sci. Soc. Amer. J.* 47: 922–927.

Posner, G. H. 1983. Organic reactions at alumina surfaces. *Angew. Chem. Int. Ed. Engl.* 17: 487–496.

Prinn, R., D. Cunnold, R. Rasmussen, P. Simmonds, F. Alyea, A. Crawford, P. Fraser, and R. Rosen. 1987. Atmospheric trends in methylchloroform and the global average for the hydroxyl radical. *Science* 238: 945–950.

Rabani, J., W. A. Mulac, and M. S. Matheson. 1965. The pulse radiolysis of aqueous tetranitromethane. I. Rate constants and the extinction coefficient of e^-_{aq}. *J. Phys. Chem.* 69: 53–62.

Rånby, B. and J. F. Rabek. 1975. Photodegradation, photo-oxidation and photostabilization of polymers. Wiley, London.

Rao, R. S. and E. Hayon. 1975. Redox potential of free radicals. IV. Superoxide and hydroperoxy radicals $\cdot O_2^-$ and ·HOO. *J. Phys. Chem.* 79: 397–402.

Rodgers, M. A. and P. T. Snowden. 1982. Lifetime of O_2 ($^1\Delta_g$) in liquid water as determined by time-resolved infrared luminescence measurements. *J. Amer. Chem. Soc.* 104: 5541–5543.

Russell, G. A. 1957. Deuterium-isotope effects in the autoxidation of aralkyl hydrocarbons. Mechanism of the interaction of peroxy radicals. *J. Amer. Chem. Soc.* 79: 3871–3877.

Saito, I., M. Imuta, and T. Matsuura. 1972. Photoinduced reactions. LIX. Reactivity of singlet oxygen toward methoxybenzenes. *Tetrahedron* 28: 5307–5311.

Saito, I., T. Matsuura, M. Nakagawa, and T. Hino. 1977. Peroxidic intermediates in photosensitized oxygenation of tryptophan derivatives. *Acc. Chem. Res.* 10: 346–352.

Sakugari, H., G.-I. Furusawa, K. Ueno, and K. Tokamura. 1982. Generation of singlet oxygen on irradiation of contact charge-transfer pairs of 1,2,3,4-tetramethylnaphthalene with oxygen. *Chem. Lett.* 1213–1216.

Sakugawa, H., I. R. Kaplan, W. Tsai, and Y. Cohen. 1990. Atmospheric hydrogen peroxide. *Environ. Sci. Technol.* 24: 1452–1462.

Sawnhey, B. L., R. K. Kozloski, P. J. Isaacson, and M. P. N. Gent. 1984. Polymerization of 2,6-dimethoxyphenol on smectite surfaces. *Clays Clay Min.* 32: 108–114.

Sawyer, D. T. and E. J. Nanni, Jr. 1981. Redox chemistry of dioxygen species and their chemical reactivity. In M. A. J. Rodgers and E. L. Powers, Eds. *Oxygen and Oxy-Radicals in Chemistry and Biology*. (New York, NY: Academic Press), pp. 15–44.

Schofield, P. J., B. J. Ralph, and J. H. Green. 1964. Mechanisms of hydroxylation of aromatics on silica surfaces. *J. Phys. Chem*. 68: 472–476.

Schumacher, J. N., C. R. Green, F. W. Best, and M. P. Newell. 1977. Smoke composition. An extensive investigation of the water-soluble portion of cigarette smoke. *J. Agric. Food Chem*. 25: 310–320.

Scully, F. E. and J. Hoigné. 1987. Rate constants for reactions of singlet oxygen with phenols and other compounds in water. *Chemosphere* 16: 681–694.

Shaub, W. M. and W. Tsang. 1983. Dioxin formation in incinerators. *Environ. Sci. Technol*. 17: 721–730.

Shindo, H. and P. M. Huang. 1985. Catalytic polymerization of hydroquinone by primary minerals. *Soil Sci*. 139: 505–511.

Singh, H. B. and L. J. Salas. 1983. Peroxyacetyl nitrate in the free troposphere. *Nature* 302: 326–328.

Smotkin, E. S., F. T. Moy, and W. Z. Plachy. 1991. Dioxygen solubility in aqueous phosphatidylcholine dispersions. *Biochim. Biophys. Acta* 1061: 33–38.

Solar, S., W. Solar and N. Getoff. 1986. Resolved multisite OH-attack on aqueous aniline studied by pulse radiolysis. *Radiat. Phys. Chem*. 28: 229–234.

Solomon, D. H., B. C. Loft, and J. D. Swift. 1968. Reactions catalyzed by minerals. 4. The mechanism of the benzidine blue reaction. *Clay Min*. 7: 389–397.

Soma, Y., and M. Soma. 1983. Formation of hydroxydibenzofurans from chlorophenols adsorbed on Fe-ion exchanged montmorillonite. *Chemosphere* 18: 1895–1902.

Spencer, W. F., J. D. Adams, R. E. Hess, T. D. Shoup, and R. C. Spear. 1980. Conversion of parathion to paraoxon on soil dusts and clay minerals as affected by ozone and UV light. *Agric. Food Chem*. 28: 366–371.

Stone, A. T. 1987. Reductive dissolution of manganese (III/IV) oxides by substituted phenols. *Environ. Sci. Technol*. 21: 979–988.

Stone, A. T. and J. J. Morgan. 1984. Reduction and dissolution of manganese(III) and manganese(IV) oxides by organics. 1. Reaction with hydroquinone. *Environ. Sci. Technol*. 18: 450–456.

Sunda, W. G., S. A. Huntsman, and G. R. Harvey. 1983. Photoreduction of manganese oxides in seawater and its geochemical and biological implications. *Nature* 301: 234–236.

Swartz, H. M. and N. J. F. Dodd. 1981. The role of ascorbic acid on radical reactions *in vivo*. In M. A. J. Rodgers and E. L. Powers, Eds. *Oxygen and Oxy-Radicals in Chemistry and Biology*. (New York, NY: Academic Press), pp. 161–168.

Taylor, W. I. and A. R. Battersby. 1967. *Oxidative Coupling of Phenols*. (New York: NY: Marcel Dekker).

Temple, P. J. and O. C. Taylor. 1983. World-wide ambient measurements of peroxyacetyl nitrate (PAN) and implications for plant injury. *Atmos. Environ*. 17: 1583–1587.

Theng, B. K. G. 1974. *The Chemistry of Clay-Organic Reactions*. (New York, NY: John Wiley & Sons).

Thomas, M. J. and C. S. Foote. 1978. Chemistry of singlet oxygen. XXVI. Photooxygenation of phenols. *Photochem. Photobiol.* 27: 683–693.

Tratnyek, P. G. and J. Hoigné. 1991. Oxidation of substituted phenols in the environment: a QSAR analysis of rate constants for reaction with singlet oxygen. *Environ. Sci. Technol.* 25: 1596–1604.

True, M. B. and O. C. Zafiriou. 1987. Reaction of Br_2^- produced by flash photolysis of sea water with components of the dissolved carbonate system. In Zika, R. G. and Cooper, W. J., Eds. *Aquatic Photochemistry. Amer. Chem. Soc. Sympos. Ser.* 327: 106–115.

Ulrich, H.-J. and A. T. Stone. 1989. Oxidation of chlorophenols adsorbed to manganese oxide surfaces. *Environ. Sci. Technol.* 23: 421–428.

Vaghijiani, G. L. and A. R. Ravishankara. 1991. New measurement of the rate coefficient for the reaction of OH with methane. *Nature* 350: 406–409.

Van Cauwenberghe, K., L. Van Vaeck, and J. N. Pitts, Jr. 1979. Chemical transformations of organic pollutants during aerosol sampling. *Advan. Mass Spectrom.* 8: 1499–1520.

von Sonntag, C. 1987. *The Chemical Basis of Radiation Biology*. (London: Taylor and Francis).

Voudrias, E. A. and M. Reinhard. 1986. Abiotic organic reactions at mineral surfaces: a review. In J. A. Davis and K. F. Hayes, Eds., *Geochemical processes at mineral surfaces. Amer. Chem. Soc. Sympos. Ser.* 323: 462–486.

Waite, T. D., I. C. Wrigley, and R. Szymczak. 1988. Photoassisted dissolution of a colloidal manganese oxide in the presence of fulvic acid. *Environ. Sci. Technol.* 22: 778–785.

Waldman, J. M., J. W. Munger, D. J. Jacob, R. C. Flanagan, J. J. Morgan, and M. R. Hoffmann. 1982. Chemical composition of acid fog. *Science* 218: 677–680.

Walling, C. 1975. Fenton's reagent revisited. *Acc. Chem. Res.* 12: 125–131.

Wang, T. S. C., M. C. Wang, Y. L. Ferng, and P. M. Huang. 1983. Catalytic synthesis of humic substances by natural clays, silts, and soils. *Soil Sci.* 135: 350–360.

Weschler, C. J., M. L. Mandich, and T. E. Graedel. 1986. Speciation, photosensitivity, and reactions of transition metal ions in atmospheric droplets. *J. Geophys. Res.* 91: 5189–5204.

Wilkinson, F. and J. G. Brummer. 1981. Rate constants for the decay and reactions of the lowest electronically excited singlet state of molecular oxygen in solution. *J. Phys. Chem. Ref. Data* 10: 809–999.

Wilshire, J. and D. T. Sawyer. 1979. Redox chemistry of dioxygen species. *Acc. Chem. Res.* 12: 105–110.

Yasuhara, A. and M. Morita. 1988. Formation of chlorinated aromatic hydrocarbons by thermal decomposition of vinylidene chloride polymer. *Environ. Sci. Technol.* 22: 646–650.

Ye, M. and R. H. Schuler. 1990. Determination of oxidation products in radiolysis of halophenols with pulse radiolysis, HPLC, and ion chromatography. *J. Liq. Chromatogr.* 13: 3369–3387.

Yokley, R. A., A. A. Garrison, E. L. Wehrly, and G. Mamantov. 1986. Photochemical transformation of pyrene and benzo[a]pyrene vapor-deposited on eight coal stack ashes. *Environ. Sci. Technol.* 20: 86–90.

Yoshizumi, K., K. Aoki, I. Nouchi, T. Okita, T. Kobayashi, S. Kamukara, and M. Tajima. 1984. Measurements of the concentration in rainwater and of the Henry's law constant of hydrogen peroxide. *Atmos. Environ.* 18: 395–401.

Zafiriou, O. C., J. Joussot-Dubien, R. G. Zepp, and R. G. Zika. 1984. Photochemistry of natural waters. *Environ. Sci. Technol.* 18: 358A-371A.

Zafiriou, O. C., M. B. True, and E. Hayon. 1987. Consequences of OH radical reaction in sea water: formation and decay of Br_2^- ion radical. In Zika, R. G. and Cooper, W. J., eds. *Aquatic photochemistry. Amer. Chem. Soc. Sympos. Ser.* 327: 89–105.

Zepp, R. G., N. L. Wolfe, G. L. Baughmann, and R. C. Hollis. 1977. Singlet oxygen in natural waters. *Nature* 267: 421–423.

Zepp, R. G., A. M. Braun, J. Hoigné, and J. A. Leenheer. 1987a. Photoproduction of hydrated electrons from natural organic solutes in aquatic environments. *Environ. Sci. Technol.* 21: 485–490.

Zepp, R. G., J. Hoigné, and H. Bader. 1987b. Nitrate-induced photooxidation of trace organic chemicals in water. *Environ. Sci. Technol.* 21: 443–450.

Zepp, R. G., Y. I. Skurlatov, and J. T. Pierce. 1987c. Algal-induced decay and formation of hydrogen peroxide in water: its possible role in oxidation of anilines by algae. In Zika, R. G. and Cooper, W. J., Eds. *Aquatic photochemistry, Amer. Chem. Soc. Sympos. Ser.* 327: 215–224.

Zhou, X. and K. Mopper. 1990. Determination of photochemically produced hydroxyl radicals in seawater and freshwater. *Mar. Chem.* 30: 71–88.

Zika, R., E. Saltzman, W. L. Chameides, and D. D. Davis. 1982. H_2O_2 levels in rainwater collected in south Florida and the Bahama Islands. *J. Geophys. Res.* 87: 5015–5017.

CHAPTER 5

REACTIONS WITH DISINFECTANTS

In the past 15 to 20 years there has been growing interest in the organic chemistry of potable water. Although most finished drinking waters have extremely low organic carbon concentrations, the large absolute amount of water (about 50,000 liters) ingested by an individual over a lifetime means that human exposure to chemical contaminants may be significant. Water disinfection has been quite successful in practically eliminating many acute waterborne diseases from the countries where it has been widely practiced, but its possible roles in the development of chronic health disorders, such as cancer, have not been fully scrutinized. Potential biological hazards that may arise during drinking water treatment have been summarized (Larson, 1988).

In addition, disinfecting agents are used for the treatment of industrial wastewaters to remove undesirable colors, odors, and toxic organic compounds as well as to reduce the number of potentially harmful microorganisms in the effluent. Because of the higher concentrations of organic matter typically found in these waters, the likelihood of organic-disinfectant reactions occurring at significant rates is enhanced.

A. FREE AQUEOUS CHLORINE (HOCl)

1. Chlorine in Water

Aqueous reactions of molecular chlorine have been reviewed by several authors (see, e. g., White, 1972). When chlorine gas is bubbled into pure water, a rapid hydrolysis to hydrochloric and hypochlorous (HOCl) acids results:

$$Cl_2 + H_2O \rightarrow HCl + HOCl \quad (5.1)$$

At millimolar levels of molecular chlorine, the reaction goes to 99% completion in a matter of a few seconds. Above pH 4.4, essentially no molecular chlorine remains in aqueous solution. Hypochlorous acid is a weak acid (pKa approximately 7.5) and thus, near neutrality, both the protonated form and the anion occur at appreciable levels. HCl-free solutions may be prepared by adding salts such as sodium hypochlorite (NaOCl, commercially available as a stabilized 5.25% [0.7 M] solution as a fabric bleach). Either chlorine gas or hypochlorite solutions can be used in large-scale water chlorination applications. In addition to water treatment, chlorine is also used as a disinfectant for beef, pork, and poultry carcasses and also as a bleaching agent for paper pulp and cake flour (Wei et al., 1985). The disinfecting ability of aqueous chlorine is closely associated with its vigorous oxidant character: the redox potential for the reactions

$$ClO^- + 2\,H_2O + 2\,e^- \rightarrow Cl^- + 2\,OH^- \quad (5.2)$$

$$HOCl + H^+ + 2\,e^- \rightarrow Cl^- + H_2O \quad (5.3)$$

are 0.9 V and 1.49 V, respectively (Masschelein, 1979).

In solution, HOCl and hypochlorite concentrations are commonly determined together by standard water analysis methods (AWWA, 1989). The sum of their concentrations is referred to as **free available chlorine** and is usually reported in parts per million (1 ppm free available chlorine = 1.4×10^{-5} M). Detectable levels of free and **combined** (nitrogenous) available chlorine in treated water constitute a **chlorine residual**. In normal practice in North America, water utilities attempt to adjust chlorination levels to a small chlorine residual (2×10^{-5} M or less) that is sufficient to survive throughout the drinking water distribution system all the way to the user's water taps. To achieve this level in the distribution lines, it may be necessary to add as much as 10 times higher amounts to the water in the plant, depending on the amount of reactive material in the source water and in the lines.

Available chlorine disappears from treated water by numerous pathways. In addition to reacting with the organic matter (including living organisms) in water, it also reacts readily with inorganic reductants such as ferrous and manganous cations, and sulfide, bromide, iodide, and nitrite anions. When exposed to sunlight, these species also underwent photolysis at varying rates (Oliver and Carey, 1977: Nowell and Crosby, 1985). Hypochlorite anion has an absorption maximum at 292 nm (ε = 400) and can decompose either to O^- and Cl· (favored at 313 nm) or Cl^- and ·O· (favored at 365 nm). Either of these potent radicals can then initiate free-radical reactions by abstraction or other mechanisms (Ogata et al., 1979). In the presence of light, NaOCl attacked such normally unreactive substrates as ethanol, *n*-butanol, nitrobenzene, acetic, benzoic, and propionic acids (Oliver and Carey, 1977: Ogata et al., 1979; Nowell and Crosby, 1985), giving oxygenated and chlorinated derivatives. Short-wavelength solar UV (UV-B) converted HOCl to HO·, a potent oxidant (Molina et al., 1980).

2. Oxidation Reactions

Reactions in which HOCl acts as an oxidant, for example,

$$RCHO + HOCl \rightarrow RCOOH + Cl^- + H^+ \quad (5.4)$$

have not received as much attention from environmental scientists as have the substitution and addition reactions discussed in the following sections. Obviously, organochlorine compounds are of much greater public health concern at the present than are most oxidized organic derivatives. Nevertheless, it seems incontrovertible that much of the chlorine applied to treated waters is not incorporated into organic molecules, but simply converted to chloride ion. For example, Sigleo et al. (1980) treated estuarine water with 0.14 m*N* NaOCl and obtained 6–37% yields of CO_2 (based on the amount of chlorine applied), suggesting that oxidation, with or without the incorporation of chlorine into aromatic compounds, was a major pathway for hypochlorite consumption. The species responsible for the reduction of chlorine in hypochlorite are not fully defined, but presumably include reduced metal ions, aldehydes, ketones, alcohols (especially secondary alcohols; Nwaukwa and Keehn, 1982), amino acids, and sulfur compounds.

Acetone, under some conditions, can give rise to lactic acid (Guthrie and Cossar, 1986: Equation 5.5), and some phenyl ketones were similarly oxidized in the α-position to various acidic products, some of which were proposed to be formed by benzilic acid rearrangements (Guthrie et al., 1991: Figure 5.1).

$$CH_3C(=O)CH_3 \xrightarrow{OCl^-} CH_3CH(OH)CO_2H \quad (5.5)$$

Carbohydrates are also reported by a few authors to be oxidized by HOCl, but the mechanisms are rather uncertain. In one study, some starch slurries were extensively depolymerized when treated with aqueous chlorine, apparently due to cleavage of the 1,4-bonds between glucose residues and attack at the C_2 and C_3 carbons of the D-glucose residues (Whistler and Schweiger, 1957).

Thiols were oxidized by HOCl to the corresponding sulfonic acids; thus, cysteine afforded cysteic acid (Equation 5.6: Ingols et al., 1953). This is in contrast to the

$$(H_2N)(HO_2C)CH{-}CH_2SH \xrightarrow{OCl^-} (H_2N)(HO_2C)CH{-}CH_2SO_3^- \quad (5.6)$$

behavior of NH_2Cl, which oxidized cysteine only to the disulfide. Thiophenes were chlorinated in the nucleus as well as undergoing partial oxidation of the sulfur atom (Equation 5.7: Morton, 1946), whereas more highly condensed sulfur heterocycles such as benzo[b]thiophene (**1**) and dibenzothiophene (**2**) were oxidized to the sulfone

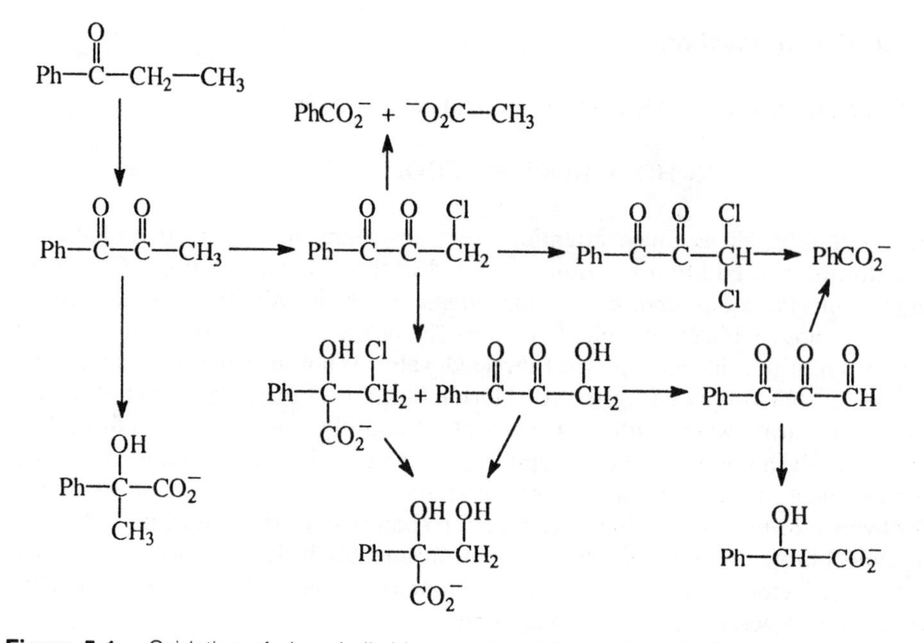

Figure 5.1. Oxidation of phenyl alkyl ketones to acidic products by hypochlorite.

and sulfoxide almost to the exclusion of chlorination (Lin and Carlson, 1984). Tetrachlorothiophene (**3**), 3-formyl-2,4,5-trichlorothiophene (**4**), and 3-acetyl-2,4,5-trichlorothiophene (**5**) were identified in a spent liquor from the bleaching of a softwood kraft pulp (McKague et al., 1989); the precursors of these chemicals were not conjectured upon.

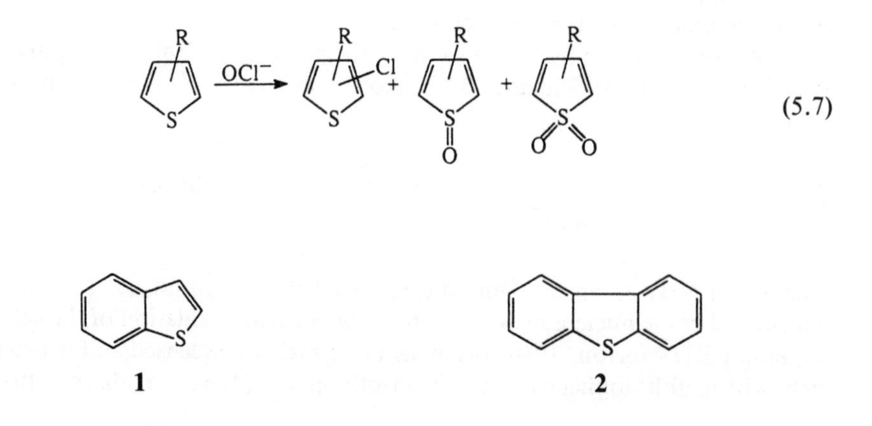

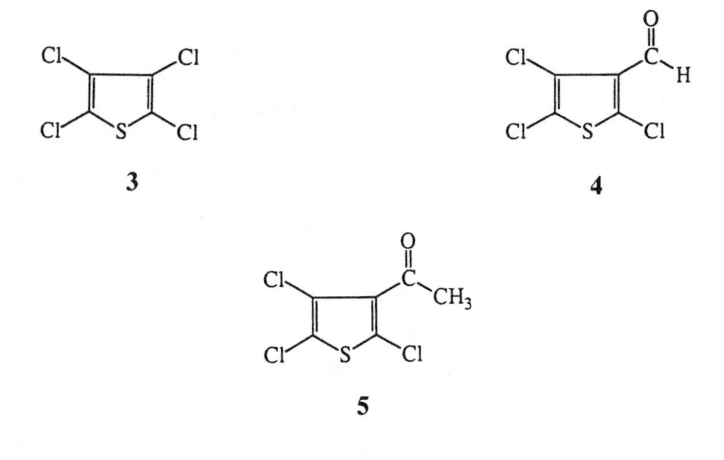

3. Substitution and Addition Reactions

Although conceivably HOCl could undergo electrophilic attack either at the chlorine atom (with displacement of hydroxyl, leading to chlorination) or at the hydroxyl group (with loss of chlorine), the former would (theoretically) be expected from the relative electronegativity of the two elements; oxygen, with electronegativity 3.5, is more electronegative than chlorine (electronegativity 3.0: Pauling, 1960). In any event, chlorination by HOCl has been widely observed in an extensive range of organic structures.

Phenols

The reactions of HOCl with phenolic compounds in drinking water has long been known to produce malodorous and unpalatable chlorophenols by substitution reactions (Streeter, 1929). Phenols are not chlorinated as rapidly as ammonia (the second-order rate constant for phenol is smaller by approximately three orders of magnitude at pH 8.3); hence substantial chlorination would be expected to occur only in the presence of free chlorine (Lee and Morris, 1962). The kinetics of chlorination of phenols clearly show that the rate constants are inversely correlated with the acidity of the phenol (Figure 5.2), with the more acidic phenols being chlorinated more slowly (Soper and Smith, 1926); the correlation is clearly with the electron density on the ring, since this would contribute both to acceleration of the rate of an electrophilic reaction such as chlorination and to the extent of ionization of a substituted phenol. Phenol itself is chlorinated most rapidly at pH 8.5, which is roughly the arithmetic mean of the pKa's of HOCl and phenol (and would thus be the pH where the product of HOCl and the phenolate ion would be maximal). The chlorination of phenol proceeds via a typical electrophilic substitution pathway, with the *o* and *p* positions being attacked first to give a mixture or 2- and 4-chlorophenols (Equation 5.8). These compounds were also readily substituted to the expected products—2,4- and 2,6-dichlorophenols and 2,4,6-trichlorophenol—

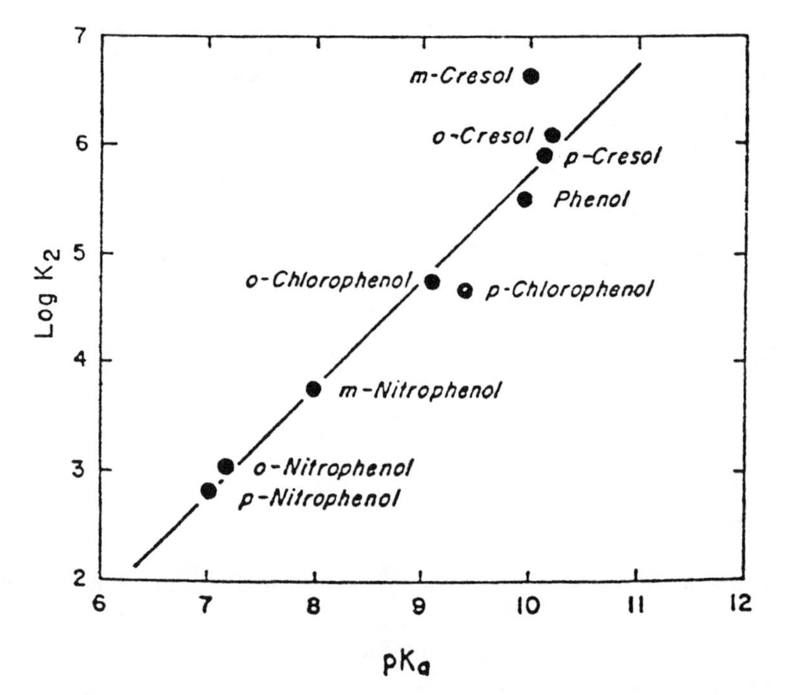

Figure 5.2. Effect of phenol acidities on the rate constants of their reactions with HOCl. From Soper and Smith (1926). Reprinted by permission of the Royal Society of Chemistry (U.K.).

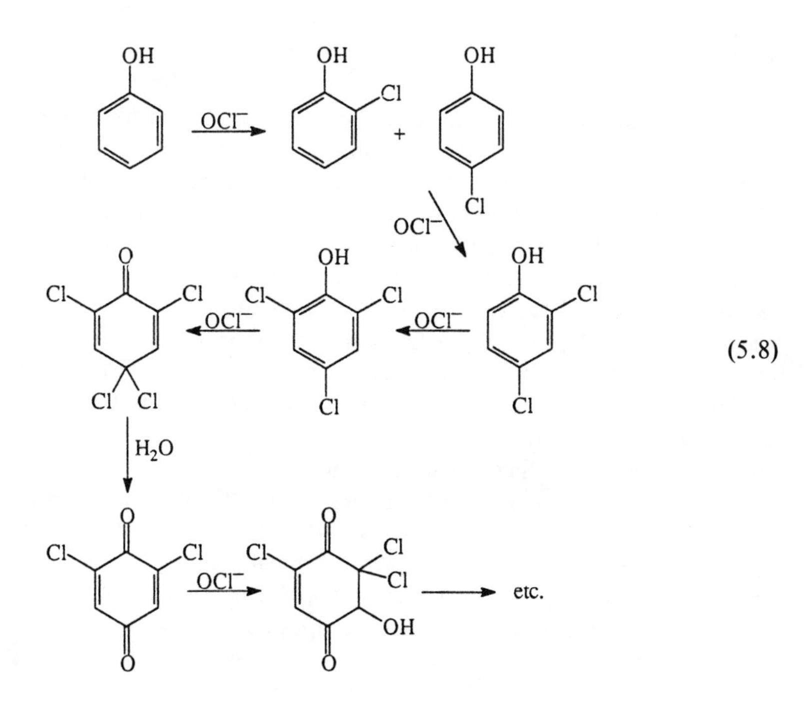

(5.8)

although the rates decreased with increasing chlorine substitution (Burttschell et al., 1959; Lee and Morris, 1962). 2,4,6-Trichlorophenol was attacked slowly via *ipso* substitution at the 4-position (Everly and Traynham, 1979: Fischer and Henderson, 1979) to give the cyclohexadienone **6**. Further attack of this compound also occured, possibly via S_N1 nucleophilic displacement and elimination of HCl, to the benzoquinone **7** (Smith et al., 1975), and eventually cleavage of the ring occurred, with production of such ultimately stable compounds as trichloroacetic and 2-chloromaleic acids (Onodera et al., 1984).

Other 4-substituted phenols also reacted with HOCl to give *ipso* cyclohexadienones, either as stable compounds or as intermediates in the displacement of side chains by chlorine (Dence and Sarkanen, 1960: Larson and Rockwell, 1979). For example, *p*-hydroxybenzoic acid reacted rapidly with HOCl at environmentally realistic concentrations to give a mixture of substitution and decarboxylation products (Larson and Rockwell, 1979). It is plausible that such decarboxylation and side-chain cleavage products are at least partially responsible for the occurrence of chlorinated phenol derivatives in drinking water and in paper pulp bleaching effluents.

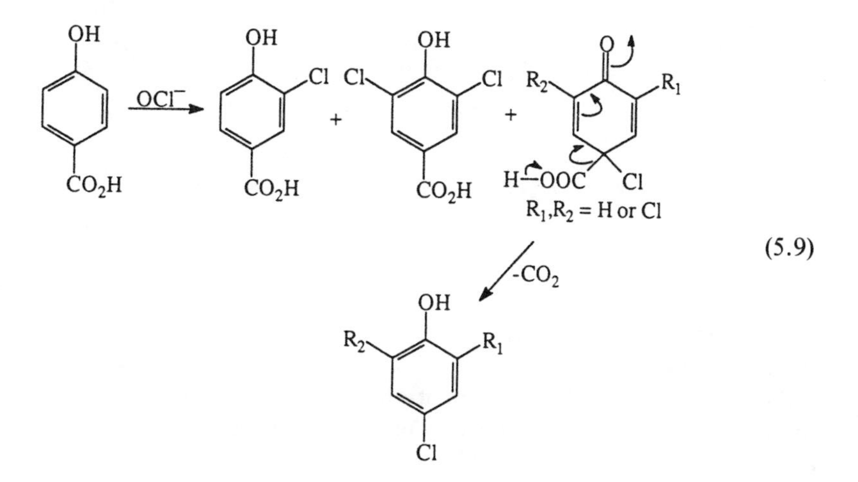

(5.9)

Dimeric products have been observed in some studies of the chlorination of phenol. On the basis of chromatographic and spectroscopic considerations, these compounds have been tentatively identified as diphenyl ethers with 2 to 5 chlorine atoms (**8** and **9**: Onodera et al., 1984). The highest yields of these compounds occurred at low pH values, although small amounts were formed at as high as pH 8. Although the authors did not speculate on the mechanisms of the interconversions,

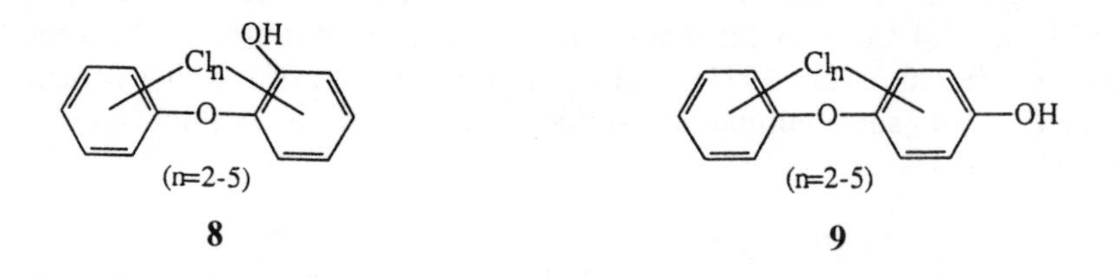

the occurrence of these compounds suggests that free-radical coupling reactions between phenolate radicals and phenols may be occurring in these solutions (Taylor and Battersby, 1967). Similar coupling reactions took place on the surface of activated carbon treated with disinfectants (see Section 5.E). Free radical mechanisms in solutions of HOCl might be expected to occur if the following reactions were important (Fonouni et al., 1983):

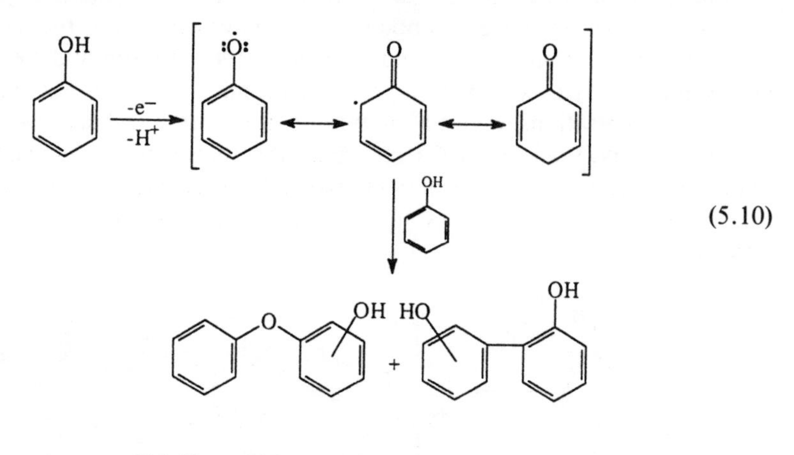

(5.10)

$$HOCl + ClO^- \rightarrow Cl_2O + HO^- \tag{5.11}$$

$$Cl_2O \rightarrow Cl\cdot + ClO\cdot \tag{5.12}$$

As in atmospheric chemistry, Cl· and ClO· could be important chain-carrying or -initiating radicals.

Another possibility for the formation of free radical species from hypochlorite is through its reactions with transition metal ions. Thus, Guilmet and Meunier (1980) reported a manganese-promoted epoxidation of olefins such as styrene (Equation 5.13) and cyclohexene in a two-phase dichloromethane-water solvent mixture. The epoxide oxygen was derived from HOCl, not from air, but no mechanistic details were speculated upon. Further evidence needs to be obtained on the possibility of free-radical reactions in water and wastewater chlorination.

$$\mathrm{Ph(H)C{=}CH_2} \xrightarrow[\mathrm{Mn^{2+}}]{\mathrm{OCl^-}} \mathrm{Ph(H)\overset{O}{C{-}CH_2}} \qquad (5.13)$$

Several studies have dealt with the chlorination of other types of phenols. Nitrophenols, which may be produced when phenols are oxidized in the presence of nitrite, have been shown to be converted to chloropicrin, Cl_3CNO_2, by chlorination (Thibaud et al., 1987). The mechanism of this reaction needs further study. The product has been detected in some chlorinated drinking water samples. Anisole (phenyl methyl ether) was readily chlorinated to the expected *o* and *p* isomers (de la Mare et al., 1954). Creosol (2-methoxy-4-methylphenol) was converted to a variety of compounds including 6-chlorocreosol (the major product), 5-chlorocreosol, 5,6-dichlorocreosol, *o*-benzoquinones, and ring cleavage products (Gess, 1971).

Phenolic Acids

Norwood et al. (1980) chlorinated a considerable number of phenolic acids of both benzoic and cinnamic acid types and looked for haloform production. Concentrations were from 0.5–6 m*M* and approximately a twofold excess of chlorine was added. For most of the acids tested, except those with 1,3-dihydroxy (resorcinol-type) substituents, chloroform yields were low (0–4%) except in the case of 3-methoxy-4-hydroxycinnamic acid (ferulic acid, **10**), where about an 11% yield was

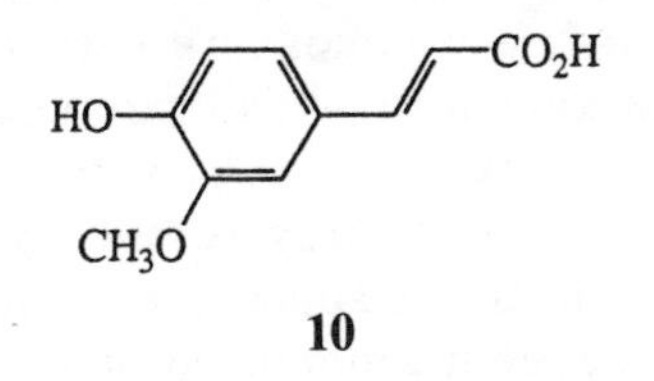

noted. It is probable that this $CHCl_3$ originated from the side chain, although no data were presented to establish this hypothesis. Previous work has shown that the alkene linkage of cinnamic acid derivatives such as *p*-coumaric acid (**11**) was susceptible to HOCl attack to give chlorohydrins (*vide infra*). The formation of compounds such as **12** can be rationalized by chlorohydrin formation followed by

hypochlorite-induced decarboxylation (Equation 5.14). Further oxidative side-chain degradation may also occur with these compounds, resulting in the formation of stable chlorinated quinones like **13** (Shimizu and Hsu, 1975), and, conceivably, haloforms.

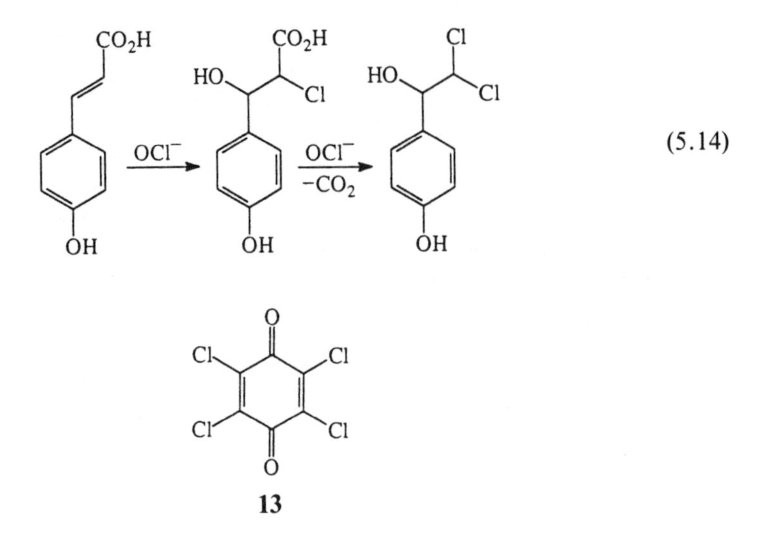

Aromatic Hydrocarbons

Unactivated aromatic rings such as benzene show little tendency to chlorinate under water treatment conditions. Biphenyl has been studied somewhat intensively because of concerns that polychlorinated biphenyls might form in wastewaters containing the parent compound, a common dye carrier particularly in the carpet industry. Wastewaters may contain this hydrocarbon at up to 2 mg/L (Gaffney, 1977). Biphenyl would not be expected to be very reactive toward electrophilic attack on empirical grounds, since the phenyl substituent has poor electron-donating characteristics (Table 1.3). In model aqueous chlorination practice, only under rather extreme conditions does even a monochlorobiphenyl isomer form, and these compounds were virtually inert to further substitution (Carlson and Caple, 1978: Snider and Alley, 1979).

Highly alkylated benzenes might be expected to acquire sufficient electron density in the ring for substitution to take place; evidence that this does occur comes from the identification of various isomers of chloro- and bromocymenes (**14**) from the spent bleaching liquors of a sulfite pulp mill (Eklund et al., 1978). These workers also identified some chlorinated and brominated tetrahydronaphthalene derivatives that appeared to be of sesquiterpene origin, and, interestingly, a chlorotoluene, which was a minor product. Dehydroabietic acid (**15**), a sesquiterpene found in wood pulp whose structure incorporates a trisubstituted aromatic ring, was found to incorporate 1-3 chlorine atoms into various positions of this ring when chlorinated in the laboratory under controlled conditions, and some of the same compounds were identified in a kraft mill caustic extraction effluent (Equation 5.15) (Thakore and Oehlschlager, 1977). Chlorine was also incorporated into some hydrocarbons in the aromatic fraction of a diesel fuel. 1,3,5-Trimethylbenzene was particularly reactive, affording the monochloro adduct (Reinhard et al., 1976).

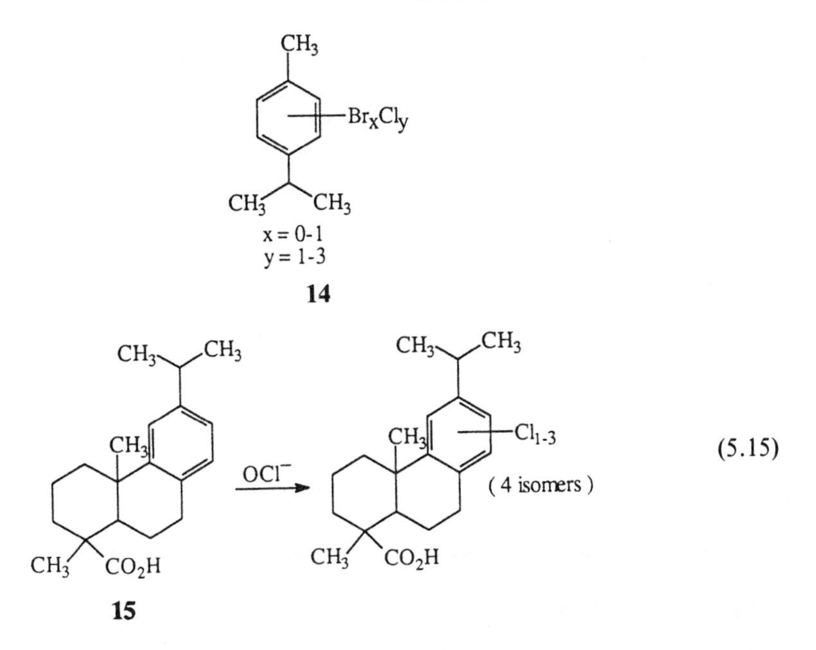

Polycyclic aromatic hydrocarbons contain sites with high electron density, and scattered evidence suggests that they react with aqueous chlorine. For example, naphthalene and alkylnaphthalenes reacted to form chloronaphthalenes, phthalic anhydride, *o*- and *p*-quinone derivatives, and chlorinated naphthols. Fluorene and phenanthrene likewise afforded monochloro derivatives, and benzo[a]pyrene and anthracene gave a mixture of chlorine-substituted products and carbonyl compounds (Reinhard et al., 1976; Oyler et al., 1978, 1983; Onodera et al., 1986). Phenanthrene-9,10-oxide (**16**) was the major product of phenanthrene chlorination at pHs above 4 in one study (Oyler et al., 1983); at low pH, phenanthrene-9,10-dione and 9-chlorophenanthrene were the principal products. Most of the compounds studied were more reactive at low pH, suggesting a possible requirement for molecular chlorine in their substitution reactions (Carlson et al., 1978). In realistic situations such as natural waters that contain more reactive substrates such as humic materials, significant chlorination of polycyclic aromatic hydrocarbons does not appear to occur (with the exceptions of anthracene and benzo[a]pyrene), except at unrealistically high chlorine concentrations (Johnsen and Gribbestad, 1988).

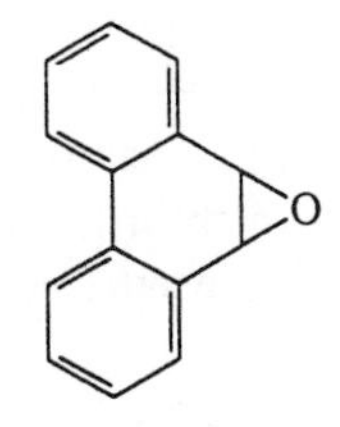

16

Alben (1980) chlorinated a coal-tar leachate that contained many polycyclic aromatic hydrocarbons and demonstrated that many hydrocarbons were significantly reduced in concentration. Oxygenated compounds, such as dibenzofuran (**17**), were formed, together with low concentrations of chlorinated and brominated derivatives of naphthalene and fluorene. Chlorinated polycyclic compounds have been demonstrated to occur in drinking water samples (Shiriashi et al., 1985), including mono- and dichloro derivatives of naphthalene, fluorene, dibenzofuran, phenanthrene, fluoranthene, and fluorenone (Figure 5.3).

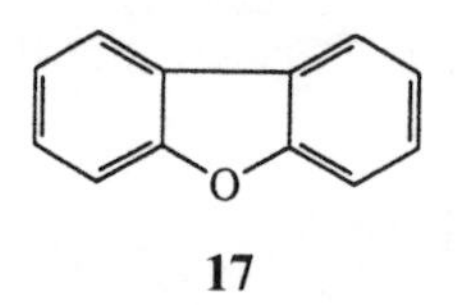

17

Enolizable Carbonyl Compounds: the Haloform Reaction

In the middle 1970s, several groups of chemists reported the then-startling news that chloroform and several other trihalomethanes (THMs: **18**) were almost ubiquitously present in chlorinated drinking water (Rook, 1974: Bellar et al., 1974: Symons et al., 1975). Initial suggestions that the compounds were due to industrial contamination or impurities in commercial chlorine gas were rapidly discounted, and it was clearly demonstrated by many groups of workers that the source of these compounds was naturally occurring organic matter in water. Disputes almost immediately arose over the functional groups responsible for the THM formation, and to some extent these are still continuing. Although chloroform is almost always the most abundant THM in chlorinated drinking water, brominated THMs are usually present (Figure 5.4).

$$CHBr_xCl_y$$

$$\left.\begin{matrix} x = 0\text{-}3 \\ y = 0\text{-}3 \end{matrix}\right\} (x+y = 3)$$

18

The classical organic reaction for the synthesis of THMs, the so-called "haloform reaction," is actually a series of reactions of enolizable compounds whose rate is usually determined by the rate of enolization of a precursor molecule. It is outlined in Figure 5.5.

Formally, it involves an electrophilic addition to the α-carbon of an enolizable carbonyl compound. Introduction of a chlorine atom at this position helps to accelerate enolization by virtue of the increased electron-withdrawing character of Cl relative to H, so further addition of chlorine atoms is highly favored. Although simple ketones such as acetone are theoretically capable of enolization, only a vanishingly small fraction of the compound is present in the form which is reactive

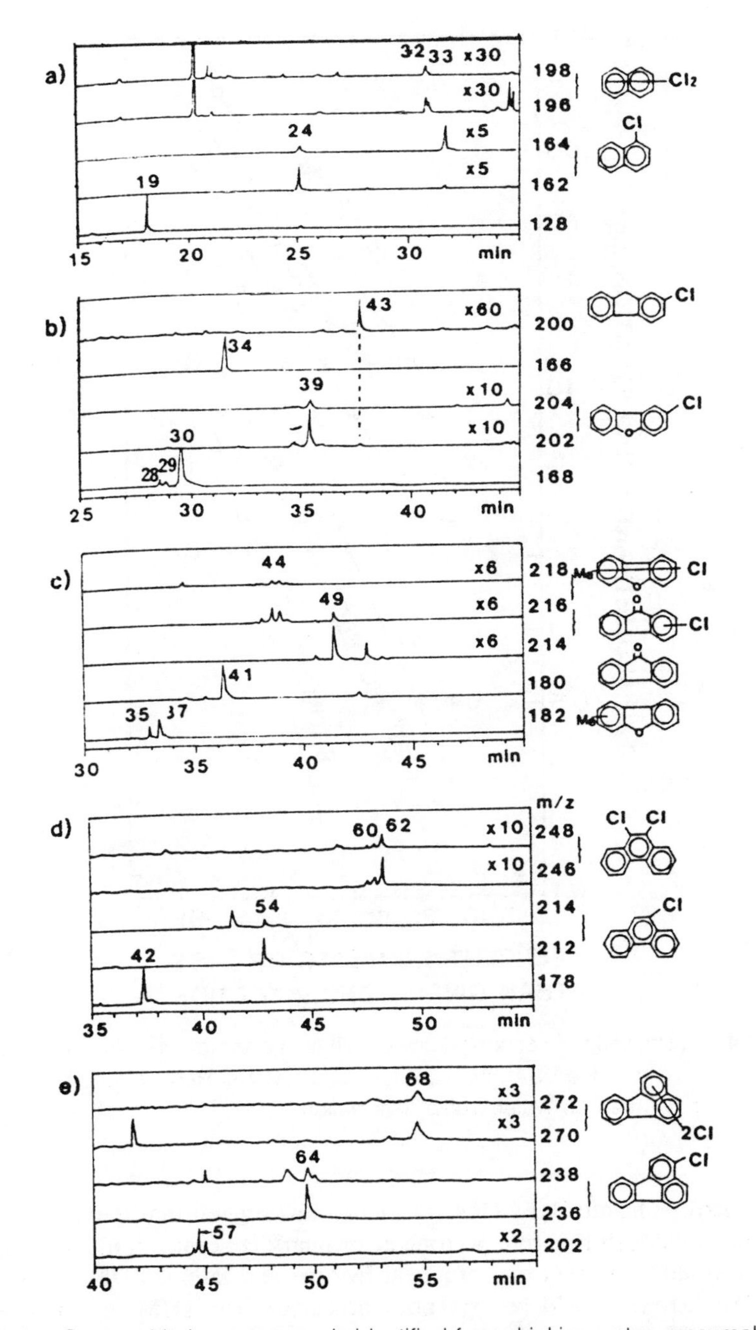

Figure 5.3. Organochlorine compounds identified from drinking water, presumably formed by hypochlorite chlorination of polycyclic compounds. From Shiriashi et al. (1985). Reprinted by permission of the American Chemical Society.

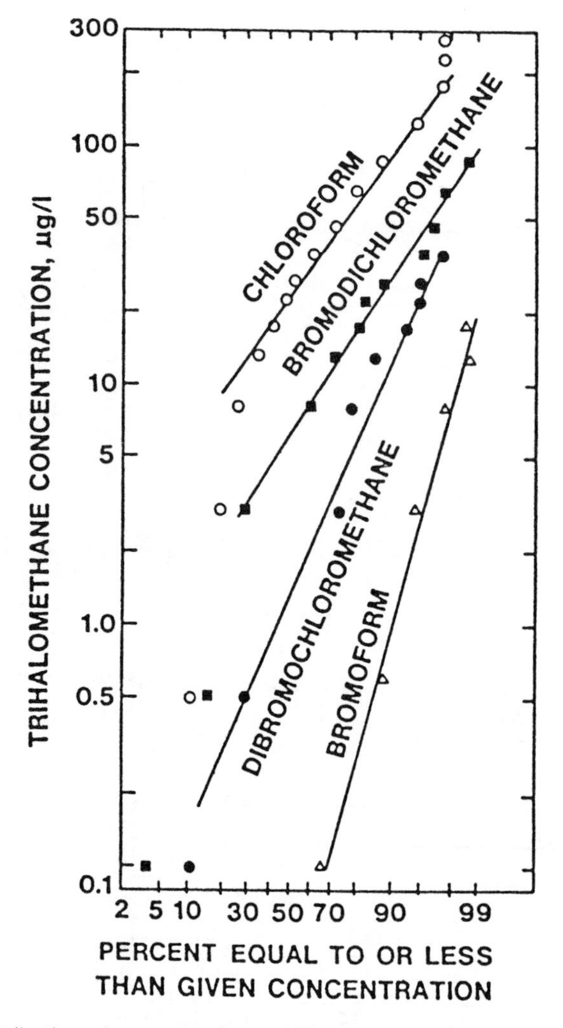

Figure 5.4. Distribution of concentrations of THMs in a survey of chlorinated drinking waters in the USA. From Symons et al. (1975). Reprinted by permission of the American Water Works Association.

toward electrophilic attack by HOCl. Calculations suggest that the half-life for the formation of chloroform from acetone at ordinary concentrations, pHs, and temperatures would be of the order of a year (Morris and Baum, 1978). Thus, acetone and similar ketones would be very poor precursors for THMs even if they were present in sufficient quantities in raw water to give rise to the concentrations of haloforms (often approximately 100 μg/L, or around 10^{-9} M) observed in finished drinking water. Attention has been focused on more readily enolizable carbonyl compounds (including β-diketones such as acetylacetone (**19**), in which the enol

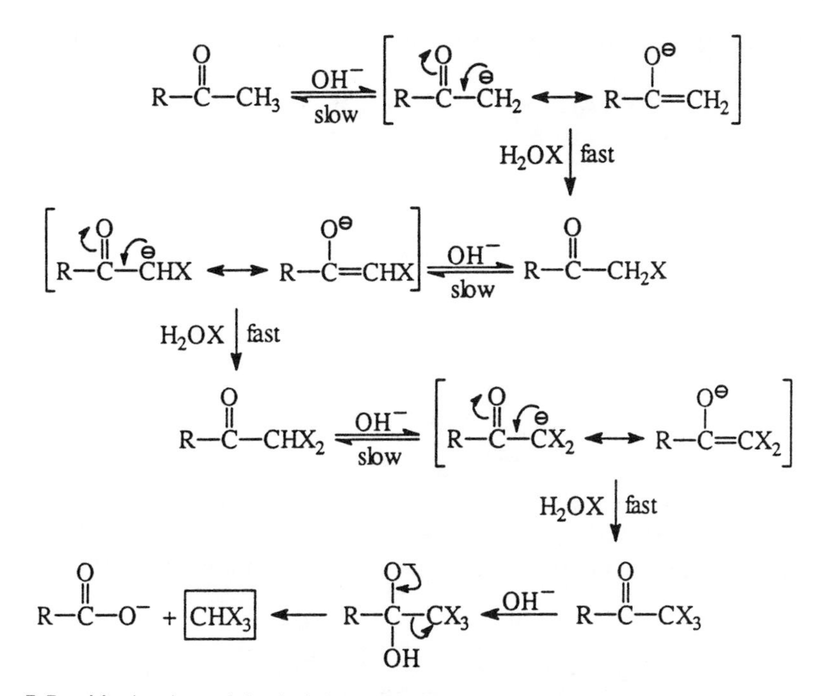

Figure 5.5. Mechanism of the haloform reaction with methyl ketones. From Morris and Baum (1978). Reprinted by permission of Lewis Publishers.

form is strongly stabilized by hydrogen bonding) and on phenols, which may be considered to be extremely stable enol forms of cyclohexenones.

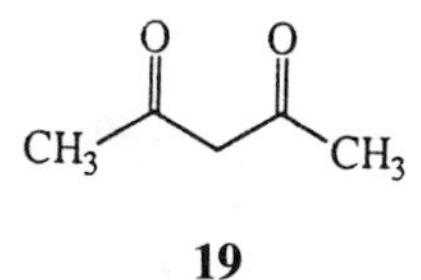

Rook (1976) demonstrated that cyclic 1,3-diketones such as ninhydrin (**20**), dimedone (**21**), and 1,3-cyclohexanedione were efficient precursors of chloroform

(50–100% yields with 3 mM starting concentrations of chlorine). In a later publication, he (Rook, 1977) found that *m*-dihydroxybenzenes also gave very high yields of chloroform. Substances such as resorcinol (**22**), 3,5-dihydroxybenzoic acid, and several flavonoids such as hesperetin (**23**) contain "masked" β-diketones (cf. Figure

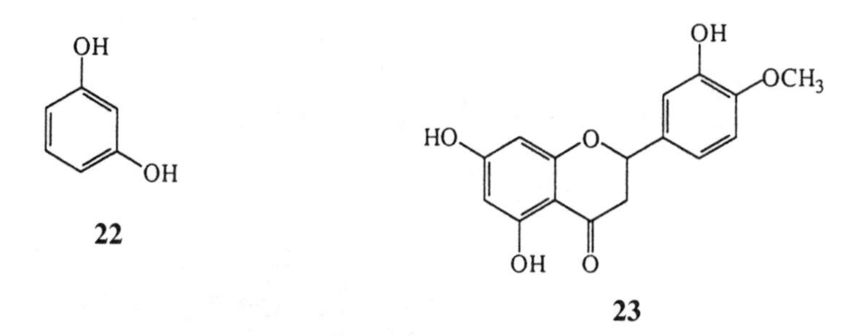

5.6), and Rook suggested that the 2-position of such compounds might be expected to be chlorinated readily and to react rapidly as haloform precursors. This suggestion was confirmed by an elegant experiment using isotopically labeled resorcinol

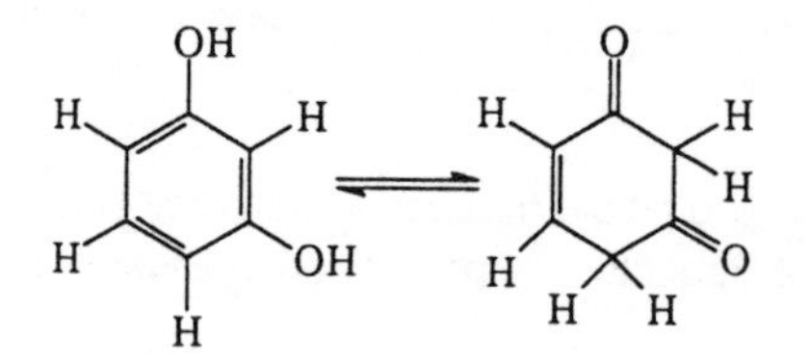

Figure 5.6. Representation of the structure of resorcinol as a "masked" diketone.

(**22**) (Boyce and Hornig, 1983). The detailed mechanisms of formation of $CHCl_3$ from this position are not completely understood, but a reasonable pathway invokes a triketocyclohexene (**24**): ring opening of this intermediate could give rise to the

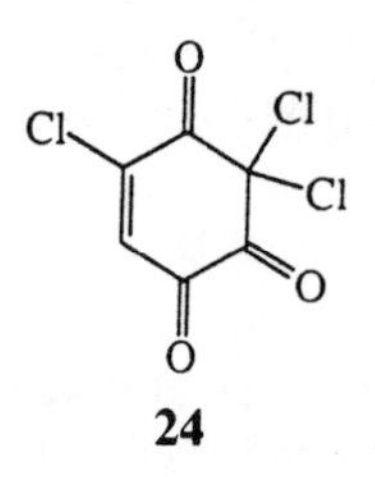

principal observed products, chloroform, CO_2, 2-chlorobutenedioic acid, and di- and trichloroacetic acids (Boyce and Hornig, 1983: Lin et al., 1984: De Leer and Erkelens, 1988). Christman et al. (1978) also chlorinated resorcinol, showing that during chloroform formation it underwent ring contraction to produce the unusual cyclopentenedione **25**. Larson and Rockwell (1979) confirmed Rook's and Christman's observations that resorcinol derivatives were active haloform precursors, and also found that even if the 2-carbon of resorcinol was blocked by a carboxyl group (cf. **26**), yields of chloroform were still high. (However, an -OH or $-CH_3$ substituent reduced the yields to <10%: De Leer and Erkelens, 1988). The reaction of the 2-carboxylated resorcinol derivative is an example of hypochlorite-induced decarboxylation, which is also an important reaction with aliphatic carboxylic acids.

Some β-keto acids (or compounds that could readily be converted to β-keto acids by decarboxylation in the presence of hypochlorite) were also shown to be efficient precursors of chloroform. Thus, citric acid at pH 7.0 gave a 78% yield of $CHCl_3$ in a model experiment, presumably by the mechanism shown in Equation 5.16. The

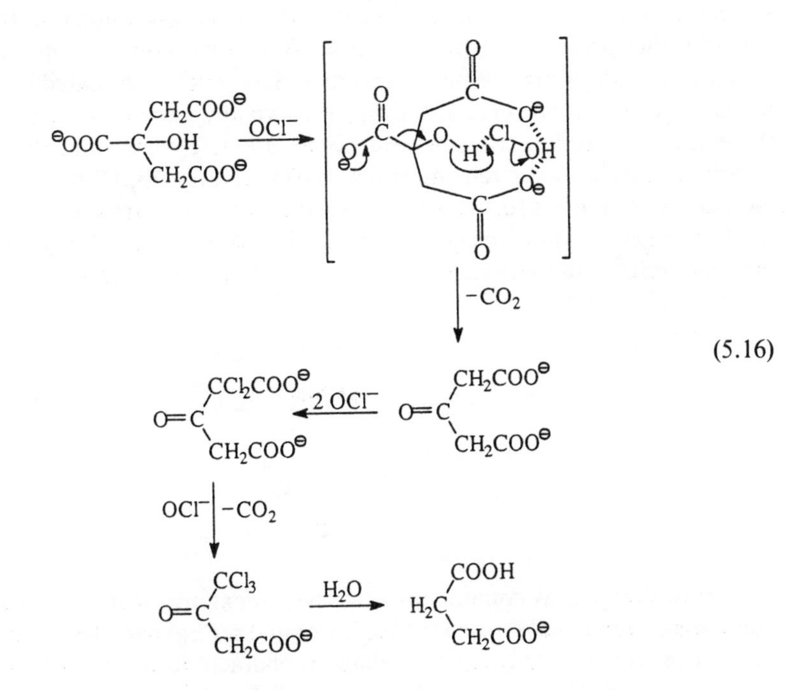

(5.16)

assumed product of the initial hypochlorite-induced decarboxylation, acetone-1,3-dicarboxylic acid (**27**), reacted very rapidly with HOCl to give high yields of chloroform (Larson and Rockwell, 1979). Although citric acid has been demonstrated to

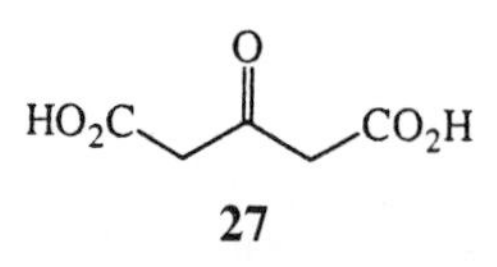

27

occur at low levels in natural waters, it is more likely that in the chlorination of natural organic matter other types of β-keto acid groupings would represent the active sites for haloform production by this mechanism. Citric acid itself is also used in coagulant aid formulations that may be used in drinking water treatment (Streicher et al., 1986). Another possibility is that citric acid might be exuded from killed microbial cells during the chlorination process. Thus, it was shown that $CHCl_3$ was produced only from the "strong acid" fraction (which included citric acid) when microorganisms isolated from a stream sediment and grown on acetate as a sole carbon source were ultrasonically disrupted, fractionated, and chlorinated (Larson and Rockwell, 1979). Thus, the relative importance of phenolic and β-ketonic sources of haloforms, as well as contributions from other types of structures, remains unresolved. An experiment such as that of Oliver and Lawrence (1979), who demonstrated significant decreases in haloform yield when natural waters are treated with alum, may someday be of predictive value when the structures of humic materials are better understood, assuming that the alum interacts in some specific way with particular functional groups to deactivate them.

Although many other phenolic derivatives have been chlorinated and examined as potential haloform precursors, few have shown any particular proclivity to react in this particular fashion. Possible exceptions are a few chlorinated phenols having chlorine atoms in a *meta* relationship to the OH group (**28**; Hirose et al., 1982). The mechanistic rationales for the formation of haloforms from these compounds have not become clear. Many phenols, particularly phenolic acids, do, however, undergo other interesting and unusual chlorination reactions (*vide supra*).

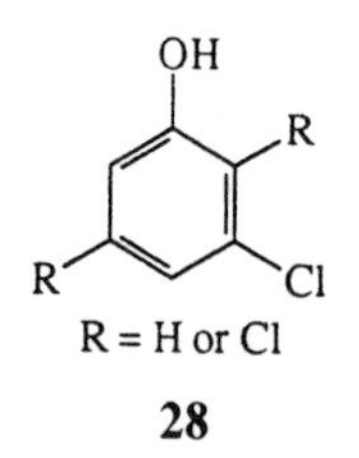

28

Several nitrogenous compounds have been demonstrated to be good haloform precursors. Morris and Baum (1978) chlorinated several pyrroles, which were anticipated to exist with considerable carbanion character at the position α to the ring nitrogen. Naturally occurring substances such as chlorophyll were among the pyrroles that gave high yields of haloforms. Scully et al. (1988) demonstrated that model proteins, hydrolyzed proteins, and protein fractions isolated from natural waters were rather efficient precursors of trihalomethanes and estimated that about 10% of the observed potential for the formation of trihalomethanes might be due to protein-HOCl reactions. These observations may be related to studies that show correlations between haloform concentrations in drinking water with algal populations in source waters (Hoehn et al., 1978) or demonstrate that haloforms are produced by chlorination of extracellular metabolites of algae (Crane et al., 1980: Oliver and Shindler, 1980: Wachter and Andelman, 1984: van Steenderen et al., 1988).

$$\text{pyrrole-R (N-H)} \xrightarrow[\text{pH}=9.5]{OCl^-} CHCl_3\ (30\ \%) \qquad (5.17)$$

Several attempts have been made to examine the ^{13}C-NMR spectra and other properties of natural organic matter before and after chlorination to infer the structural types that give rise to haloforms. Norwood et al. (1987) chlorinated a lake fulvic acid and, on the basis of decreases in the amounts of certain phenols in the copper oxide oxidation products of the chlorinated material, postulated that some of these phenols were precursors for organohalogen compounds. The total quantity of phenols identified by this means, however, was quite low (about 7 mg/g of C), and other potential precursor structures could not be ruled out. Reckhow et al. (1990) attempted to correlate trihalomethane production with the "activated (defined as electron-rich) aromatic" content of ten different humic materials. The activated aromatic content was calculated by an algorithm including the aromatic carbon content of the humic substance as determined by ^{13}C-NMR, the phenolic content as assessed by a difference titration method, and the total nitrogen content, which was assumed to contribute to at least a portion of the activated aromatic reactivity. In another study that utilized pyrolysis-GC-MS and ^{13}C-NMR spectroscopy as probes of the structure of five aquatic humic materials before and after chlorination, the principal observable changes were decreases in intensity in the aromatic, methoxy, and ketonic carbon regions and increases in the relative amounts of aliphatic, alkyl-oxygen, and carboxyl carbons (Hanna et al., 1991). These observations would be consistent with a model for humic chlorination in which either aromatic rings or carbonyl compounds were the active sites of hypochlorite attack.

The source of bromine-containing haloforms has been established by a number of investigators to be the rapid oxidation of bromide by hypochlorite to hypobromite, BrO^-, and the participation of this species in the haloform reaction with dissolved organic matter. Hypobromite appears to be a more reactive halogenating agent, but a less reactive oxidant, than HOCl (Bunn et al., 1975: Macalady et al., 1977: Rook et al., 1978). Some of the factors influencing brominated haloform production in natural waters include pH, temperature, bromide and ammonia concentration, and chlorine dose (Minear and Bird, 1980: Amy et al., 1984).

Intermediates of the types expected to persist under some conditions of the haloform reaction have been isolated in aqueous chlorination experiments. Suffet et al. (1976) identified 1,1,1-trichloroacetone in treated river water samples from the Philadelphia area. This compound is unstable to hydrolysis at pH greater than about 5.

The same compound and several α-dichloroketones were also detected by Giger et al. (1976) in chlorinated water from Lake Zurich. Chlorinated carbonyl compounds have been demonstrated to occur in other mixtures where chlorination is carried out at low pH (Glaze et al. 1978: Kringstad et al., 1981: McKague et al., 1981: Meier et al., 1986: Streicher et al., 1986).

Early suggestions that chloroform, like the carbon tetrachloride occasionally observed in finished drinking water (Shackelford and Keith, 1976), was a contaminant in gaseous chlorine have been shown to be correct in a few instances (Otson et al., 1986). However, the amount of chloroform produced by reactions of chlorine with organic matter in water normally far outweighs the amount introduced in this fashion. Incidentally, hexachloroethane, hexachlorobenzene, and a few other chlorinated species have also been identified as trace contaminants of chlorine gas (White, 1972: NAS, 1980: Otson et al., 1986).

Alkenes

Unsaturated compounds add HOCl to form chlorohydrins. Although halohydrin formation has been rather well-studied in preparative and mechanistic organic chemistry, not many mechanistically directed investigations of these reactions have been conducted under typical water-treatment conditions. However, there is little reason to doubt that the same reactivity constraints prevail at low concentration; primary addition of the equivalent of Cl^+ to the less-substituted end of the double bond to afford a more stable α-chloro carbocation, for example R_2C^+-CH_2Cl, followed by attack of water or HO^- to give the chlorohydrin, should still apply. Seven different chlorinated derivatives of monoterpenes containing both HO and Cl groups were identified by Stuthridge et al. (1990) in a pulp bleaching effluent. At least five of these were formally chlorohydrins. In laboratory investigations of the chlorination of aqueous or aqueous-acetone mixtures of the monounsaturated terpene, camphene (**29**) under acidic conditions, the major products were a few isomers containing one or two chlorine atoms (Buchbauer et al., 1984; Larson and Marley, 1988), although on prolonged standing or in the presence of light, more highly chlorinated mixtures were formed. Product mixtures whose GC-MS patterns had a striking resemblance to those of the insecticide, toxaphene (**30**), developed when an aqueous suspension of the reactants was allowed to stand in sunlight (Larson and Marley, 1988). Interestingly, a toxaphene-like mixture was also isolated from the spent bleach liquor of a pine kraft pulp mill in Finland (Pyysalo and Antervo, 1985).

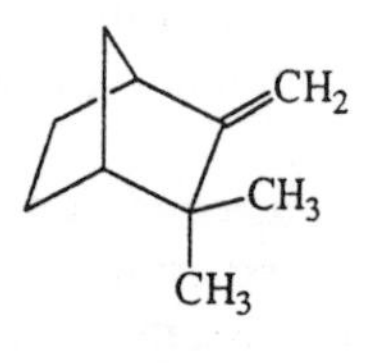

29

$Cl_{6\text{-}9}$

CH_3

CH_3 (> 300 isomers)

CH_3

30

Lipids containing various unsaturated fatty acids in combined (triglyceride) form reacted in an aqueous suspension with radioactive $HO^{36}Cl$ to form low to moderate yields (1–30%) of labeled derivatives, with those lipids containing the highest proportion of multiply unsaturated fatty acids showing the highest incorporation (Ghanbari et al., 1983). These experiments, however, were done with unrealistically high HOCl concentrations (5×10^{-3} *M*, 180 mg/L), at least from a water-treatment perspective. In more controlled laboratory experiments, unsaturated fatty acids such as oleic acid (**31**) were converted to a mixture of chlorohydrins (Carlson and Caple, 1978: Gibson et al., 1986) and it has been suggested that these compounds may eliminate HCl under alkaline conditions to produce, for example, **32**, the epoxide of

stearic acid (Naudet and Desnuelle, 1950), a possible mutagen that has been identified in pulp mill effluents (Leach and Thakore, 1977). Epoxides were also produced at high pH from HOCl and such common terpenes as α-terpineol (**33**: Kopperman et al., 1976), though the major products in acidic solution were the *cis* and *trans* diols **34** and **35**.

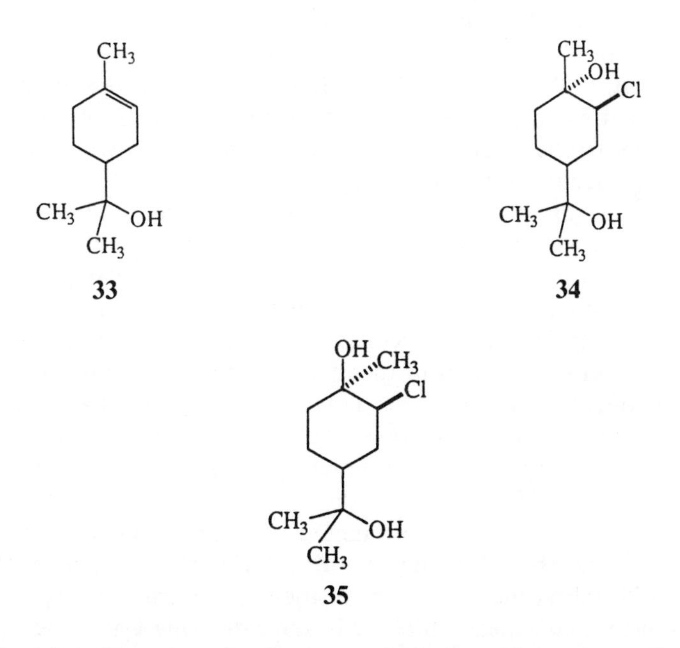

As mentioned earlier, the tricyclic aromatic hydrocarbon phenanthrene, when treated with HOCl at pHs above 4, was converted mainly to the epoxide, phenanthrene-9,10-oxide (**16:** Oyler et al., 1983). The mechanism of formation of the epoxide under these conditions is not clear.

Humic Polymers and Natural Waters

The majority of studies of chlorination of humic materials have been undertaken in order to demonstrate that trihalomethanes are indeed formed from these naturally occurring polymers. The earliest experiment along these lines was that of Rook (1974), who showed that an extract of peat, when chlorinated, gave rise to these compounds. Subsequent investigations with humic compounds from a variety of sources have amply confirmed the original findings (e.g., Stevens et al., 1976: Hoehn et al., 1977: Oliver and Visser, 1980).

Chlorine appears to be incorporated into most humic materials to the extent of about 1–2%, with a variety of products besides trihalomethanes being produced (Rook, 1976: Johnson et al., 1982: De Leer and Erkelens, 1988). Trichloroacetaldehyde (chloral), halogenated acetic acids (principally di- and trichloroacetic acid), and 2,2-dichlorobutanedioic acid (**36**) have been among the major products isolated

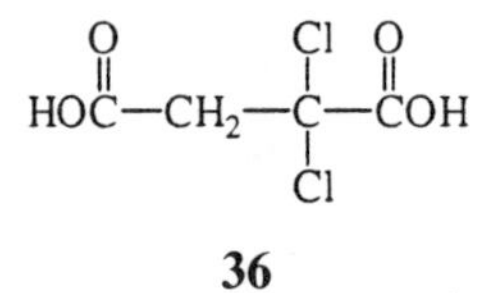

36

from the chlorination of solutions of humic substances or of natural waters high in humic material (Keith et al., 1976: Quimby et al., 1980: Christman et al., 1983: Miller and Uden, 1983: DeLeer et al., 1985). Chlorinated carboxylic acids were also formed when softwood pulp was bleached (Lindström and Österberg, 1986); although the chloroacetic acids predominated, chlorinated derivatives of acrylic, maleic, 2-keto-3-pentenoic, and fumaric acids were also produced. These compounds, plus the haloforms, usually account for approximately half of the chlorinated products derived from aqueous humic substances. It appears that in some instances, nonchlorinated aldehydes (octanal and nonanal, in particular), were formed by chlorination of raw river water by some unspecified oxidative mechanism (Lykins et al., 1986). Among the minor products identified are a variety of compounds listed in Table 5.1. The molecular details of the reactions leading to most of these products are still uncertain, although De Leer et al. (1985) have proposed a mechanism for the formation of the principal chlorination products, based on cleavage of putative 1,3-dihydroxybenzene groups within the humic polymer structure to carboxylic acid intermediates (Figure 5.7).

Bioassay-directed fractionation of humic acid chlorination products has revealed the presence of a number of mutagens with chlorinated γ-lactone and lactol structures such as **37**, (also called MX), one of the most potent bacterial mutagens ever tested, and one which appears to occur in some drinking waters (Hemming et al., 1986: Kronberg et al., 1988: Coleman et al., 1988). Compounds of this type also were formed in the waste liquors from the bleaching of wood pulp (McKague and Kringstad, 1988). The lactol structure of **37** suggests that it might be formed by cyclization of the half-aldehyde **38**, a close relative of the chlorinated C_4-diacids that

Table 5.1. Major Chlorination Products of Humic Materials

Volatile products	$CHCl_3$, CCl_3–CHO, $CHCl_2$–CN	
Nonvolatile products		
Chlorinated monobasic acids	$CHCl_2$–COOH, CCl_3–COOH, CH_3–CCl_2–COOH, CCl_2 = CCl–COOH	
Chlorinated dibasic acids	HOOC–CCl_2–COOH, HOOC–CHCl–CH_2–COOH, HOOC–CCl_2–CH_2–COOH, HOOC–CCl = CH–COOH, HOOC–CCl = CCl–COOH	
Chlorinated tribasic acids	HOOC–CCl = C = $(COOH)_2$	
Chloroform intermediates	CCl_3–CO–CCl = C = $(COOH)_2$, CCl_3–CHOH–CCl_2–CHCl–COOH	
Cyanoalkanoic acids	NC–CH_2–CH_2–COOH, NC–CH_2–CH_2–CH_2–COOH	
Alkanoic acids	CH_3–$(CH_2)_n$–COOH	n = 6 – 22
Alkanedioic acids	HOOC–$(CH_2)_n$–COOH	n = 0 – 2
Benzene carboxylic acids	Phenyl–$(COOH)_n$	n = 1 – 6

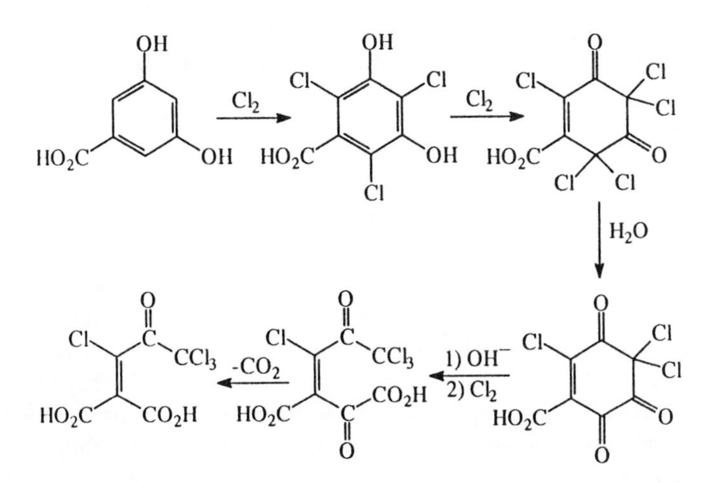

Figure 5.7. Proposed mechanism for the formation of ring cleavage products from 3,5-dihydroxybenzoic acid, a presumed humic model compound. Reprinted by permission from DeLeer et al., *Environ. Sci. Technol.* 19, 522–522. Copyright 1985 American Chemical Society.

are common products of humic acid chlorination (Table 5.1). The ultimate source of these diacids is uncertain, although a wide variety of phenols, when treated with excess HOCl at low pH, produce at least a little MX (Horth et al., 1990: Långvik et al., 1991). The highest yields (around 2.5%) were reported from 3,4,5-trimethoxybenzaldehyde (**39**). No mechanistic speculations have yet appeared in the literature. Similar compounds have also been identified from amino acid chlorination (De Leer et al., 1987).

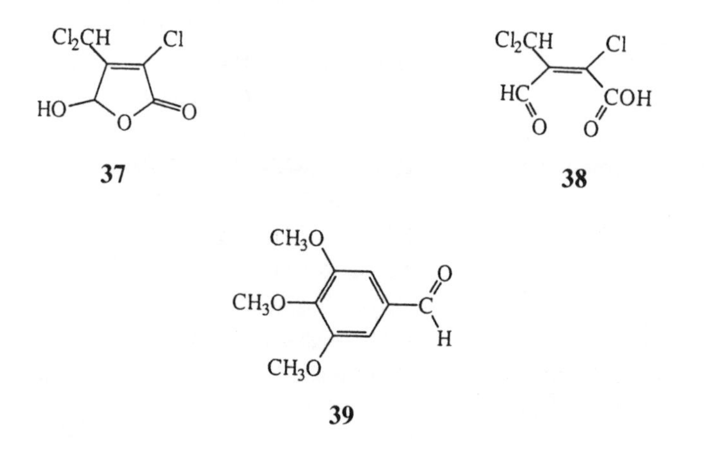

A certain fraction of chlorine appears to be incorporated into macromolecular regions of humic material (Sander et al., 1977), especially at low pH. For example, a fulvic acid from an Ontario creek gave about 70% high molecular-weight organochlorine compounds at pH 5 (30% volatile), whereas volatile compounds predominated at pH 7 and especially pH 11 (Figure 5.8). The same approximate amount of chlorine was taken up by the material at all pHs tested (Oliver, 1978).

Other Polymers

The reactions of aqueous chlorine with the biopolymers and other constituents of the human digestive system have received surprisingly little attention. The issue has been reviewed by Scully and White (1991). Saliva, a largely aqueous solution, contains enzymes and glycoproteins, free amino acids, urea, and low levels of lipids; gastric fluids are much more complex and variable mixtures whose composition depends on diet. Whereas saliva has a pH of ca. 6.5, stomach fluid pH can be as low as 1.3. (At such low pHs, chlorination by Cl_2 becomes an important factor.) Based on kinetic considerations, it is likely that the principal chlorine-consuming reactions within the human digestive tract will be those of proteins and amino acids, with reactions of salivary proteins perhaps playing an important role. Proteins have been shown to react with HOCl to form modified derivatives that contain a variety of N-chloro (perhaps lysine-derived) side-chain adducts as well as oxidative transformation derivatives (Ingols et al., 1953: Stoward, 1975). Tyrosine residues have also been shown to be chlorinated in proteins at the pH of stomach fluid (ca. 2; Nick-

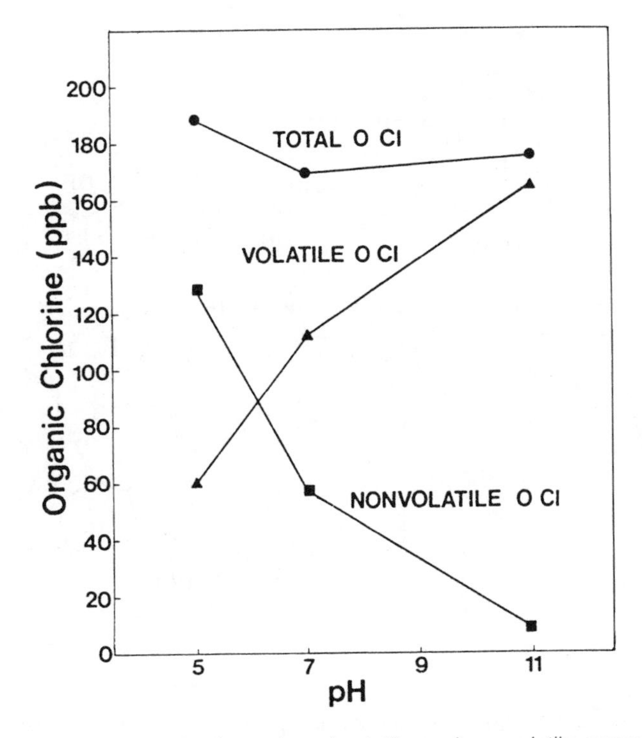

Figure 5.8. Effect of pH on the formation of volatile and nonvolatile organochlorine compounds from chlorination of fulvic acid. From Oliver (1978). Reprinted by permission of MacLean-Hunter, Ltd.

elsen et al., 1991). The destruction of catalytically active thiol (cysteine) and thio ether (methionine) groups in enzymes has long been suggested to be one of the principal mechanisms for the bactericidal effects of chlorine (Knox et al., 1948), and perhaps also occurs in the gastrointestinal tract of humans as well. Very few data on this suggestion have been obtained, although it has been shown that a critical methionine residue in the protein α_1-antiproteinase is specifically converted to a sulfoxide by HOCl, leading to loss of activity (Matheson and Travis, 1985). Furthermore, externally added thiols such as N-acetylcysteine, glutathione, and lipoic acid were able to protect another protein, elastase, from HOCl inactivation (Haenen and Bast, 1991). Reactions of proteins that produce trihalomethanes have also been described (although such reactions are not likely to take place in the digestive tract upon ingestion of water containing chlorine at the concentrations normally used for disinfection; Scully et al., 1988).

Lignin, one of the major polymeric constituents of wood, is a highly crosslinked phenolic polymer derived from the free-radical coupling of substituted phenylpropanoids (Figure 5.9). It reacts with aqueous chlorine to form oxidized and chlorinated species. This is a key reaction in the bleaching of wood pulp by the kraft process

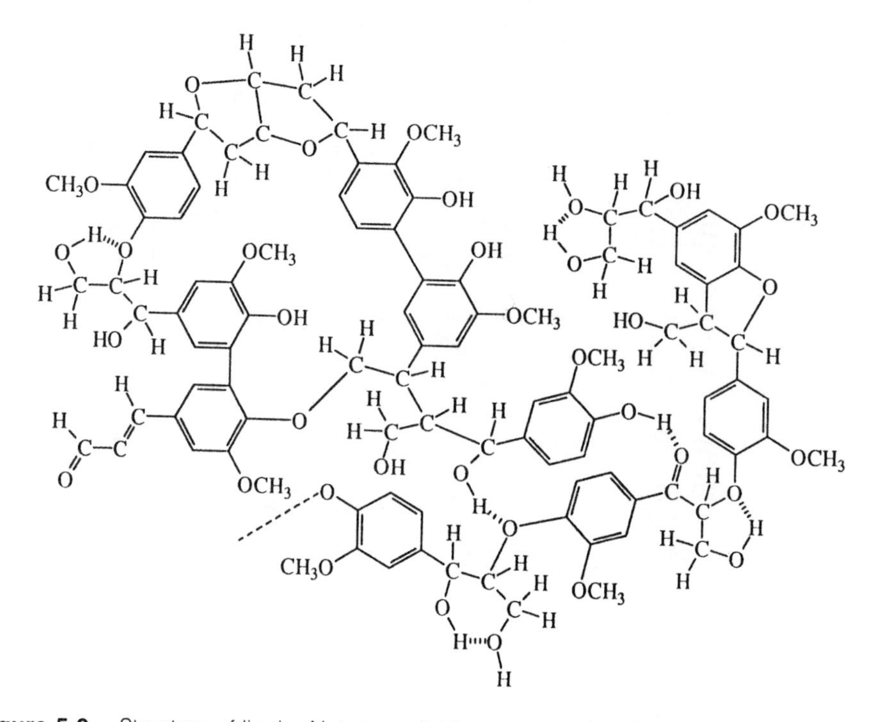

Figure 5.9. Structure of lignin. Note crosslinking due to adventitious radical reactions of C_6–C_3 cinnamylphenol precursors. From K. Haider, S. Lim, and W. Flaig, *Holzforschung* 18, 81–88 (1964). Reprinted by permission of Walter de Gruyter & Co.

(Van Buren and Dence, 1970). Both high molecular-weight "chlorolignins" and smaller molecules such as chlorinated aliphatic carboxylic acids, chloromuconic acids (**40**), chlorinated phenolic derivatives, and chlorobenzoquinones are formed (Das et al., 1969: Leach and Thakore, 1977: Erickson and Dence, 1976: Österberg and Lindström, 1985: McKague et al., 1987, 1989). The conversion of the lignin chlorination product tetrachlorocatechol to the unusual product 2,2,4,5-tetrachlorocyclopenten-1,3-dione (**41**) has been rationalized by the mechanism in Figure 5.10.

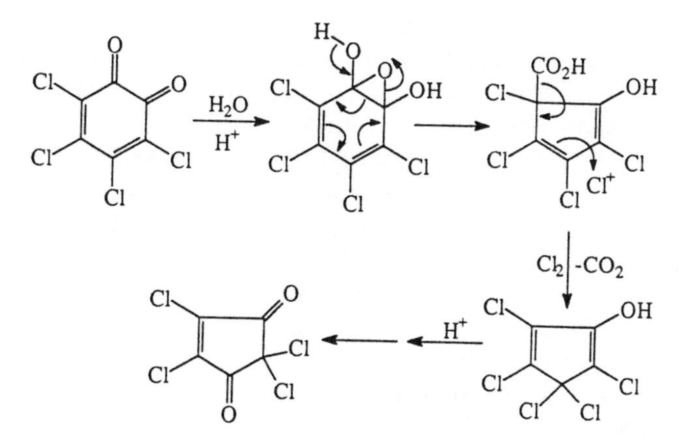

Figure 5.10. Proposed mechanism for formation of a cyclopentenedione product from an *ortho*-quinone derived from tetrachlorocatechol. From McKague et al. (1987). Reprinted by permission of Walter de Gruyter & Co.

B. COMBINED AQUEOUS CHLORINE (CHLORAMINES)

The chemistry of chloramines has been reviewed by Kovacic et al. (1970). In water, depending on pH and concentration, monochloramine (NH_2Cl), dichloramine ($NHCl_2$), or trichloramine (NCl_3), may be present, in addition to organic chloramines derived from amines.

1. Formation of Chloramines

When water containing ammonia or ammonium ions is treated with chlorine gas, chloramines are rapidly formed:

$$HOCl + NH_3 \leftrightharpoons NH_2Cl + HOCl \leftrightharpoons NHCl_2 + HOCl \leftrightharpoons NCl_3 \quad (5.18)$$

This reaction can be considered as a nucleophilic displacement on chlorine by the free electron pair of nitrogen.

Monochloramine predominates at $pH > 7.5$, whereas dichloramine prevails down to pH 5, except at high ammonium ion concentrations, where it is destroyed by the following reaction (Fair et al., 1948):

$$NH_4^+ + NHCl_2 \leftrightharpoons 2\,NH_2Cl + H^+ \quad (5.19)$$

Trichloramine occurs in appreciable concentrations only at $pH < 3.5$. At very high relative HOCl concentrations, combined chlorine species disappear by a complex process referred to as the **breakpoint reaction**, characterized by decreasing concentrations of combined chlorine, concomitant formation of inorganic nitrogen compounds, and the eventual appearance of free chlorine in solution, which occurs at a hypochlorite–to–ammonia mole ratio of approximately 2. During this incompletely understood process (Weil and Morris, 1974), chloramines are further oxidized to intermediate forms that may include hydroxylamine (NH_2OH), nitric oxide (NO), and nitrite (NO_2^-). The ultimate products are nitrogen gas and nitrate, NO_3^-.

Monochloramine is not a particularly reactive agent for bringing about substitution reactions; chlorination by NH_2Cl is about four orders of magnitude slower than the corresponding HOCl reaction (Morris, 1967), except at low pH (<2) where its conjugate acid, $^+NH_3Cl$, is present; this species is a potent chlorinating agent. In many cases, observed reactions may be due to the slow hydrolysis of NH_2Cl to HOCl. Monochloramine does not appear to participate to a significant extent in haloform reactions (Stevens et al., 1978), which has made it attractive to drinking water plant personnel as a disinfectant, since these compounds are regulated by the Safe Drinking Water Act.

2. Formation and Reactions of Chloramines

Aromatic Compounds

Phenol reacts with monochloramine at neutral pH in a multi-step reaction to form a colored compound, indophenol blue (**42**: AWWA, 1989). The sequence of its formation appears to be initiated by amination at the *p*-position of phenol, followed by oxidation of 4-aminophenol to an iminoquinone and condensation to form the ultimate product (Equation 5.20). The reaction is used in a standard procedure for the determination of ammonia in natural waters. Amination reactions of a few other compounds by NH_2Cl have been observed under some conditions (for example, phenolic ethers are converted to aniline derivatives in methanol: Kovacic et al., 1970), but it is uncertain to what extent these would occur under environmental conditions.

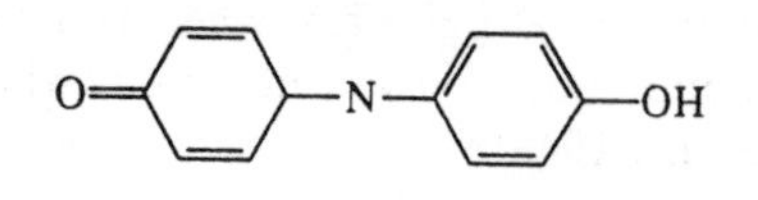

42

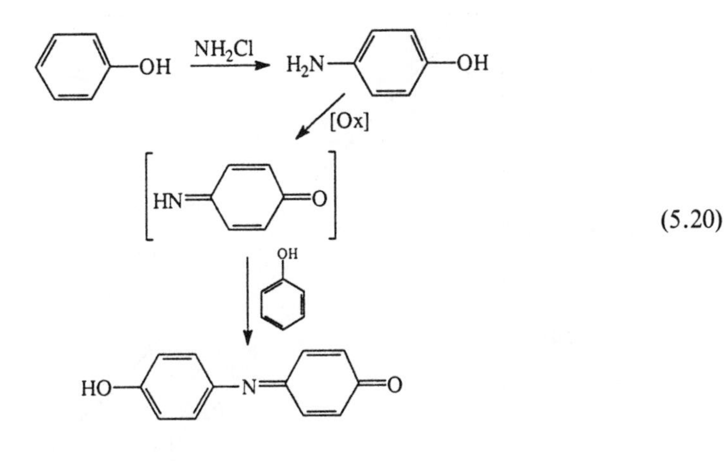

(5.20)

Aromatic amines, which are far less basic than the aliphatic forms, tend to react by substitution on the ring rather than by N-chlorination; the amino group is a powerful electron donor (cf. Chapter 1). Thus, aniline gives rise to *p*-chloroaniline as the principal product of reaction with HOCl (Barnhart and Campbell, 1972; Hwang et al., 1990). However, benzidine (**43**) reacts rapidly with HOCl to form a deeply colored polymer without ring chlorine substitution. It was suggested that an oxidized polymer similar to the common dye, aniline black (**44**) had been produced (Jenkins et al., 1978).

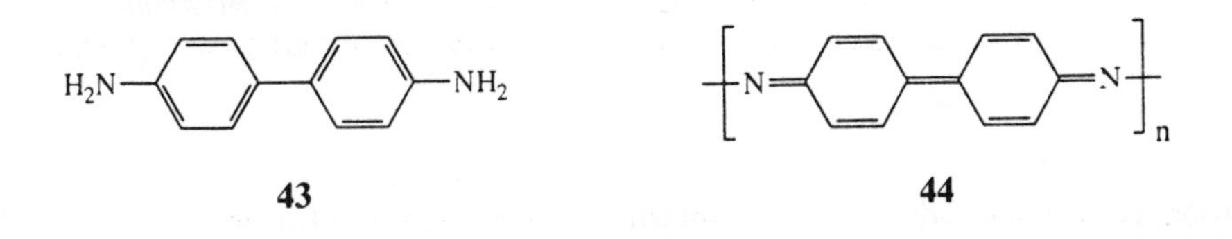

Aliphatic Compounds

Thiols are oxidized by NH_2Cl to disulfides (Ingols et al., 1953). Monochloramine reacts with aldehydes (although in many cases the reaction is slow: Hauser and Hauser, 1930; Conyers and Scully, 1993) to form N-chloroimines (R-CH = NCl) which then undergo further reactions (elimination of HCl to the nitrile; hydrolysis to aldehydes; polymerization; etc.)

Organic aliphatic amines may react rapidly with HOCl to form N-chloro compounds. Methylamine, for example, reacts nearly 100 times as fast as ammonia with HOCl (Morris, 1967). In general, the rates of these reactions parallel the basicities of the amines, although steric and other effects undoubtedly play important roles. Morris (1967) has shown excellent correlation of reaction rate with pK_b over many orders of magnitude of the latter parameter (Figure 5.11). Tertiary amines were converted to secondary N-chloro compounds and carbonyl compounds derived from side-chain cleavage (Ellis and Soper, 1954). The mechanism of this reaction

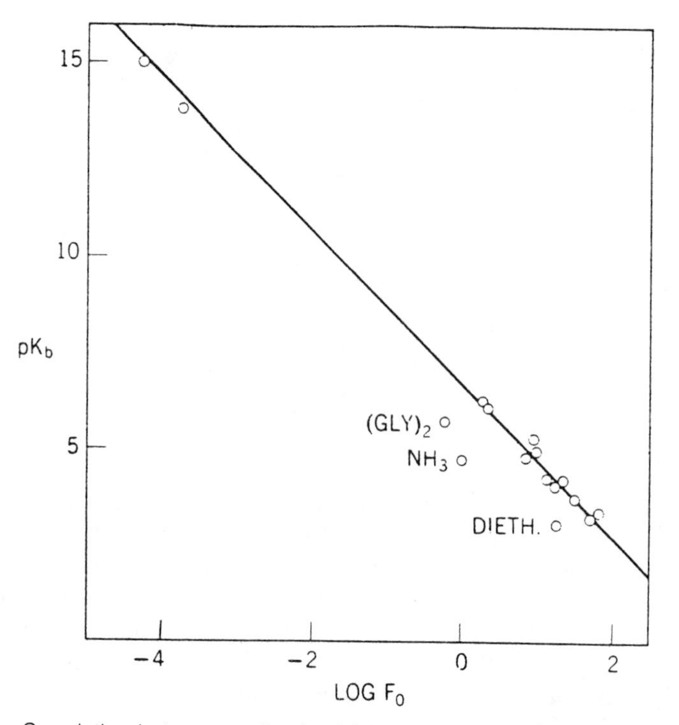

Figure 5.11. Correlation between amine basicities and their rates of reaction with HOCl. From S. D. Faust and J. V. Hunter, Eds. *Principles and Applications of Water Chemistry*, John Wiley & Sons, 1967. Reprinted by permission.

probably involves hydrolysis of an intermediate iminium ion formed by loss of HCl from a quaternary N-chloronium intermediate.

$$R_1CH_2{-}N\begin{matrix}R_2\\R_3\end{matrix} \xrightarrow{HOCl} R_1CH_2{-}\overset{\displaystyle R_2}{\underset{\displaystyle Cl}{\overset{|}{\underset{|}{N^{\oplus}}}}}{-}R_3$$

$$\downarrow HCl$$

$$\begin{matrix}R_1\\ \end{matrix}\!\!\!\!C{=}O \;\; (H) + HN\begin{matrix}R_2\\R_3\end{matrix} \xleftarrow{H_2O} R_1CH{=}\overset{\oplus}{N}\begin{matrix}R_2\\R_3\end{matrix} \qquad (5.21)$$

The detergent additive nitrilotriacetic acid (NTA, **45**), although practically inert to chlorination under environmentally realistic conditions, does react at pH 11 with 0.7 M NaOCl (essentially, undiluted laundry bleach) to give a 40% yield of the side-chain cleavage product, N-formyliminodiacetic acid (**46**: Spanggord and Tyson, 1979).

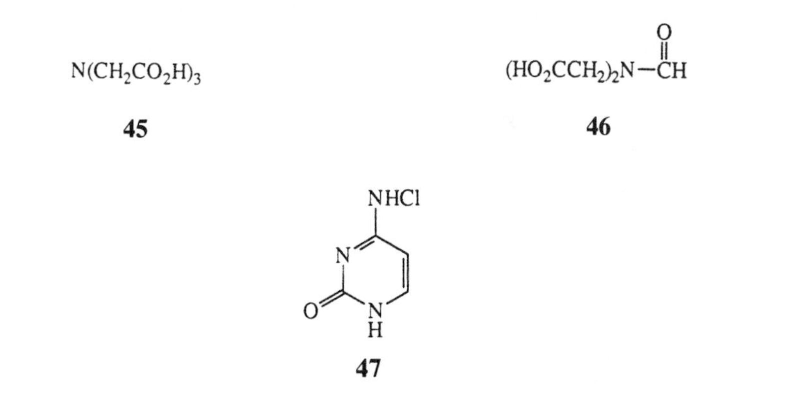

N-Chlorocompounds are themselves capable of reacting with some organic compounds to produce oxidation or chlorination products. 4-N-Chlorocytosine (**47**), for example, reacts with phenylalanine to give a low yield of phenylacetaldehyde (Patton et al., 1972). Presumably, the N-chloro amino acid is formed as an intermediate by direct chlorine atom transfer between **47** and phenylalanine.

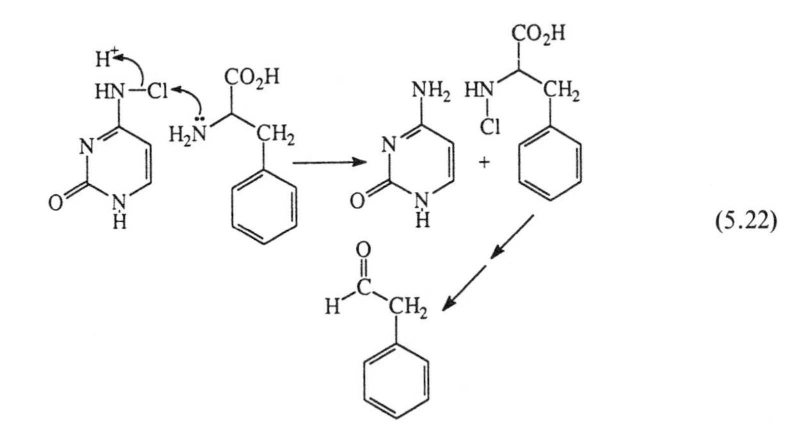

(5.22)

N-chlorination of amides, although it can be observed (Mauger and Soper, 1946), is very much slower than amine chlorination and is not likely to occur under normal water treatment conditions. N-Methylacetamide reacts more than a million times slower than ammonia (Morris, 1967). The unreactivity of the N–H bond in amides is further illustrated by the fact that acetanilide chlorinates at the ring 4-position rather than the amide nitrogen (Houben and Weyl, 1962).

Amino Sugars

Aqueous HOCl reacts with amino sugars to produce the corresponding N-chloro compounds (Sandford et al., 1971). The reaction has been used as the basis for a colorimetric determination of these sugars. Glucosamine (**48**) undergoes an interesting reaction with alkaline NaOCl in which a chain shortening to arabinose (**49**) occurs, with concomitant production of ammonia and HCOOH (Matsushima, 1951). The mechanism may proceed by the route indicated in Equation 5.23.

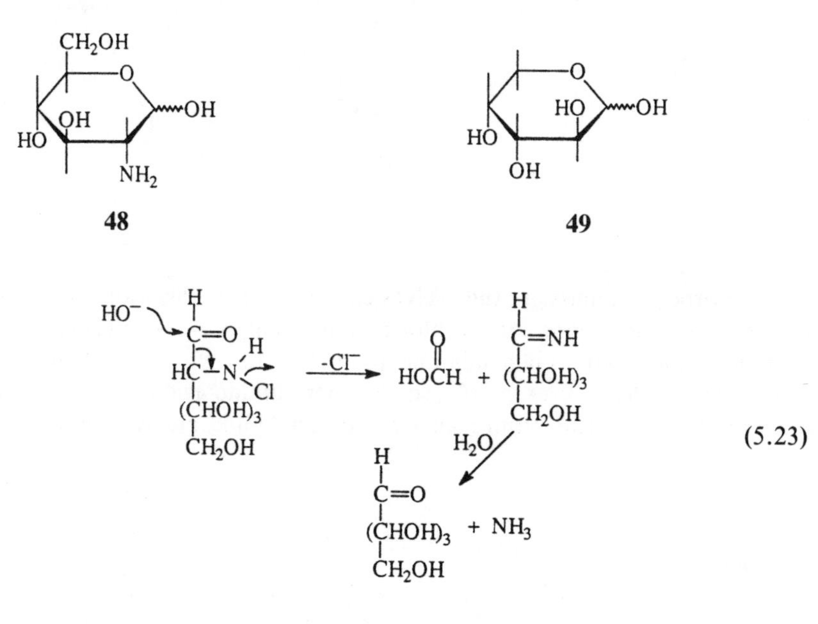

(5.23)

Amino Acids

The hypochlorite chlorination of amino acids generally begins with the rapid formation of an N-chloro intermediate, which can then undergo further reactions leading to α-keto acids, nitriles, and aldehydes, depending on conditions (Friedman and Morgulis, 1936: Stanbro and Smith, 1979: Tan et al., 1987). It has been estimated that N-chloroalanine will degrade in a few hours to pyruvic acid, acetaldehyde, ammonia, CO_2, and Cl^-; the rate is strongly dependent on temperature (Stanbro and Smith, 1979). The basic mechanistic pathway is outlined in Figure 5.12.

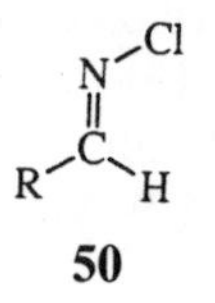

50

Under certain conditions, chlorinated aldimines (**50**) which are surprisingly stable can be formed from isoleucine and perhaps other amino acids (Nweke and Scully, 1989). The mechanism proposed for this reaction is the elimination of Cl^- and CO_2 from a hypothetical dichloramino acid (Equation 5.24).

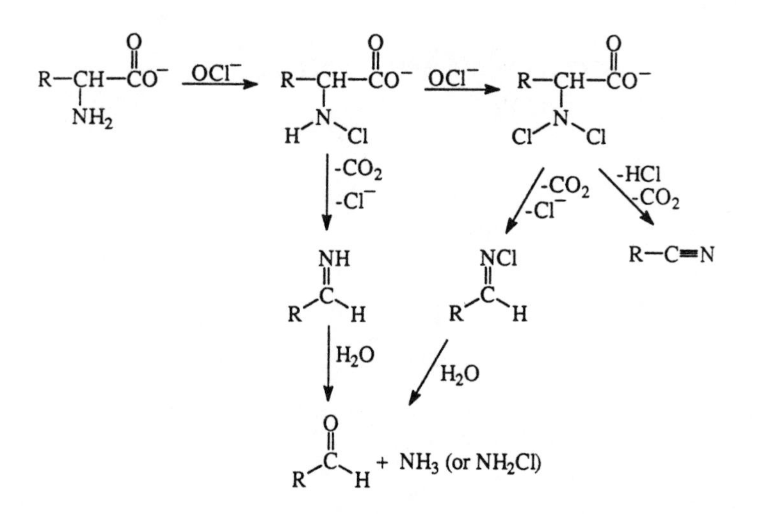

Figure 5.12. Pathways for formation of aldehydes and nitriles from amino acids and HOCl. Modified from Stanbro and Smith (1979). Reprinted by permission. Copyright 1979 American Chemical Society.

$$\mathrm{R{-}CH(NH_2){-}CO_2^-} \xrightarrow{2\,\mathrm{HOCl}} \mathrm{R{-}CH(NCl_2){-}CO_2^-} \xrightarrow[-\mathrm{HCl}]{-\mathrm{CO_2}} \mathrm{R_1{-}CH{=}NCl} \qquad (5.24)$$

The formation of halogenated acetonitriles, known constituents of drinking water (McKinney et al., 1976: Oliver, 1983), was shown to be probably due to reactions of amino acids such as aspartic acid, tyrosine, and tryptophan (Trehy and Bieber, 1981; Trehy et al., 1988). Experiments by Oliver (1983) confirmed that either natural organic matter (fulvic acid) or algal products yielded these compounds under conditions similar to those used in water treatment. In addition to the nitrile derivatives, these amino acids also yield chloral and chloroform by variants of the basic pathway (Figures 5.13 and 5.14).

The formation of haloacetonitriles from tyrosine must involve an *ipso* ring-chlorination step at some stage in order to cleave off the side chain at the benzylic position. Nuclear chlorination of the activated aromatic ring of tyrosine has also been shown to occur, with the formation of chlorinated phenylacetonitrile derivatives such as **51** and one or more chlorinated phenylacetaldehyde derivatives

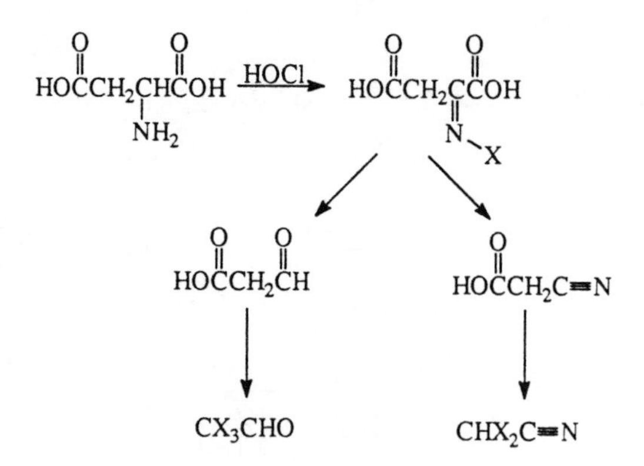

Figure 5.13. Halogenation of aspartic acid. From Trehy et al. (1988). Reprinted by permission of Lewis Publishers.

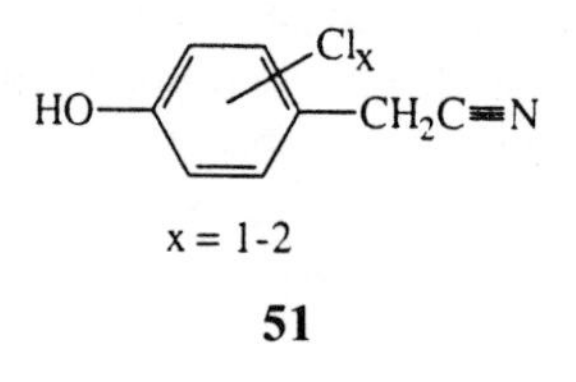

(Pereira et al., 1973: Burleson et al., 1978; Shimizu and Hsu, 1975). In alkaline solution (pH 10.5), some nitriles reacted by formally adding the elements of HOCl

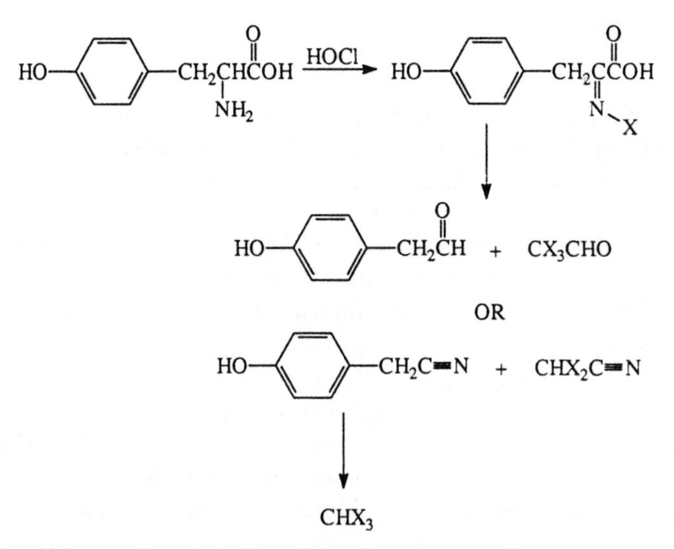

Figure 5.14. Halogenation of tyrosine. From Trehy et al. (1988). Reprinted by permission of Lewis Publishers.

to the CN triple bond to produce isomeric chloroamides (Equation 5.25) (Peters et al., 1990).

$$Cl_2CH{-}C{\equiv}N \xrightarrow{HOCl} Cl_2CH{-}C({=}N{-}Cl){-}OH \longrightarrow Cl_2CH{-}C({=}O){-}NHCl \qquad (5.25)$$

Aldehydes from amino acid chlorination, such as 2-methylpropanal (derived from valine), 2- and 3-methylbutanal (derived from leucine and isoleucine, respectively), and phenylacetaldehyde (derived from phenylalanine), have been found to be potent contributors to taste-and-odor problems in the drinking water supply of Edmonton, Alberta. The incidents occur in the spring following snowmelt and concomitant increases in raw water color and TOC (Hrudey et al., 1988). Interestingly, the aldehydes were also formed during monochloramine disinfection.

$$HC(NH_2)(CO_2H){-}(CH_2)_2{-}CO_2H \xrightarrow{HOCl} N{\equiv}C{-}(CH_2)_2{-}CO_2H + CO_2 \qquad (5.26)$$

Chlorination of glutamic acid at pH 7.2 gave a high yield of the cyano acid **52** (Equation 5.26). Proline reacted to give the dichlorodiacid **53** (De Leer et al., 1987), which represents one of the major structural types found in humic acid chlorination products (*vide supra*). Tryptophan reacted with HOCl under acidic conditions to

$$HO_2C{-}CH_2{-}CH_2C{\equiv}N \quad \textbf{52}$$

$$HO_2C{-}CCl_2{-}CH_2CO_2H \quad \textbf{53}$$

give an oxindole and a C-chlorinated oxindole (Equation 5.27) (Burleson et al., 1978). These products may arise from addition of HOCl to the pyrrole ring double bond. In another investigation, conducted at pH 7, several highly mutagenic compounds were isolated from HOCl-tryptophan reactions, including a dichloroquinoline, a trichloroacetone, and a tetrachloracetone. Mechanisms for formation of these products were not discussed (Owusu-Yaw et al., 1990).

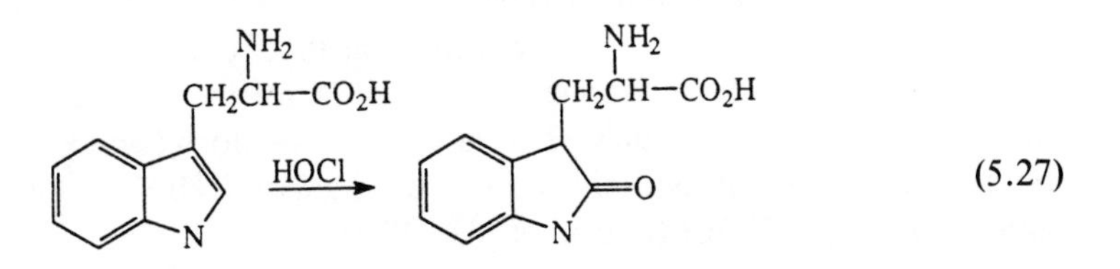

(5.27)

Glycine reacted with HOCl to form HCN and ClCN, depending on pH and molar excess of HOCl (Sawamura et al., 1982).

Heterocyclic Nitrogen Compounds

The reactions of aqueous solutions of free chlorine with nucleic acid bases has been studied by many groups of workers. They are generally very complex. The free pyrimidine base uracil (**54**), for example, gave a variety of products when chlorinated. The initial product, 5-chlorouracil (**55**: Hayatsu et al., 1971: Hoyano et al., 1973), was converted by excess HOCl to di- and trichloro derivatives, to chlorohydrins, and finally to ring destruction products including CO_2, NCl_3, and Cl_3CCOOH. At high pH, significant quantities of $CHCl_3$ were produced. Formaldehyde and formic acid were not observed, and the mechanism of the ring cleavage remains obscure (Dennis et al., 1978). The closely related compound, uridine monophosphate (**56**), was surprisingly unreactive toward HOCl relative to the other RNA nucleoside monophosphates. Cytidine and adenosine monophosphates (**57**,

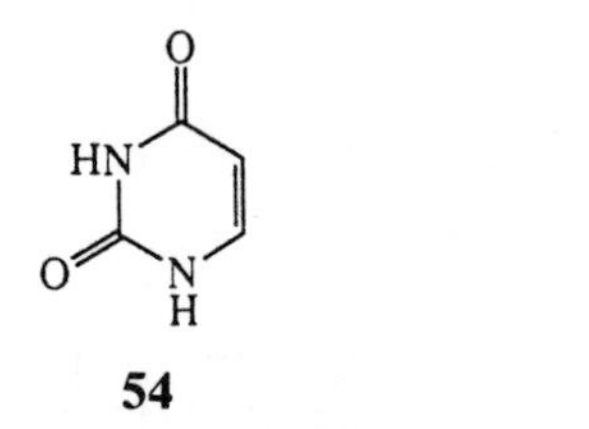

54

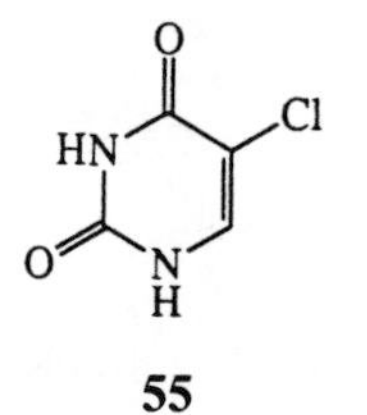

55

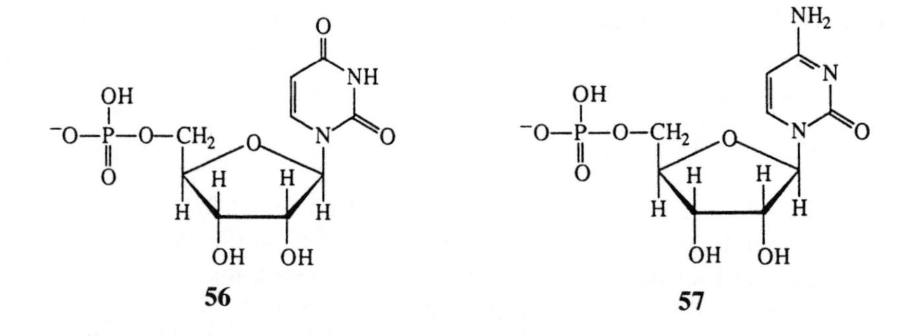

56 **57**

58) were most reactive as measured by UV spectroscopic changes. Products of the reactions were not determined (Dennis et al., 1979). Cytosine (**59**) and cytidine appeared to chlorinate more or less cleanly at the 5-position (Hayatsu et al., 1971); other mono- and dichloro isomers have been identified depending on chlorine dose and pH (Reynolds et al., 1988). The side-chain N-chloro adduct may be an intermediate, and chlorohydrins derived from double-bond HOCl addition arise at high concentrations of HOCl (Patton et al., 1972).

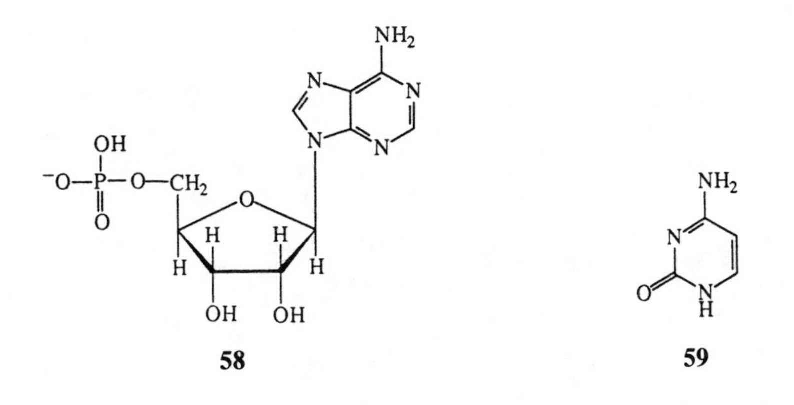

58 **59**

Purine bases are rather less reactive to HOCl than pyrimidines. Thus, in one study, treatment of guanine, adenine, or xanthine (**60–62**) with 1 or 2 equivalents of HOCl resulted only in high yields of starting material recovery. After prolonged exposure, variable yields of parabanic acid (**63**) were obtained (Hoyano et al., 1973). This compound has been demonstrated to be formed from both pyrimidines and purines in other high-energy oxidative processes (see, e. g., LeRoux et al. [1969]), but the details of its formation from purines by the HOCl process are not known.

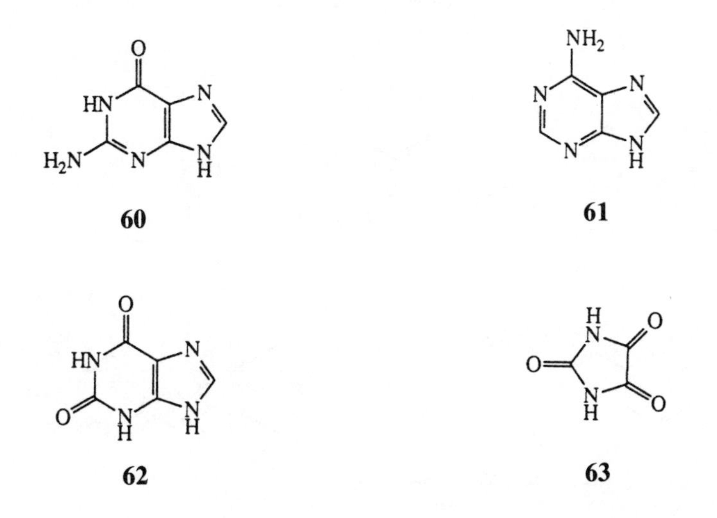

60 **61** **62** **63**

Indole (**64**) reacted with aqueous chlorine under mildly acidic conditions (Babenzian et al., 1963: Lin and Carlson, 1984). The products are shown in Equation 5.28. The initial product formed is presumably N-chloroindole (DeRosa and Alonso, 1978). Carbazole (**65**) also reacted with free chlorine to produce a complex mixture of Cl_1–Cl_4 substitution products (Lin and Carlson, 1984).

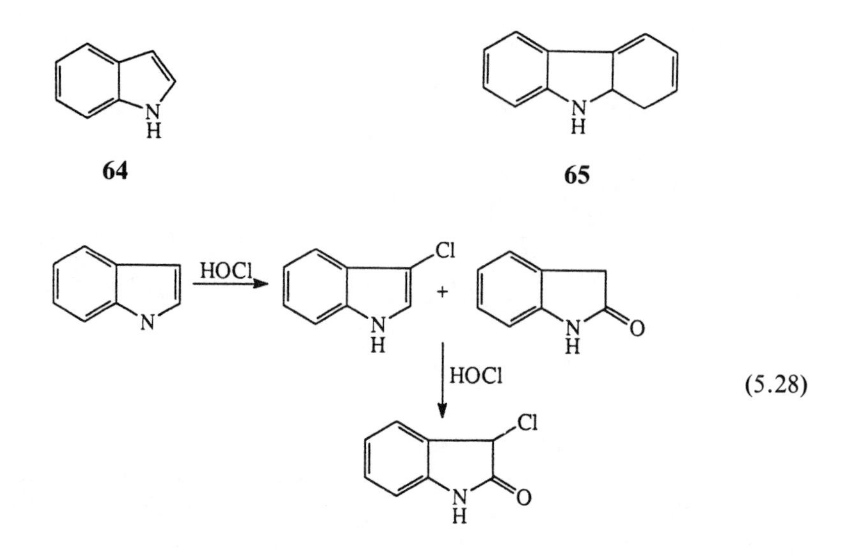

Pyridine reacted much less readily than even benzene toward HOCl, in keeping with its behavior toward other electrophilic reagents (Lin and Carlson, 1984), but quinoline (**66**) did react to form both oxidation and substitution products.

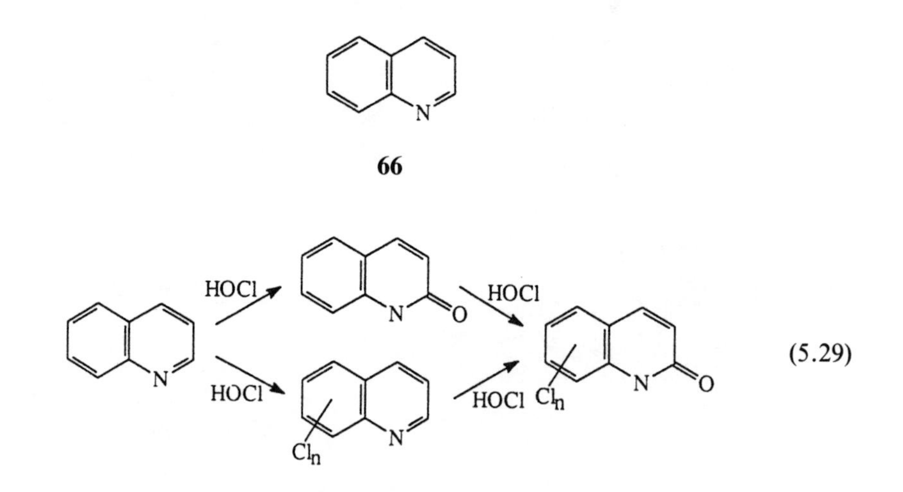

When treated with large excesses of hypochlorite at alkaline pHs, much chloroform was produced from the nitrogen heterocycles pyrrole, proline, hydroxyproline, tryptophan, and chlorophyll (Morris and Baum, 1978).

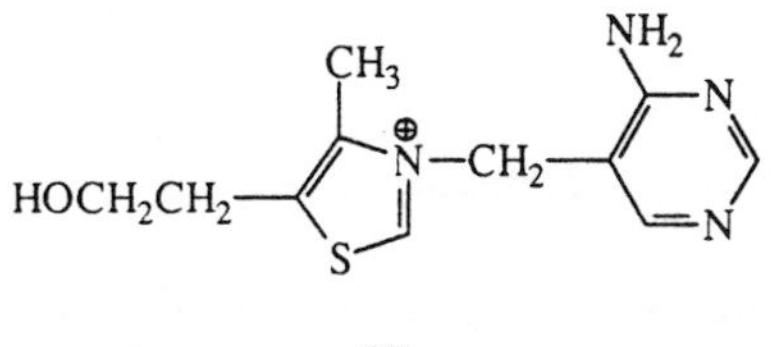

67

The vitamin, thiamine (**67**), was rapidly destroyed by HOCl above pH 5; the half-life of a 3μM solution in the presence of an equimolar amount of HOCl was only 2 hr (Yagi and Itokawa, 1979). Products were not identified.

C. OZONE

Ozone, O_3, is one of the most potent oxidants known, having a standard redox potential of about 2.1 V in water. It is readily generated by passing an electric discharge through air or oxygen, and has been widely used in synthetic organic chemistry for many decades. Because of its great water solubility and its extremely high reactivity toward microorganisms and soluble organic compounds, it has been widely employed in water treatment, particularly in Europe; in North America, however, because it does not persist for long periods in the distribution lines, it has normally been used in combination with free or combined chlorine for drinking water treatment. The added costs associated with the onsite generation of ozone have also been a consideration that has militated against its use, relative to the cheapness and convenience of chlorine. Recently, however, there has been a renewed consciousness in the U.S. and Canada of ozone treatment of drinking water, as the concentrations of more and more chlorinated organic compounds are regulated by governmental bodies (Glaze, 1987).

Drinking water plants typically use ozone doses of about 1–5 mg/L (2×10^{-5} – 1×10^{-4} *M*) with contact times of around 10 min. In wastewater applications, much higher levels may be applied. These concentrations are normally sufficient for virtually complete microbial inactivation, but are not adequate to exhaustively oxidize organic compounds. Thus, partial oxidation products of susceptible organic compounds will remain after treatment. Although the chemistry of ozonolysis in nonaqueous solvents is rather well understood due to its widespread use as a degradative reagent, less is known about the course of its reactions in water, particularly concerning the highly polar and labile products formed when it reacts with trace constituents of the dissolved organic material.

1. Ozone in Water

Decomposition Mechanisms of Aqueous Ozone

The lifetime of ozone in aqueous systems depends principally on the pH. Ozone decomposition is catalyzed by HO^- in a bimolecular process whose rate is dependent on both $[O_3]$ and $[HO^-]$. However, the kinetics are complicated by the existence of a radical chain process involving $HO\cdot$ and other intermediate odd-electron species. Forni et al. (1982) indicated that the initiation step for the decomposition reaction appeared to be an oxygen atom transfer to form molecular oxygen and the anion of H_2O_2:

$$O_3 + HO^- \rightarrow HOO^- + O_2 \tag{5.30}$$

However, other evidence (Bühler et al., 1984) suggested that the products of the above reaction were $HOO\cdot$ and $O_2\cdot^-$. In any event, subsequent reactions of either HOO^- or $HOO\cdot$ would result in the formation of $HO\cdot$:

$$HOO^- + O_3 \rightarrow HOO\cdot + O_3^-\cdot \tag{5.31}$$

$$HOO\cdot \rightarrow H^+ + O_2\cdot^- \qquad (pKa = 4.8) \tag{5.32}$$

$$O_2\cdot^- + O_3 \rightarrow O_2 + O_3^-\cdot \tag{5.33}$$

$$O_3^-\cdot + H_2O \rightarrow HO\cdot + HO^- + O_2 \tag{5.34}$$

This radical takes part in further chain decomposition reactions leading to ozone loss:

$$HO\cdot + O_3 \rightarrow HOO\cdot + O_2 \tag{5.35}$$

In pure water, therefore, the operation of the above mechanisms would lead smoothly to conversion of ozone to molecular oxygen. However, actual waters almost always contain substrates which can compete efficiently for either ozone or the intermediate free radicals such as $\cdot OH$ that are formed during its decomposition. In particular, in seawater molecular ozone reacts most rapidly with dissolved I^-, despite its low concentration of about 10^{-7} *M* (Haag and Hoigné, 1983); in moderately alkaline freshwater with a typical degree of hardness (roughly millimolar in carbonate alkalinity), it is usually the carbonate or (with lower effectiveness) bicarbonate anion that reacts with the ozone-derived $HO\cdot$:

$$HO\cdot + CO_3^{2-} \rightarrow HO^- + CO_3^-\cdot \tag{5.36}$$

This process is described more fully below. Moreover, $\cdot OH$ also rapidly attacks dissolved organic compounds (see also Section 4.B.2), with rate constants for most ranging from a little more than a tenth to a little less than a hundredth of diffusion control (10^8 - 6×10^9 L/mol sec; Farhataziz and Ross, 1977). At ordinary water treatment pHs (>6) it is probable that ·OH is by far the principal oxidizing species present in ozonized water (Hoigné and Bader, 1975).

In sufficient concentrations, carbonate anion, by scavenging one of the major chain carriers of the ozone decomposition pathway, can slow down the decay rate of ozone significantly (Hoigné and Bader, 1975, 1979). The second-order rate constant for reaction of carbonate anion (4×10^8 L/mol sec; 1.5×10^7 for bicarbonate anion) with HO· is about an order of magnitude less than that for reaction of ozone (or for most organic compounds) with HO·, so in waters where carbonate levels exceed that for applied ozone (or dissolved organic matter), a significant fraction of the HO· will be scavenged by carbonate, inhibiting the chain reaction for ozone decomposition. The bicarbonate anion is, however, about an order of magnitude less reactive than the carbonate anion. The further fate of the carbonate radical in natural waters is uncertain; it is a very selective species that only reacts rapidly with quite electron-rich organic compounds such as aniline derivatives (Larson and Zepp, 1988).

2. Reactions of Ozone

Kinetics

Most of the known rate data for reactions of organic solutes in water have been determined by Hoigné and Bader (1983a-b). These authors measured the losses of ozone in solutions having a large stoichiometric excess of the potential organic reactant, thus obtaining pseudo-first-order rate law terms for O_3. The kinetic experiments were usually performed at low pHs to avert contributions from ·OH. These studies confirm the high selectivity of aqueous ozone as an oxidant (Table 5.2, Table 5.3, and Table 5.4). Hammett plots such as Figure 5.15 show that the direct attack of ozone on the aromatic ring is highly selective and dependent on electron density.

Polycyclic hydrocarbons are extremely reactive with ozone. For example, naphthalene reacted with ozone at a rate about 1500 times faster than benzene (Hoigné and Bader, 1983b), and higher polycyclic hydrocarbons such as phenanthrene, pyrene, and benzo[a]pyrene were also extremely reactive (Butkovic et al., 1983). The experiments of Hoigné and Bader also indicate that the rate constants for reaction of ozone with aromatic hydrocarbons in water were about 100 times greater than in nonpolar solvents such as CCl_4. However, aliphatic compounds did not show such a profound solvent effect.

Hydrocarbons

In organic solvents, ozone undergoes a classic reaction with olefins, reacting rapidly and almost quantitatively with most of these compounds. Carbonyl compounds and carboxylic acids are the products of this reaction. The traditional Criegee (cf. Bailey, 1958) mechanism is generally agreed to provide the best rationaliza-

Table 5.2. Reaction-Rate Constants of Substituted Benzenes with Ozone[a,b]

Solute	mM	pH	Scavenger	mM	k_o (M^{-1} s^{-1})
Nitrobenzene	5/10	1	*t*-BuOH	50–1000	0.09 ± 0.02
Benzene sulfonate ion	10–800	1.7–2	*t*-BuOH	50–800	*0.23 ± 0.05*
1,2,4-Trichlorobenzene	0.05	2	*t*-BuOH	10	<1.6
1,4-Dichlorobenzene	0.05	2	*t*-BuOH	0.5	<3
Chlorobenzene	0.8–3	2	PrOH	1	0.75 ± 0.2
Methyl benzoate	1000	2	*t*-BuOH	100	1.1 ± 0.3
Benzoate ion	10–100	5	*t*-BuOH	100–1000	*1.2 ± 0.2*
Benzene	1–10	1.7–3	—	—	2 ± 0.4
Benzaldehyde	2–10	1.7	—	—	2.5 ± 0.5
Iso-Propylbenzene	0.14	2.0	*t*-BuOH	100	11 ± 3
Toluene	0.4–4	1.7	—	—	14 ± 3
4-Nitrophenol[c]	0.01–14	1.7	PrOH		<50
Ethylbenzene	0.25–1	2.0	*t*-BuOH	100	14 ± 4
o-Xylene	0.03–0.8	1.7–5	—	—	90 ± 20
m-Xylene	0.2–0.5	2.0	*t*-BuOH	1	94 ± 20
p-Xylene	0.2–0.5	2.0	*t*-BuOH	1	140 ± 30
Anisole	0.05–0.8	2	PrOH	1	290 ± 50
1,2,3-Trimethylbenzene	0.06–0.2	1.7	—	—	400 ± 100
1,3,5-Trimethylbenzene	0.05–0.1	1.7	—	—	700 ± 200
4-Chlorophenol	0.1–0.5	2	PrOH	1	600 ± 100
Phenol[c]	0.01–0.1	1.7–2	PrOH	1	1300 ± 300
Salicylic acid[c]	0.1–1	1.3–3	*t*-BuOH	4	<600
Salicylate ion	0.03	4–7	*t*-BuOH	4	(30 ± 10)·10^3
Resorcinol[c]	0.003	2	PrOH	—	>300,000
Naphthalene	0.002–0.14	2	*t*-BuOH	1	*3000 ± 600*

[a]All measurements by indigo method.
[b]Data from Hoigné and Bader (1983a).
[c]Rates of reactions of deprotonated species becomes significant at very low pH values.
[d]Reaction-rate constants determined primarily by relative measurements (references: benzene, toluene; *o*-xylene).

tion of the observed facts concerning the reaction, and also appears to explain the pathway of the reaction in low-pH aqueous media. It is summarized in Equation 5.37.

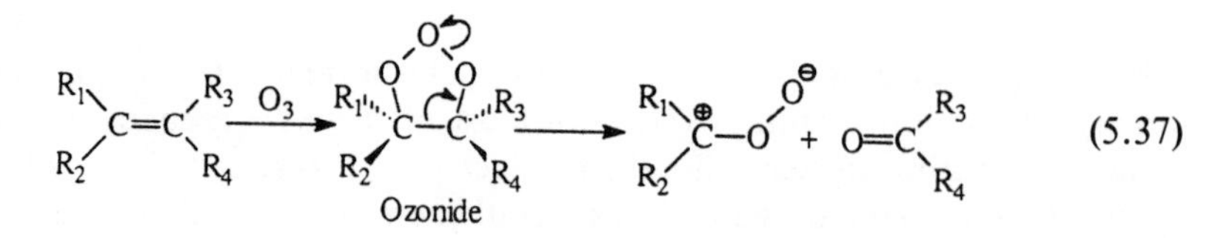

(5.37)

Table 5.3. Reaction-Rate Constants of Substituted Alkanes with Ozone[a,b]

Solute	mM	pH	k_{O_3} (M^{-1} s^{-1})
1-Propylamine		2	<0.01
1-Butylamine		2	<0.02
tert-Butanol	600	2/6	~0.003
Methanol	600	2/5	~0.024
Ethanol	6–60	2	*0.37 ± 0.04*
1-Propanol	6–60	2	*0.37 ± 0.04*
1-Octanol	0.8–1.4	2	≤0.8
1-Butanol	1–10	2	0.58 ± 0.06
2-Propanol	2–30	2/6	1.9 ± 0.2
Cyclopentanol	2–20	2	2.0 ± 0.2
Formaldehyde	70–600	2	0.1 ± 0.03
Acetaldehyde	20–100	2	1.5 ± 0.2
Propanal	30–300	2	2.5 ± 0.4
n-Octanal	0.1–0.2	1.8–5	8 ± 0.8

[a]All measurements by direct u.v. method.
[b]Data from Hoigné and Bader (1983a).

The initial product of the reaction between ozone and most olefins appears to be the 1,2,3-trioxolane ("ozonide"), which then collapses to a carbonyl compound and a zwitterionic intermediate, **68**. In the presence of water, the intermediate hydroxy-hydroperoxide, **69**, is produced and decomposes to additional carbonyl-containing compounds.

68 **69**

Although the majority of aromatic compounds will react with ozone at low pH in water, the relative rates are slower than those of olefins. In general, the order of reactivity is (Bailey, 1958):

olefins > tricyclic and higher > bicyclic > monocyclic

As expected, electron-donating groups enhance the reactivity of the aromatic compound toward ozone, whereas electron-withdrawing groups inhibit it.

Table 5.4. Reaction-Rate Constants of Substituted Ethylenes with Ozone[b]

Solute	mM	pH	Method[a]	Scavenger	mM	k_{O_3} (M^{-1} s^{-1})
Tetrachloroethylene	0.7	2.0	u.v.	—	—	*<0.1*
Trichloroethylene	0.06–0.6	2.0	u.v.	—	—	*17 ± 4*
cis-Dichloroethylene[c]	0.06–0.2	2.0	u.v.	—	—	<800
1,1-Dichloroethylene	0.04–0.4	2.0	u.v.	—	—	*110 ± 20*
Maleic acid	0.1–0.7	2	I	*t*-BuOH	1.5	$\sim 10^3$
Fumaric acid	0.1–1	2	I	*t*-BuOH	1.5	$\sim 6 \times 10^3$
trans-*Dichloroethylene*	0.03–0.1	2.0	u.v.	*t*BuOH	1	*5.7 ± 1.0 × 10³*
1-Hexane-3-ol	3×10^{-3}	2.0	u.v.	*t*-BuOH	4	$\sim 1.0 \times 10^5$
Allylbenzene	$3/4 \times 10^{-3}$	2.0	I	PrOH	2	$1.2 \pm 0.4 \times 10^5$
1-Hexane-4-ol	3×10^{-3}	2.0	u.v.	*t*-BuOH	4	$\sim 1.8 \times 10^5$
Styrene	7×10^{-3}	2.0	I	*t*-BuOH	1	$\sim 3 \times 10^5$

[a]Determination of ozone by: I = indigo method; u.v. = u.v. absorbance at 258 nm.
[b]From Hoigné and Bader (1983a).
[c]Contaminated with about 5% *trans*-dichloroethylene.

Benzene reacted with three equivalents of ozone with rupture of the ring to give the expected products, glyoxal, glyoxylic acid, and oxalic acid, and alkylbenzenes gave rise to analogous compounds (Bailey, 1958; Jurs, 1966; Cerkinsky and Trahtman, 1972.)

Naphthalenes have been ozonized in water by at least three groups and the products reported have been rather different. Naphthalene is a particularly active substrate for ozone, kinetically being approximately 1500 times more reactive than benzene (Hoigné and Bader, 1983a); at low pH it is even more reactive than phenol. After application of 2 molar equivalents of ozone, Legube et al. (1986) demonstrated that naphthalene entirely disappeared and that the principal products were phthalic acid (**70**), 2-formylbenzoic acid (**71**), hydrogen peroxide, and phthalalde-

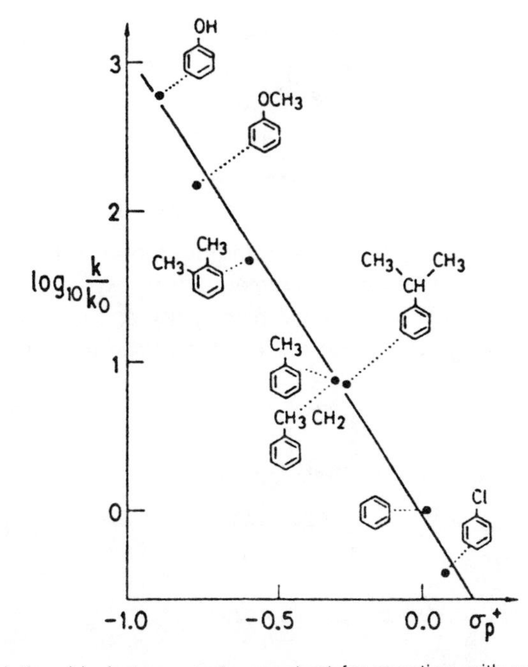

Figure 5.15. Relationship between rate constant for reaction with ozone and Hammett σ^+ constant for a series of substituted benzenes. From Hoigné and Bader (1983a). Reprinted by permission of Pergamon Press Ltd.

hyde (**72**). The authors also tentatively identified an unprecedented cyclic peroxide (**73**), but did not speculate on possible mechanisms whereby it could have

been formed. Further treatment with additional ozone led to the isolation of formic and oxalic acids. Somewhat different products were reported by Marley et al. (1987), who ozonized naphthalene in buffered solutions. In pH 5 acetate buffer, when one equivalent of ozone was used, the major products were the two isomers of 2-formylcinnamaldehyde (**74–75**), whereas at pH 9 (bicarbonate buffer), only one

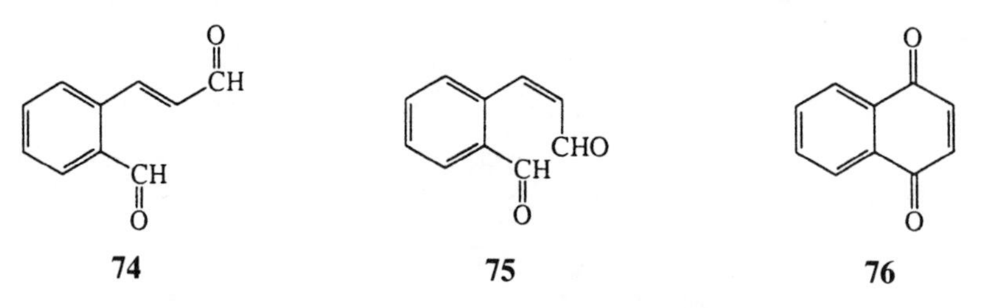

isomer (**74**) was detected along with 1,4-naphthoquinone (**76**). The possible role of carbonate radicals in the pH 9 reaction remains to be explored.

Phenanthrene (**77**) reacted with ozone principally at the 9,10 double bond, to give biphenyl derivatives such as the diacid **78** and the hemialdehyde **79** (Bailey,

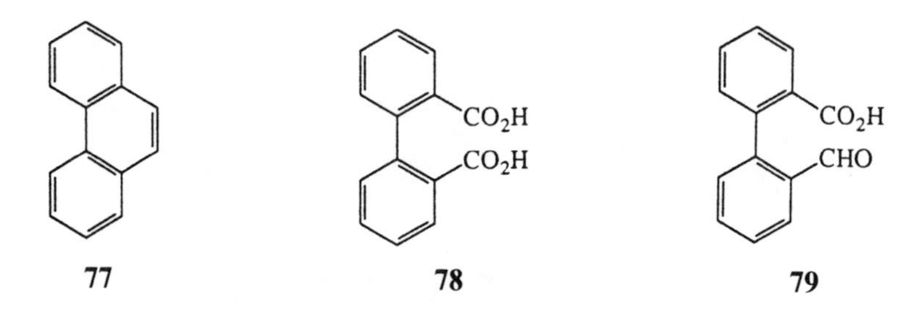

1956; Sturrock et al., 1963). Anthracene appeared to be attacked in the central ring by a mechanism similar to 5.38; anthraquinone and phthalic acid were the

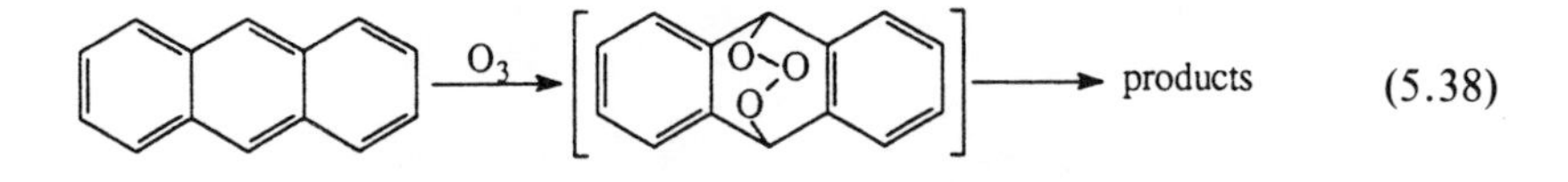

major products isolated (Bailey et al., 1964). Acenaphthylene (**80**) and acenaphthene (**81**) were ozonized in water, hexane, and methanol by Chen et al. (1979), who identified a variety of compounds including acids, an anhydride, acetals, lactols, carbonyl compounds, and an epoxide (Figure 5.16).

Notably, the epoxide was isolated from an aqueous ozonation reaction mixture.

Several polycyclic aromatic hydrocarbons that are genotoxic were ozonized in acetone-water or dimethylsulfoxide-water solvents. In most cases, mutagenic or carcinogenic activity was entirely lost after a brief exposure to ozone, but in one case (7,12-dimethylbenz[a]anthracene, **82**) a small number of animal tumors developed,

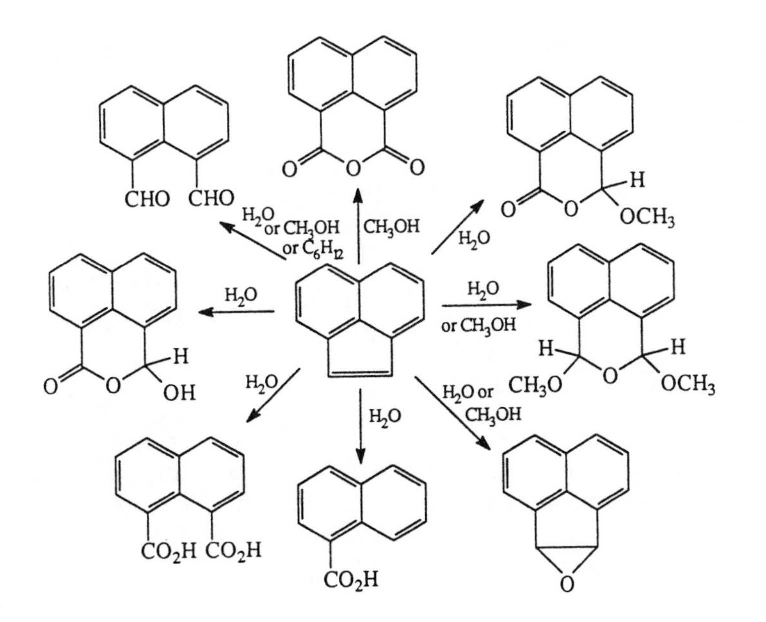

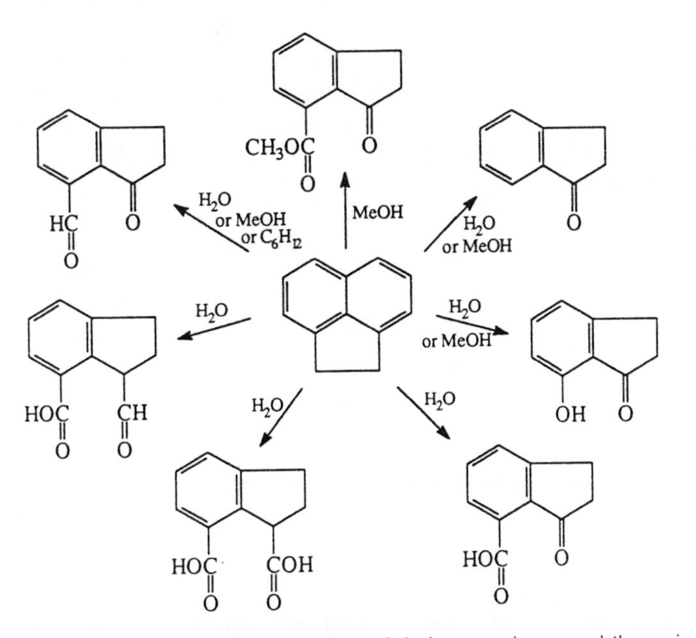

Figure 5.16. Ozonation products from acenaphthylene and acenaphthene in various solvents. Reprinted with permission from Chen et al. Reactions of organic pollutants. 1. Ozonation of acenaphthylene and acenaphthene. *Environ. Sci. Technol.* 13, 451–454. Copyright 1979 American Chemical Society.

possibly due to products of ozonolysis. No products, however, were identified in this study (Burleson et al., 1979).

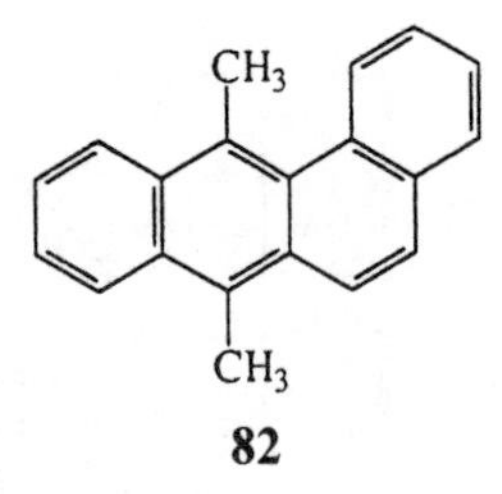

82

Fatty Acids

Inhaled ozone is known to initiate free-radical autooxidation of unsaturated fatty acids in animal pulmonary lipids (Pryor et al., 1981). These reactions lead to the formation of such typical autooxidation products as conjugated dienes and short-chain alkanes like ethane and pentane. Whether these reactions also occur in water treatment is uncertain. Glaze et al. (1988) showed that 9-hexadecenoic acid (**83**) reacted readily in aqueous solution to form the expected C_7 and C_9 aldehydes and acids. Linoleic acid (**84**) was converted to a mixture of aldehydes and acids (Carlson and Caple, 1977); notably, 3-nonenal (**85**) was among the products. Isolation of an unsaturated aldehyde is significant because of the high reported toxicity of these compounds. Carlson and Caple (1977) also implied that the epoxide of stearic acid was formed when an aqueous solution of oleic acid was ozonized; the product probably derives from an indirect attack on the double bond by peracids or peroxy radicals (Equation 5.39). Nevertheless, it is conceivable that similar reactions could occur in natural waters.

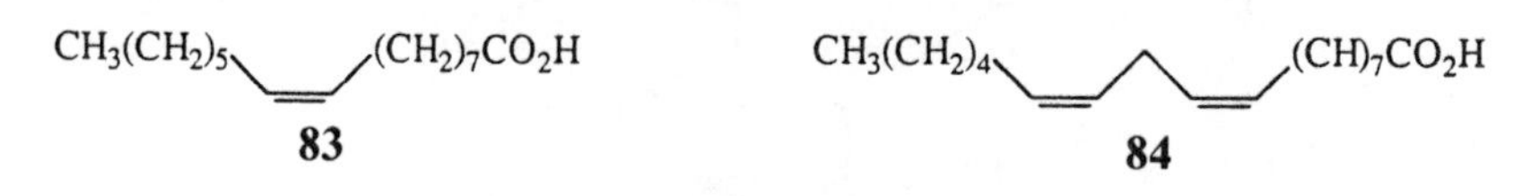

$$CH_3(CH_2)_4CH{=}CHCH_2CHO$$

85

$$CH_3(CH_2)_7CH{=}CH(CH_2)_7CO_2H \xrightarrow{O_3} CH_3(CH_2)_7HC\overset{O}{\frown}CH(CH_2)_7CO_2H \quad (5.39)$$

Phenols

At low pH, aqueous ozone appears to act as a typical electrophile with aromatic compounds; ozone reacts with the electron-rich compound, phenol, more than 600 times faster than it does with benzene (Hoigné, 1982). Fundamental studies of the kinetics of ozone-phenol reactions have revealed the expected correlation between electron density, as measured by the Hammett σ substituent parameter, and reaction rate (Figure 5.17; Hoigné, 1982; Gurol and Nekoulnalni, 1984). Of course, at higher pHs (>6), where decomposition of ozone to hydroxyl radical becomes important, much less selectivity among substrates would be expected because of the nearly indiscriminate character of HO·. In addition, at these higher pHs phenols begin to ionize, and rates of reaction of electrophiles with phenolate anions are typically greater than the rates with the protonated forms. The effects of pH on the rate constants of reactions of ozone with several phenols have been tabulated by Hoigné and Bader (1983b) and their data largely confirm the expected increases of reaction rate with pH and degree of dissociation. The measured rate constants for some phenolates approached the diffusion-controlled limits.

Molecular ozone appears to attack the phenolic ring by initially hydroxylating it, as in the Eisenhauer (1968) mechanism outlined below, and then further oxidizing the dihydric phenol to either an *o*- or *p*-quinone. These mechanistic steps have been supported by the isolation of hydroquinone and catechol from the reaction of ozone with phenol (Doré et al., 1978: Gould and Weber, 1976). Additional equivalents of ozone lead to ring cleavage and to the identification of CO_2, formic acid, and C_2 - C_6 dicarboxylic acids and aldehydes (Gilbert, 1976: Yamamoto et al., 1979). Chloride and nitrate ions were observed during the ozonolysis of chlorophenols or nitro-

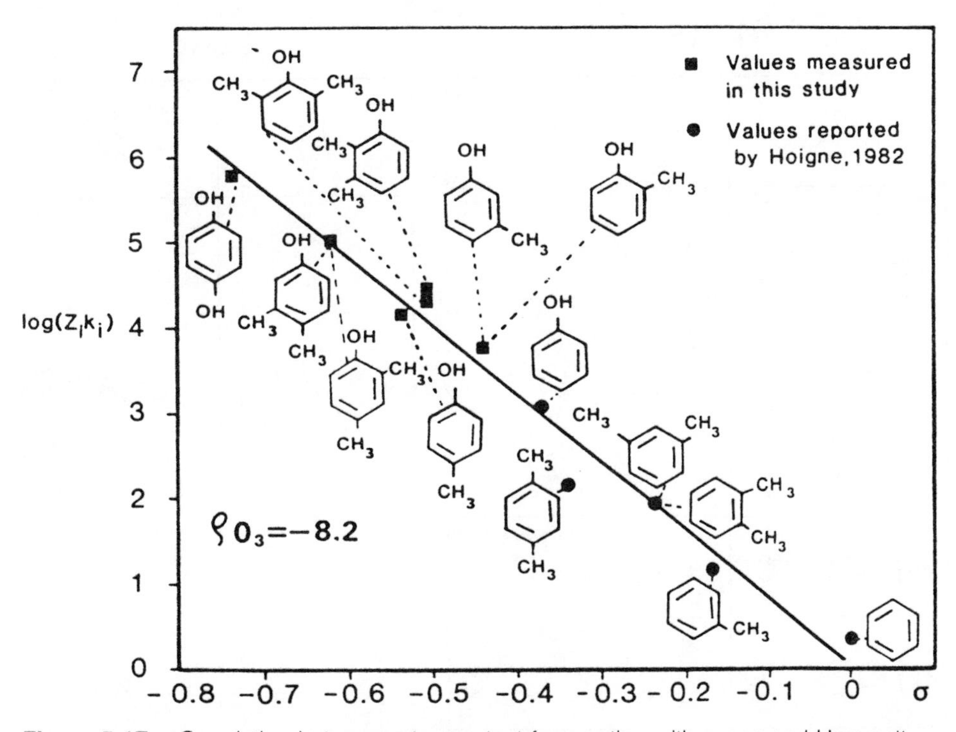

Figure 5.17. Correlation between rate constant for reaction with ozone and Hammett σ constant for substituted phenols. Reprinted with permission from Gurol and Nekoulnalni (1984). Copyright American Chemical Society.

phenols, respectively (Gilbert, 1978). In the case of 4-chloro-*o*-cresol, the yield of chloride was 100% at a point when all of the starting material had reacted, and similar results were observed for the formation of nitrate from 2-nitro-*p*-cresol. Duguet et al. (1988) treated 2,4-dichlorophenol with ozone and demonstrated the formation of a variety of oxidation products (Figure 5.18). Notably, chlorinated derivatives of dibenzofurans and dibenzodioxins were among the identified compounds, together with polymers containing at least six phenol units. Several of these coupling products were partially dechlorinated.

$$\text{OH} \xrightarrow{O_3} \text{OH, OH} \xrightarrow{O_3} \text{O, O} \qquad (5.40)$$

Several other groups have observed oxidative coupling reactions of phenols (see Equation 5.10) in the presence of ozone. Chrostowski et al. (1983) exposed several polyphenolic compounds to ozone at various pHs and observed changes in color; for example, catechol (**86**) was observed to develop first a red and then a brown color,

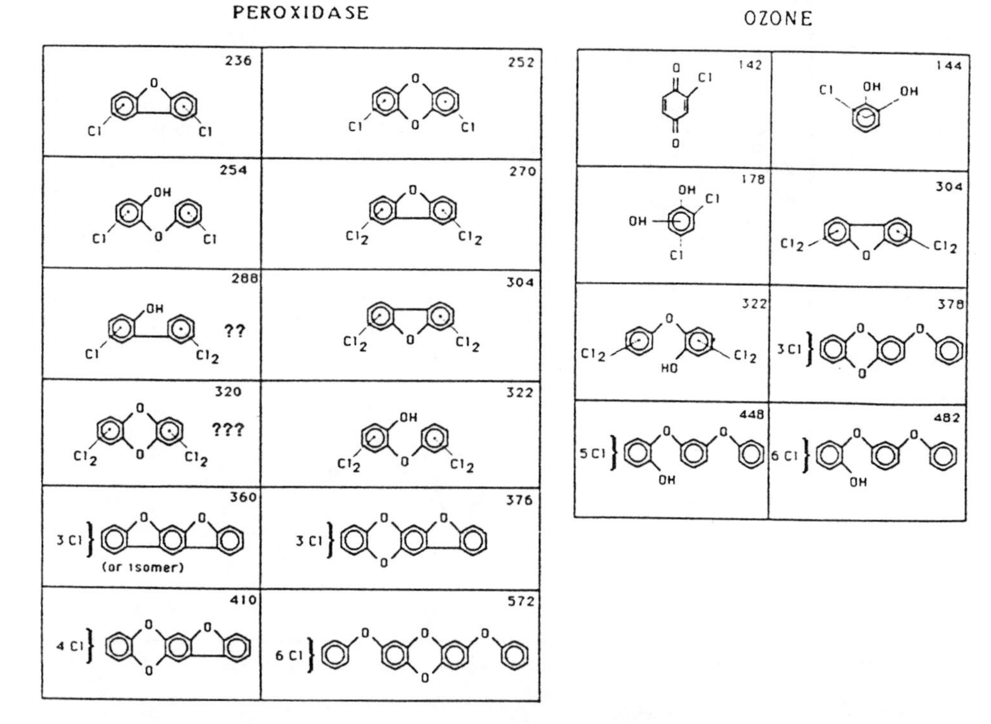

Figure 5.18. Reaction products identified from the ozonation of 2,4-dichlorophenol. From Duguet et al. (1988). Reprinted by permission of Lewis Publishers.

and then eventually to become colorless as more and more ozone was passed through its solutions. Size exclusion chromatography data indicted that the colored reaction products were of greater molecular size than the starting materials. Phenolate anions and radicals as well as quinones were postulated as likely intermediates in the reaction, but no intermediates were isolated or characterized. The increases in reaction rates with pH were cited as evidence that the oxidizing agent carrying out the coupling reactions was the hydroxyl radical.

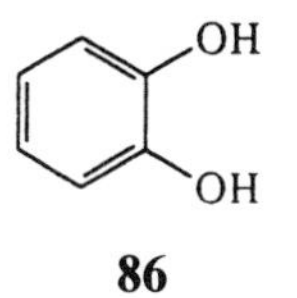

86

Nitrogen Compounds

The kinetic studies of Hoigné and Bader (1983b) indicate that protonated aliphatic amines do not react with ozone; however, as the concentration of free amine increases in solution, its reaction rate also increases. Whereas the rate of reaction of ozone with ammonia is quite slow (except at high pH where HO· reactions become important), successive substitution of its hydrogens by aliphatic groups increases the rate by orders of magnitude. Thus, dimethylamine reacted approximately 10^6 times as fast as ammonia. A variety of products are formed by ozone-amine reactions, including nitro compounds, nitroxides, and products apparently resulting from attack on the carbon chain (Bailey et al., 1972). The course of the reaction is consistent with an intermediate zwitterion (Bailey, 1982) resulting from electrophilic attack of ozone on nitrogen:

$$R_3N + O_3 \rightarrow R_3N^+\text{-O-O-O}^- \qquad (5.41)$$

Some amino acids are quite rapidly attacked, especially those containing sulfur such as cystine, cysteine, and methionine (Hoigné and Bader, 1983b: Pryor et al., 1984: Menzel, 1971), and attack on enzymes containing these compounds is probably one of the principal mechanisms by which ozone exerts its antibacterial activity. In addition to the above-mentioned compounds, the aromatic amino acids tryptophan, tyrosine, histidine, and phenylalanine are also susceptible to destruction by ozone, as they are with other electrophilic oxidizing agents (Mudd et al., 1969). As in the case of other organic amines, the rates of reaction are proportional to the amount of unprotonated amine present. Few data are available on the reaction products of ozone with these compounds. Methionine sulfoxide was almost the sole product of methionine oxidation at pH 7.2 in phosphate buffer, and cysteine gave a mixture of cystine and cysteic acid, whereas ammonia was a major product of histidine ozonolysis under these conditions (Mudd et al., 1969). Ozonolysis of tyrosine gave phenol-coupled dimer, *o*, *o*′-dityrosine (**87**), as well as 3,4-

dihydroxyphenylalanine (DOPA, **88**) and polymeric material. In addition, phenylalanine was partly converted to tyrosine and *o, o'*-dityrosine by ozone. These products are best explained by HO•-initiated free-radical reactions (Verweij et al., 1982). Anilines reacted with aqueous ozone to form a series of oxidation products (azobenzenes, azoxybenzenes, and benzidines: **89–91**) that appeared largely to result from attack of •OH on either the amine group or the aromatic ring, followed by radical-radical coupling and (in some cases) further oxidation (Chan and Larson, 1992).

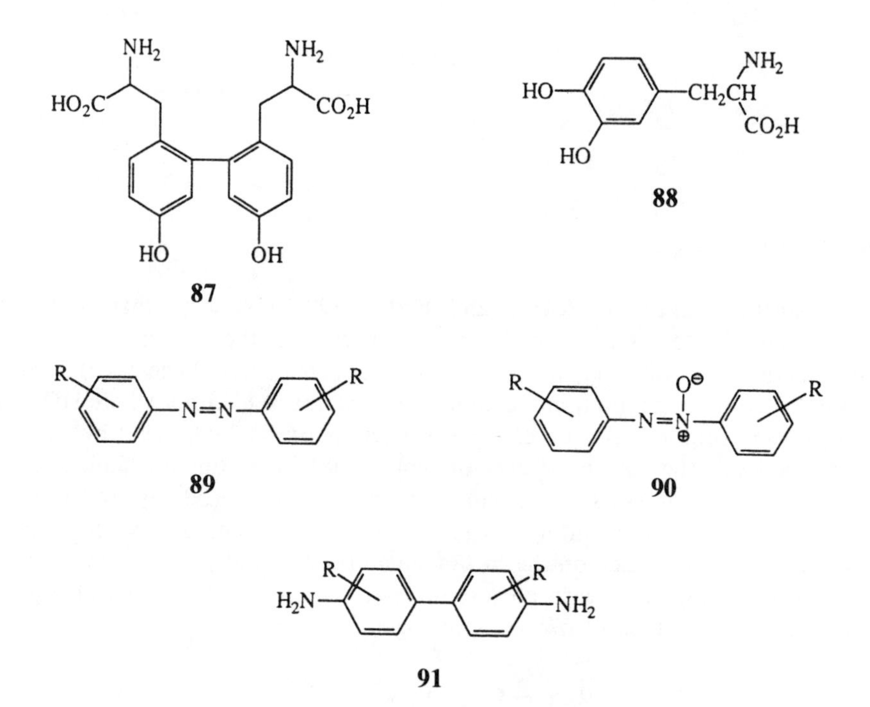

Reactions of ozone with nucleic acid constituents have not been much studied. Uracil (**54**) gave a variety of products including acylated hydroxyhydantoins such as **92**, whose structure was established by X-ray diffraction (Matsui et al., 1989). Ring-opened products were also identified; a mechanism is shown in Equation 5.42. Caffeine (**93**) was attacked by ozone in the imidazole ring to give two prin-

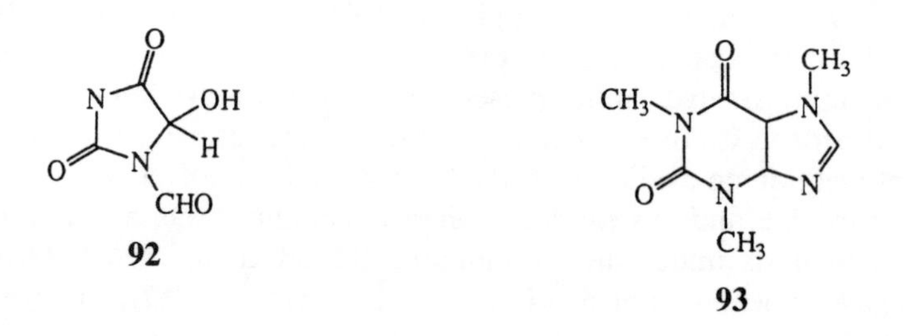

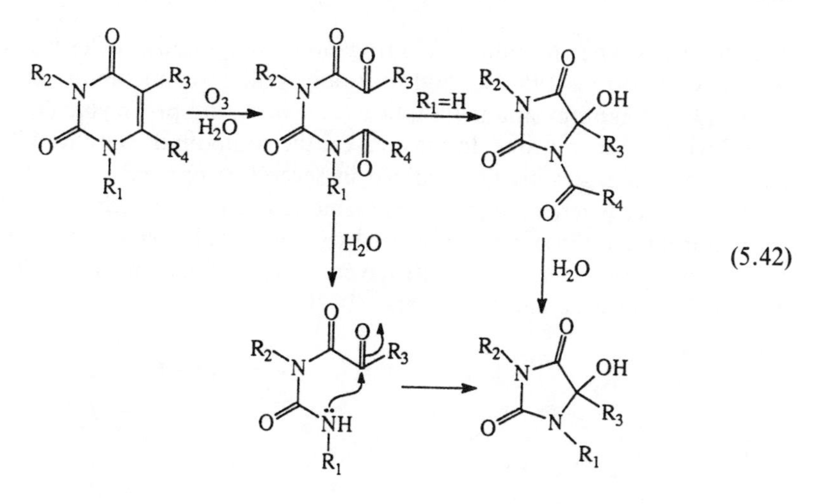

cipal products, **94** and **95**. No mechanism was postulated for the formation of the former compound, which was produced in 44% yield. The latter compound, dimethylparabanic acid, is also formed by reaction of caffeine with many other oxidants (Shapiro et al., 1978).

Pyridine is relatively stable to ozonolysis under nonradical conditions, giving the N-oxide as almost the only product (Andreozzi et al., 1991). The pyridinium cation is practically inert to ozonation. At higher pHs many products, including ring cleavage compounds, ammonia, and nitrate are produced (Equation 5.43).

Quinoline was converted in good yield to nicotinic acid (Sturrock et al., 1960). With indoles, the pyrrole ring opens to give aniline derivatives, as in the formation of 2-aminobenzaldehyde and anthranilic acid from indole (Equation 5.44) (Jurs, 1966).

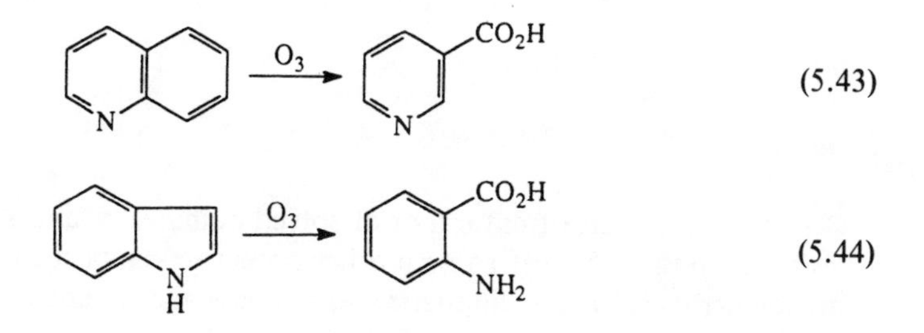

A number of known mutagenic nitrogenous compounds, including several pesticides, two hydrazines (**96–97**), and four polycyclic aromatic amine mutagens, acriflavine (**98**), β-naphthylamine (**99**), benzidine (**43**), and proflavine (**100**), have been treated with ozone in water. In most cases, the compounds were rapidly destroyed, but some mutagenic activity remained in several compounds, including benzidine and proflavine, whereas dimethylhydrazine (**101**), maleic hydrazide (**102**), and N-aminomorpholine (**103**) were converted to new, highly active mutagens. Reaction products were not determined (Cotruvo et al., 1977: Burleson et al., 1979: Caulfield et al., 1979: Burleson and Chambers, 1982).

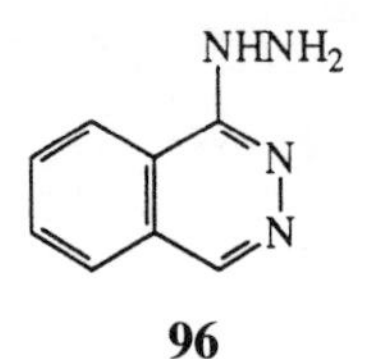

96

$H_2N{-}NH{-}CH_2CH_2OH$

97

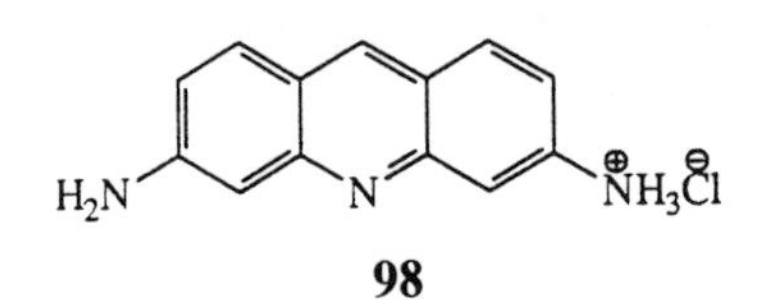

98

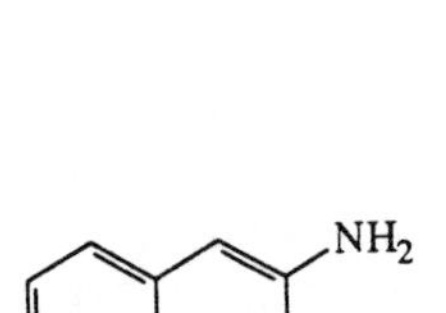

99

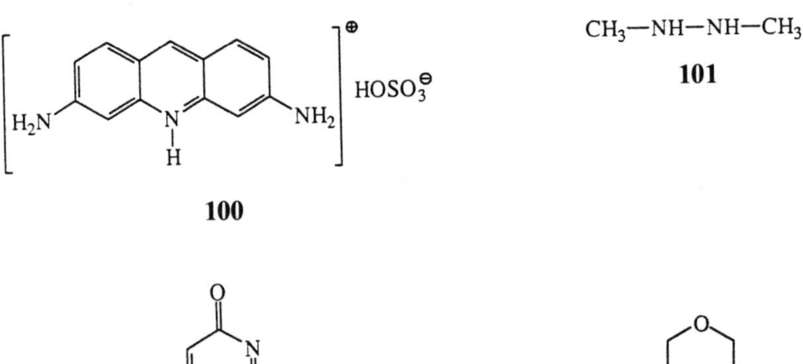

100

$CH_3{-}NH{-}NH{-}CH_3$

101

O
N
N
O

102

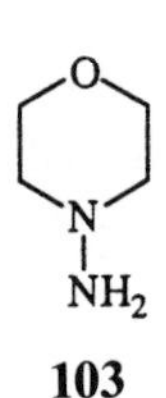

103

Humic Materials: Natural Waters

Ozonation products of natural or dissolved humic substances appear to be generally similar to those identified with other potent oxidants such as permanganate or hydrogen peroxide; mainly aliphatic and aromatic mono- and dicarboxylic acids and carbonyl compounds (Lawrence et al., 1980: Killops, 1986: Glaze et al., 1988; Xie and Reckhow, 1992).

The ozonation of a variety of surface waters as well as wastewater have indicated that several aldehydes (including formaldehyde, glyoxal, methylglyoxal and medium-chain length saturated isomers such as hexanal, heptanal, octanal and nonanal) were major identifiable products (Sievers et al., 1977: Lykins et al., 1986: Glaze et al., 1988: Yamada and Somiya, 1989). Longer-chain aldehydes (up to C_{15}) and other complex carbonyl compounds have been detected. Keto acids have also been reported as major products of ozonation of drinking waters and fulvic acid solutions (Xie and Reckhow, 1992).

Ozone is capable of oxidizing bromide ion to hypobromous acid, HOBr, although it does not oxidize chloride. Formation of this oxidant means that some organobromine compounds may be formed when ozone is added to waters high in bromide; and, in fact, Helz et al. (1978) demonstrated that bromoform can be detected in ozonized seawater. More recently, many other brominated compounds, such as brominated acetic acids, haloforms, bromopicrins, and bromohydrins, have been identified in freshwater or drinking water to which bromide and ozone are added (Cavanagh et al., 1992). Presumably, HOBr is the active brominating species in the formation of all these compounds. Further oxidation of HOBr to bromate, BrO_3, by ozone can occur if the hypobromite anion is present (pKa for HOBr is 8.8: Haag and Hoigné, 1983).

Advanced Oxidation: Wastewater Treatment

Ozone is a key ingredient in a family of relatively recent treatment methods, called "advanced oxidation processes" (AOPs), that show promise for virtually complete removal of organic matter from contaminated waters (Kuo et al., 1977: Glaze, 1987). These techniques, which include ozone-UV, ozone-H_2O_2, metal ion-ozone and H_2O_2-UV combinations, fundamentally rely on the production of large fluxes of hydroxyl radicals which, in the presence of other potent oxidants, bring about (in favorable cases) complete mineralization of organic matter. Externally supplied short-wave (254-nm) ultraviolet light is needed to generate HO· by UV techniques, since the key intermediate, H_2O_2, only weakly absorbs longer wavelengths such as those in sunlight.

Hydroxyl radicals, of course, react with organic compounds either by abstraction (of a hydrogen atom or electron) or addition pathways. Both of these routes result in the formation of new free-radical species that are usually quite reactive toward molecular oxygen, invariably present in AOP systems. The intermediate peroxy radicals participate together with the other reactive intermediates in the system to establish a mechanistic relationship (Staehelin and Hoigné, 1985: Peyton et al., 1988) that is at least partly cyclic and self-sustaining in nature, as indicated in Figure 5.19. In this figure, HRH represents an organic compound capable of forming a peroxy radical HRO_2· that can eliminate superoxide. Such an elimination produces a species that can react very rapidly with ozone (see *Decomposition Mechanisms of Aqueous Ozone*, p. 314) to generate HO· and thus sustain a chain reaction that

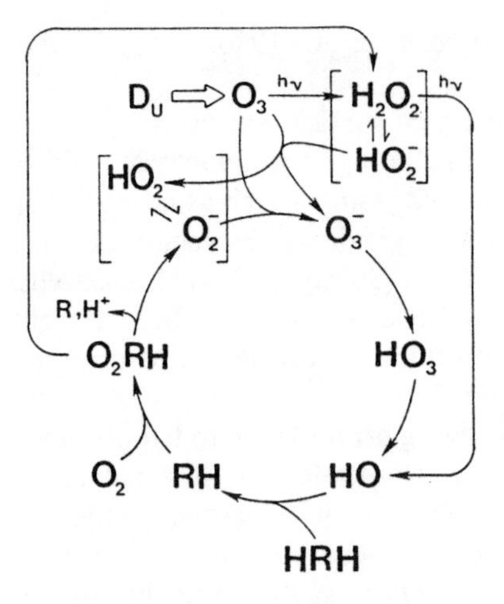

Figure 5.19. Cyclic mechanism for ozone photolysis ("advanced oxidation") in the presence of a generalized oxidizable substrate HRH. D_u = "utilized" ozone dose. From Peyton et al. (1988). Reprinted by permission of Lewis Publishers.

destroys more organic compound. An intermediate peroxy radical that is not capable of losing $O_2{\cdot}^-$ directly may break down by other routes, such as the Russell tetroxides (ROO-OOR; see also Chapter 4), which are known to decompose to alkoxy radicals and (in some cases) H_2O_2, which also reacts rapidly with ozone:

$$\text{HROO-OORH} \rightarrow 2\ \text{RO}\cdot + H_2O_2 \tag{5.45}$$

$$\text{ROO-OOR} \rightarrow 2\ \text{RO}\cdot + O_2 \tag{5.46}$$

$$H_2O_2 + O_3 \rightarrow O_3^- + HO_2\cdot + H^+ \tag{5.47}$$

Polar intermediates such as carbonyl compounds and carboxylic acids, whose reactions with HO· are comparatively slow, tend to accumulate in AOP systems. The possibility that aldehydes might react with H_2O_2 (commonly present in high concentrations in these systems) to form hydroxyhydroperoxides has been considered by Peyton et al. (1988). The general reaction is summarized in Equation 5.48.

$$\text{RCHO} \xrightarrow{H_2O_2} \text{RCH(OH)OOH} \tag{5.48}$$

When formaldehyde was mixed with H_2O_2 at millimolar concentrations, a rapid loss of H_2O_2 from solution occured as the intermediate **104** formed; on standing, H_2O_2 was slowly liberated into solution as other compounds were formed by secondary reactions. The implications for this phenomenon for water treatment by AOPs are that more or less stable organic peroxides may be formed which can either persist in effluents or release H_2O_2 over a period of time. Peroxidic compounds, or at least oxidants, have also been observed to form in waters treated with ozone alone. Carlson and Caple (1977), for example, reported that residual oxidants remained in water long after the ozone itself disappeared.

$$HOCH_2C(=O)OH$$

104

Catalytic ozonation has been shown to be highly effective in degrading such recalcitrant substrates as ethanol (Figure 5.20: Namba and Nakayama, 1982).

In pure water, organochlorine compounds such as chloroform and tetrachloroethylene, which are very resistant to degradation by many other treatment methods, disappeared rapidly in ozone-UV or other AOP systems (Peyton et al., 1982). In natural waters, dissolved organic matter or other solutes inhibited the reaction to some extent, probably by scavenging reactive free radicals or other potent oxidants.

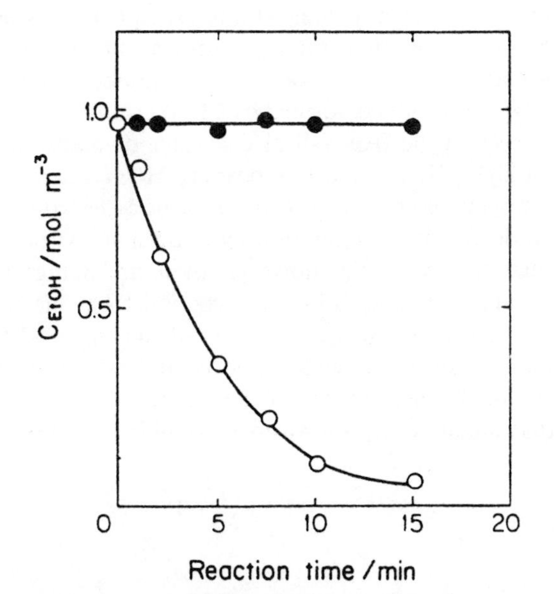

Figure 5.20. Effect of hydrogen peroxide on the ozonolytic decomposition of ethanol in water. Open circles, H_2O_2 present at 5×10^{-4} M; closed circles, H_2O_2 absent. From Namba and Nakayama (1982). Reprinted by permission of the Chemical Society of Japan.

D. CHLORINE DIOXIDE

Because of the objectionable reactions of organic matter in raw water with hypochlorite to form chlorinated organic compounds, many studies during the last 15 years on reactions of a possible alternative disinfectant, chlorine dioxide (ClO_2), have been undertaken. The compound, like chlorine, is a potent disinfectant and numerous industries, notably the paper processing industry, have used it as a bleaching agent in large quantitites, but it was used for only a brief period in the middle 1940s and early 1950s in the U.S. domestic water treatment industry before largely falling into disuse. In parts of Europe, it is still applied on a limited scale.

Chlorine dioxide is a bent (O–Cl–O bond angle 117°), relatively stable free-radical species with a solubility in water about an order of magnitude greater than chlorine gas. Both the O and Cl atoms have some radical density and may enter into one-electron redox reactions. The redox potential for addition of an electron to aqueous ClO_2 is 0.936 V, very close to that for OCl^- (Merényi et al., 1988), but lower than that of HOCl. The compound is stable for long periods in acidic solution, but decomposes in alkali by disproportionation to chlorite (ClO_2^-) and chlorate (ClO_3^-):

$$2\ ClO_2 + 2\ OH^- \leftrightarrows H_2O + ClO_2^- + ClO_3^- \tag{5.49}$$

Many other valence states are accessible to ClO_2 as it reacts with various reagents; some of these will be discussed below.

The principal reason for recent enhanced interest in ClO_2 has been that, in general, it acts as an oxidant rather than a halogenating agent. In particular, little or no trihalomethane formation is observed when water is treated with it. The often-heard suggestion that the type of products formed by ClO_2 reactions are similar to those of ozonolyses has its basis in the free-radical character of many of the reactions of ClO_2. Like ozone at high pH, which decomposes to HO· (*vide supra*), the ClO_2 free radical reacts with organic compounds to produce odd-electron intermediates that subsequently can undergo further oxidation by molecular oxygen or other endogenous oxidants (Equation 5.50). Still, however, there are numerous reports of the formation of chlorinated compounds in ClO_2 reactions. The mechanisms of chlorination by ClO_2 are obscure. Thus, many aspects of the chemical behavior of ClO_2 with organic compounds in water remain poorly understood. In particular, there have been no studies of the mechanisms of these reactions using ^{18}O-labeled water, ClO_2, or oxygen that might clarify the various possible oxidation pathways.

$$\cdot ClO_2 + RH \longrightarrow ClO_2^{\ominus} + H^{\oplus} + R\cdot \xrightarrow{O_2} ROO\cdot \longrightarrow \text{products} \tag{5.50}$$

In the presence of light, ClO_2 (which has an absorbance maximum at 360 nm, ε about 1400, well within the solar UV range), decomposes to O_2, ClO_3^-, hypochlorite (or Cl_2), and other products (Bowen and Cheung, 1932: Gordon et al., 1972; Zika et al., 1985). In full sunlight, it has a half-life of about 25 min, depending on water depth and organic matter content (Zika et al., 1985). The effects of the transient intermediates that form during its decomposition on the organic material in treated waters are not clearly comprehended as yet.

Hydrocarbons

Few reactions of ClO_2 with hydrocarbons have been reported except under conditions very far removed from water treatment practice. Saturated alkanes and alkyl side chains appear almost inert in its presence except where unusually stable radicals may result (benzylic hydrocarbons, e. g.). In these cases, typical products of autooxidation (see Chapter 4) have been isolated, probably by attack of O_2 on intermediate free radicals formed by electron transfer (Ozawa and Kwan, 1984: Rav-Acha and Choshen, 1987: Merényi et al., 1988) or (less likely) hydrogen atom abstraction (Chen et al., 1982). A few polycyclic hydrocarbons have been shown to be partially converted to chlorinated derivatives and quinones by ClO_2 (Thielemann, 1972a: Taymaz et al., 1979).

Olefinic bonds afford a great variety of oxidation products with ClO_2. It is the allylic position of the molecule that is normally attacked fastest. Cyclohexene, for example, gave the allylic alcohol, ketone, and chloro derivatives. In addition, the double bond was attacked in a substitutive fashion, with formation of the dichloride, chlorohydrin, and epoxide (Lindgren et al., 1965). Styrene was converted to its epoxide and to a chlorohydrin at pH 6; at lower pH, the epoxide was hydrolyzed to a diol (Kolar and Lindgren, 1982). Many of these products could be formed by free-radical pathways (Rav-Acha and Choshen, 1987), though an alternative mechanism involving oxygen atom transfer to form the epoxides has been advanced by Lindgren and Nilsson (1975):

$$ClO_2 + R_1CH{=}CHR_2 \longrightarrow ClO^- + R_1HC\overset{O}{\frown}CHR_2 \quad (5.51)$$

The chlorine monoxide intermediate is then supposed to undergo electron transfer with ClO_2, forming hypochlorous acid and chlorate, ClO_3^-. This mechanism has the merit of explaining the formation of a chlorinating agent, HOCl, in reactions involving ClO_2. The hydrolysis of chlorite esters, ROClO (formed by addition of ClO_2 to a double bond) could also give rise to HOCl (Lindgren, 1971). It is also possible that, since ClO_2^- also reacts with peroxides such as H_2O_2 (Masschelein, 1979),

$$ClO_2^- + H_2O_2 \rightarrow ClO^- + O_2 + H_2O \quad (5.52)$$

that organoperoxy intermediates could convert chlorite (or ClO_2) to hypochlorite. A still further alternative species that may have chlorinating ability is Cl_2O_2 (Taube and Dodgen, 1949: Gordon et al., 1972).

Stilbene was converted to an interesting series of products with ClO_2 in an aqueous-organic solvent mixture (Equation 5.53: Lindgren and Nilsson, 1974). Electron-poor olefins such as crotonic, maleic, fumaric, and cinnamic acids, however, have been reported to be unreactive with ClO_2 (Masschelein, 1979).

Although cleavage of the olefinic double bond is not normally observed, as is the case with ozone, a possible cleavage reaction with unsaturated material in natural waters was indicated by the isolation of C_2–C_8 aliphatic aldehydes from Ohio River water treated with ClO_2 (Stevens et al., 1978). Aldehydes, however, are reportedly converted readily to carboxylic acids by ClO_2 (Somsen, 1960), so the detailed pathway of this process remains to be worked out.

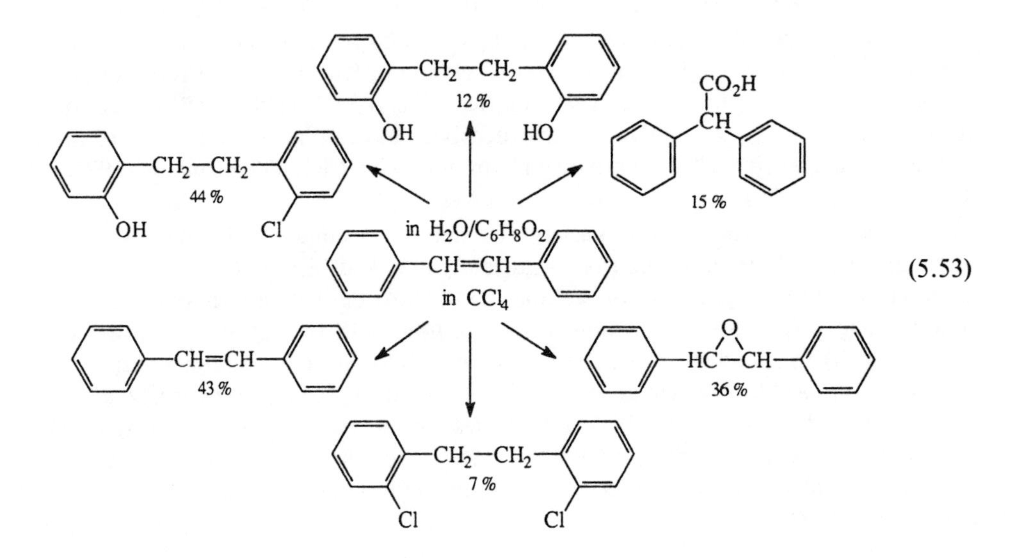

(5.53)

Phenols

The reactions of ClO_2 with phenols have been somewhat comprehensively studied, and the kinetics of its reactions with variously substituted phenols indicate that the phenolate anion is the reactive species (Hoigné and Bader, 1982). Phenol itself, for example, gave a mixture of hydroquinone, chlorinated phenols, benzoquinone, 2-chlorobenzoquinone, and 2,6-dichlorobenzoquinone (Equation 5.54). When phenol was in stoichiometric excess, chlorinated products were formed, but if ClO_2 was in excess, benzoquinone appeared to be the major product. If ClO_2 is in great excess, destructive oxidation of the ring occurs, with oxalic, fumaric, and maleic acids as the ultimate stable products. The mechanism is complex: benzoquinone apparently does not react with ClO_2. Trichlorophenols have been postulated as intermediates, but little can yet be said with certainty about the course of the reaction (Gordon et al., 1972: Stevens, 1982: Brueggeman et al., 1982).

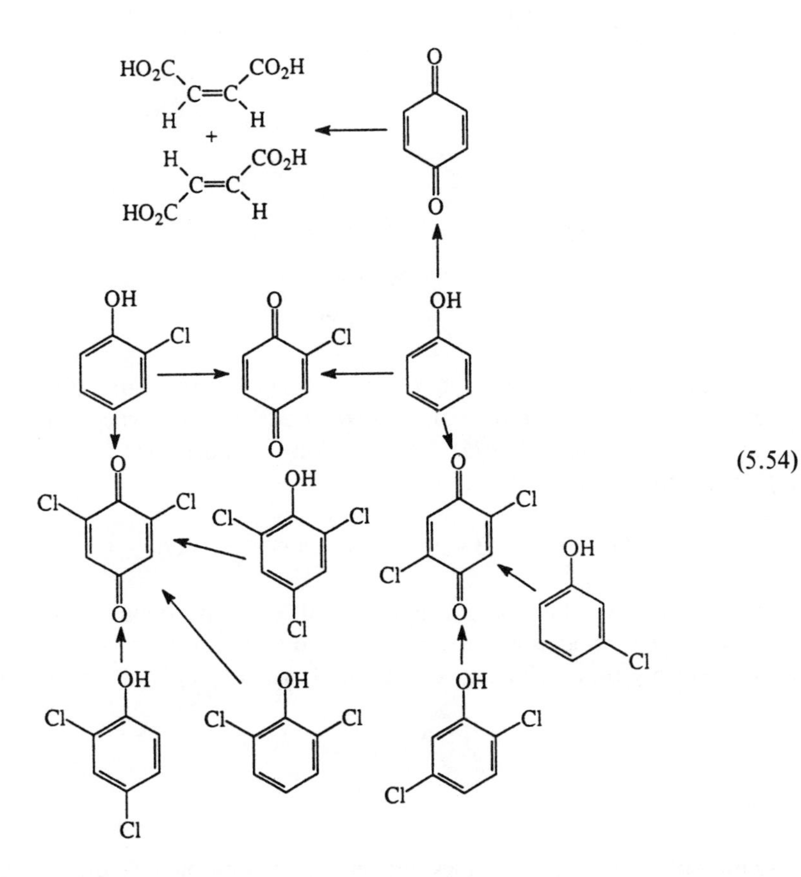

(5.54)

Phenols with *p*-substituents often lose them during ClO_2 oxidation. For example, *p*-Cl, O_2N-, -COOH, and -CHO groups were lost during the conversion of phenols to quinones (NAS, 1980: Carlson and Lin, 1985: Lin et al., 1984). The details of the mechanisms are not clear, but analogous *ipso* substitutions have been observed with HOCl and with peroxy radical reactions of phenols (Everly and Traynham, 1979; Larson and Rockwell, 1979; Bickel and Gersman, 1959). Electron-poor phenols such as di- and trinitro derivatives are unreactive toward ClO_2 (Paluch, 1964), as are most other aromatic compounds that are not electron-rich.

Resorcinol (**22**) was converted to 2-hydroxybenzoquinone (**105**: Lin et al., 1984). Phenols related to lignin were partly converted to quinones, but also to ring-opened products such as muconic acid derivatives, and to polymeric materials (Masschelein, 1979). For example, guaiacol (**106**) afforded methoxybenzoquinone, monomethyl-*cis*-muconate (**107**), *o*-benzoquinone, and polymer (Lindgren, 1971). Quinones are also formed by the reactions of ClO_2 and either 1- or 2-naphthol (Thielemann, 1972b).

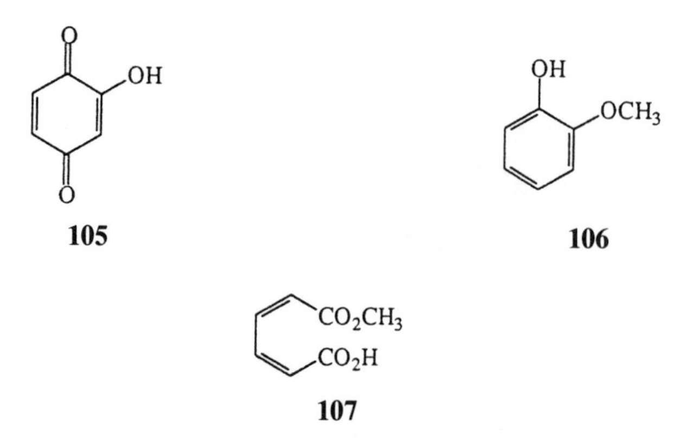

It should be noted that chlorite (ClO_2^-), a major product of ClO_2 reduction, also reacts rapidly with some phenols, α-diketones, and aldehydes. Chlorite is rather nucleophilic (Rosenblatt, 1975). The reaction of chlorite with aldehydes obeys the following stoichiometry (White et al., 1942):

$$RCHO + H^+ + 3\ ClO_2^- \rightarrow RCOO^- + 2\ ClO_2 + Cl^- + H_2O \qquad (5.55)$$

The mechanism of this reaction is not at all obvious. It has been suggested that HOCl may be formed as an intermediate, and that it could accelerate oxidations by ClO_2^-, possibly by reoxidizing it to ClO_2 (Lindgren, 1971: Lindgren and Nilsson, 1973).

Amines

Quite unlike HOCl, ClO_2 does not react with ammonia or primary amines, and attacks most secondary amines slowly. However, some benzylic and aliphatic tertiary amines in their free base forms reacted with ClO_2, leading ultimately to oxidative dealkylation, with production of the corresponding secondary amine and an aldehyde derived from one of the N-alkyl substituents (Gordon et al., 1972: Rosenblatt, 1978).

$$(CH_3CH_2)_3N \xrightarrow{ClO_2} (CH_3CH_2)_2N + CH_3\text{-}\overset{\overset{O}{\|}}{C}\text{-}H \qquad (5.56)$$

Electron transfer from the amine nitrogen to ClO_2, with production of chlorite and the radical cation derived from the amine (Ozawa and Kwan, 1984), appears to predominate over hydrogen abstraction from the carbon adjacent to the amino nitrogen by ClO_2. This latter mechanism, however, appears to prevail in the reaction of benzylic secondary amines with ClO_2, as confirmed with studies of deuterated amines (Gordon et al., 1972). Sterically hindered tertiary amines, such as quinuclidine (**108**) did not undergo cleavage when treated with ClO_2, but were rather oxidized to the N-oxide (Dennis et al., 1967).

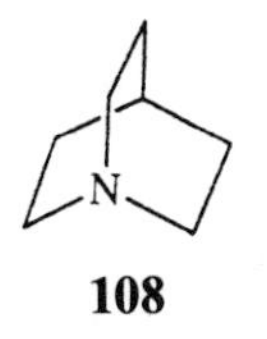

108

Although the majority of amino acids are more or less inert to ClO_2 under typical water treatment conditions, histidine and tryptophan underwent ring or side-chain cleavage reactions (Fujii and Ukita, 1957; Noss et al., 1986), and the phenolic ring of tyrosine was attacked with the formation of *o*-quinone (dopaquinone) derivatives (Hodgen and Ingols, 1954). Proline and hydroxyproline were quite reactive but apparently no information exists on reaction products (Tan et al., 1987). Cysteine and cystine appeared to be very reactive from kinetic measurements (Hoigné and Bader, 1982: Noss et al., 1986); the products from cysteine were the disulfoxide and cysteic acid (Equation 5.57). Methionine (**109**) similarly experienced oxidation at the thio ether linkage to the sulfoxide and sulfone (Masschelein, 1979). Disulfide linkages in the protein, keratin, were converted in a similar fashion to sulfonic acids (Rosenbusch, 1965). The reactivity of proteins toward ClO_2 in intact bacterial cells was much greater than that of nucleic acids (Roller et al., 1980) and is probably responsible for the inactivation of bacteria during disinfection with ClO_2.

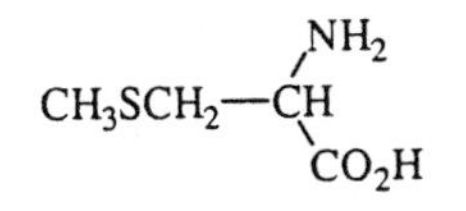

109

$$HSCH_2-CH(NH_2)(CO_2H) \xrightarrow{ClO_2} HO-S(=O)_2-CH_2-CH(NH_2)(CO_2H) + \left((H_2N)(HO_2C)HC-CH_2-S(=O)- \right)_2 \quad (5.57)$$

Other Compounds

Studies of ClO_2 reactions with nucleic acid constituents showed that it is virtually inert toward nucleotides of adenine, cytosine, or uracil, but that guanosine monophosphate (**110**) was rapidly attacked at pH 7.1 (less rapidly at lower pHs).

The products were not reported (Hauchman et al., 1986). Bacterial DNA was not significantly damaged by ClO_2 treatment (Roller et al., 1980).

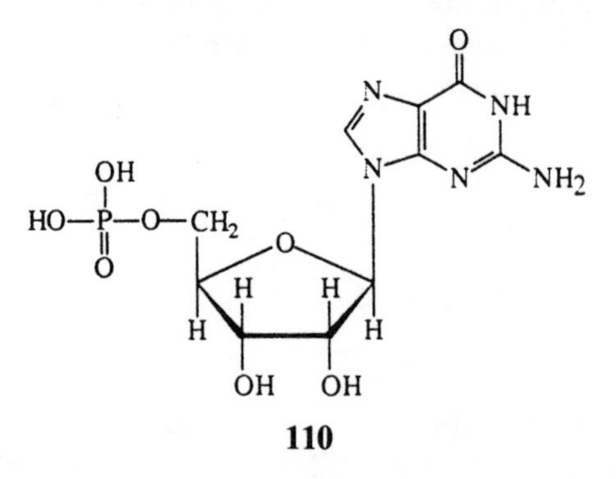

110

Indole was attacked by ClO_2 at the 2,3-double bond to form a diketone and an anhydride (Equation 5.58: Lin and Carlson, 1984). Carbazole (**65**) underwent an interesting reaction with low concentrations of ClO_2 to produce the quinonoid compound **111**; at higher concentrations, a colored solution was formed from which no identifiable compounds could be isolated (Lin and Carlson, 1984).

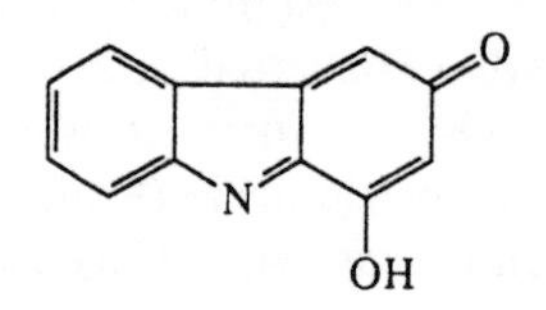

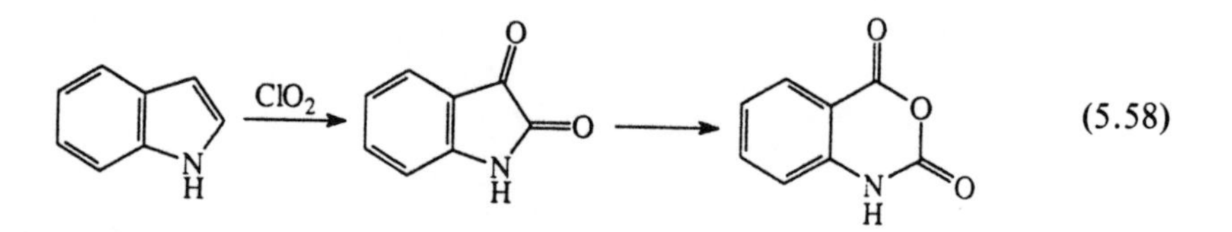

(5.58)

E. SURFACE REACTIONS OF DISINFECTANTS

With the exception of activated carbon, not much work has been reported on the reactions of disinfectants at surfaces of the types expected to be encountered in water treatment, although reactions of free chlorine with the organic matter at the surfaces of several natural sediments have been shown to produce trihalomethanes (Uhler and Means, 1985). Activated carbons have long been known to be potent catalysts of redox (cf. Austin et al., 1980) and other reactions such as eliminations

(Boehm et al., 1984). These carbons have a fairly high density of surface free radicals, whose nature is not well understood (Donnet, 1982).

Free chlorine, in high concentrations (2 g chlorine per g carbon) reacts with a virgin granular activated carbon (GAC) to produce a high molecular weight (> 100,000) brown-black product in addition to a few volatile compounds. The polymeric product is water-soluble and is not observed at more "normal" chlorine doses (Snoeyink et al., 1981). At lower concentrations, however, some sites on the GAC surface are converted by free chlorine to what appear to be odd-electron oxidants, capable of reacting with electron-rich substances such as phenols and converting them to oxidized products, including many that are unlike the compounds found when chlorine is applied in solution. For example, vanillic acid (**112**) in aqueous

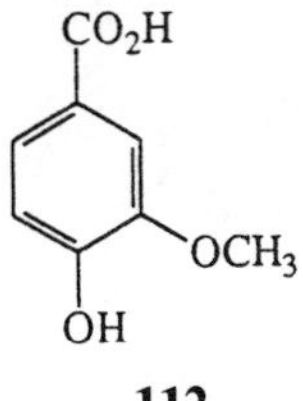

112

solution was chlorinated to only three products (Equation 5.59), but after adsorption on GAC and chlorine application, a considerable number of oxidized compounds including quinones, hydroxylated and decarboxylated phenols, and demethylated isomers were detected (Figure 5.21; McCreary et al., 1982).

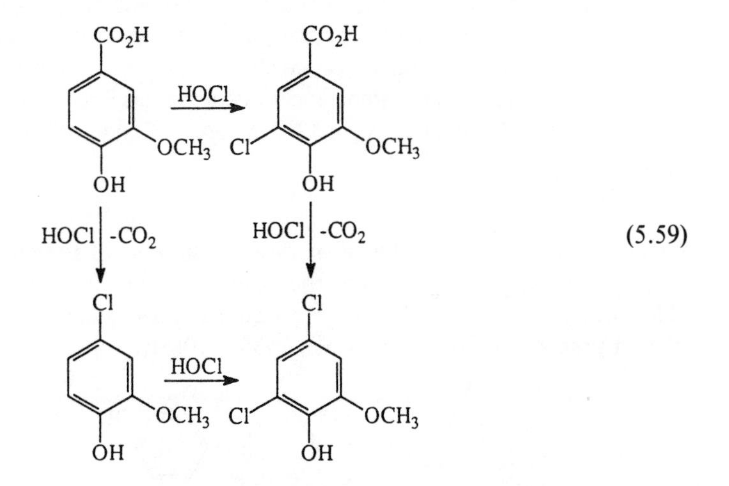

(5.59)

Other phenolic compounds underwent similar reactions. GAC after application of either free or combined chlorine became capable of promoting reactions such as hydroxylation of the aromatic ring, oxidation to phenols, chlorine substitution, carboxylation, and oxidative coupling (dimer and trimer formation). The formation of chlorohydroxybiphenyls ("hydroxylated PCBs:" **113**) by dimerization of chloro-

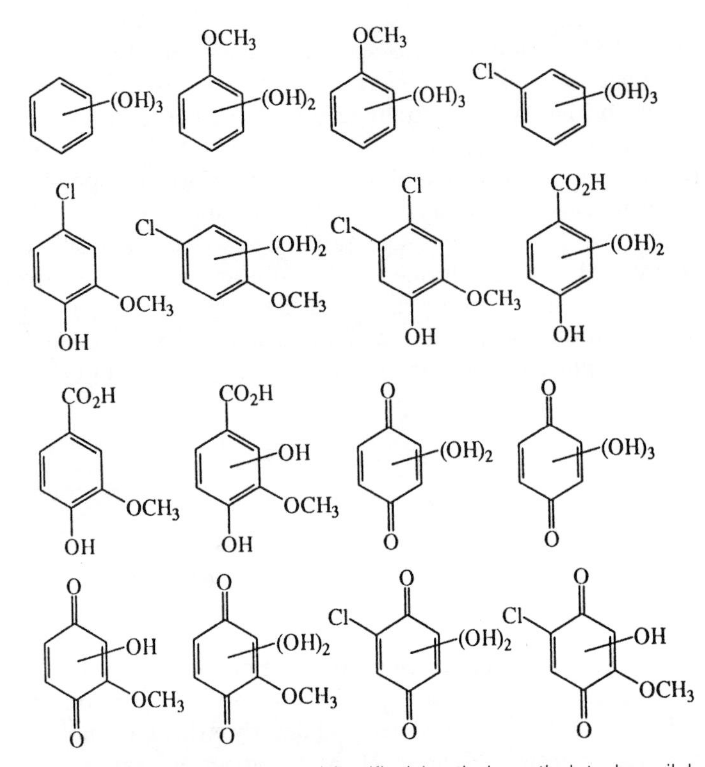

Figure 5.21. Structures of compounds identified (as their methylated or silylated derivatives) in the extract of a carbon receiving hypochlorite and vanillic acid. Reprinted with permission from McCreary et al. Comparison of the reaction of aqueous free chlorine with phenolic acids in solution and adsorbed on granular activated carbon. *Environ. Sci. Technol.* 16, 339–344. Copyright 1982 American Chemical Society.

phenols is of particular interest in view of the known toxicity of these substances. Although these compounds are the main products from the surface reactions of chlorophenols on GAC, they were produced in lesser amounts from nonchlorinated phenol precursors (Voudrias et al., 1985a, 1985b).

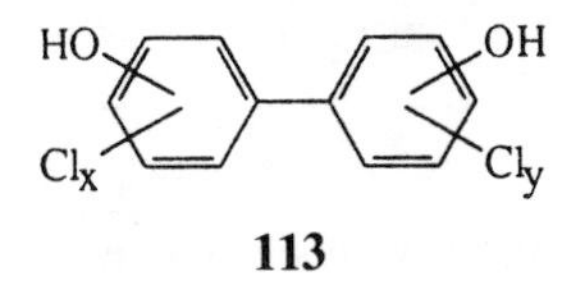

113

The free-radical character of GAC was further demonstrated by reactions of resorcinol (**22**). After absorption to GAC columns and treatment with either HOCl or ClO_2, resorcinol reacted to afford a single major product, the dichlorocyclopentenedione **114**. This compound and related isomers were the principal products formed when ClO_2 was allowed to react with resorcinol in solution, but these compounds were found in lesser yield in HOCl solution reactions. Apparently, similar free radical sites are produced on the GAC surface by reaction with both oxidants, and these sites bring about reactions that are similar to the radical reactions initiated by ClO_2 in solution (Jackson et al., 1987).

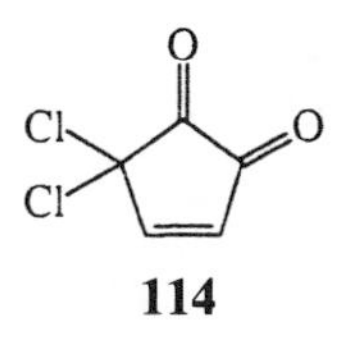

114

The only report on the reactions of GAC-sorbed nitrogen compounds (Hwang et al., 1990) similarly indicated that sorbed anilines were converted by HOCl to oxidative coupling products such as azobenzenes (**89**), unlike solution reactions with HOCl that led to ring substitution. Additional compounds such as formanilides and acetanilides were also identified in some cases; the mechanisms of formation of these N-substituted derivatives are unknown, although similar N-acyl derivatives have been reported to occur during the metabolism of anilines (Freitag et al., 1984).

REFERENCES

Alben, K. 1980. Gas chromatographic-mass spectrometric analysis of chlorination effects on commercial coal-tar leachate. *Anal. Chem.* 52, 1825–1828.

Amy, G. L., P. A. Chadlik, P. H. King, and W. J. Cooper. 1984. Chlorine utilization during trihalomethane formation in the presence of ammonia and bromide. *Environ. Sci. Technol.* 18: 781–786.

Andreozzi, R., A. Insola, V. Caprio, and M. G. D'Amore. 1991. Ozonation of pyridine in aqueous solution: mechanistic and kinetic aspects. *Water Res.* 25: 655–659.

Austin, J. M., T. Groenewald, and M. Spiro. 1980. Heterogeneous catalysis in solution. Part 18. The catalysis by carbons of oxidation-reduction reactions. *J. Chem. Soc. Dalton Trans* 854–859.

AWWA (American Water Works Association). 1989. Standard methods for the examination of water and wastewater. 16th ed. American Public Health Association, Washington, DC.

Babenzian, H. D., A. Schwartz, and W. Schwartz. 1963. Über Chlorschwund im Trinkwasser. II. Einfluss von Bakterien und von organischen Verbindungen auf den Chlorschwund. *Zentralbl. Bakteriol. Parasitenkr. Infektionskr. Abt. I* 185: 398–415.

Bailey, P. S. 1956. The ozonolysis of phenanthrene in methanol. *J. Amer. Chem. Soc.* 78: 3811–3816.

Bailey, P. S. 1958. The reactions of ozone with organic compounds. *Chem. Rev. 58*: 925–1010.

Bailey, P. S. 1982. *Ozonation in Organic Chemistry*. Vol. 2. (New York, NY: Academic Press).

Bailey, P. S., P. Kolsaker, B. Sinha, J. B. Ashton, F. Dobinson, and J. E. Batterbee. 1964. Competing reactions in the ozonation of anthracene. *J. Org. Chem.* 29: 1400–1409.

Bailey, P. S., T. P. Carter, Jr., and L. M. Southwick. 1972. Ozonation of amines. VI. Primary amines. *J. Org. Chem.* 37: 2997–3004.

Barnhart, E. L. and G. R. Campbell. 1972. Effect of chlorination on selected organic chemicals. EPA Report EPA-12020 EXG PB 211–160.

Bellar, T. L., J. J. Lichtenberg, and R. C. Kroner. 1974. The occurrence of organohalides in chlorinated drinking water. *J. Amer. Water Works Assoc.* 66: 703–706.

Bickel, A. F. and H. R. Gersman. 1959. Autoxidation of hindered phenols in alkaline media. Part 1. *J. Chem. Soc.* 2711–2716.

Boehm, H. P., G. Mair, T. Stoehr, A. R. de Rincon, and B. Terczi. 1984. Carbon as catalyst in oxidation reactions and hydrogen halide elimination reactions. *Fuel* 63, 1061–1063.

Bowen, E. J. and W. M. Cheung. 1932. The photodecomposition of chlorine dioxide solutions. *J. Chem. Soc.* 1200–1208.

Boyce, S. D. and J. F. Hornig. 1983. Reaction pathways of trihalomethane formation from the halogenation of dihydroxyaromatic model compounds for humic acid. *Environ. Sci. Technol.* 17: 202–211.

Brueggeman, E., J. E. Wajon, C. W. R. Wade, and E. P. Burrows. 1982. Analysis of unquenched mixtures of chlorine dioxide and phenols by reversed-phase high-performance liquid chromatography. *J. Chromatogr.* 237: 484–488.

Buchbauer, G., S. Freudenreich, C. Hampl, E. Haslinger, and W. Robien. 1984. Zur Reaktion von Camphen mit unterchloriger Säure. Synthesen in der Isocamphenreihe, 23. Mitt. *Monatsh. Chem.* 115, 509–517.

Bühler, R. E., J. Staehelin, and J. Hoigné. 1984. Ozone decomposition in water studied by pulse radiolysis. 1. HO_2/O_2^- and HO_3/O_3^- as intermediates. *J. Phys. Chem.* 88: 2560–2564.

Bunn, W. W., B. B. Haas, E. R. Deane, and R. D. Kleopfer. 1975. Formation of trihalomethanes by chlorination of surface waters. *Environ. Lett.* 10: 205–213.

Burleson, G. R. and T. M. Chambers. 1982. Effect of ozonation on the mutagenicity of carcinogens in aqueous solution. *Environ. Mutagenesis* 4: 469–476.

Burleson, G. R., M. J. Caulfield, and M. Pollard. 1979. Ozonation of mutagenic and carcinogenic polyaromatic amines and polyaromatic hydrocarbons in water. *Cancer Res.* 39: 2149–2154.

Burleson, J. L., G. R. Peyton, and W. H. Glaze. 1978. Chlorinated tyrosine in municipal waste treatment plant products after superchlorination. *Bull. Environ. Contam. Toxicol.* 19: 724–728.

Burttschell, R. H., A. A. Rosen, F. M. Middleton, and M. B. Ettinger. 1959.

Chlorine derivatives of phenol causing taste and odor. *J. Amer. Water Works Assoc.* 51: 205–213.

Butkovic, V., L. Klasinc, M. Orhanovic, J. Turk, and H. Güsten. 1983. Reaction rates of polynuclear aromatic hydrocarbons with ozone in water. *Environ. Sci. Technol.* 17: 546–548.

Carlson, R. M. and R. Caple. 1977. Chemical-biological implications of using chlorine and ozone for disinfection. US EPA Report EPA/600/3-77/066.

Carlson, R. M. and R. Caple. 1978. Organochemical implications of water chlorination. In R. L. Jolley, ed. *Water Chlorination: Environmental Impact and Health Effects.* Vol. 1. (Ann Arbor, MI: Ann Arbor Sci. Pub.), pp. 65–75.

Carlson, R. M. and S. Lin. 1985. Characterization of the products from the reaction of hydroxybenzoic and hydroxycinnamic acids with aqueous solutions of chlorine, chlorine dioxide, and chloramine. In *Water Chlorination: Chemistry, Environmental Impact and Health Effects.* Vol. 5. R. L. Jolley, R. J. Bull, W. P. Davis, S. Katz, M. H. Roberts, Jr., and V. A. Jacobs, Eds., (Chelsea, MI: Lewis Pub.), pp. 835–842.

Carlson, R. M., R. Caple, A. R. Oyler, K. J. Welch, D. L. Bodenner, and R. Liukonnen. 1978. Aqueous chlorination products of polynuclear aromatic hydrocarbons. In *Water Chlorination: Environmental Impact and Health Effects.* Vol. 2. R. L. Jolley, H. Gorchev, and D. H. Hamilton, Jr., Eds. (Ann Arbor, MI: Ann Arbor Sci. Pub.), pp. 59–65.

Caulfield, M. J., G. R. Burleson, and M. Pollard. 1979. Ozonation of mutagenic and carcinogenic alkylating agents, pesticides, aflatoxin B^1, and benzidine in water. *Cancer Res.* 39: 2155–2159.

Cavanagh, J. E., H. S. Weinberg, A. Gold, R. Sangalah, D. Marbury, W. H. Glaze, T. W. Collette, S. D. Richardson, and A. D. Thruston Jr. 1992. Ozonation byproducts: identification of bromohydrins from the ozonation of natural waters with enhanced bromide levels. *Environ. Sci. Technol.* 26, 1658–1662.

Cerkinsky, S. N. and N. Trahtman. 1972. The present status of research on the disinfection of drinking water in the USSR. *Bull. World Health Org.* 46, 277–283.

Chan, W. F. and R. A. Larson. 1992. Formation of mutagens from the aqueous reactions of ozone and anilines. *Water Res.* 25, 1529–1538.

Chen, A. S.-C., R. A. Larson, and V. L. Snoeyink. 1982. Reactions of chlorine dioxide with hydrocarbons: effects of activated carbon. *Environ. Sci. Technol.* 16: 268–273.

Chen, P. N., G. A. Junk, and H. J. Svec. 1979. Reactions of organic pollutants. 1. Ozonation of acenaphthylene and acenaphthene. *Environ. Sci. Technol.* 13: 451–454.

Christman, R. F., J. D. Johnson, J. R. Haas, F. K. Pfaender, W. T. Liao, D. L. Norwood, and H. J. Alexander. 1978. Natural and model aquatic humics: reaction with chlorine. In *Water Chlorination: Environmental Impact and Health Effects.* Vol. 2. R. L. Jolley, H. Gorchev, and D. H. Hamilton, Jr., Eds. (Ann Arbor, MI: Ann Arbor Sci. Pub.), pp. 15–28.

Christman, R. F., D. L. Norwood, D. S. Millington, and J. D. Johnson. 1983.

Identity and yields of major halogenated products of aquatic fulvic acid chlorination. *Environ. Sci. Technol.* 17: 625–628.

Chrostowski, P. C., A. M. Dietrich, and I. H. Suffet. 1983. Ozone and oxygen induced oxidative coupling of aqueous phenolics. *Water Res.* 17: 1627–1633.

Coleman, W. E., J. W. Munch, P. A. Hodakievic, F. C. Kopfler, J. R. Meier, R. P. Streicher, and H. Zimmer. 1988. GC/MS identification of mutagens in aqueous chlorinated humic acid and drinking waters following HPLC fractionation of strong acid extracts. In Larson, R. A., Ed. *Biohazards of Drinking Water Treatment.* (Chelsea, MI: Lewis Pub.), pp. 107–121.

Conyers, B., and F. E. Scully, Jr. 1993. *N*-Chloroaldimines. 3. Chlorination of phenylalanine in model solutions and in a wastewater. *Environ. Sci. Technol.* 27: 261–266.

Cotruvo, J. A., V. F. Simmon, and R. J. Spanggord. 1977. Investigation of mutagenic effects of products of ozonation reactions in water. *Ann. N.Y. Acad. Sci.* 298: 124–140.

Crane, A. M., S. J. Erickson, and C. E. Hawkins. 1980. Contribution of marine algae to trihalomethane production in chlorinated estuarine water. *Estuar. Coast. Mar. Sci.* 11: 239–249.

Das, B. S., S. G. Reid, J. L. Betts, and K. Patrick. 1969. Tetrachloro-o-benzoquinone as a component in bleached kraft chlorination effluent toxic to young salmon. *J. Fish. Res. Bd. Canada* 26: 3055–3067.

De la Mare, P. B. D., A. D. Ketley, and C. A. Vernon. 1954. The kinetics and mechanisms of aromatic halogen substitution. I. Acid-catalyzed chlorination by aqueous solutions of hypochlorous acid. *J. Chem. Soc.* 1290–1297.

De Leer, E. W. B. and C. Erkelens. 1988. Pathways for the production of organochlorine compounds in the chlorination of humic materials. In Larson, R. A., Ed. *Biohazards of Drinking Water Treatment.* (Chelsea, MI: Lewis Pub.), pp. 97–106.

De Leer, E. W. B., J. S. Sinninghe Damsté, C. Erkelens, and L. de Galan. 1985. Identification of intermediates leading to chloroform and C-4 diacids in the chlorination of humic acid. *Environ. Sci. Technol.* 19: 512–522.

De Leer, E. W. B., T. Baggerman, C. Erkelens, and L. de Galan. 1987. The production of cyano compounds on chlorination of humic acid. A comparison between terrestrial and aquatic material. *Sci. Total Environ.* 62: 329–334.

Dence, C. and K. Sarkanen. 1960. A proposed mechanism for the chlorination of softwood lignin. *Tappi* 43: 87–96.

Dennis, W. H., Jr., L. A. Hull, and D. H. Rosenblatt. 1967. Oxidation of amines. IV. Oxidative fragmentation. *J. Org. Chem.* 32: 3783–3787.

Dennis, W. H., Jr., V. P. Olivieri, and C. W. Krusé. 1978. Reaction of uracil with hypochlorous acid. *Biochem. Biophys. Res. Commun.* 83: 168–171.

Dennis, W. H., Jr., V. P. Olivieri, and C. W. Krusé. 1979. The reaction of nucleotides with aqueous hypochlorous acid. *Water Res.* 13: 357–362.

DeRosa, M. and J. L. T. Alonso. 1978. Studies of the mechanism of chlorination of indoles. Detection of N-chloroindole and 3-chloro-3H-indole as intermediates. *J. Org. Chem.* 43: 2639–2643.

Donnet, J.-B. 1982. Structure and reactivity of carbons; from carbon black to carbon composites. *Carbon* 20: 266–282.

Doré, M., B. Langlais, and B. Legube. 1978. Ozonation des phénols et des acides phénoxyacetiques. *Water Res.* 12: 413–425.

Duguet, J.-P., A. Bruchet, B. Dussert, and J. Mallevialle. 1988. Formation of aromatic polymers during the ozonation or enzymatic oxidation of waters containing phenolic compounds. In Larson, R. A., Ed. *Biohazards of Drinking Water Treatment*. (Chelsea, MI: Lewis Pub.), pp. 171–184.

Eisenhauer, H. R. 1968. The ozonisation of phenolic wastes. *J. Water Pollut. Contr. Fed.* 40: 1887–1899.

Eklund, G., B. Josefsson and A. Bjørseth. 1978. Determination of chlorinated and brominated lipophilic compounds in spent bleach liquors from a sulphite pulp mill. Glass capillary column gas chromatography-mass spectrometry-computer analysis and identification. *J. Chromatogr.* 150: 161–169.

Ellis, A. J. and F. G. Soper. 1954. Studies of N-halogeno compounds. VI. The kinetics of chlorination of tertiary amines. *J. Chem. Soc.* 1750–1755.

Erickson, M. and C. W. Dence. 1976. Phenolic and chlorophenolic oligomers in chlorinated pine kraft pulp and in bleach plant effluents. *Svensk Papperstidn.* 10: 316–322.

Everly, C. R. and J. G. Traynham. 1979. Formation and rearrangement of ipso intermediates in aromatic free-radical chlorination reactions. *J. Org. Chem.* 44: 1784–1787.

Fair, G. M., J. C. Morris, S. L. Chang, I. Weil, and R. P. Burden. 1948. The behavior of chlorine as a water disinfectant. *J. Amer. Water Works Assoc.* 40: 1051–1061.

Farhataziz and A. B. Ross. 1977. Selected specific rates of reactions of transients from water in aqueous solution. III. Hydroxyl radical and perhydroxyl radical and their radical ions. NSRDS-NBS Publication #59. US Dept. of Commerce, Washington, DC.

Fischer, A. and G. N. Henderson. 1979. Ipso chlorination of 4-alkylphenols. Formation of 4-alkyl-4-chlorocyclohexa-2,5-dienones. *Can. J. Chem.* 57: 552–557.

Fonouni, H. E., S. Krishnan, D. G. Kuhn, and G. A. Hamilton. 1983. Mechanisms of epoxidations and chlorinations of hydrocarbons by inorganic hypochlorite in the presence of a phase-transfer catalyst. *J. Amer. Chem. Soc.* 105: 7672–7676.

Forni, L., D. Bahnemann, and E. J. Hart. 1982. Mechanism of the hydroxide ion initiated decomposition of ozone in aqueous solution. *J. Phys. Chem.* 86: 255–259.

Freitag, D., L. Scheunert, W. Klein, and F. Korte. 1984. Long-term fate of 4-chloroaniline-^{14}C in soil and plants under outdoor conditions. A contribution to terrestrial ecotoxicology of chemicals. *J. Agric. Food Chem.* 32, 203–207.

Friedman, A. H. and S. Morgulis. 1936. The oxidation of amino acids with sodium hypobromite. *J. Amer. Chem. Soc.* 58: 909–913.

Fujii, M. and M. Ukita. 1957. Mechanism of wheat protein coloring by chlorine dioxide. *Nippon Nogei Kagaku Kaishi* 31: 101–109.

Gaffney, P. E. 1977. Chlorobiphenyls and PCBs: formation during chlorination. *J. Water Pollut. Contr. Fed.* 49: 401–404.

Gess, J. M. 1971. Reactions of creosol with aqueous chlorine. Ph.D. Thesis, State Univ. Coll. Forestry (Syracuse, NY), 1971. *Diss. Abstr.* B31: 5258.

Ghanbari, H. A., W. B. Wheeler, and J. R. Kirk. 1983. Reactions of aqueous chlorine and chlorine dioxide with lipids: chlorine incorporation. *J. Food Sci.* 47: 482–485.

Gibson, T. M., J. Haley, M. Righton, and C. D. Watts. 1986. Chlorination of fatty acids during water treatment disinfection: reactivity and product identification. *Environ. Technol. Lett.* 7: 365–372.

Giger, W., M. Reinhard, C. Schaffner, and F. Zurcher. 1976. Analyses of organic constituents in water by high-resolution gas chromatography in combination with specific detection and computer-assisted mass spectrometry. In L. H. Keith, Ed. *Identification and Analysis of Organic Pollutants in Water*. (Ann Arbor, MI: Ann Arbor Sci. Pub.), pp. 433–452.

Gilbert, E. 1976. Über den Abbau von organischen Schadstoffen in Wasser durch Ozon. *Vom Wasser* 43: 275–290.

Gilbert, E. 1978. Reactions of ozone with organic compounds in dilute aqueous solution: identification of their oxidation products. In R. G. Rice and J. A. Cotruvo, Eds. *Ozone-Chlorine Dioxide Oxidation Products of Organic Materials*. (Cleveland, OH: Ozone Press Intl.), pp. 227–242.

Glaze, W. H. 1987. Drinking-water treatment with ozone. *Environ. Sci. Technol.* 21: 224–230.

Glaze, W. H., J. E. Henderson IV, and G. Smith. 1978. Analysis of new chlorinated organic compounds formed by chlorination of municipal wastewater. In R. L. Jolley, Ed. *Water Chlorination: Environmental Impact and Health Effects*. Vol. 1. (Ann Arbor, MI: Ann Arbor Sci. Pub.), pp. 139–159.

Glaze, W. H., M. Koga, E. C. Ruth, and D. Cancilla. 1988. Application of closed loop stripping and XAD resin adsorption for the determination of ozone by-products from natural water. In Larson, R. A., Ed. *Biohazards of Drinking Water Treatment*. (Chelsea, MI: Lewis Pub.), pp. 201–210.

Gordon, G., R. G. Kieffer, and D. H. Rosenblatt. 1972. The chemistry of chlorine dioxide. *Progr. Inorg. Chem.* 15: 201–286.

Gould, J. P. and W. J. Weber Jr. 1976. Oxidation of phenols by ozone. *J. Water Pollut. Contr. Fed.* 48: 47–60.

Guilmet, E. and B. Meunier. 1980. A new catalytic route for the epoxidation of styrene with sodium hypochlorite activated by transition metal complexes. *Tetrahedron Lett.* 21: 4449–4450.

Gurol, M. D. and S. Nekoulnalni. 1984. Kinetic behavior of ozone in aqueous solutions of substituted phenols. *Industr. Eng. Chem. Fundamentals* 23: 54–60.

Guthrie, J. P. and J. Cossar. 1986. The chlorination of acetone: a complete kinetic analysis. *Can. J. Chem.* 64: 1250–1266.

Guthrie, J. P., J. Cossar, and J. Lu. 1991. Dihydroxyacids from the chlorination of ketones: an unexpected process. *Can. J. Chem.* 69: 1904–1908.

Haag, W. R. and J. Hoigné. 1983. Ozonation of bromide-containing waters: kinetics of formation of hypobromous acid and bromate. *Environ. Sci. Technol.* 17: 261–267.

Haenen, G. R. M. M. and A. Bast. 1991. Scavenging of hypochlorous acid by lipoic acid. *Biochem. Pharmacol.* 42: 2244–2246.

Hanna, J. V., W. D. Johnson, R. A. Quezada, M. A. Wilson, and L. Xiao-Qiao. 1991. Characterization of aqueous humic substances before and after chlorination. *Environ. Sci. Technol.* 25: 1160–1164.

Hauchman, F. S., C. I. Noss, and V. P. Olivieri. 1986. Chlorine dioxide reactivity with nucleic acids. *Water Res.* 20: 357–361.

Hauser, C. R. and M. L. Hauser. 1930. Researches on chloramines. I. Orthochlorobenzalchlorimine and anisalchlorimine. *J. Amer. Chem. Soc.* 52: 2050–2054.

Hayatsu, H., S. K. Pan, and E. T. Ukita. 1971. Reaction of sodium hypochlorite with nucleic acids and their constituents. *Chem. Pharm. Bull.* 19: 2189–2192.

Helz, G. R., R. Y. Hsu, and R. M. Block. 1978. Bromoform production by oxidative biocides in marine waters. In R. G. Rice and J. A. Cotruvo, Eds. *Ozone—Chlorine Dioxide Oxidation Products of Organic Materials.* (Cleveland, OH: Ozone Press Intl.), pp. 68–76.

Hemming, J., B. Holmbom, M. Reunanen, and L. Kronberg. 1986. Determination of the strong mutagen 3-chloro-4-(dichloromethyl)-5-hydroxy-2(5H)-furanone in chlorinated drinking and humic waters. *Chemosphere* 15: 549–556.

Hirose, Y., T. Okitsu and S. Kanno. 1982. Formation of trihalomethanes by reaction of halogenated phenols or halogenated anilines with sodium hypochlorite. *Chemosphere* 11: 81–87.

Hodgen, H. W. and R. S. Ingols. 1954. Direct colorimetric method for the determination of chlorine dioxide in water. *Anal. Chem.* 26: 1224–1226.

Hoehn, R. C., C. W. Randall, F. A. Bell, Jr., and P. T. B. Shaffer. 1977. Trihalomethanes and viruses in a water supply. *J. Environ. Eng. Div., Amer. Soc. Chem. Eng.* 103: 803–814.

Hoehn, R. C., R. P. Goode, C. W. Randall, and P. T. B. Shaffer. 1978. Chlorination and treatment for minimizing trihalomethanes in drinking water. In *Water Chlorination: Environmental Impact and Health Effects.* Vol. 2. R. L. Jolley, H. Gorchev, and D. H. Hamilton, Jr., Eds. (Ann Arbor, MI: Ann Arbor Sci. Pub.), pp. 519–535.

Hoigné, J. 1982. Mechanisms, rates, and selectivities of oxidations of organic compounds initiated by ozonation of water. In *Handbook of Ozone Technology and Applications.* R. G. Rice and A. Netzer, Eds. (Ann Arbor, MI: Ann Arbor Sci. Pub.), pp. 341–379.

Hoigné, J. and H. Bader. 1975. Ozonation of water: role of hydroxyl radicals as oxidizing intermediates. *Science* 190: 782–783.

Hoigné, J. and H. Bader. 1979. Ozonation of water: "oxidation-competition values" of different types of waters used in Switzerland. *Ozone Sci. Eng.* 1: 357–372.

Hoigné, J. and H. Bader. 1982. Kinetik typischer Reaktionen von Chlordioxid mit Wasserinhaltsstoffen. *Vom Wasser* 59: 253–267.

Hoigné, J. and H. Bader. 1983a. Rate constants of reactions of ozone with organic and inorganic compounds in water. I. Non-dissociating organic compounds. *Water Res.* 17: 173–183.

Hoigné, J. and H. Bader. 1983b. Rate constants of reactions of ozone with organic and inorganic compounds in water. II. Dissociating organic compounds. *Water Res.* 17: 185–194.

Horth, H., M. Fielding, H. A. James, M. J. Thomas, T. Gibson, and P. Wilcox. 1990. Production of organic chemicals and mutagens during chlorination of amino acids in water. In *Water Chlorination: Chemistry, Environmental Impact and Health Effects*. Vol. 6. R. L. Jolley, L. W. Condie, J. D. Johnson, S. Katz, R. A. Minear, J. S. Mattice, and V. A. Jacobs, Eds. (Chelsea, MI: Lewis Pub.), pp. 107–124.

Houben, J. and T. Weyl. 1962. Methoden der organischen Chemie ("Houben-Weyl"). 4. Aufl., B. 5, p. 793. Thieme Verlag, Stuttgart.

Hoyano, Y., V. Bacon, R. E. Simmons, W. E. Pereira, B. Halpern, and A. M. Duffield. 1973. Chlorination studies. IV. The reaction of aqueous hypochlorous acid with pyrimidine and purine bases. *Biochem. Biophys. Res. Commun.* 53: 1195–1199.

Hrudey, S. E., A. Gac, and S. A. Daignault. 1988. Potent odor-causing chemicals arising from drinking water disinfection. *Water Sci. Technol.* 20: 55–61.

Hwang, S.-C., R. A. Larson, and V. L. Snoeyink. 1990. Reactions of free chlorine with substituted anilines in aqueous solution and on granular activated carbon. *Water Res.* 24: 427–432.

Ingols, R. S., H. A. Wyckoff, T. W. Kethley, H. W. Hodgen, E. L. Fincher, J. C. Hildebrand, and J. E. Mandel. 1953. Bacterial studies of chlorine. *Industr. Eng. Chem.* 45: 996–1000.

Jackson, D. E., R. A. Larson, and V. L. Snoeyink. 1987. Reactions of chlorine and chlorine dioxide with resorcinol in aqueous solution and adsorbed on granular activated carbon. *Water Res.* 231: 849–857.

Jenkins, R. L., J. E. Haskins, L. G. Carmona, and R. B. Baird. 1978. Chlorination of benzidine and other aromatic amines in aqueous environments. *Arch. Environ. Contam. Toxicol.* 7: 301–315.

Johnsen, S. and I. S. Gribbestad. 1988. Influence of humic substances on the formation of chlorinated polycyclic aromatic hydrocarbons during chlorination of polycyclic aromatic hydrocarbon polluted water. *Environ. Sci. Technol.* 22: 978–981.

Johnson, J. D., D. L. Norwood, and R. F. Christman. 1982. Reaction products of aquatic humic substances with chlorine. *Environ. Health Perspect.* 46: 63–71.

Jurs, R. H. 1966. Die Wirkung des Ozons auf wassergelöste Stoffe. *Fortschr. Wasserchem.* 4: 40–64.

Keith, L. H., A. W. Garrison, F. R. Allen, M. H. Carter, T. L. Floyd, J. D. Pope, and A. D. Thruston Jr. 1976. Identification of organic compounds in drinking water from thirteen U.S. cities. In L. H. Keith, Ed. *Identification and Analysis of Organic Pollutants in Water*. (Ann Arbor, MI: Ann Arbor Sci. Pub.), pp. 329–373.

Killops, S. D. 1986. Volatile ozonation products of aqueous humic material. *Water Res.* 20: 153–165.

Knox, W. E., P. K. Stumpf, D. E. Green, and V. H. Auerbach. 1948. The inhibition of sulfhydryl enzymes as the basis of the bactericidal action of chlorine. *J. Bacteriol.* 55: 451–458.

Kolar, J. J. and B. O. Lindgren. 1982. Oxidation of styrene by chlorine dioxide and by chlorite in aqueous solutions. *Acta Chem. Scand.* B36: 599–605.

Kopperman, H. L., R. C. Hallcher, A. Riehl, R. M. Carlson, and R. Caple. 1976. Aqueous chlorination of α-terpineol. *Tetrahedron* 32: 1621–1626.

Kovacic, P., M. Lowery, and K. W. Field. 1970. Chemistry of N-bromamines and N-chloramines. *Chem. Rev.* 70: 639–665.

Kringstad, K. P., P. O. Ljungquist, F. De Sousa, and L. M. Strömberg. 1981. Identification and mutagenic properties of some chlorinated aliphatic compounds in the spent liquor from kraft pulp chlorination. *Environ. Sci. Technol.* 15: 562–566.

Kronberg, L., B. Holmbom, M. Reunanen, and L. Tikkanen. 1988. Identification and quantification of the Ames mutagenic compound 3-chloro-4-(dichloromethyl)-5-hydroxy-2(5H)-furanone and of its geometric isomer (E)-2-chloro-3-(dichloromethyl)-4-oxobutenoic acid in chlorine-treated humic water and drinking water extracts. *Environ. Sci. Technol.* 22: 1097–1103.

Kuo, P. P. K., E. S. Chian, and B. J. Chang. 1977. Identification of end products resulting from ozonation and chlorination of organic compounds commonly found in water. *Environ. Sci. Technol.* 11: 1177–1181.

Långvik, V.-A., O. Hormi, L. Tikkanen, and B. Holmbom. 1991. Formation of the mutagen 3-chloro-4-(dichloromethyl)-5-hydroxy-2(5H)-furanone and related compounds by chlorination of phenolic compounds. *Chemosphere* 22, 547–555.

Larson, R. A., Ed. 1988. *Biohazards of Drinking Water Treatment.* (Chelsea, MI: Lewis Pub.).

Larson, R. A. and K. A. Marley. 1988. Sunlight photochlorination and dark chlorination of monoterpenes. *Sci. Total Environ.* 77: 245–252.

Larson, R. A. and A. L. Rockwell. 1979. Chloroform and chlorophenol production by decarboxylation of natural acids during aqueous chlorination. *Environ. Sci. Technol.* 13: 325–329.

Larson, R. A. and R. G. Zepp. 1988. Reactivity of the carbonate radical with aniline derivatives. *Environ. Toxicol. Chem.* 7: 265–274.

Lawrence, J., H. Tosine, F. I. Onuska, and M. E. Comba. 1980. The ozonation of natural waters: product identification. *Ozone Sci. Eng.* 2: 55–64.

Leach, J. M. and A. N. Thakore. 1977. Compounds toxic to fish in pulp mill waste streams. *Progr. Water Technol.* 9: 787–798.

Lee, G. F. and J. C. Morris. 1962. Kinetics of chlorination of phenol: chlorophenolic tastes and odors. *Internat. J. Air Water Pollut.* 6: 419–431.

Legube, B., S. Guyon, H. Sugimitsu, and M. Doré. 1986. Ozonation du naphthalene en milieu aqueux. I. Consommation d'ozone et produits de reaction. *Water Res.* 20: 197–208.

LeRoux, Y., J. C. Ginisty, and C. Nofre. 1969. Dégradation comparée de la xanthine sous l'effet de l'oxygene singulet excitée et du radical libre hydroxyle. *Compt. Rend. Acad. Sci. Paris* 269C: 744–747.

Lin, S. and R. M. Carlson. 1984. Susceptibility of environmentally important heterocycles to chemical disinfection: reactions with aqueous chlorine, chlorine dioxide, and chloramine. *Environ. Sci. Technol.* 18: 743–748.

Lin, S., R. J. Liukkonen, R. E. Thorn, J. G. Bastian, M. T. Lukasewycz, and R. M. Carlson. 1984. Increased chloroform production from model compounds of aquatic humus and mixtures of chlorine dioxide/chlorine. *Environ. Sci. Technol.* 18: 932–935.

Lindgren, B. O. 1971. Chlorine dioxide and chlorite oxidations of phenols related to lignin. *Svensk Papperstidn.* 74: 57–63.

Lindgren, B. O., C. M. Svahn, and G. Widmark. 1965. Chlorine dioxide oxidation of cyclohexene. *Acta Chem. Scand.* 19: 7–13.

Lindgren, B. O. and T. Nilsson. 1973. Preparation of carboxylic acids from aldehydes (including hydroxylated benzaldehydes) by oxidation with chlorite. *Acta Chem. Scand.* 27: 888–890.

Lindgren, B. O. and T. Nilsson. 1974. Oxidation of lignin model compounds with chlorine dioxide and chlorite. Reactions with stilbenes. *Acta Chem. Scand.* 28: 847–852.

Lindgren, B. O. and T. Nilsson. 1975. Chlorate formation during the reaction of chlorine dioxide with lignin model compounds. *Svensk Papperstidn.* 78: 66–68.

Lindström, K. and F. Österberg. 1986. Chlorinated carboxylic acids in softwood kraft pulp spent bleach liquors. *Environ. Sci. Technol.* 20: 133–138.

Lykins, B. W., W. Koffskey, and R. G. Miller. 1986. Chemical products and toxicologic effects of disinfection. *J. Amer. Water Works Assoc.* 78: 66–75.

Macalady, D. L., J. H. Carpenter, and C. A. Moore. 1977. Sunlight-induced bromate formation in chlorinated seawater. *Science* 195: 1335–1337.

Marley, K. A., R. A. Larson, P. L. Stapleton, W. J. Garrison, and C. L. Klodnycky. 1987. Ozonolysis of naphthalene derivatives in water and in kerosene films. *Ozone Sci. Eng.* 9: 23–36.

Masschelein, W. J. 1979. Chlorine dioxide: chemistry and environmental impact of oxychlorine compounds. (Ann Arbor, MI: Ann Arbor Science).

Matheson, N. R. and T. J. Travis. 1985. Differential effects of oxidizing agents on human plasma α_1-antiproteinase inhibitor and human neutrophil myeloperoxidase. *J. Biochem.* 24, 1941–1945.

Matsui, M., H. Nakazumi, K. Kamiya, C. Yatome, K. Shibata, and H. Muramatsu. 1989. Ozonolysis of uracils in water. *Chem. Lett.* 723–724.

Matsushima, Y. 1951. Studies on amino-hexoses. I. A new method for preparing crystalline D-arabinose. *Bull. Chem. Soc. Japan* 24: 7–20.

Mauger, R. P. and F. G. Soper. 1946. Acid catalysis in the formation of chloramides from hypochlorous acid. N-chlorination by hypochlorite ions and by acyl hypochlorite. *J. Chem. Soc.* 71–75.

McCreary, J. J., V. L. Snoeyink, and R. A. Larson. 1982. Comparison of the reaction of aqueous free chlorine with phenolic acids in solution and adsorbed on granular activated carbon. *Environ. Sci. Technol.* 16: 339–344.

McKague, A. B. and K. P. Kringstad. 1988. Some lipophilic compounds formed in the chlorination of pulp lignin and humic acids. In Larson, R. A., Ed. *Biohazards of Drinking Water Treatment*. (Chelsea, MI: Lewis Pub.), pp. 123–131.

McKague, A. B., E. G.-H. Lee, and G. R. Douglas. 1981. Chloroacetones: mutagenic constituents of bleached kraft chlorination effluent. *Mutat. Res.* 91: 301–306.

McKague, A. B., F. de Sousa, L. M. Strömberg, and K. P. Kringstad. 1987. Formation of 2,2,4,5-tetrachlorocyclopentene-1,3-dione in the chlorination of kraft lignin, kraft pulp, tetrachlorocatechol and tetrachloro-o-benzoquinone. *Holzforschung* 41: 191–193.

McKague, A. B., M.-C. Kolar, and K. P. Kringstad. 1989. Nature and properties of some chlorinated, lipophilic, organic compounds in spent liquors from pulp bleaching. 2. *Environ. Sci. Technol.* 23: 1126–1129.

McKinney, J. D., R. R. Maurer, J. R. Hass, and R. O. Thomas. 1976. Possible factors in the drinking water of laboratory rats causing reproductive failure. In L. H. Keith, Ed. *Identification and Analysis of Organic Pollutants in Water.* (Ann Arbor, MI: Ann Arbor Sci. Pub.), pp. 417–432.

Meier, J. R., H. P. Ringhand, W. E. Coleman, K. M. Schenk, J. W. Munch, R. P. Streicher, W. H. Kaylor, and F. C. Kopfler. 1986. Mutagenic by-products from chlorination of humic acid. *Environ. Health Perspect.* 69: 101–107.

Menzel, D. B. 1971. Oxidation of biologically active reducing substances by ozone. *Arch. Environ. Health* 23: 149–153.

Merényi, G., J. Lind, and X. Shen. 1988. Electron transfer from indoles, phenol, and sulfite (SO_3^{2-}) to chlorine dioxide ($ClO_2\cdot$). *J. Phys. Chem.* 92: 134–137.

Miller, J. W. and P. C. Uden. 1983. Characterization of nonvolatile aqueous chlorination products of humic substances. *Environ. Sci. Technol.* 17: 150–156.

Minear, R. A. and J. Bird. 1980. Trihalomethanes: impact of bromide ion concentration on yield, species distribution, rate of formation and influence of other variables. In *Water Chlorination: Environmental Impact and Health Effects.* Vol. 3. R. L. Jolley, Ed. (Ann Arbor, MI: Ann Arbor Sci. Pub.), pp. 151–160.

Molina, M. J., T. Ishiwata, and L. T. Molina. 1980. Production of OH from photolysis of HOCl at 307–309 nm. *J. Phys. Chem.* 84: 821–826.

Morris, J. C. 1967. Kinetics of reactions between aqueous chlorine and nitrogen compounds. In S. D. Faust and J. V. Hunter, Eds. *Principles and Applications of Water Chemistry*. (New York, NY: John Wiley & Sons), pp. 22–53.

Morris, J. C. and B. Baum. 1978. Precursors and mechanisms of haloform formation in the chlorination of water supplies. In *Water Chlorination: Environmental Impact and Health Effects.* Vol. 2. R. L. Jolley, H. Gorchev, and D. H. Hamilton, Jr., Eds. (Ann Arbor, MI: Ann Arbor Sci. Pub.), pp. 29–48.

Morton, A. A. 1946. *The Chemistry of Heterocyclic Compounds.* McGraw-Hill, New York.

Mudd, J. B., R. Leavitt, A. Ongun, and T. T. McManus. 1969. Reaction of ozone with amino acids and proteins. *Atmos. Environ.* 3: 669–682.

Namba, K. and S. Nakayama. 1982. Hydrogen peroxide-catalyzed ozonation of refractory organics. 1. Hydroxyl radical formation. *Bull. Chem. Soc. Japan* 55: 3339–3340.

NAS (US National Academy of Sciences). 1980. *Drinking Water and Health*. Vol. 2. (Washington, DC: National Academy Press).

Naudet, M. and P. Desnuelle. 1950. Sur la formation des acides dihydroxy et époxy-stéariques su cours du traitement de l'acide chlorhydroxystéarique par divers agents alcalins. *Bull. Soc. Chim. France* 845–848.

Nickelsen, M. G., A. Nweke, F. E. Scully, Jr., and H. P. Ringhand. 1991. Reactions of aqueous chlorine in vitro in stomach fluid from the rat: chlorination of tyrosine. *Chem. Res. Toxicol.* 4: 94–101.

Norwood, D. L., J. D. Johnson, R. F. Christman, J. R. Hass, and M. J. Bobenreith. 1980. Reactions of chlorine with selected aromatic models of humic material. *Environ. Sci. Technol.* 14: 187–190.

Norwood, D. L., R. F. Christman, and P. G. Hatcher. 1987. Structural characterization of aquatic humic material. 2. Phenolic content and its relationship to chlorination mechanism in an isolated aquatic fulvic acid. *Environ. Sci. Technol.* 21: 791–798.

Noss, C. I., F. S. Hauchman, and V. P. Olivieri. 1986. Chlorine dioxide reactivity with proteins. *Water Res.* 20: 351–356.

Nowell, L. H. and D. G. Crosby. 1985. Photodegradation of water pollutants in chlorinated water. In *Water Chlorination: Chemistry, Environmental Impact, and Health Effects*. Vol. 5. R. L. Jolley, R. J. Bull, W. P. Davis, S. Katz, M. H. Roberts, Jr., and V. A. Jacobs, Eds., (Chelsea, MI: Lewis Pub.), pp. 1055–1062.

Nwaukwa, S. O. and P. M. Keehn. 1982. The oxidation of alcohols and ethers using calcium hypochlorite. *Tetrahedron Lett.* 23: 35–38.

Nweke, A. and F. E. Scully, Jr. 1989. Stable N-chloroaldimines and other products of the chlorination of isoleucine in model solutions and in a wastewater. *Environ. Sci. Technol.* 23, 989–994.

Ogata, Y., Y. Suzuki, and K. Takagi. 1979. Photolytic oxidation of aliphatic acids by aqueous sodium hypochlorite. *J. Chem. Soc. Perkin Trans. II*, 1715–1719.

Oliver, B. G. 1978. Chlorinated non-volatile organics produced by the reaction of chlorine with humic materials. *Can. Res.* 11: 21–22.

Oliver, B. G. 1983. Dihaloacetonitriles in drinking water: algae and fulvic acid as precursors. *Environ. Sci. Technol.* 17: 80–83.

Oliver, B. G. and J. H. Carey. 1977. Photochemical production of chlorinated organics in aqueous solutions containing chlorine. *Environ. Sci. Technol.* 11: 893–895.

Oliver, B. G. and J. Lawrence. 1979. Haloforms in drinking water: a study of precursors and precursor removal. *J. Amer. Water Works Assoc.* 71: 161–164.

Oliver, B. G. and D. B. Shindler. 1980. Triahlomethanes from the chlorination of aquatic algae. *Environ. Sci. Technol.* 14: 1502–1505.

Oliver, B. G. and S. A. Visser. 1980. Chloroform production from the chlorination of aquatic humic material: the effect of molecular weight, environment and season. *Water Res.* 14: 1137–1141.

Onodera, S., K. Yamada, Y. Yamaji, and S. Ishikura. 1984. Chemical changes of organic compounds in chlorinated water. IX. Formation of polychlorinated phenoxyphenols during the reaction of phenol with hypochlorite in dilute aqueous solution. *J. Chromatogr.* 288: 91–100.

Onodera, S., T. Muritani, N. Kobatake, and S. Suzuki. 1986. Chemical changes of organic compounds in chlorinated water. XII. Gas chromatographic-mass spectrometric studies of the reactions of methylnaphthalenes with hypochlorite in dilute aqueous solution. *J. Chromatogr.* 370: 259–274.

Österberg, F. and K. Lindström. 1985. Characterization of the high molecular mass chlorinated matter in spent bleach liquors (SBL). 3. Mass spectrometric interpretation of aromatic degradation products in SBL. *Org. Mass Spectrom.* 20: 515–524.

Otson, R., G. L. Polley, and J. L. Robertson. 1986. Chlorinated organics from chlorine used in water treatment. *Water Res.* 20: 775–779.

Owusu-Yaw, J., J. P. Toth, W. B. Wheeler, and C. I. Wei. 1990. Mutagenicity and identification of the reaction products of aqueous chlorine or chlorine dioxide with L-tryptophan. *J. Food Sci.* 55: 1714–1720.

Oyler, A. R., D. L. Bodenner, K. J. Welch, R. J. Liukkonen, R. M. Carlson, H. L. Kopperman, and R. Caple. 1978. Determination of aqueous chlorination reaction products of polynuclear aromatic hydrocarbons by reversed phase high performance liquid chromatography—gas chromatography. *Anal. Chem.* 50: 837–842.

Oyler, A. R., R. J. Liukkonen, M. T. Lukasewycz, K. E. Heikkila, D. A. Cox, and R. M. Carlson. 1983. Chlorine "disinfection" chemistry of aromatic compounds. Polynuclear aromatic hydrocarbons: rates, products, and mechanisms. *Environ. Sci. Technol.* 17: 334–342.

Ozawa, T. and T. Kwan. 1984. Electron spin resonance studies on the reactive character of chlorine dioxide (ClO_2) radical in aqueous solution. *Chem. Pharm. Bull.* 32: 1587–1589.

Paluch, K. 1964. The reaction of chlorine dioxide with phenols. II. Hydroquinone, chloro derivatives of hydroquinone, and nitrophenols. *Roczniki Chem.* 38: 43–46.

Patton, W., V. Bacon, A. M. Duffield, B. Halpern, Y. Hoyano, W. Pereira, and J. Lederberg. 1972. Chlorination studies. I. The reaction of aqueous hypochlorous acid with cytosine. *Biochem. Biophys. Res. Commun.* 48: 880–884.

Pauling, L. 1960. *The Nature of the Chemical Bond.* (Ithaca, NY: Cornell Univ. Press).

Pereira, W. E., Y. Hoyano, R. E. Simmons, V. A. Bacon, and A. M. Duffield. 1973. Chlorination studies. II. The reaction of aqueous hypochlorous acid with α-amino acids and dipeptides. *Biochim. Biophys. Acta* 313: 170–180.

Peters, R. J. B., E. W. de Leer, and L. de Galan. 1990. Chlorination of cyanoethanoic acid in aqueous medium. *Environ. Sci. Technol.* 24: 81–86.

Peyton, G. R., F. Y. Huang, J. L. Burleson, and W. H. Glaze. 1982. Destruction of pollutants in water with ozone in combination with ultraviolet radiation. 1. General principles and oxidation of tetrachloroethylene. *Environ. Sci. Technol.* 16: 448–453.

Peyton, G. R., C. S. Gee, M. A. Smith, J. Bandy, and S. W. Maloney. 1988. Byproducts from ozonation and photolytic ozonation of organic pollutants in water: preliminary observations. In Larson, R. A., Ed. *Biohazards of Drinking Water Treatment*. (Chelsea, MI: Lewis Pub.), pp. 185–200.

Pryor, W. A., D. G. Prier, and D. F. Church. 1981. Radical production from the interaction of ozone and PUFA as demonstrated by electron spin resonance spin-trapping techniques. *Environ. Res.* 24: 42–52.

Pryor, W. A., D. H. Giamalva, and D. F. Church. 1984. Kinetics of ozonation. 2. Amino acids and model compounds in water and comparisons to rates in nonpolar solvents. *J. Amer. Chem. Soc.* 106: 7094–7100.

Pyysalo, H. and K. Antervo. 1985. GC profiles of chlorinated terpenes (toxaphene) in some Finnish environmental samples. *Chemosphere* 14: 1723–1728.

Quimby, B. D., M. F. Delaney, P. C. Uden, and R. M. Barnes. 1980. Determination of the aqueous chlorination products of humic substances by gas chromatography with microwave emission detection. *Anal. Chem.* 52: 259–263.

Rav-Acha, C. and E. Choshen. 1987. Aqueous reactions of chlorine dioxide with hydrocarbons. *Environ. Sci. Technol.* 21: 1069–1074.

Reckhow, D. A., P. C. Singer, and R. L. Malcolm. 1990. Chlorination of humic materials: byproduct formation and chemical interpretations. *Environ. Sci. Technol.* 24, 1655–1664.

Reinhard, M., V. Drevenkar, and W. Giger. 1976. Effects of aqueous chlorination on the aromatic fraction of diesel fuel. Analysis by computer-assisted gas chromatography-mass spectrometry. *J. Chromatogr.* 116: 43–51.

Reynolds, G. L., H. A. Filaderli, A. E. McIntyre, N. J. D. Graham, and R. Perry. 1988. Isolation and identification of reaction products arising from the chlorination of cytosine in aqueous solution. *Environ. Sci. Technol.* 22: 1425–1429.

Roller, S. D., V. P. Olivieri, and K. Kawata. 1980. Mode of bacterial inactivation by chlorine dioxide. *Water Res.* 14: 635–641.

Rook, J. J. 1974. Formation of haloforms during chlorination of natural waters. *Water Treat. Exam.* 23: 234–243.

Rook, J. J. 1976. Haloforms in drinking water. *J. Amer. Water Works Assoc.* 68: 168–172.

Rook, J. J. 1977. Chlorination reactions of fulvic acids in natural waters. *Environ. Sci. Technol.* 11: 478–482.

Rook, J. J., A. A. Gras, B. G. van der Heidjen, and J. de Wee. 1978. Bromide oxidation and organic substitution in water treatment. *J. Environ. Sci. Health* 13A: 91–116.

Rosenblatt, D. H. 1975. Chlorine and oxychlorine species reactivity with organic substances. In J. D. Johnson, Ed. *Disinfection: Water and Wastewater*. (Ann Arbor, MI: Ann Arbor Sci. Pub.), pp. 249–276.

Rosenblatt, D. H. 1978. Chlorine dioxide: chemical and physical properties. In R. G. Rice and J. A. Cotruvo, Eds. *Ozone—Chlorine Dioxide Oxidation Products of Organic Materials*. (Cleveland, OH: Ozone Press Intl.), pp. 332–343.

Rosenbusch, K. 1965. Neure Erfahrungen bei der Oxydativ-Enthaarung. *Leder* 16: 237–248.

Sander, R., W. Kühn, and H. Sontheimer. 1977. Untersuchungen zur Umsetzung von Chlor mit Huminsubstanzen. *Z. Wasser-Abwasser Forsch.* 10: 155–158.

Sandford, P. A., A. J. Nafziger, and A. Jeanes. 1971. Reaction of sodium hypochlorite with amines and amides: a new method for quantitating amino sugars in monomeric form. *Anal. Biochem.* 42: 422–436.

Sawamura, R., E. Sakurai, M. Yamamoto, M. Tachikawa, and A. Hasegawa. 1982. The reaction of hypochlorite with glycine. I. Decomposition of glycine and formation of cyanogen chloride. *Eisei Kagaku* 28: 267–273.

Scully, F. E., Jr., and W. N. White. 1991. Reactions of chlorine, monochloramine in the GI tract. *Environ. Sci. Technol.* 25: 820–828.

Scully, F. E., Jr., G. D. Howell, R. Kravitz, J. T. Jewell, V. Hahn, and M. Speed. 1988. Proteins in natural waters and their relation to the formation of chlorinated organics during water disinfection. *Environ. Sci. Technol.* 22: 537–542.

Shackelford, W. M. and L. H. Keith. 1976. Frequency of organic compounds identified in water. EPA Report EPA-600/4-76-062.

Shapiro, R. H., K. J. Kolonko, P. M. Greenstein, R. M. Barkley, and R. E. Sievers. 1978. Ozonation products from caffeine in aqueous solutions. In R. G. Rice and J. A. Cotruvo, Eds. *Ozone—Chlorine Dioxide Oxidation Products of Organic Materials*. (Cleveland, OH: Ozone Press Intl.), pp. 284–290.

Shimizu, Y. and R. Y. Hsu. 1975. Interaction of chlorine and selected plant phenols in water. *Chem. Pharm. Bull.* 23: 2179–2181.

Shiriashi, H., N. H. Pilkington, A. Otsuki, and K. Fuwa. 1985. Occurrence of chlorinated polynuclear aromatic hydrocarbons in tap water. *Environ. Sci. Technol.* 19: 585–590.

Sievers, R. E., R. M. Barkley, G. Z. Eiceman, L. P. Haack, R. H. Shapiro, H. F. Walton, K. J. Kolonko, and I. R. Field. 1977. Environmental trace analysis of organics in water by glass capillary column chromatography and ancillary techniques. *J. Chromatogr.* 142: 745–754.

Sigleo, A. C., G. R. Helz, and W. H. Zoller. 1980. Organic-rich colloidal material in estuaries and its alteration by chlorination. *Environ. Sci. Technol.* 14: 673–679.

Smith, J. G., S.-F. Lee, and A. Netzer. 1975. Chlorination in dilute aqueous systems: 2,4,6-trichlorophenol. *Environ. Lett.* 10: 47–52.

Snider, E. H. and F. C. Alley. 1979. Kinetics of the chlorination of biphenyl under conditions of waste treatment processes. *Environ. Sci. Technol.* 13: 1244–1248.

Snoeyink, V. L., R. R. Clark, J. J. McCreary, and W. F. McHie. 1981. Organic compounds produced by the aqueous free-chlorine-activated carbon reaction. *Environ. Sci. Technol.* 15: 188–192.

Somsen, R. A. 1960. Oxidation of some simple organic molecules with aqueous chlorine dioxide. II. Reaction products. *TAPPI* 43: 157–160.

Soper, F. G. and G. F. Smith. 1926. The halogenation of phenols. *J. Chem. Soc.* 1582–1591.

Spanggord, R. J. and C. A. Tyson. 1979. N-Formyliminodiacetic acid, a new compound from the reaction of nitrilotriacetic acid and chlorine. *Science* 204: 1081–1082.

Staehelin, J. and J. Hoigné. 1985. Decomposition of ozone in water in the presence of organic solutes acting as promoters and inhibitors of radical chain reactions. *Environ. Sci. Technol.* 19: 1206–1213.

Stanbro, W. D. and W. D. Smith. 1979. Kinetics and mechanism of the decomposition of N-chloroalanine in aqueous solution. *Environ. Sci. Technol.* 13: 446–451.

Stevens, A. A. 1982. Reaction products of chlorine dioxide. *Environ. Health Perspect.* 46: 101–110.

Stevens, A. A., C. J. Slocum, D. R. Seeger, and G. G. Robeck. 1976. Chlorination of organics in drinking water. In R. L. Jolley, Ed. *Water Chlorination: Environmental Impact and Health Effects*. Vol. 1. (Ann Arbor, MI: Ann Arbor Sci. Pub.), pp. 77–104.

Stevens, A. A., D. R. Seeger, and C. J. Slocum. 1978. Products of chlorine dioxide treatment of organic materials in water. In R. G. Rice and J. A. Cotruvo, Eds. *Ozone—Chlorine Dioxide Oxidation Products of Organic Materials*. (Cleveland, OH: Ozone Press Intl.), pp. 383–399.

Stoward, P. J. 1975. A histochemical study of the apparent deamination of proteins by sodium hypochlorite. *Histochemistry* 45: 213–226.

Streeter, H. W. 1929. Chlorophenol tastes and odors in water supplies of Ohio River cities. *Publ. Health Rep.* 44: 2149–2156.

Streicher, R. P., H. Zimmer, J. P. Bercz, and W. E. Coleman. 1986. The interactions of aqueous solutions of chlorine with citric acid. A source of mutagens. *Anal. Lett.* 19: 681–696.

Sturrock, M. G., E. L. Cline, K. R. Robinson, and K. A. Zercher. 1960. Pyridine carboxylic acids. US Patent 2964529.

Sturrock, M. G., E. L. Cline, and K. R. Robinson. 1963. The ozonation of phenanthrene with water as a participating solvent. *J. Org. Chem.* 28: 2340–2343.

Stuthridge, T. R., A. L. Wilkins, A. G. Longdon, K. L. Mackie, and P. N. Macfarlane. 1990. Identification of novel chlorinated monoterpenes formed during kraft pulp bleaching of *Pinus radiata*. *Environ. Sci. Technol.* 24: 903–908.

Suffet, I. H., L. Brenner, and B. Silver. 1976. Identification of 1,1,1-trichloroacetone (1,1,1-trichloropropanone) in two drinking waters: a known precursor in the haloform reaction. *Environ. Sci. Technol.* 10: 1273–1275.

Symons, J. M., T. A. Bellar, J. K. Carswell, J. Demarco, K. L. Kropp, G. G. Robeck, D. R. Seeger, C. J. Slocum, B. L. Smith, and A. A. Stevens. 1975. National organics reconaissance survey for halogenated organics. *J. Amer. Water Works Assoc.* 67: 634–647.

Tan, H., A. C. Sen, W. B. Wheeler, J. A. Cornell, and C. I. Wei. 1987. A kinetic study of the reaction of aqueous chlorine and chlorine dioxide with amino acids, peptides, and proteins. *J. Food Sci.* 52: 1706–1711.

Taube, H. and H. Dodgen. 1949. Application of radioactive chlorine to the study of the mechanisms of reactions involving changes in the oxidation state of chlorine. *J. Amer. Chem. Soc.* 71: 3330–3336.

Taylor, W. I. and A. R. Battersby. 1967. *Oxidative Coupling of Phenols*. (New York, NY: Marcel Dekker).

Taymaz, K., D. T. Williams, and F. M. Benoit. 1979. Chlorine dioxide oxidation of aromatic hydrocarbons commonly found in water. *Bull. Environ. Contam. Toxicol.* 23: 398–404.

Thakore, A. N. and A. C. Oehlschlager. 1977. Structures of toxic constituents in kraft mill caustic extraction effluents from ^{13}C and ^{1}H nuclear magnetic resonance. *Can. J. Chem.* 55: 3298–3303.

Thibaud, H., J. DeLaat, N. Merlet, and M. Doré. 1987. Formation de chloropicrine en milieu aqueux: influence des nitrites sur la formation de precurseurs par oxydation de composes organiques. *Water Res.* 21: 813–821.

Thielemann, H. 1972a. Über die Einwirkung von Chlordioxid auf einige polycyklische aromatischen Kohlenwasserstoffe. *Mikrochim. Acta* 575–577.

Thielemann, H. 1972b. Effect of chlorine dioxide on 1-and 2-naphthol containing test solutions. *Mikrochim. Acta* 669–671.

Trehy, M. L. and T. I. Bieber. 1981. Detection, identification, and quantitative analysis of dihaloacetonitriles in chlorinated natural waters. In L. H. Keith, Ed. *Advances in the Identification and Analysis of Organic Pollutants in Water.* Vol. 2. (Ann Arbor, MI: Ann Arbor Sci. Pub.), pp. 941–975.

Trehy, M. L., R. A. Yost, and C. J. Miles. 1988. Amino acids as model compounds for halogenated by-products formed on chlorination of natural waters. In Larson, R. A., Ed. *Biohazards of Drinking Water Treatment.* (Chelsea, MI: Lewis Pub.), pp. 133–140.

Uhler, A. D. and J. C. Means. 1985. Reaction of dissolved chlorine with surficial sediment: oxidant demand and production of trihalomethanes. *Environ. Sci. Technol.* 19: 340–344.

Van Buren, J. B. and C. W. Dence. 1970. Chlorination behavior of pine kraft lignin. *TAPPI* 53: 2246–2253.

van Steenderen, R. A., W. E. Scott, and D. I. Welch. 1988. *Microcystis aeruginosa* as an organohalogen precursor. *Water SA* 14: 59–62.

Verweij, H., K. Christianse, and J. van Steveninck. 1982. Different pathways of tyrosine oxidation by ozone. *Chemosphere* 11: 721–725.

Voudrias, E. A., R. A. Larson, and V. L. Snoeyink. 1985a. Effects of activated carbon on the reactions of free chlorine with phenols. *Environ. Sci. Technol.* 19: 441–449.

Voudrias, E. A., R. A. Larson, and V. L. Snoeyink. 1985b. Effects of activated carbon on the reactions of combined chlorine with phenols. *Water Res.* 19: 909–915.

Wachter, J. K. and J. B. Andelman. 1984. Organohalide formation on chlorination of algal extracellular products. *Environ. Sci. Technol.* 18: 811–817.

Wei, C.-I., D. L. Cook, and J. R. Kirk. 1985. Use of chlorine compounds in the food industry. *Food Technol.* 39: 107–115.

Weil, I. W. and J. C. Morris. 1974. Dynamics of breakpoint chlorination. In A. J. Rubin, Ed., *Chemistry of Water Treatment, Supply, and Distribution.* (Ann Arbor, MI: Ann Arbor Sci. Pub.), pp. 297–332.

Whistler, R. L. and R. Schweiger. 1957. Oxidation of amylopectin with hypochlorite at different hydrogen ion concentrations. *J. Amer. Chem. Soc.* 79: 6460–6464.

White, G. C. 1972. *Handbook of Chlorination*. (New York: NY: Van Nostrand Reinhold).

White, J. F., M. C. Taylor, and G. P. Vincent. 1942. The chemistry of chlorites. *Industr. Eng. Chem.* 34: 782–792.

Xie, Y. and D. A. Reckhow. 1992. Formation of ketoacids in ozonated drinking water. *Ozone Sci. Eng.* 14: 269–275.

Yagi, N. and Y. Itokawa. 1979. Cleavage of thiamine by chlorine in tap water. *J. Nutr. Sci. Vitaminol.* 25: 281–287.

Yamada, H. and I. Somiya. 1989. The ozonation of natural water: identification of carbonyl products and their behavior. *Ozone Sci. Eng.* 20: 125–141.

Yamamoto, Y., E. Niki, H. Shiokawa, and Y. Kamiya. 1979. Ozonation of organic compounds. 2. Ozonation of phenol in water. *J. Org. Chem.* 44: 2137–2142.

Zika, R. G., C. A. Moore, L. T. Gidel, and W. J. Cooper. 1985. Sunlight-induced photodecomposition of chlorine dioxide. In *Water chlorination: chemistry, environmental impact, and health effects*, Vol. 5, R. L. Jolley, R. J. Bull, W. P. Davis, S. Katz, M. H. Roberts, Jr., and V. A. Jacobs, Eds., (Chelsea, MI: Lewis Pub.), pp. 1041–1053.

CHAPTER 6

ENVIRONMENTAL PHOTOCHEMISTRY

The uptake of light energy (quanta) by organic compounds may cause subsequent photophysical or photochemical events to occur. Photophysical processes include emission of radiant energy (light or heat), whereas photochemical changes produce new compounds by transformations that include isomerization, bond cleavage, rearrangement, or intermolecular chemical reactions. In the environment, photochemical reactions have been reported to occur in the gas phase (troposphere, stratosphere), aqueous phase (atmospheric aerosols or droplets, surface waters, land-water interfaces) and in the solid phase (plant tissue exteriors, soil and mineral surfaces). All of these possibilities need to be taken into account when fates of organic compounds in nature are considered. Although in many cases the contribution of photochemical processes is negligible, in others photolysis is by far the dominant pathway for loss of a chemical.

A. SUNLIGHT

With the exception of photochemical reactions carried out using artificial sources of light such as mercury lamps, environmental photochemistry is exclusively sunlight-driven; therefore, it is imperative to understand the nature and variability of sunlight.

The sun is an immense ball of gases heated by internal nuclear reactions that produce radiant energy. The sun's electromagnetic radiation is emitted into space in all directions and is constantly bombarding the earth. A continuous input of sunlight energy, to the extent of about 1.3 kW/m^2 (19 kcal/min; the "solar flux"), reaches the earth's upper atmosphere. Approximately 50% of this energy is reflected back into space by the atmosphere; the rest reaches the earth's surface either directly or after intermediate scattering by clouds, smoke, or other particulate materials. Sunlight, as we experience it at the earth's surface, is made up of a mixture of visible and invisible radiation characterized by different wavelengths. The distribution of solar energy by wavelength is charted in Figure 6.1. The maximum intensity occurs at about 500 nm, in the visible region.

Visible light is usually taken to include those wavelengths covering the spectral range of 400–760 nm, whereas shorter wavelengths (290–400 nm) are in the so-called ultraviolet or UV region, and longer wavelengths (760–2000 nm) are considered to be infrared (IR) or heat rays. The earth's atmospheric constituents (especially ozone) are responsible for removing even shorter-wavelength UV radiation and, in addition, they prevent some IR wavelengths (those absorbed by CO_2 and water, especially) from reaching the surface.

Changes in the frequency distribution and intensity of sunlight vary with time of day and season, of course, and also with elevation, ozone concentration, and presence or absence of clouds, smoke, and dust particles in the atmosphere. In theory, clouds should transmit UV radiation to a greater extent than visible light, but the actual relationship is complex and depends to a great extent on cloud thickness. Tropospheric clouds strongly affect photochemical reaction rates, also, by their ability to differentially scavenge water-soluble reactive species from the gas phase, increasing their concentrations by orders of magnitude and also providing a solvent that can facilitate reactions involving polar species (Lelieveld and Crutzen, 1991).

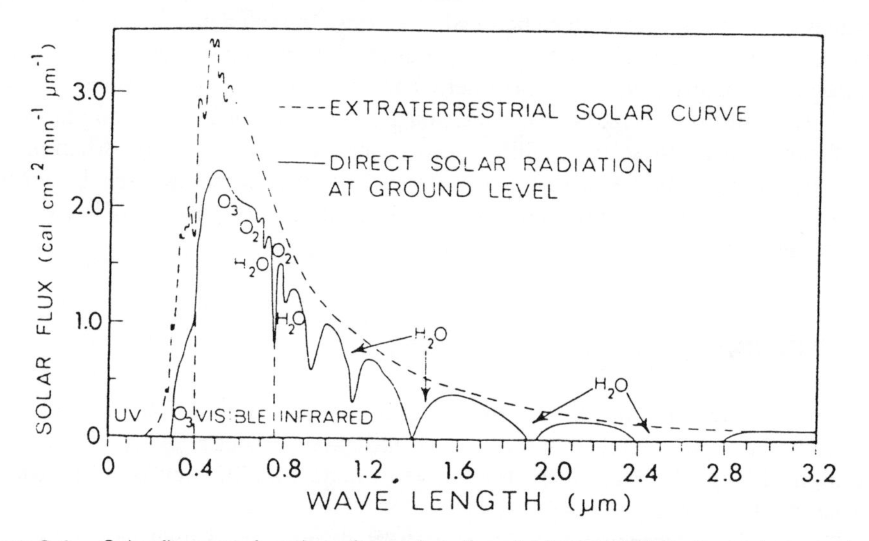

Figure 6.1. Solar flux as a function of wavelength outside the earth's atmosphere and at ground level, showing the atmospheric chemical species responsible for absorption in certain wavelength regions. From R. G. Wetzel, *Limnology* (1975). Reprinted by permission of W. B. Saunders.

Even the distance from the earth to the sun and the level of sunspot activity can affect sunlight, especially the UV wavelengths, significantly.

The concept of the wave-particle duality of light, conceived by Planck and Einstein at the turn of the 20th century, underlies our current understanding of photochemistry. We speak of the wavelength of light as though it was analogous to an oscillating mechanical medium such as water in a trough, and monochromatic light can be understood as having characteristic wave behavior that can be measured on a peak-to-peak basis. Although light does indeed behave in other ways as if it were made up of waves (it can be reflected, diffracted, and focused), it also has many of the characteristics of a chemical reagent or catalyst; it can be measured in stoichiometric terms and used to bring about chemical changes requiring an external source of energy, in almost the same way as a thermal reaction can be caused to occur by heating a reaction mixture. In the latter view, light is assumed to consist of virtually massless, but energy-bearing, particles or packets of energy called photons. The radiation energy associated with a photon varies with the wavelength of light according to the equation

$$E \text{ (in kJ/mol)} = \frac{120{,}000}{\lambda \text{ (in nm)}} \tag{6.1}$$

A "mole" of photons of a given wavelength (Avogadro's number) has the energy calculated by the above expression. This quantity is called an **Einstein**.

In chemical terms, the energies contained in visible and UV photons are comparable to those observed in the covalent bonds of organic molecules. A typical carbon-carbon single bond, for example, has a strength of about 350 kJ/mol, corresponding to photons of wavelengths around 340 nm and thus falling within the solar emission spectrum. However, we do not observe wholesale random bond fission in organic compounds exposed to sunlight because light energy absorption is localized within certain substructures (**chromophores**) and is dissipated via a variety of mechanisms, many of which do not result in chemical alterations. Clearly, however, solar ultraviolet wavelengths are more energetic than visible or IR radiation, and thus they take on importance far in excess of the 4% of the energy they constitute in terms of radiation reaching the ground.

The absorption of light of a given wavelength by a dilute solution of a single chromophoric molecule is described by the Beer-Lambert law,

$$E \text{ or } A = \varepsilon c l \tag{6.2}$$

where E (the extinction) or A (the absorbance) of the solution (also sometimes called its optical density) is equal to the product of the molar concentration, c, of the chromophoric molecule; the path length, l, of the solution; and a constant, ε, called the extinction coefficient, which is characteristic for each molecule. Absorbance is inversely related to transmittance T (transparency) of a solution by the equation,

$$A = \log (1/T) \tag{6.3}$$

Accordingly, a solution with an absorbance of 0 at a given wavelength would be completely transparent (T = 1) to light of that wavelength, whereas a solution with an absorbance of 1 would transmit only 10% of the incident light.

B. CHROMOPHORES AND EXCITED STATES

1. Photophysics of Light Absorption

The initiating event for any photochemical reaction is the absorption of a photon by an organic molecule and the conversion of that molecule to an electronically excited state; that is, a new electronic configuration of the molecule that has a higher potential energy than the lower, or ground, state. In order for the uptake of a photon to occur, the molecule must possess an absorption band in its UV-visible spectrum that includes the wavelength of that photon (that is, the difference between the energy levels must be equal to the quantum energy of the photon). In other words, since sunlight UV photons have a minimum wavelength value of around 290 nm, organic molecules must absorb light to at least some extent above 290 nm in order to be able to undergo photolysis. The lowest energy for electronic excitation of an organic molecule lies near 800 nm, corresponding to light at the fringes of the red/infrared region.

After the absorption of photon energy by a molecule, one of its electrons from a low-lying orbital is promoted to a new, higher-energy, previously unoccupied orbital. Thus, an electronically excited molecule, with its new and different electronic configuration from its precursor, can be looked at as either a distinct isomer of the ground state, or, perhaps more productively, as a completely different substance with its own properties and reactivities.

Singlet and Triplet States

Most stable organic molecules have an even number of electrons that exist in spin-paired configurations; such molecules are said to be **singlets**, since in the presence of an external magnetic field they remain in one distinct state. When such molecules absorb photon energy, the initial form of the excited state is almost always a different, quantum-chemically allowed singlet also. In diagrammatic terms, the process is shown in Figure 6.2. In the figure S^o represents the ground-state singlet and S^1 represents an excited-state singlet resulting from the additional photon energy. Such singlets are usually characterized by extremely short lifetimes* (on the order of a few microseconds or less). Their fates include:

*Defined as the time required for the excited species to decline to $1/e$ (about 37%) of its initial concentration. Often designated by the symbol τ.

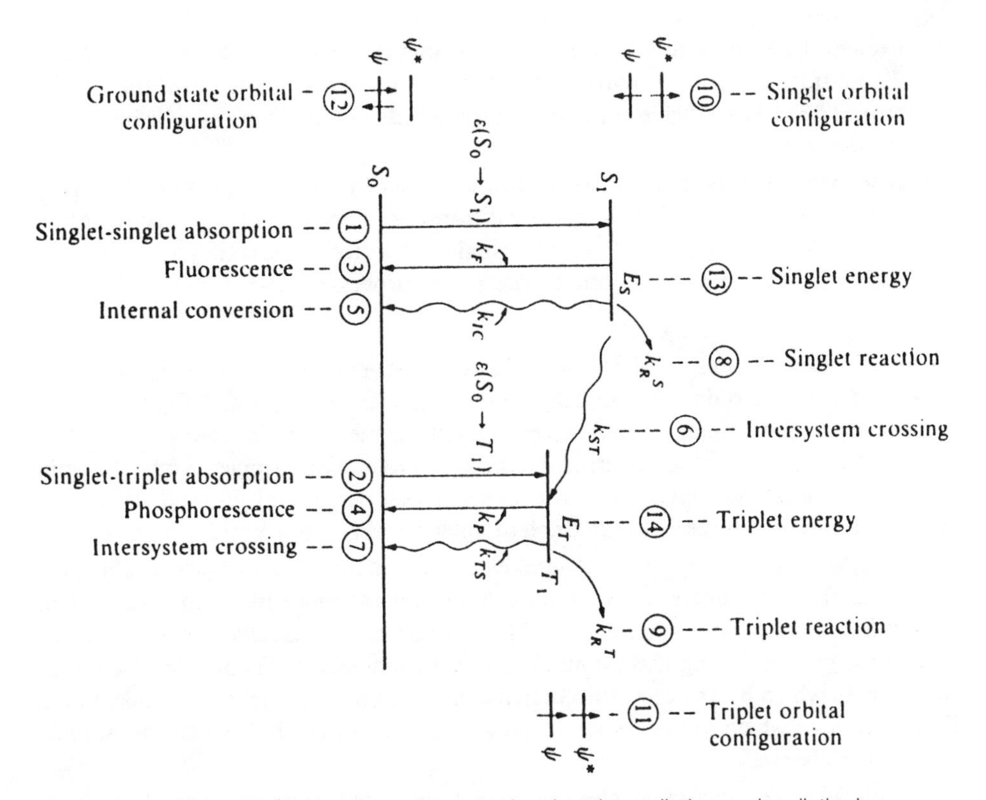

Figure 6.2. Photophysical energy diagram showing the radiative and radiationless interconversions among excited and ground states. 1 = singlet-singlet absorption, 2 = singlet-triplet absorption, 3 = singlet-singlet emission (fluorescence), 4 = triplet-singlet emission (phosphorescence), 5 = radiationless singlet-singlet transition (internal conversion), 6 = radiationless singlet-triplet transition (intersystem crossing), 7 = radiationless triplet-singlet transition, 8 = chemical reaction from the singlet state, 9 = chemical reaction from the triplet state. From Turro (1978). Reprinted by permission of Benjamin-Cummings Pub. Co.

- **Internal conversion,** thermal reversion of the excited singlet to the ground state with the release of heat to surrounding molecules such as the solvent ($S_1 \rightarrow S_0$ + heat). Because these two states are of like multiplicity, the transformation is allowed in terms of quantum theory and is often a very favorable process with a rate constant close to diffusion control
- **Fluorescence,** emission of visible or ultraviolet radiation ($S_1 \rightarrow S_0 + h\nu$). The emitted photon is always of a longer wavelength than the absorbed photon. This is also a quantum mechanically allowed process and usually occurs rapidly.
- **Intersystem crossing,** a quantum-forbidden transition that produces a new

excited state with unpaired electrons, a **triplet** ($S_1 \rightarrow T_1$ + heat: Figure 6.2). These transitions are usually 100 or more times slower than diffusion control. Triplet states have some degree of diradical character, as will be discussed later.

- **Chemical reaction**. Because of the low instantaneous concentrations of most excited singlets, bimolecular chemical transformations involving them (such as reaction with a solvent) are not usually important unless they are kinetically very fast. Normally, one of the above processes predominates.

Excited singlet states (that is, those having spin-paired but orbitally unpaired electrons) can in certain cases undergo intersystem crossing ("ISC," Figure 6.2) to the corresponding triplet excited states. Quantum mechanics predicts that triplet states are always lower in electronic energy content than singlets, but the spin inversion required to convert one to another is theoretically disallowed. However, experimentally it is observed that singlet-triplet conversions are quite common.

The triplet state, having spin-unpaired electrons, differs chemically in its chemical properties from the singlet state. Triplet states are far more likely to take part in chemical reactions than singlets are. Their lifetimes are usually many orders of magnitude greater, giving them a much higher probability of encountering another species with which it can react. Photochemical reactions of triplet species take many forms, some of which will be discussed under the reactions of individual compounds later in the chapter.

A somewhat unproductive photoprocess that usually involves triplet states is quenching, in which a molecule of the solvent or another dissolved species collides with the triplet and returns it to the ground state. In the process the excess energy of the triplet is dissipated thermally or vibrationally; no further photochemistry takes place. Quenchers vary greatly in their quantum efficiency of triplet deactivation. Another common reaction of triplets is triplet-triplet energy transfer, a special type of quenching in which an excited triplet $*T_1$ donates some of its excess excited-state energy to an acceptor in the ground state, A_o, with production of a new triplet, $*A_1$, and the ground state of the donor:

$$*T_1 + A_o \rightarrow S_o + *A_1 \quad (6.4)$$

Such reactions may be photochemically productive if the new triplet possesses sufficient energy to take part in other processes.

Conversion of a triplet to the corresponding ground-state singlet (without the intervention of a quencher) is another quantum-mechanically forbidden and usually inefficient ISC process, called **phosphorescence** when it is accompanied by emission of radiation. Phosphorescence and fluorescence are examples of luminescence phenomena.

Quantum Yield

Chemical reactions of an excited state are in competition, kinetically speaking, with all of the other possible modes of its deactivation. The **quantum yield,** ϕ, of any photochemically induced process is defined as that fraction of the species converted to an excited state that undergoes a particular fate. Thus we may speak, for example, of quantum yields for fluorescence, triplet formation, disappearance of starting material, or formation of a reaction product. A quantum yield of 1 means that every excited-state molecule proceeds to the same outcome; quantum yields of 1 or close to it have been observed, for example, for fluorescence of stilbene derivatives (DeBoer and Schlessinger 1968), but many quantum yields, particularly for chemical reaction, are very small. Therefore, even molecules that absorb sunlight strongly may be surprisingly resistant to photodegradation if other pathways of excited-state deactivation are more favorable (have a higher quantum yield). Methods for determining quantum yields of disappearance are discussed by Zepp (1982) and Leifer (1988).

2. Chromophores

The extent of absorption of UV radiation by any organic compound is related to its molecular structure. Almost all organic compounds that significantly absorb UV and visible wavelengths in the solar spectrum have one or (usually) more double bonds involving carbon, nitrogen, and oxygen. Saturated compounds containing these groups are usually transparent to solar UV. Chromophoric structures thus include alkenes, aromatic and heterocyclic compounds, aldehydes and ketones, nitro compounds, azo dyes, and many others. The absorption spectra of these compounds are usually quite characteristic (Table 6.1).

Electronic transitions in unsaturated molecules are of two principal types, named after the valence electrons that take part in the transition. Electrons that lie in **sigma** orbitals, such as those of single covalent bonds, are not usually involved in environmental photochemistry, since their absorption maxima lie far below the solar UV minimum of 290 nm. However, π electrons (those involved in the multiple bonding of C=C and C=O groups, for example) and some **n** electrons (like the nonbonding pairs associated with the oxygen atoms of carbonyl groups) often undergo excitation processes that can be initiated by sunlight photons. In practically all of the reactions that are important in environmental photochemistry, either π or n electrons undergo promotion to higher-energy, antibonding π^* ("π-star") orbitals.

An n $\rightarrow$ π^* transition can be visualized in a simplified fashion as summarized in Figure 6.3. After photon absorption, one n electron associated with the unshared pairs of the carbonyl oxygen occupies an antibonding π^* orbital. The excited state now contains two spin-paired but orbitally unpaired electrons and accordingly has diradical character. In addition, since the partial negative character of the oxygen atom is decreased by the removal of one of its electrons, the excited state becomes much more electrophilic and therefore much more able to act as an oxidizing agent.

Table 6.1. Typical Chromophoric Values for Selected Organic Groupings

Chromophore	λ_{max} (nm)	ϵ (approx)	Transition
C–H or C–C	<180	1000	$\sigma \rightarrow \sigma^*$
C=C	180	10,000	$\pi \rightarrow \pi^*$
C=C–C=C	220	20,000	$\pi \rightarrow \pi^*$
Benzene	260	200	$\pi \rightarrow \pi^*$
Naphthalene	310	200	$\pi \rightarrow \pi^*$
Anthracene	350	10,000	$\pi \rightarrow \pi^*$
Phenol	275	1500	$\pi \rightarrow \pi^*$
Aniline	290	1500	$\pi \rightarrow \pi^*$
RS–SR	300	300	$n \rightarrow \sigma^*$
C=O	280	20	$n \rightarrow \pi^*$
Benzoquinone	370	500	$n \rightarrow \pi^*$
C=C–C=O	320	50	$n \rightarrow \pi^*$
C=N	<220	20	$n \rightarrow \pi^*$
N=N	350	50	$n \rightarrow \pi^*$
N=O	300	100	$n \rightarrow \pi^*$
Ar–NO_2	280	7000	$\pi \rightarrow \pi^*$
Indole	290	5000	$\pi \rightarrow \pi^*$

Source: Scott (1964) and Turro (1978).

Geometrically, with the destruction of the π bonding orbital, the molecule tends toward a pyramidal rather than a planar configuration. Spectroscopic determinations show also that the C=O bond length changes in the excited state toward that of a C–O single bond.

Similarly, a $\pi \rightarrow \pi^*$ transition entails the orbital unpairing of the electrons of the C–C double bond and the formation of a diradical-like intermediate that is capable of free rotation around the former double bond (Figure 6.4), unless, of course, the new antibonding state is constrained in a ring. A $\pi \rightarrow \pi^*$ transition usually has a higher extinction coefficient (is more probable) than an $n \rightarrow \pi^*$ transition because the π-orbitals of the former transition occupy much the same region in space, whereas the latter transition requires the electron being promoted to move to a new position.

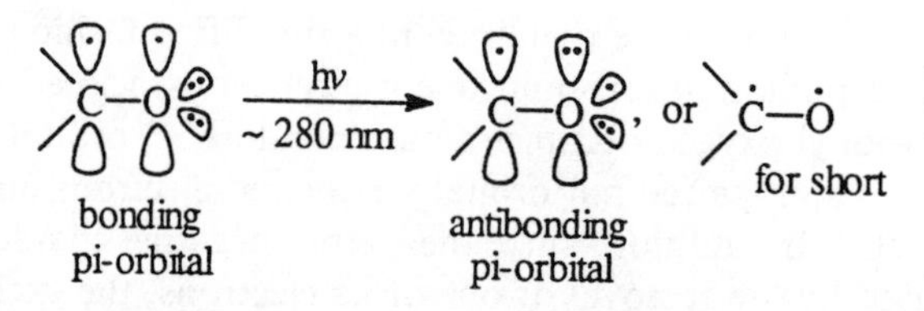

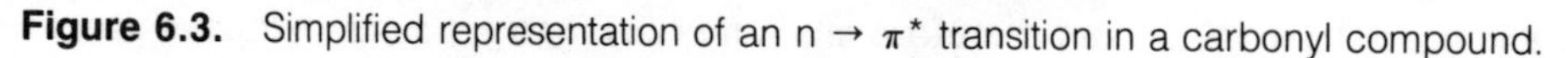
Figure 6.3. Simplified representation of an $n \rightarrow \pi^*$ transition in a carbonyl compound.

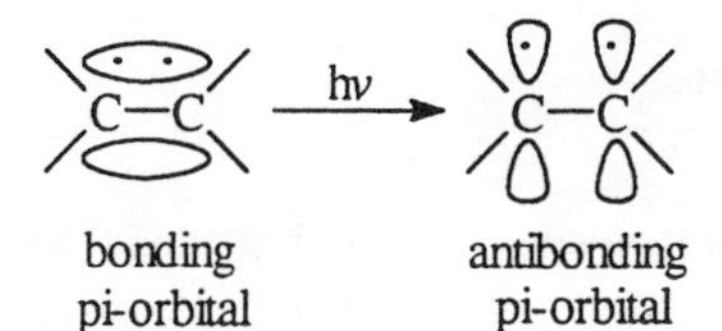

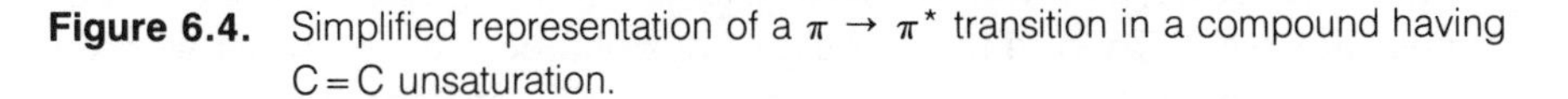
Figure 6.4. Simplified representation of a $\pi \rightarrow \pi^*$ transition in a compound having C = C unsaturation.

C. PHOTOCHEMICAL REACTION PRINCIPLES

Since excited states are so different electronically from ground states, it is not surprising that they have very dissimilar chemistry. The atypical molecular orbital interactions available to the highest occupied orbitals of excited singlets and triplets allow many reaction pathways that are unfavorable in the ground state to proceed readily. In fact, some reactions that are virtually impossible to carry out in the dark occur readily in the excited state.

It should be realized that a photochemically induced reaction may have a multistep mechanism in which, perhaps, only one step may involve light absorption. For example, an excited molecule may transfer an electron to some acceptor molecule in its ground state to produce two odd-electron species. Both these free radicals may then take part in subsequent "dark" reactions. Although it is often stated that photochemistry is relatively insensitive to temperature, this is strictly only correct for the initial, light absorbing step and the rapid internal rearrangements of the excited state. Subsequent processes may be very susceptible to temperature effects.

1. Direct Photolysis

It is a law of photochemistry that only the light that is absorbed by a molecule can bring about a photochemical reaction. Intuitively, it seems obvious that a molecule that is incapable of gaining energy cannot react. Complications occur when the absorbing molecule, after formation, is able to transfer some of its energy or some portion of its structure, such as an electron, to a nonabsorbing species; but many environmentally important compounds do undergo so-called direct photolysis (photochemical reactions that result from the uptake of solar quanta "directly" by a substrate, followed by rearrangements or other reactions of the excited state). These reactions often are kinetically simple and easily modeled, especially if the absorption spectrum of the compound and its quantum yield of disappearance are known or can be measured. Other species are transparent to solar radiation, however, and must of necessity react "indirectly." Many examples of both types of photoreactions will be given later in the chapter.

2. Sensitized Photolysis

In addition to direct photolysis, the possibility of indirect, or sensitized, photolysis must be taken into account whenever a light-absorbing chemical can transfer energy, a hydrogen atom, a proton, or an electron to another species. Early forms of this phenomenon were observed in the field of dye photochemistry, when workers observed that "photosensitizing dyes" were capable of bringing about reactions with other species in solution with them. Photosensitization is now usually defined as the transfer of energy from a photochemically excited molecule to an acceptor, often oxygen, to form a reactive, transient form of oxygen, singlet oxygen ($^1O_2$). Many photosensitizers of both natural and synthetic origin have been described; those of environmental importance include humic materials, tetrapyrroles, flavins, polycyclic aromatic hydrocarbons, and mineral surfaces. The photophysics of photosensitization are interesting in that the triplet excited state of the sensitizer interacts with the singlet ground state of molecular oxygen to produce the singlet excited state. The lowest singlet of oxygen lies only 92 kJ/mol above the ground state; accordingly, sensitizers that absorb in the visible region can be extremely efficient in carrying out the transition. Visible dyes such as methylene blue (λ_{max} 668 nm, E_T 135 kJ/mol) and rose bengal (λ_{max} 549 nm, E_T 175 kJ/mol) are examples of such sensitizers.

The reactions of $^1O_2$ have been discussed in Section 4.A.3. The fate of singlet oxygen in water is predominantly to be quenched back to the ground state by water, with a lifetime of 3–4 μsec (Rodgers and Snowden, 1982). Scully and Hoigné (1987) calculated that, in aqueous medium, only the most reactive compounds toward $^1O_2$ attack could compete with such fast thermal quenching. However, when $^1O_2$ is generated in hydrophobic media such as petroleum surface films or cellular membranes, it is probable that its longer lifetime in such environments, together with the much higher concentration of potentially reactive species, makes it an important factor in the aging or weathering of these materials (Larson et al., 1979: Lichtenthaler et al., 1989).

3. Radical-Producing Photochemical Reactions

Especially in aquatic photochemistry, the photochemical excitation of a chromophore is often followed by electron, hydrogen atom, or proton transfer. This form of indirect photolysis is usually distinguished on theoretical grounds from photosensitization, or energy transfer; but often both processes may be occurring simultaneously, or it may be otherwise difficult to distinguish electron transfer from other mechanistic possibilities. In any event, many well-studied examples of electron transfer are recognized in environmental photochemistry. The transfer of an electron from an excited molecule to oxygen, for example, produces the dioxygen radical anion $\cdot O_2^-$, usually called superoxide.

4. Kinetics

The kinetics of aquatic photolysis reactions have been discussed by Leifer (1988). The rate of a photochemical reaction involving a single light-absorbing species can be expressed as the product of the quantum yield (ϕ) of the reaction and the rate of light absorption (I_a) of the reacting substance. (Quantum yields are usually wavelength-independent if only one chromophore is involved.) I_a basically can be calculated from the overlap between the absorption spectrum of the reactant and the emission spectrum of the light source,

$$\text{rate} = I_a\phi \tag{6.5}$$

such as sunlight. At low reactant concentrations, first-order kinetics are observed since I_a is proportional to concentration (Beer-Lambert Law). An excellent computer program, GCSOLAR, is available for calculating rates of photolysis of organic compounds in sunlight when the absorption spectrum is known and the quantum yield of disappearance is either known or can be estimated (Zepp and Cline, 1977). Laboratory photolysis data derived in distilled water or in water-polar solvent mixtures can usually be used to estimate rates of direct photolysis; often, however, complications occur due to the occurrence of dissolved substances in natural waters that either quench or promote photochemical reactions. Some of these interferences will be discussed later in this chapter.

Quantum yield data for organic chemicals are rather scarce. Selected data are shown in Table 6.2. Clearly, photochemical reactions have the potential to rapidly remove some compounds from water, and even compounds with low reaction quantum yields can disappear rapidly if they absorb sunlight efficiently.

The kinetics of sensitized reactions are complicated by the need to consider not only the rate of light absorption by the sensitizer but also the efficiency of formation

Table 6.2. Disappearance Quantum Yields (ϕ_d) for Selected Environmental Photoreactions

Chemical	ϕ_d
Naphthalene	0.0015
Anthracene	0.0030
Benz[a]anthracene	0.0033
Benzo[a]pyrene	0.00089
Pyrene	0.0022
Naphthacene	0.013
3,3′-Dichlorobenzidine	—
Ferrocyanide complex	0.14
Trifluralin	0.0020
N-Nitrosoatrazine	0.30

Source: Zepp (1982).

of the sensitizer triplet (usually the excited state involved in a sensitized reaction) and the efficiency of energy transfer from the sensitizer to the substrate. Because the steady-state concentrations of sensitizer triplets should, at low concentration, be proportional to sensitizer concentration, rate expressions for sensitized reactions normally have to include sensitizer concentration as a kinetic term. In natural waters, particularly, the concentration of the naturally occurring sensitizers cannot be determined from the absorption spectrum alone. Although the effect of acceptor concentration on triplet energy transfer quantum yield could be variable, at low acceptor concentrations the quantum yield can be assumed to be directly proportional to the concentration of the acceptor. The net result of these assumptions is that, at constant sensitizer concentration, the rate expression is in a first-order form,

$$\text{rate} = k[A] \tag{6.6}$$

where [A] is the acceptor concentration and k is a proportionality constant that includes the light absorption rate and the concentration of the sensitizer as well as triplet quantum yield and triplet energy transfer terms. If the sensitizer concentration changes during the experiment, the rate expressions become much more complex (Larson et al., 1989).

D. ATMOSPHERIC PHOTOCHEMISTRY

Practically all atmospheric chemistry is sunlight-driven, and virtually every important atmospheric reaction is an oxidation. Important chromophores in the lower atmosphere (troposphere) include nitrogen oxides, organic substances such as aromatic hydrocarbons and carbonyl compounds, and complexed metal ions. These compounds are present only in very low amounts in unpolluted regions, but are much more abundant in industrial and urbanized areas. In the upper atmosphere (stratosphere), light uptake by ozone, oxygen, and organohalogens also become significant on a global basis. The principal oxidative reactants in the troposphere and stratosphere are hydroxyl radicals (•OH), hydroperoxyl radicals (•OOH) and ozone (O_3). The formation of these species and their reactions with organic compounds are analyzed in detail in Chapter 4.

E. NATURAL WATER PHOTOCHEMISTRY

The intensity of light at a given depth in a water column depends on three factors—the degree of light transmittance through the atmosphere, the transmittance of the air-water interfacial region, and the optical characteristics of the water body. Light can theoretically penetrate into pure water to relatively great depths. The absorption spectrum of water (Figure 6.5) reveals that transmittance is greatest at wavelengths in the blue region of the visible (450–500 nm), which can be detected in clear mid-ocean seawater at depths in excess of 140 m. ("Water is really a blue, not

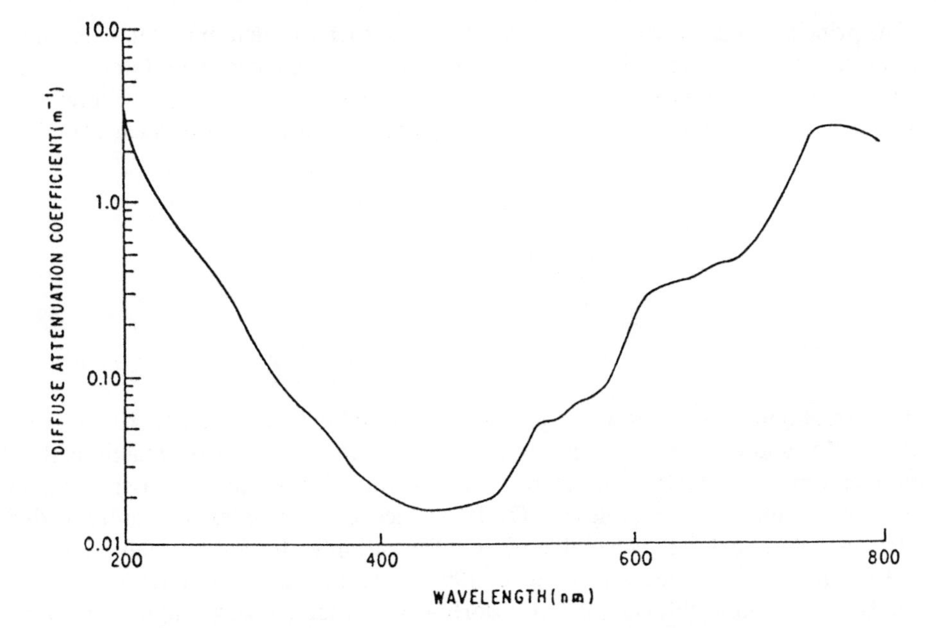

Figure 6.5. Transmittance of sunlight through pure water, showing the maximal penetration of wavelengths in the 400–500 nm region. From Weiner and Goldberg (1985). Reprinted by permission of Gordon and Breach Science Publishers.

a colourless, liquid:" Kirk, 1977.) All natural waters, however, contain dissolved and suspended materials that limit the irradiance detectable below the surface. A model devised by Baker and Smith (1982) for marine waters includes absorbance contributions from water, humic materials, chlorophyll derivatives, and inorganic species; it accounts quite well with field observations of light intensities at different ocean depths. Fresh waters are intrinsically more difficult to model because of their higher and more variable concentrations of organic compounds; qualitatively, based on the few measurements available, it can be safely assumed that except for exceptionally clear and oligotrophic lakes, light penetration, especially in the solar UV region, is greatly reduced relative to marine waters (Calkins, 1975; Kirk, 1977).

1. Inorganic Chromophores

Most of the abundant anions and cations in both natural and marine waters—chloride, bromide, carbonate, sulfate, sodium, potassium, calcium, etc.—are transparent to solar radiation. Only a few trace metal cations, nitrite, and nitrate show any absorbance at all and their contribution to the total extinction is usually negligible in comparison to the organic species.

The principal divalent cations of natural waters, such as calcium and magnesium, however, take part in complexation reactions that may enhance the photodegradability of some compounds; for example, Landymore and Antia (1978) attributed the increased rate of photolysis of pteridines such as leucopterin (**1**) to such complex

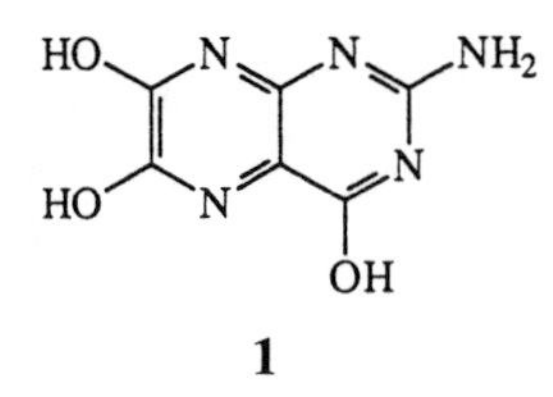

1

formation. Some complexes of ferric iron absorb sunlight significantly (Knight and Sylva, 1975) and can take part in a number of one-electron-transfer reactions that could lead to the formation of reactive free radicals (Balzani and Carassiti, 1970). Thus, for example, the complex $Fe(OH)^{2+}$, which predominates in slightly acidic waters at around pH 4–5, is capable of HO· formation with a rather high quantum yield (Equation 6.7; Faust and Hoigné, 1990). Complexes of iron(III) with other ligands, for example chloride, often absorb at much longer wavelengths and produce other radicals, such as ·Cl, that can induce vinyl polymerization and take part in other radical-initiated reactions (Rabek, 1968).

$$Fe(OH)^{2+} \xrightarrow{h\nu} Fe^{2+} + HO\cdot \quad (6.7)$$

Both nitrate (λ_{max} 303, ε = 7) and nitrite (λ_{max} 355, ε = 22) absorb solar UV and undergo homolysis to produce free radicals via Equations 6.8 to 6.10 (Wagner et al., 1980; Zafiriou, 1983; Zepp et al., 1987a). It seems probable that nitrate photolysis is the dominant process for hydroxyl radical formation in typical freshwaters (Haag and Hoigné, 1985).

$$NO_3^- + H_2O + h\nu \rightarrow \cdot NO_2 + HO\cdot + OH^- \quad (6.8)$$

$$\rightarrow NO_2^- + \cdot O\cdot \quad (6.9)$$

$$NO_2^- + H_2O + h\nu \rightarrow \cdot NO + HO\cdot + OH^- \quad (6.10)$$

The quantum yield for ·OH production at 313 nm in lake water was about 0.015 (Zepp et al., 1987a). Consistent with the production of ·OH, the radicals derived from nitrite and nitrate photolysis readily take part in organic reactions; a variety of compounds are photolyzed at faster rates in the presence of these ions (Kotzias et al., 1982). Biphenyl in aqueous nitrate solution was converted by photolysis (using 254 nm light) to nitrobiphenyl ethers (Suzuki et al., 1985). Other polycyclic aromatic compounds were converted to nitro products in aqueous nitrite solution with longer-wavelength (Pyrex-filtered) light, however (Suzuki et al., 1987). In the presence of nitrite and air, naphthols were photolyzed to mixtures that included mutagenic compounds. The reaction mixtures included naphthoquinones, nitroso and nitronaphthols, and isocoumarin (Suzuki et al., 1988). Naphthoxy radicals were proposed as initial intermediates (Figure 6.6).

Particulate matter such as sediment particles and microorganisms suspended in a water column may scatter incident light, greatly reducing penetration of light beneath the surface. Photochemical processes may still occur in turbid waters, how-

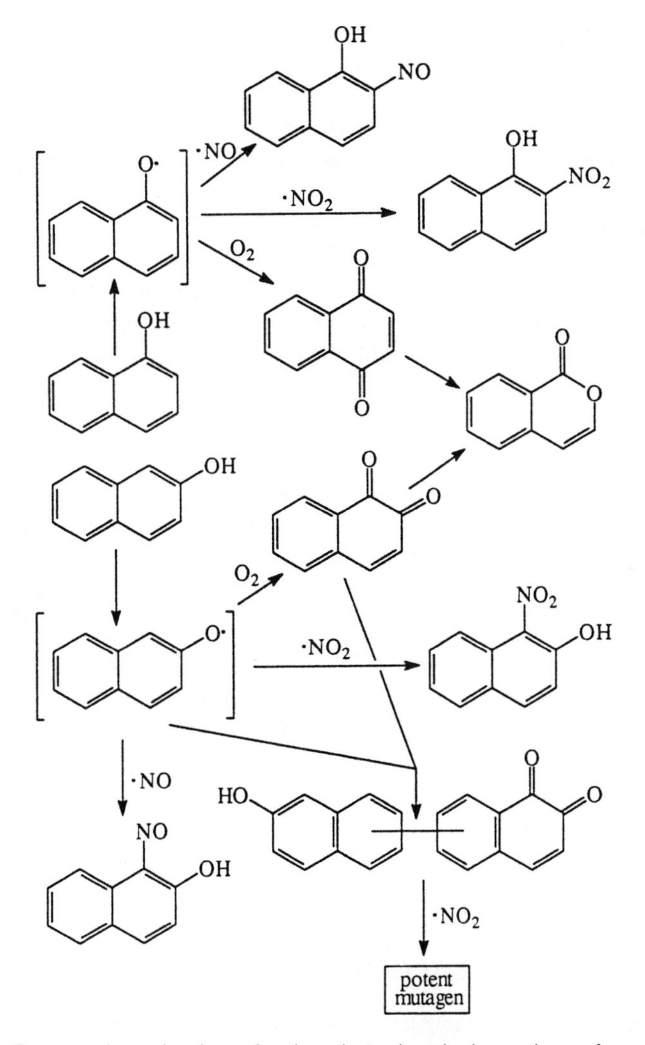

Figure 6.6. Proposed mechanisms for the photochemical reactions of α– and β–naphthol in the presence of nitrite ion and oxygen. From Suzuki et al. (1988). Reprinted by permission of the Pharmaceutical Society of Japan.

ever, if the transmittance of the particles permits passage of light in some regions of the spectrum. Miller and Zepp (1979), for example, demonstrated that UV-transparent clays actually increased the rate of photolysis of a butyrophenone derivative (Figure 6.7). The increased rates were attributed to longer UV pathlengths due to light scattering by the clay particles. In other cases, sorption protects substrates from photolysis, possibly by competitive light attenuation, by migration of the substrate into regions of the particle where light does not penetrate, or by quenching of the excited states of substrates by constituents of the particles (Hautala, 1978).

2. Organic Chromophores

The principal light-absorbing component of both marine and freshwaters is dissolved organic matter. The visibly colored fraction of this heterogeneous composite approximately corresponds to the humic and fulvic acid fraction ("aquatic humus"). A typical absorption spectrum for this material (Figure 6.8) manifests a rising, almost monotonic absorbance, extending from the visible into and beyond the solar UV region. As such, it is an important filter for UV wavelengths near the water surface. Zepp et al. (1981a,b) have recorded the optical properties for a variety of natural waters and have calculated that, when adjusted to the same optical density, samples are remarkably constant in their photosensitizing efficiencies. Absorption

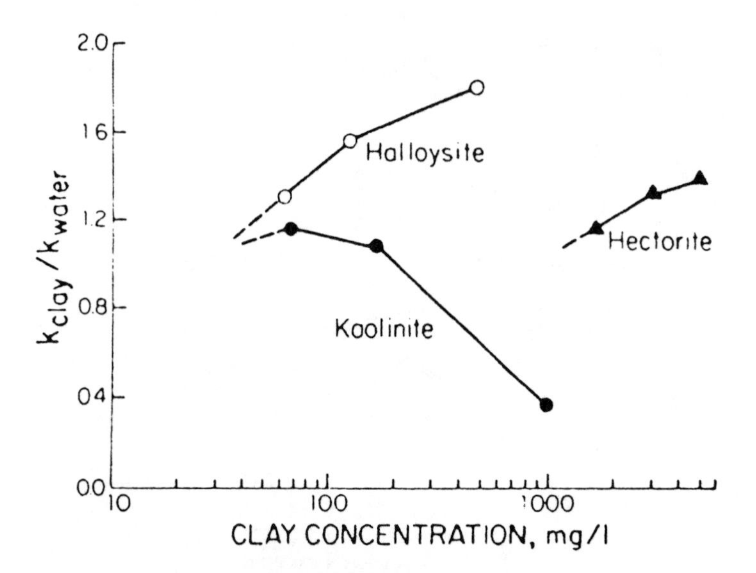

Figure 6.7. Relative rate constants for photodegradation of a butyrophenone derivative in the presence of varying amounts of UV-transparent (halloysite, hectorite) and UV-absorbing (kaolinite) clay minerals. From Miller and Zepp (1979). Reprinted by permission of Pergamon Press Ltd.

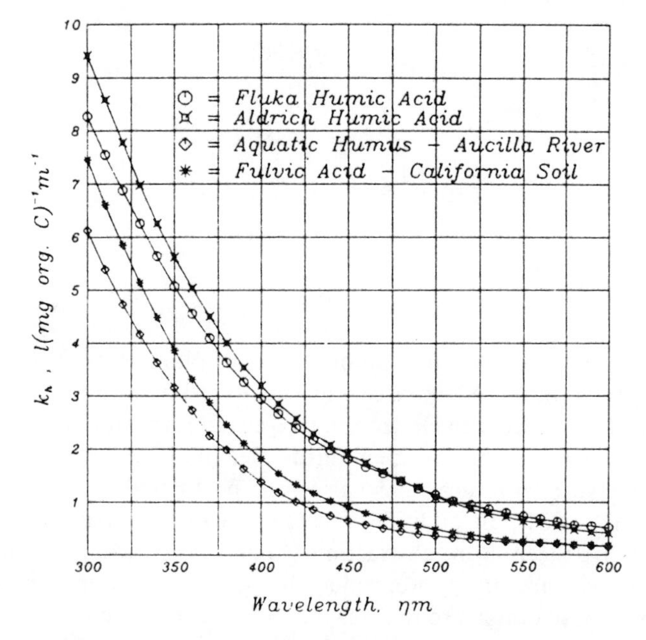

Figure 6.8. UV-visible absorption spectra (at pH 6.0) of some commercial and naturally occurring humic materials. From Zepp et al. (1981a). Reprinted by permission of Pergamon Press Ltd.

of light by natural organic matter may greatly affect the rates of either direct or indirect photolysis of a dissolved compound. Direct photochemical reactions can be slowed by the shielding effect, and sensitized reactions may be promoted by indirect light-induced reactions of the natural dissolved materials (Zepp et al., 1977: Ross and Crosby, 1985). In highly colored waters, the photic zone may be only a few centimeters in depth, but in this relatively narrow region, the high concentration of humic chromophores may be promoting sensitized reactions at accelerated rates.

In a few eutrophic waters, absorbance due to pigments derived from algae or other microorganisms may become consequential. Chlorophylls and their degradation products are readily detectable because of their characteristic absorbance and fluorescence spectra. For example, chlorins (magnesium-free, tetrapyrrolic derivatives of chlorophyll) were detected in Mackenzie River (Canada) water at concentrations of up to 100 ng/L (Peake et al., 1972) and in other, more contaminated waters these classes of compounds may occur at higher levels.

Riboflavin (**2**) and some related flavins are pigments of potential importance for aquatic photochemistry that exist at low concentrations in natural waters. These compounds have broad, moderately strong absorption bands in the visible region at about 440 nm and show high quantum yields for triplet formation (Heelis, 1982).

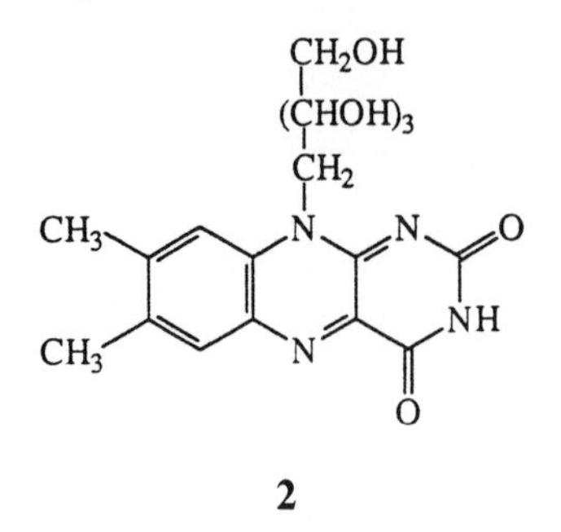

2

Flavin excited states take part in a complex array of photophysical and photochemical processes. In addition to fluorescing efficiently, riboflavin produces singlet oxygen ($^1O_2$) and also transfers an electron to oxygen, to form superoxide radical ($\cdot O_2^-$), when irradiated with visible or solar ultraviolet light in the presence of oxygen (Joshi, 1985). Hydroxyl radicals have also been postulated in some of its reactions (Ishimitsu et al., 1985). In addition, excited flavins take part in energy-transfer processes and hydrogen atom abstraction (Chan, 1977). Mopper and Zika (1987) detected riboflavin and its photoproduct lumichrome at approximately 200 pM concentrations in seawater from Biscayne Bay, Florida and implicated them in abiotic aquatic redox processes such as the photogeneration of hydrogen peroxide, the photooxygenation of amino acids (particularly methionine), and the formation of organic free radicals that can undergo further oxidation by molecular oxygen.

Riboflavin undergoes quite rapid direct photolysis, with full or partial loss of the ribityl side chain, probably due to self-sensitized photooxidation (Smith and Metzler, 1963). Riboflavin esters, such as tetraacetylriboflavin, are much more photochemically stable (Larson et al., 1992a).

In general, it has been found that flavins are photochemically reactive toward electron-rich aromatic compounds. Thus, in reactions of substituted phenylalanines, electron-donating substituents increase the rate of photodecomposition, whereas electron-withdrawing groups deactivate the ring to attack (Rizzuto et al., 1986). Larson et al. (1989) showed that riboflavin, when added at 5 μM to solutions of phenols or anilines, greatly accelerated the rate of their loss in the presence of light from a Pyrex-filtered mercury lamp. For example, aniline had a half-life in direct photolysis of 180 min under these conditions; added riboflavin, however, decreased the first half-life to less than 2 min. The sensitized photolysis rates increased further in the absence of oxygen, suggesting a mechanism involving direct energy or electron transfer between flavin excited states and target molecules.

Riboflavin has been shown to react with dissolved organic matter in filtered seawater to produce carbonyl compounds such as formaldehyde, glyoxal, glyoxylic acid and pyruvic acid (Mopper and Zika, 1987). Berends and Posthuma (1962) showed that riboflavin photodegraded the polyene fungicide, pimaricin (**3**) in aqueous solution. The reaction proceeded more rapidly in the absence of oxygen and was quenched by paramagnetic metal ions such as Cu^{2+} or Co^{3+}, suggesting a triplet energy transfer mechanism.

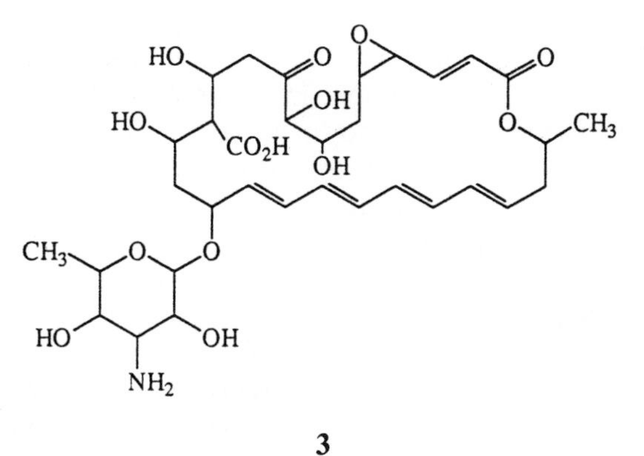

3

F. INTERFACIAL PHOTOCHEMISTRY

1. The Air-Water Interface

Natural Surface Films

Organic surface layers (microlayers) have been demonstrated to exist on most natural water bodies. Many classes of compounds have been found to be enriched at this interface, including those with surface-active (detergent-like) properties (see Section 1.B.2d). The environment at the surface therefore tends to resemble an organic solvent or a mixed water-solvent environment rather than the more "purely aqueous" character typical of water at depth. The chemical constituents of microlayers are exposed to more photochemically active radiation than are compounds in the water column, and because of the higher concentrations of organic materials in this region, reactions of all kinds may be kinetically favored. The constituents of surface layers are in a state of flux as the input of new material is balanced by microbial degradation and photodecomposition reactions.

Many studies have shown that unsaturated fatty acids, common constituents of aquatic organisms, are unstable to sunlight in a surface environment. For example, Wheeler (1972) showed in an early investigation that soft glass-filtered (>320 nm) wavelengths promoted the formation of oxidation products from linoleic acid (**4**), linolenic acid (**5**), and the lipid extract from the diatom *Thalassiosira fluviatilis*. The

$$CH_3(CH_2)_4CH{=}CH{-}CH_2{-}CH{=}CH(CH_2)_7CO_2H$$

4

$$CH_3CH_2CH{=}CH{-}CH_2{-}CH{=}CH{-}CH_2{-}CH{=}CH{-}(CH_2)_7CO_2H$$

5

mechanisms of transformation of lipids in surface films have not been fully elucidated. A possible function for metal ions in the process has been indicated by the experiments of Morita et al. (1976), who showed that iron-containing catalysts such as heme or Fe^{2+} greatly accelerated methyl linoleate photooxidation. The authors postulated that the catalysts reacted with trace quantitites of hydroperoxides at the lipid-water interface, supplying radicals that could take part in chain reactions involving molecular oxygen. A potential role for ozone in surface microlayer destruction is suggested by the experiments of Srisankar and Patterson (1979), who demonstrated the rapid collapse of oleic or linoleic acid monolayers exposed to low (0.03 ppm) levels of ozonated air. The rate of destruction did not depend on whether an O_2 or N_2 atmosphere was used, indicating that oxidative chain reactions were probably much less important than direct attack of ozone on double bonds.

Effects of microlayer constituents or other surface-active agents on photochemical reactions in aqueous media have not been widely studied. Zadelis and Simmons (1983) demonstrated that the photodecomposition of naphthalene in Lake Michigan microlayer material was slower by a factor of about 2 relative to that in lake water. Larson and Rounds (1987) also showed that a 30 mM concentration of a surface-active material (sodium dodecyl sulfate) significantly decreased the photolysis rate of 1-naphthol at pH 7. In contrast, Epling et al. (1988) showed that borohydride-promoted photodechlorination of polychlorinated biphenyls was significantly increased in the presence of ca. 100 mM concentrations of the surface-active agents Brij-58® (a polyethoxyethanol) and sodium dioctyl succinate. The mechanisms for these observed rate effects are unknown.

Oil Spills

Photochemical aging or weathering of petroleum on a water surface leads to great changes in concentration of some oxygen-containing and water-soluble products. The subject has been reviewed by Payne and Phillips (1985). For example, the low-boiling fraction of a Libyan crude oil, when exposed to short-wave UV, was principally converted to a mixture of carboxylic acids, including alkylated benzoic, salicylic, and phthalic acids (Hansen, 1975). A distillate fuel oil, however, when irradiated with longer-wavelength UV under conditions designed to simulate solar radiation, did not afford as much acidic material; the major products identified were phenols and peroxides. The phenols formed by irradiation were high-boiling, highly alkylated one-, two-, and three-ring compounds which were almost absent from the starting oil; the peroxides included alkylated derivatives of tetrahydronaphthalene-1-hydroperoxide (Larson et al., 1977, 1979).

The mechanism of photooxidation of petroleum probably includes significant contributions from both radical and singlet oxygen pathways. Petroleum contains polycyclic aromatic hydrocarbons and other compounds with known photosensitizing activity; Lichtenthaler et al. (1989) demonstrated quantum yields for $^1O_2$ formation ranging from 0.5 to 0.8 for a number of crude oils. The photooxidation of a #2 fuel oil was inhibited by β-carotene, an excellent quencher of $^1O_2$ and free radicals, at concentrations $>10^{-4}$ M (Larson and Hunt, 1978). A mechanism incorporating both $^1O_2$ and radical steps (Larson et al., 1979) is illustrated in Figure 6.9. In this mechanism, singlet oxygen, generated by the interactions of triplet oxygen and photoexcited triplet species in the oil, reacts with an efficient acceptor such as 1,2-dimethylindene to form a hydroperoxide. Direct or sensitized photolysis would afford alkoxy or hydroxyl radicals, or both, which could go on to attack reactive hydrocarbons such as substituted tetrahydronaphthalenes; these intermediate carbon-centered radicals would be likely to react with oxygen to form peroxy radi-

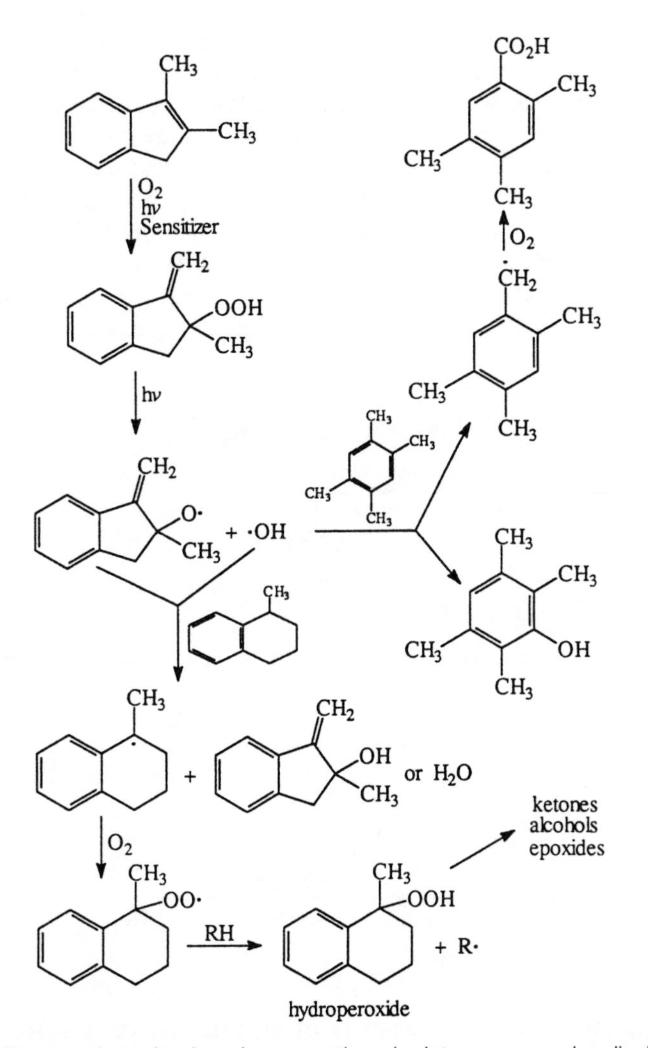

Figure 6.9. Proposed mechanism, incorporating singlet oxygen and radical steps, for the photodecomposition of a refined petroleum containing alkylbenzenes, indenes, and tetrahydronaphthalenes. From Larson et al. (1979). Reprinted by permission. Copyright 1979 American Chemical Society.

cals, which would be expected to be efficient chain carriers in a typical autooxidation (see Section 4.A.1). Termination steps would lead straightforwardly to the observed carboxylic acids, phenols, carbonyl compounds, and alcohols.

2. Solid-Water and Solid-Air Interfaces

Soils and Mineral Boundaries

Phototransformations in the presence of soils and clays have been reported by several authors. The penetration of light into soils and sediments is greatly inhibited below the first 0.2 mm or so (Hebert and Miller, 1989), but often surface photoreactions can proceed at significant rates. For example, sediment-sorbed DDE (**6**) photolyzes faster than when it is dissolved, and the product mixture differs in a way that suggests an environment rich in H-donors, probably organic matter (Miller and Zepp, 1979; Zafiriou et al., 1984). Parathion (**7**) has been shown to be rapidly photolyzed on the surfaces of dust and soil particles to paraoxon (**8**), a more toxic substance (Spencer et al., 1980). The pyrethroid insecticide fenpropathrin (**9**) was rapidly photolyzed ($t_{1/2}$ about 1 day) on soils high in organic carbon (Takahashi et al., 1985).

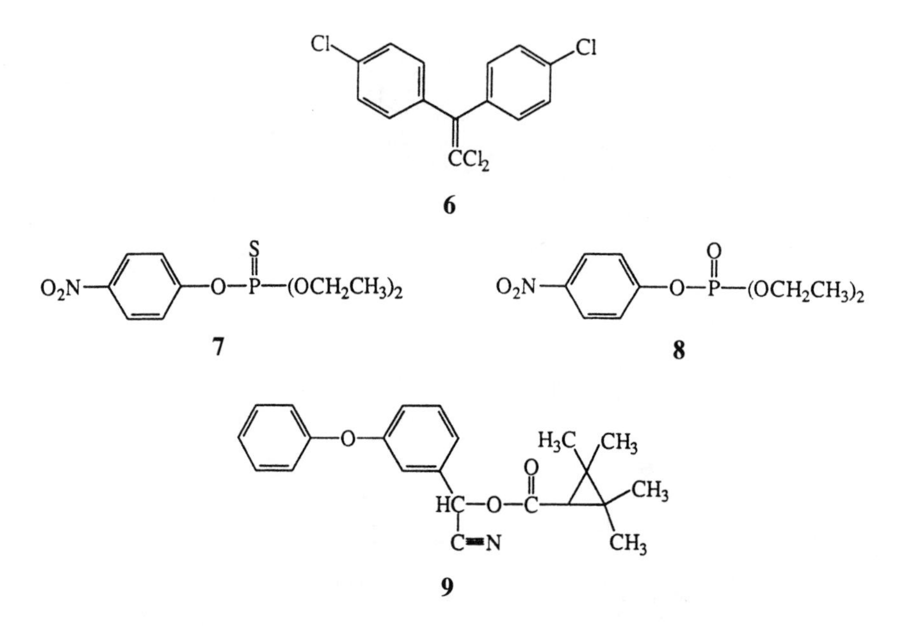

The photophysics and photochemistry of organic molecules adsorbed on some mineral surfaces have been reviewed by Oelkrug et al. (1986). The mobility of adsorbed substrates is usually restricted, and therefore photochemical transformations that require bond shifts or group rotations may be impeded; on the other hand, if concentrations are high enough the closer proximity of adsorbed molecules may facilitate intermolecular reactions. In many cases, the electronic structures, absorption and emission spectra, and excited state lifetimes of sorbed compounds are much different from their solution properties, making it very difficult to predict what effects may result from sorption. Leermakers and Thomas (1965) and Nicholls and Leermakers (1971) showed that aromatic hydrocarbons adsorbed on silica exhibited red shifts, relative to their spectra in solution, for their $\pi \rightarrow \pi^*$ bands; carbonyl compounds, in contrast, displayed pronounced $n \rightarrow \pi^*$ blue shifts. Nitrogen-containing compounds often show different spectra when sorbed on clay minerals compared to solution spectra; sometimes this is due to protonation-deprotonation effects, and at other times to poorly understood charge-transfer phenomena or excited-state energy alterations (Bailey and Karickhoff, 1973). Triplet lifetimes for many aromatic compounds are greatly extended when they are adsorbed on silica or alumina, which means they may be much more susceptible to photochemical reactions (Oelkrug et al., 1986) Also, the fluorescence of many compounds is almost completely quenched by adsorption on some solids such as TiO_2 (Chandrasekaran and Thomas, 1985).

Particles suspended in water are often responsible for a major fraction of the light absorption in near-surface waters. Furthermore, many particles are at least somewhat more hydrophobic than the surrounding water phase and may act as an efficient scavenging and concentrating mechanism for nonpolar organic matter. Frequently, these particles merely scatter and attenuate the incident light, reducing the importance of photochemical processes. In other cases, however, photochemical reactions occurring on the particle surface can be very favorable. Semiconductor photochemistry, for example, may be important in waters containing metal oxides such as ZnO, MnO_2, or Fe_2O_3 (hematite) that absorb sunlight wavelengths. Semiconductor oxides, when irradiated with light of a wavelength having greater energy than their band gap energy, emit an electron to the conduction band, leaving a positively charged site ("hole") behind. If the rate of recombination of the electron-hole pair is slow enough, the electron can be transferred to a suitable acceptor, causing its reduction ($O_2 \rightarrow \cdot O_2^-$, for example) and the hole may oxidize suitable substrates by accepting an electron from them, or by reaction of the hole with water to produce $\cdot OH$. Hydrogen peroxide could theoretically be formed either by reduction of oxygen by conduction-band electrons or oxidation of water by holes; it has been shown to be formed when metal oxides (ZnO, TiO_2) or desert sand samples were illuminated with a sunlight-simulating xenon arc (Kormann et al., 1988).

The band gap energy for hematite corresponds to a wavelength of 530 nm. Colloidal solutions of hematite have been shown to efficiently oxidize iodide to iodine at 347 nm (Moser and Grätzel, 1982) and also to oxidize sulfides to sulfones and sulfoxides (Faust and Hoffmann, 1986). Carboxylic acid anions such as tartrate, citrate, formate, etc., when sorbed onto colloidal goethite (a ferric oxyhydroxide), underwent photodecarboxylation. This was not a semiconductor-initiated reaction, but rather an electron transfer from the carboxylate anion to the iron(III) cation. The resulting carboxylate radical, RCOO·, eliminates CO_2 and leaves behind an organic free radical that is capable of further Fe(III) reduction (Weiner and Goldberg, 1985; Cunningham et al., 1988). A similar mechanism was postulated in experiments which showed the production of formaldehyde from ethylene glycol on goethite (Cunningham et al., 1985). Photodecarboxylation of EDTA (**10**) and NTA (**11**) complexes of iron and copper has been demonstrated to occur in solution experiments (Trott et al., 1972; Langford et al., 1973; Lockhart and Blakely, 1975).

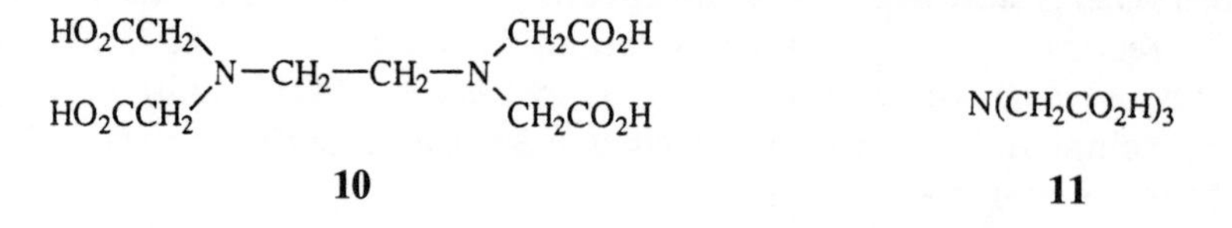

Zinc oxide (band gap equivalent to 370-nm light) is capable of photooxidizing alkanes to alkanones (Giannotti et al., 1983) and of photodegrading the herbicides molinate (**12**) and thiobencarb (**13**) in agricultural wastewaters (Draper and Crosby, 1987; Figure 6.10). The use of ZnO in the latter instance is especially promising since the oxide is used in fertilizers as a source of zinc for crop nutrition. ZnO produced

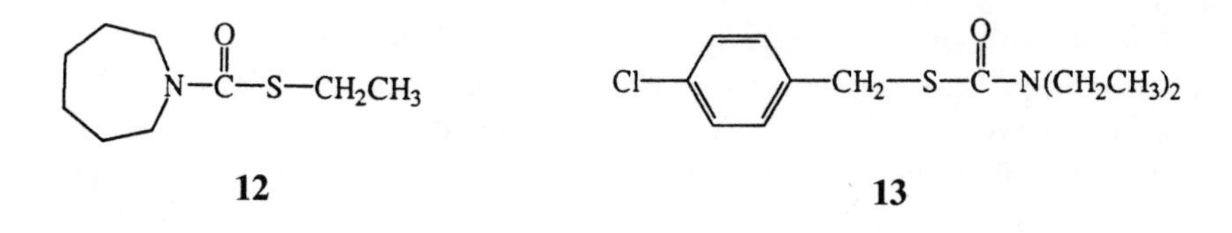

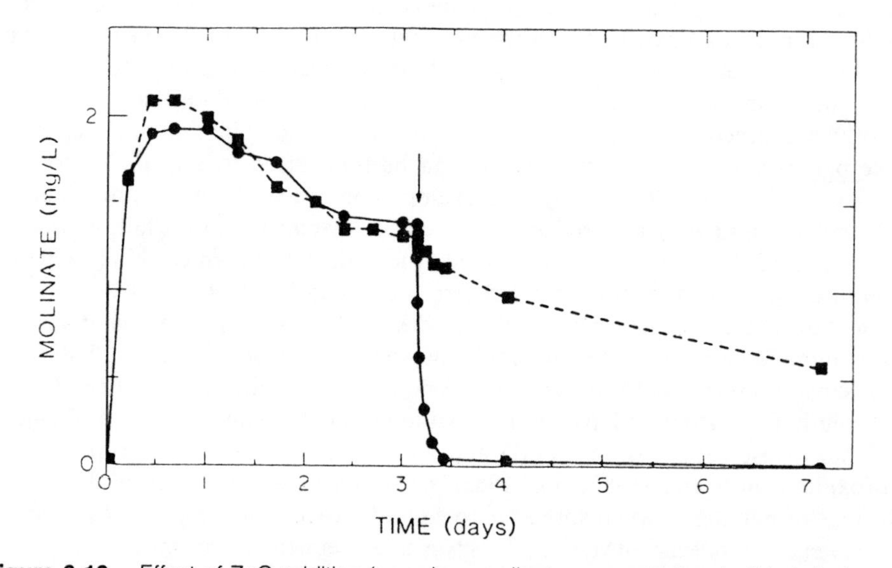

Figure 6.10. Effect of ZnO addition (arrow) on molinate concentration (solid circles). Solid squares show concentrations in untreated controls. From Draper and Crosby (1987). Reprinted by permission. Copyright 1987 American Chemical Society.

both hydrogen peroxide and (in the presence of isopropanol and acetate) organic hydroperoxides when illuminated with a xenon lamp (Kormann et al., 1988).

Although titanium dioxide (band gap equivalent to ca. 400-nm light) is not an abundant constituent of most soils, it has been widely studied as a photooxidation catalyst (Ollis et al., 1991). It is a very potent semiconductor and electron-transfer agent and has the useful property of being stable to prolonged irradiation. A classic paper by Carey et al. (1976) reported that PCBs were oxidatively dechlorinated by TiO_2 in the presence of 365-nm light absorbed by this semiconductor. Complete loss of the PCB mixture was noted after 30 min irradiation, and no products except Cl^- (in a yield of 80%) could be detected in the solution or on the surface of the oxide. After some years of neglect, the research area was revitalized in the early '80s with several independent discoveries that the oxide could generate $\cdot OH$ radicals ("photo-Fenton reaction") by way of electron transfer to O_2, forming $\cdot O_2^-$, which then went on to transient H_2O_2 and finally $\cdot OH$ (Fujihara et al., 1981; Matthews, 1983; Harbour et al., 1985). TiO_2 is an extremely potent oxidant because of this characteristic, and can convert, for example, cyclohexane to cyclohexanone (Giannotti et al., 1983), and even mineralize many organic compounds all the way to CO_2 in fair-to-excellent yields (Matthews, 1986; Ollis et al., 1991).

Other reaction pathways are also available to naturally occurring oxides. Humic materials appear to photoreduce colloidal suspensions of iron(III) oxide and manganese (IV) oxide, leading to increases in iron(II) and manganese(II) concentrations in natural waters (Sunda et al., 1983; Finden et al., 1984; Waite, 1985; Waite et al., 1988; Bertino and Zepp, 1991). In the presence of light, partial mineralization of several chlorinated organic compounds to CO_2 and Cl^- was observed using silica gel or desert sand as a support (Mansour and Parlar, 1978; Schmitzer et al., 1980). Photocatalysis of the oxidation of natural organic matter sorbed on the surface of a California beach sand containing small amounts of titanium and iron oxides has also been noted (Sancier and Wise, 1981). Death Valley, California, sand was shown to produce hydrogen peroxide when illuminated (Kormann et al., 1988) in the presence of acetate as a "hole scavenger."

Surfaces of Organisms

When pesticides are sprayed onto plants, they first contact their outer (cuticular) surfaces and may persist in this region for some time. Plant cuticles are very hydrophobic environments whose constituents include long-chain alkanes, fatty alcohols and acids, sterols, wax esters, and partly characterized polymers such as cutin. The effects of plant surfaces on photochemical reactions are not very well understood. Plant tissues may strongly attenuate light below their surfaces due to the presence of light-scattering waxy layers, as well as light-absorbing constitutents such as phenolic compounds; however, leaves may also contain compounds capable of sensitizing photochemical reactions, such as pigments and alkaloids (Tuveson et al., 1989). Miller and Zepp (1983) showed that the photolysis rate for the herbicide flumetralin (**14**) was diminished when the compound was illuminated on leaf surfaces relative to its rate in solution or as a film on a glass plate.

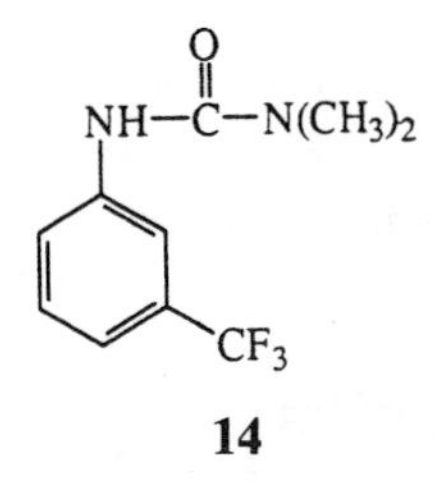

Schwack (1988) reported that DDT (**15**) and methoxychlor (**16**), when exposed to light, added to the double bond of the fatty ester, methyl oleate, to form products such as **17**, possibly explaining some of the ways by which pesticides may become "bound" to plant tissues. Berenbaum and Larson (1988) showed that singlet oxygen was produced at the surface of some leaves containing phototoxic compounds and that it could migrate for short distances in air. The authors suggested that the production of $^1O_2$ by the leaves might be a method of defense against leaf-feeding insects or other predators.

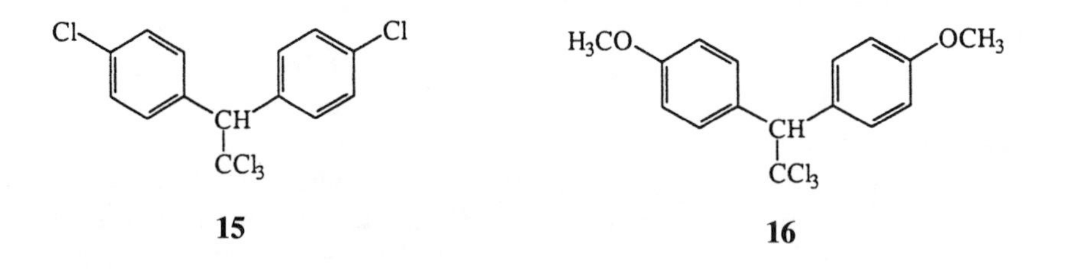

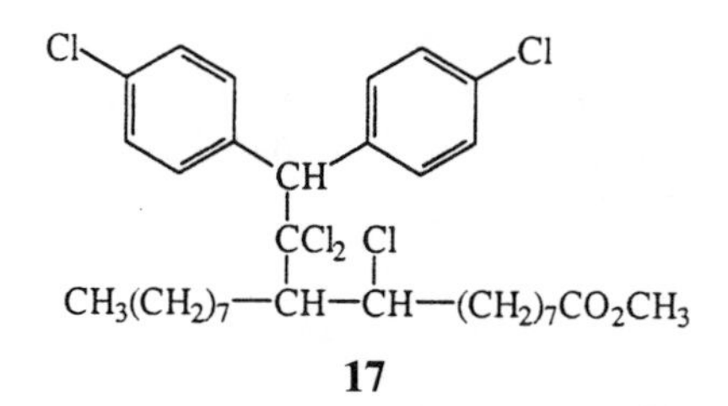

Although bacteria far outnumber algae in water columns, and contribute far more surface area to which organic matter may adsorb, algae are thought to be more important agents for promoting photochemical reactions. Being highly pigmented, they absorb much more light, and chlorophylls are known to be potent electron-transfer photocatalysts both in vivo and in vitro. Only a few experiments dealing with the photolysis of chemicals sorbed to algae have been reported. Zepp and Schlotzhauer (1983) demonstrated that the green alga *Chlamydomonas* converted sorbed aniline to azobenzene with a half-life of aniline disappearance of about 11 min. Metabolism was ruled out by conducting the experiment with killed cells, which also carried out the reaction. Algae were also shown to produce H_2O_2, which was suggested to be involved in algal-induced aniline oxidation (Zepp et al., 1987b). Wang et al. (1988) demonstrated that the photochemical polymerization of salicylic acid to a humic-like polymer was increased in the presence of a green alga, *Scenedesmus subspicatus*. Because only a small fraction of the organic compounds present in even a highly eutrophic water body can be expected to be associated with the algal biomass, it follows that only photoreactions that proceed at very much larger rates on algal surfaces than in solution will be influenced very much by the presence of algae.

G. PHOTOREACTIONS OF PARTICULAR COMPOUNDS

1. Natural Organic Matter

It has been known for a long time (Stearns, 1916; Gjessing, 1966, 1970) that humic materials undergo bleaching or color changes when exposed to light. Recently, some of the details of these light-induced processes have been studied. A number of highly reactive transient species have been shown to be formed by the photolysis of natural organic matter.

The importance of sensitized reactions in natural waters became evident with the discovery that dissolved organic matter had photosensitizing ability (Ross and Crosby 1975; Zepp et al. 1977). Many of these compounds were discovered to be producing a reactive form of oxygen, singlet oxygen ($^1O_2$: Foote et al., 1964). For details of the reactions of $^1O_2$, see Chapter 4. Flash photolysis studies of humic materials have shown that a transient species is formed whose chemical and optical properties match those of the hydrated electron (e_{aq}) observed in pulse radiolysis of water (Zepp et al., 1987c). The fate of the hydrated electron may include reaction with oxygen to form $\cdot O_2^-$, which would be expected to disproportionate rapidly to H_2O_2. Spin-trapping studies have confirmed the formation of $\cdot O_2^-$ as well as $\cdot OH$, presumably formed by Fenton-like reactions of H_2O_2 (Takahashi et al., 1988). Hydrogen peroxide is also formed when humic materials are exposed to light; see Chapter 4 for details.

Miles and Brezonik (1981) described a photolytic, iron-catalyzed, cyclic mechanism for oxygen removal. One-electron reduction of complexed Fe(III) by DOM is followed by reoxidation of the Fe(II) to Fe(III) by dissolved oxygen. During the process, the DOM is stoichiometrically decarboxylated, with the formation of two moles of CO_2 per mole of O_2 absorbed.

Carbon monoxide, CO (Conrad et al., 1982) and carbonyl sulfide, COS (Ferek and Andrae, 1984), were produced when surface seawater was illuminated. The mechanisms of formation of these compounds are unknown, although COS was formed when sulfur-containing amino acids or peptides were irradiated with 254-nm UV light (Ferek and Andrae, 1984).

2. Aromatic Hydrocarbons

Benzene derivatives as such do not absorb much light in the solar UV region; however, in the presence of oxygen some of them form complexes (Chien, 1965; Khalil and Kasha, 1978; Onodera et al., 1980; Pasternak and Morduchowitz, 1983) that are susceptible to photolysis. It appears that the absorption spectra of the complexes tail into the solar UV region (Chien, 1965) and that the quantum yield for reaction is rather high. Products of the reactions include long-chain conjugated dialdehydes from benzene itself (Wei et al., 1967) and alcohols and aldehydes from side-chain oxidation of alkyl-substituted benzenes (Pasternak and Morduchowitz, 1983). Some authors have proposed that singlet oxygen is the active species for the photooxidation, whereas others have suggested a (perhaps more likely) charge-transfer mechanism proceeding through a hydroperoxide intermediate (Figure 6.11).

Polycyclic aromatic hydrocarbons (PAHs), in contrast, absorb light strongly in the solar UV region. Some of them, e.g., naphthalene, are much more photolabile in water than in organic solvents, although their solubility, of course, decreases dramatically as the number of rings increases. Mill et al. (1981) and Zepp and

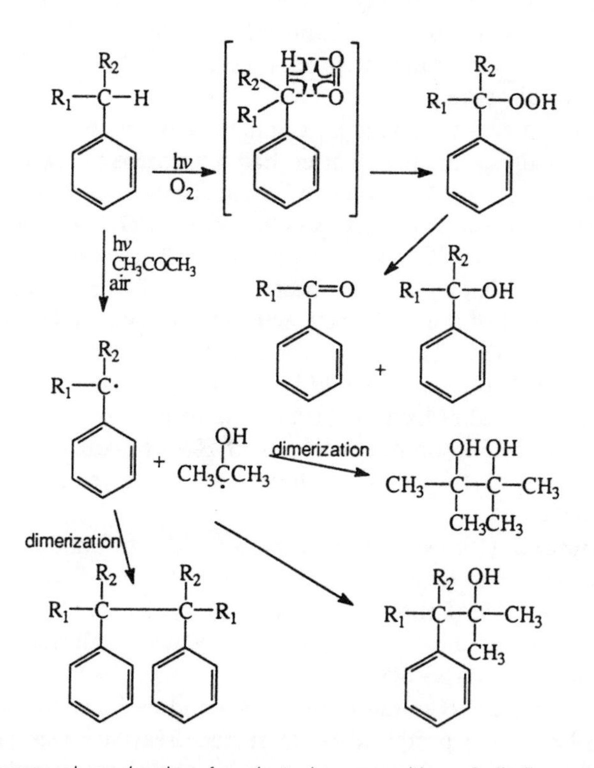

Figure 6.11. Proposed mechanism for photodecomposition of alkylbenzenes via complexes with molecular oxygen. From Pasternak and Morduchowitz (1983). Reprinted by permission of Pergamon Press Ltd.

Schlotzhauer (1979) have studied the kinetics of disappearance of several PAHs and related heterocyclic compounds in pure water. Table 6.3 summarizes the rate constants, half-lives, and quantum yields for the direct photolysis of these PAHs. It is clear that photolysis must be an important process for their removal from aqueous systems, especially considering the low rates for biodegradation of many such compounds. Notably, for many of these compounds, the presence or absence of dissolved oxygen has very little effect on their rates of disappearance. This fact indicates that reactive oxygen species such as $^1O_2$ and hydroxyl radical probably have no important roles in their photolysis, and suggests that the initial step in their photodecomposition may be photoionization to the PAH radical cation and a hydrated electron, followed by PAH-destroying reactions involving water. However, few product studies of these aqueous photoreactions have been undertaken because of solubility limitations and other problems. One study of anthracene photolysis has shown that in the presence of oxygen the primary products were anthraquinone and the endoperoxide **17A**, whereas in argon-purged solution the dimer **17B** predominated (Sigman et al., 1991). Labeling studies with ^{18}O in oxygen-deficient conditions supported a mechanism in which the cation radical of anthracene was quenched by water to a hydroxylated radical, which dimerized to the observed product.

There is evidence that the photochemical reactions of aromatic hydrocarbons in nonpolar environments, such as petroleum films, may differ greatly from those in aqueous solution. In benzene, for example, benzo[a]pyrene (**18**) was converted to a mixture of quinones (Masuda and Kuratsune, 1966).

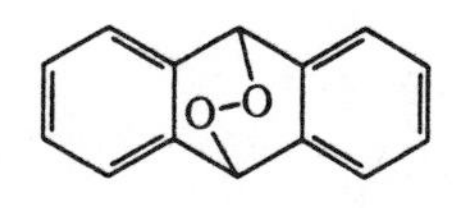

17A

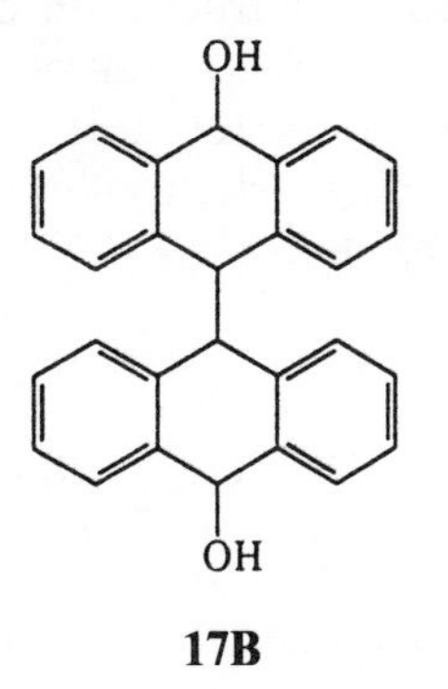

17B

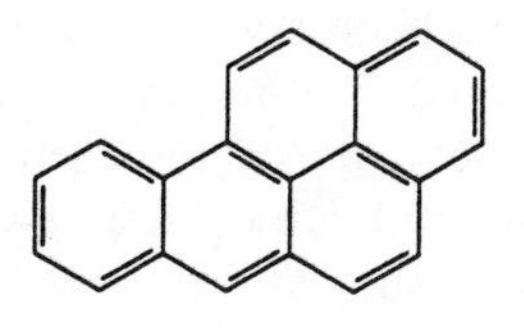

18

Table 6.3. Disappearance Quantum Yields (ϕ_d) for Selected Polycyclic Aromatic Hydrocarbons and Related Compounds in Water

Compound	ϕ_d
1-Methylnaphthalene	1.8×10^{-3}
2-Methylnaphthalene	5.3×10^{-4}
Phenanthrene	1.0×10^{-3}
9-Methylanthracene	7.5×10^{-4}
9,10-Dimethylanthracene	4.0×10^{-4}
Pyrene	2.1×10^{-4}
Chrysene	2.8×10^{-4}
Quinoline	3.2×10^{-4}
Benzo[f]quinoline	1.4×10^{-2}
Carbazole	7.6×10^{-3}
Benzo[b]thiophene	7.5×10^{-6}
Dibenzothiophene	5.0×10^{-4}

Source: Zepp and Schlotzhauer (1979) and Mill et al. (1981).

3. Halogenated Hydrocarbons

Although aliphatic monohalogenated compounds have little or no sunlight UV absorption, the introduction of two or more chlorine, e. g., atoms on the same carbon atom shifts the tail of the C-X absorption band into the solar region. The absorption process for these compounds appears to involve the promotion of one of the nonbonding *p*-electrons of the halogen atom into an antibonding σ^* orbital. This conversion weakens the C-X bond to the point where homolysis to halogen atoms

$$CCl_4 + CH_3OH \xrightarrow{h\nu} CHCl_3 + H{-}\overset{\overset{\large O}{\|}}{C}{-}H + HCl \tag{6.11}$$

and haloalkyl radicals can result. For example, carbon tetrachloride and chloroform are unstable to photodecomposition; commercial grades contain a small amount of alcohol or other preservative to intercept the free radicals that are formed in the presence of light. The alcohol is sacrificially oxidized and HCl is produced (Equation 6.11). Bromoform's absorption extends out to 380 nm, and tetrachloroethylene absorbs weakly throughout the entire solar UV region, up to ca. 420 nm. Little work has been reported on the photochemistry of these compounds under environmentally realistic conditions, however. The majority of studies so far have focused on the short-wave UV photolysis of chlorofluorocarbons under stratospheric conditions; these compounds have little absorption even at 254 nm. Photolysis of aliphatic halides in polar solvents such as water can proceed via carbocations rather than the more common free radicals obtained by homolytic cleavage (Kropp, 1984), which may make it difficult to extrapolate photolysis studies done in organic solvents to aqueous conditions. A study with 1,2-dibromo-3-chloropropane (DBCP, **19**) in water showed a dramatic rate enhancement ($t_{1/2}$ = 25 min) for hydrolysis and the production of two alcohols, **20** and **21**, which were also active to further photodehalogenation (Castro et al., 1987). Although this work was done with short-wave (254-nm) UV, in principle the results obtained could be significant for aliphatic polyhalo compounds that absorb solar UV wavelengths more strongly. Thus, for example, chloropicrin (**22**) was quantitatively converted to nitrate ion by light (Castro and Belser, 1981).

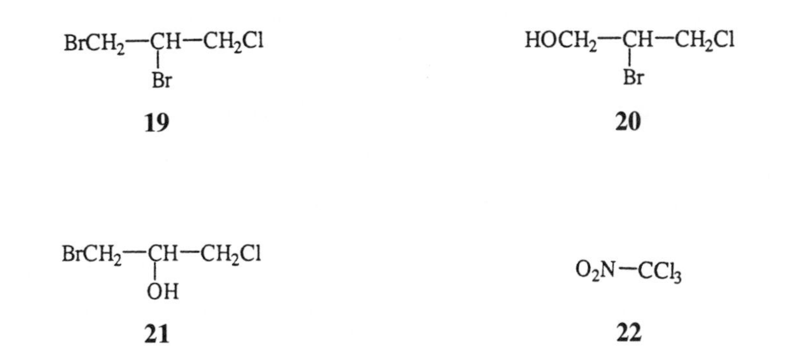

The aqueous photochemistry of simple chlorinated benzenes has been studied by only a few authors. The sunlight extinction of chlorobenzenes is quite low (ε_{297} for PhCl = 0.12 [Dulin et al., 1986]), and their solubilities (especially for the more highly chlorinated isomers) are also relatively low. Chlorobenzene triplets appear to react by homolysis to phenyl radicals and chlorine atoms; the subsequent fate of these radicals depends on the solvent used. Dulin et al. (1986) and Boule et al. (1987) indicated that monochlorobenzene was photohydrolyzed to phenol with a quantum yield of about 0.1 to 0.5. Additional very polar products were also observed, but no benzene. These quantum yields are similar to those observed in organic solvents; however, the principal fate of photolyzed chlorobenzenes in organic solvents is reduction by hydrogen donation. Dichloro- and trichlorobenzenes were similarly converted to monochloro- and dichlorophenols, respectively, in water. The quantum efficiency for trichlorobenzene photodestruction in 10% acetonitrile was about 0.03 (Choudhry and Hutzinger, 1984).

Hexachlorobenzene was very slowly photolyzed by direct or acetone-sensitized mechanisms (Choudry and Hutzinger, 1984), but it was photodecomposed much more rapidly (10–600×) in the presence of various amine and indole derivatives (Hirsch and Hutzinger, 1989). Pentachlorobenzene was the only product identified in the amine experiments. Although no mechanistic speculations were engaged in by the authors, it appears plausible that electron transfer within π-bonded complexes in the excited state could account for the observed rate enhancements.

Hexachlorocyclopentadiene (**23**) is a widely used synthetic intermediate, produced in large quantities for the manufacture of pesticides and flame retardants. In the environment, it is very reactive; photolysis is probably the major process responsible for its degradation in aquatic ecosystems. Several groups have studied its photolysis and concluded that its direct photolytic half-life in solution is between 1 to 10 min (Zepp et al., 1979; Butz et al., 1982; Chou et al., 1987). A variety of products has

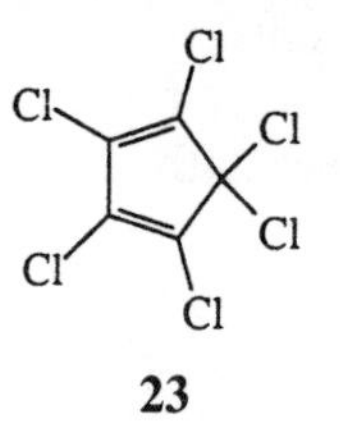

23

been tentatively identified by GC-MS, as summarized in Figure 6.12. The mechanism of these reactions is still undetermined; it would be particularly interesting to elucidate the route of the conversion of hexachlorocyclopentadiene to the rearranged, 6-chlorine ketones.

Although the maximum UV absorbance of PCBs occurs at wavelengths much shorter than those of sunlight, there is still sufficient extinction above 290 nm to permit some reactions to occur. In the presence of oxygen, these reactions are

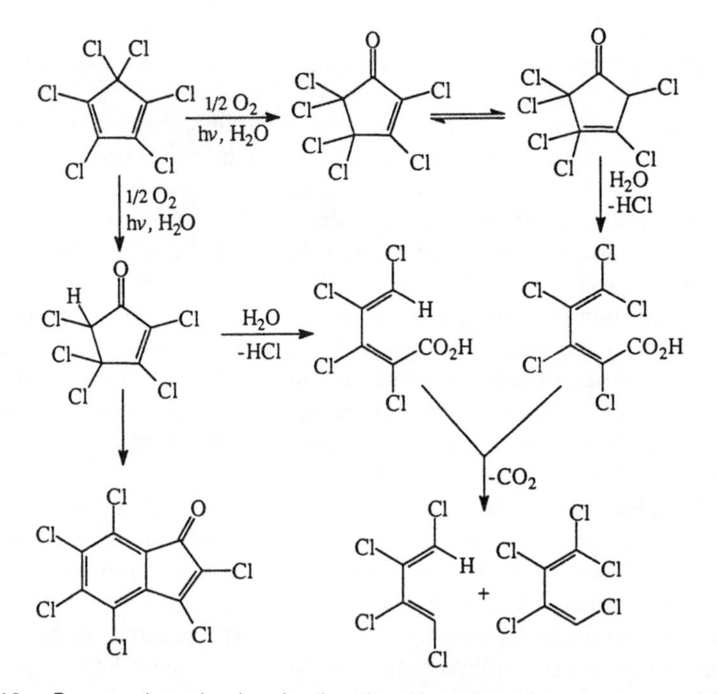

Figure 6.12. Proposed mechanism for the photodecomposition of hexachlorocyclopentadiene in water. From Chou et al. (1987). Reprinted by permission of Pergamon Press Ltd.

usually very slow. When PCBs are photolyzed in hydrocarbon solvents, homolysis of the C–Cl bonds generates an aryl radical which can rapidly abstract a hydrogen atom from a donor such as the solvent (Equation 6.12; Ruzo et al., 1974), but the high H–O bond strength in water makes this process much less likely when the compounds are photolyzed in it. For example, when 4-chlorobiphenyl was irradiated with light of >290 nm wavelengths in deoxygenated water, 3- and 4-hydroxybiphenyl were formed (Dulin et al., 1986; Moore and Pagni, 1987). The quantum yield was rather low (ϕ = 0.002). Evidence indicated that arenium ions, arynes, and addition-elimination reactions were not involved, but that the 4-chloro isomer underwent rearrangement to the 3-chloro compound which was then photohydrolyzed. 2-Chlorobiphenyl in 10% acetonitrile was photolyzed much more readily (ϕ = ca. 0.25) than the 4-isomer; the observed product, 2-hydroxybiphenyl, was even more photolabile than the starting material (Dulin et al., 1986; Orvis et al., 1991).

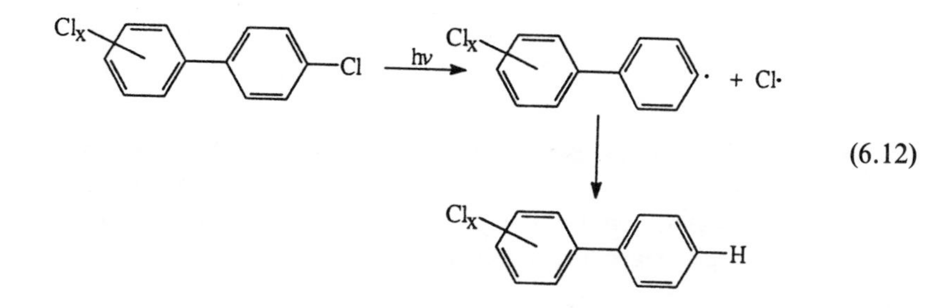

(6.12)

Epling et al. (1988) have examined the photoreduction (at 254 nm) of PCBs using sodium borohydride. The destruction of the PCBs continued smoothly to total photodechlorination of the compound (Figure 6.13). The fact that the quantum efficiencies were less than one and that the quantum efficiencies were not affected when free-radical scavengers such as BHT and acrylonitrile were added to the reacting solution supports a view that a free-radical chain mechanism for the photodechlorination of the PCBs was unlikely. However, Epling et al. (1987) did find evidence for a free-radical chain mechanism for the photodebromination of polybrominated biphenyls using sodium borohydride. Again an enhancement of photodestruction was observed when sodium borohydride was used. However, the quantum yields were greater than one, indicating a free-radical mechanism of photodebromination. Also, free-radical scavengers reduced the quantum yields of the reactions when added to the reaction mixture. Apparently, there are at least two different mechanisms by which sodium borohydride can react. One mechanism involves nonchain mechanisms which may involve electron transfer from borohydride to the excited aryl halide or a direct attack by borohydride on the excited aryl halide (see Figure 6.14).

Only a few organoiodine compounds are of environmental importance. Methyl iodide, a metabolic product of some marine and freshwater algae, produces $\cdot CH_3$ and I· by direct photolysis (Zafiriou, 1983).

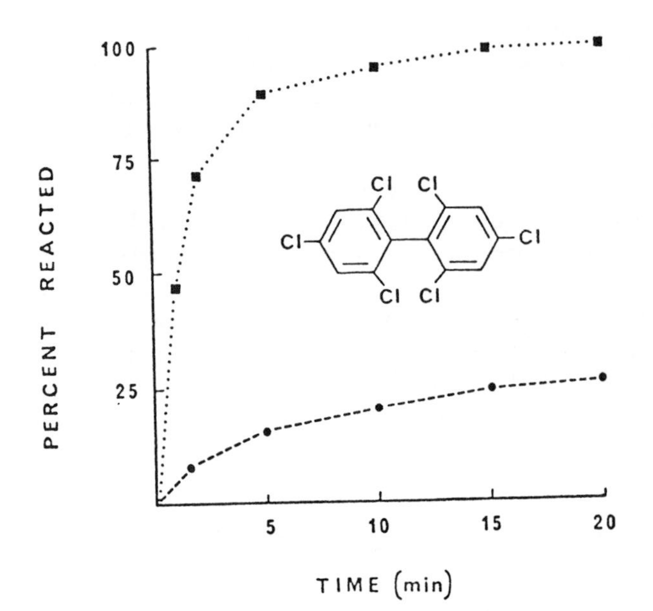

Figure 6.13. Photodecomposition (using 254-nm illumination) of hexachlorobiphenyl in the presence of sodium borohydride (solid squares) versus controls (solid circles). From Epling et al. (1988). Reprinted by permission. Copyright 1988 American Chemical Society.

4. Carbonyl Compounds

The photochemistry of aldehydes and ketones has been extensively examined (Turro, 1978). In high concentrations or in the absence of oxygen, a ketone such as acetone can fragment homolytically to an alkyl-acyl radical pair by the well-known Norrish type I cleavage mechanism:

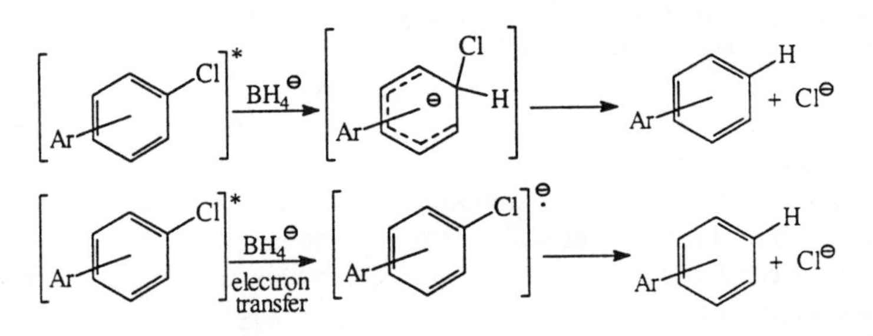

Figure 6.14. Proposed mechanism for the photolytic dechlorination of chlorinated biphenyls. From Epling et al. (1988). Reprinted by permission. Copyright 1988 American Chemical Society.

$$H_3C{-}\overset{O}{\overset{\|}{C}}{-}CH_3 \xrightarrow{h\nu} H_3C{-}\overset{O}{\overset{\|}{C}}\cdot + \cdot CH_3 \tag{6.13}$$

This reaction, however, is not usually of major importance in the environment except possibly at the low pressures and short UV wavelengths characteristic of the stratosphere. Short-chain aliphatic aldehydes such as *n*-butyraldehyde may also absorb solar UV and cleave (in a gas-phase process) to yield CO and a hydrocarbon.

A variety of other reaction pathways are available to these substances. In addition to studies of their unimolecular photoreactions, other investigations have revealed that the excited states of carbonyl compounds interact significantly with many other molecules. In some cases, they are able to efficiently promote the photodecomposition of co-solutes. Although carbonyl-containing materials are sometimes loosely called "photosensitizers," a term that on historical grounds is perhaps more properly restricted to singlet oxygen-producing species, they are capable of bringing about indirect photolysis by several mechanisms.

Absorption of light energy by the C = O group (at approximately 280–320 nm for simple aldehydes and ketones) converts the carbonyl group via n→ π^* excitation to a diradical-like intermediate:

$$>C{=}O \xrightarrow{h\nu} >\dot{C}{-}\dot{O} \tag{6.14}$$

Such diradical-like species are able to add to olefins (Paterno-Büchi reaction), producing oxetanes (Arnold, 1968; Equation 6.15). As indicated, the reaction intermediate is usually another diradical that is capable of free rotation around the newly formed single bond. Therefore, a mixture of stereoisomers usually results.

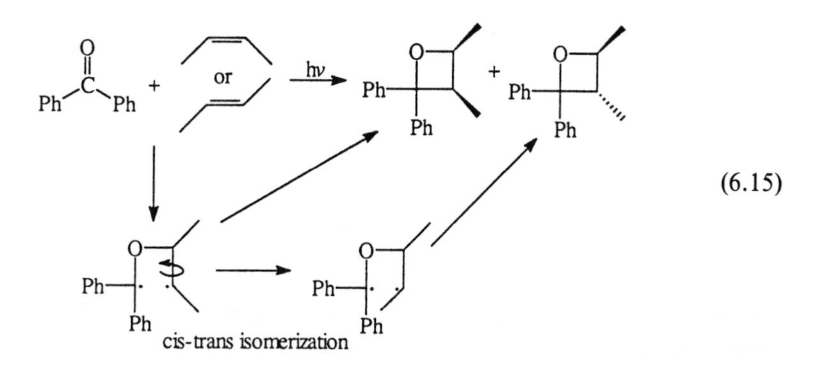

(6.15)

As described earlier in the chapter (Section 6.B.2), the excited-state oxygen of a carbonyl compound can be quite electrophilic. It can take part in oxidative processes such as hydrogen atom and electron transfer reactions. If suitably placed hydrogen atoms are available, for example, the oxygen atom of the diradical can abstract them; it is then converted to a monoradical, the so-called ketyl radical, •C-OH, and a new free radical can be produced at the abstraction site. This reaction can be intramolecular in ketones where the γ-hydrogen forms a stable six-membered transition state, or intermolecular. Hydrogen atom donation to the carbonyl excited state is especially prevalent in saturated hydrocarbons and alcohols; cyclohexane, ethanol, and isopropanol, for example, react with the triplet excited state of acetone with rate constants of 3×10^5, 4×10^5, and 10×10^5 L/mol sec, respectively (Turro, 1978). Donors with unshared electron pairs, such as amines and sulfur compounds, may donate electrons rather than H-atoms to the excited state oxygen atom, resulting in the formation of a radical cation-radical anion pair (Equation 6.16):

$$R_1C(=O)R_2 + R_3N\colon \xrightarrow{h\nu} R_1\dot{C}(O^{\ominus})R_2 + R_3\dot{N}^{\oplus} \qquad (6.16)$$

The excited state of ketones can thus initiate free-radical reactions, and this is probably the mechanism for many examples of enhanced photodecomposition of environmental pollutants sensitized by acetone or other simple carbonyl compounds. A good example of such reactions is the acetone-promoted photooxidation of atrazine (**24**) and related triazine herbicides described by Burkhard and Guth (1976). In water, atrazine absorbs almost no solar UV and was accordingly quite stable to photolysis, but in the presence of large amounts of acetone (about 0.13 M), its half-life was decreased to about 5 hr. The products were N-dealkylation products and ring-hydroxylated triazines. Similar products were also identified in riboflavin-sensitized photooxidation of triazines (Rejto et al., 1983). Presumably, a principal mechanism of photodecomposition would be H-abstraction from the N-alkyl substituents of atrazine, perhaps in conjunction with electron transfer from the unshared pairs of the nitrogen atoms.

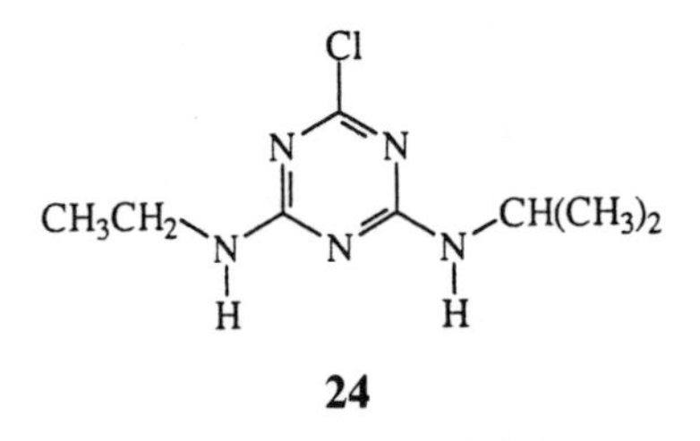

24

In the presence of oxygen, ketyl radicals and those derived from their hydrogen donors are converted to peroxy radicals. Figure 6.15 shows an example of an aldehyde photolyzed in ethanol. These peroxy radicals are often quite susceptible to loss of HOO• and subsequent production of hydrogen peroxide (see Equations 4.10 and 4.23 in the Oxidation chapter). Citral (**25**), a naturally occurring monoterpene aldehyde that is abundant in *Citrus* species and other plants, has been shown to be phototoxic, perhaps by this mechanism (Asthana et al., 1992).

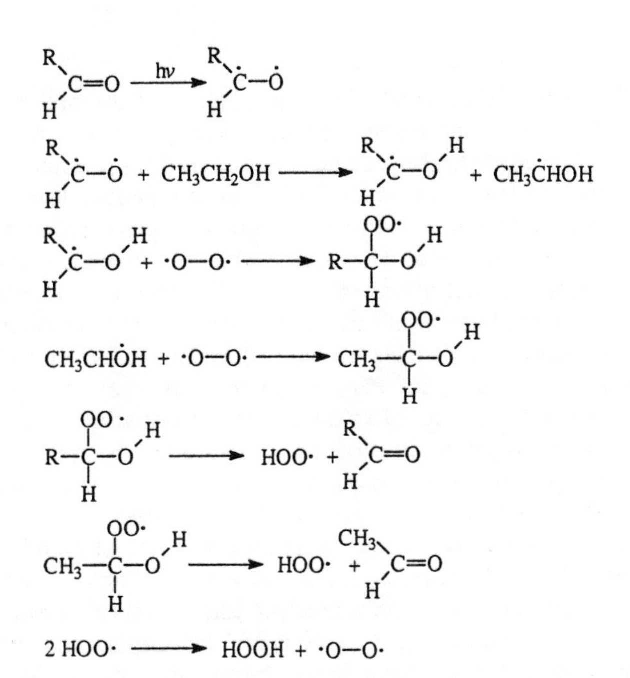

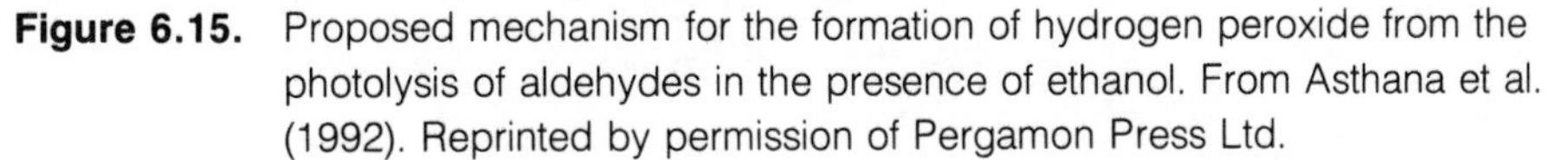

Figure 6.15. Proposed mechanism for the formation of hydrogen peroxide from the photolysis of aldehydes in the presence of ethanol. From Asthana et al. (1992). Reprinted by permission of Pergamon Press Ltd.

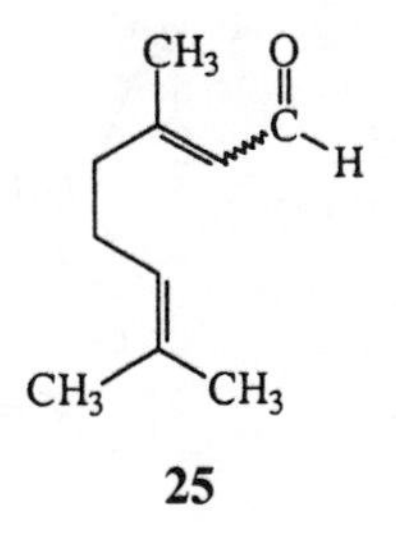

Ketone triplets may also transfer energy to acceptors:

$$^3RCOCH_3 + A \rightarrow RCOCH_3 + {}^3A \tag{6.17}$$

Acetophenone promoted the oxidative coupling of aniline to azobenzene, perhaps by such a mechanism (Anjaneyulu and Mallavadhani, 1988), but also perhaps by a ketyl radical process.

5. Phenols

Hwang et al. (1986, 1987a) have studied the kinetics of disappearance of phenol and chlorophenols in estuarine waters exposed to sunlight. Disappearance half-lives varied from <1 hr to >100 hr. In general, the rates of disappearance increased with increasing chlorine substitution from 0–3 chlorines, but pentachlorophenol was less reactive, falling between the mono- and disubstituted isomers. Solution pH was very important in governing the rate of disappearance; a pH that would allow for significant phenolate anion concentrations of a particular phenol increased its loss rate greatly. This is presumably due both to a higher rate of intrinsic photolysis for the anionic form and also to its increased light absorption characteristics. Rates of photolysis for 2,4-dichlorophenol were increased in the natural water relative to distilled water, probably because of sensitizers in the estuarine water.

The photochemistry of chlorophenols has been studied by many other authors, and depending on the conditions chosen, numerous product types have been identified. Crosby and co-workers examined the photoreactions of 2,4-dichlorophenol (Crosby and Tutass, 1966) and 2,4,5-trichlorophenol (Crosby and Wong, 1973), model compounds for the herbicides 2,4-D and 2,4,5-T. The products isolated appeared to have been formed by photonucleophilic aromatic substitution of OH groups for Cl atoms (Figure 6.16). Photolysis of 3-chlorophenol as well as other 3-halogenated phenols afforded resorcinols, presumably by similar mechanisms

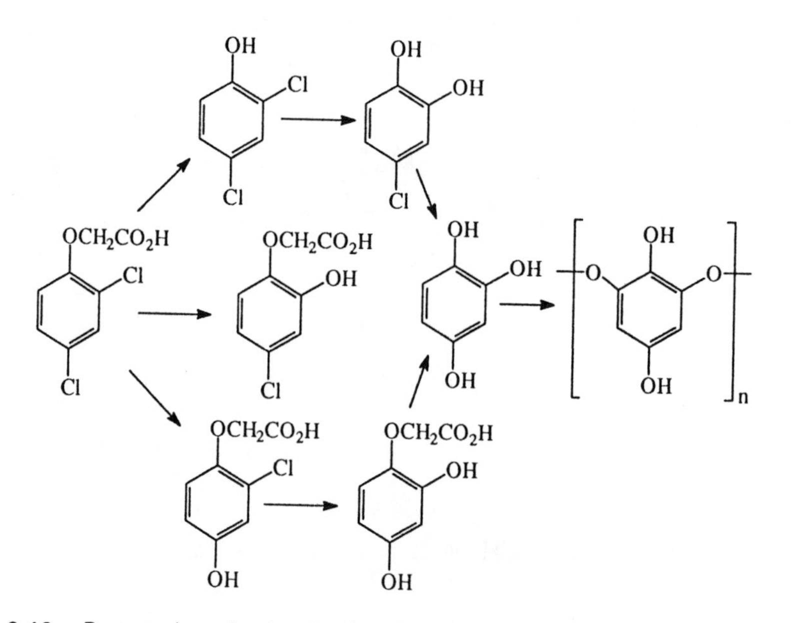

Figure 6.16. Proposed mechanism for the photodecomposition of 2,4-D. From Crosby and Tutass (1966). Reprinted by permission. Copyright 1966 American Chemical Society.

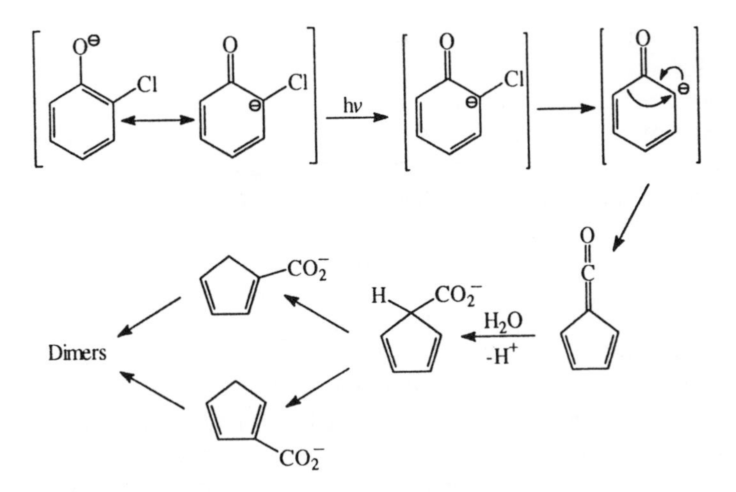

Figure 6.17. Proposed mechanism for the photochemical ring contraction of 2-chlorophenol. From Boule et al. (1987). Reprinted by permission. Copyright 1987 American Chemical Society.

(Omura and Matsuura, 1971; Boule et al., 1987). Other workers have reported different reactions for chlorophenols; for example, Boule et al. (1987) found that irradiation of 2-chlorophenol in alkaline, deaerated water yielded two unchlorinated photoproducts (see Figure 6.17) that were Diels-Alder adducts of cyclopentadiene carboxylic acids. The authors suggested a mechanism incorporating a photo-Wolff rearrangement to account for the ring contractions observed. At lower pHs (1–5.5), the acidic intermediates were still formed, together with catechol (**26**). Other *ortho*-halogenated phenols also afforded ring-contraction products, along with variable amounts of photohydrolysis products.

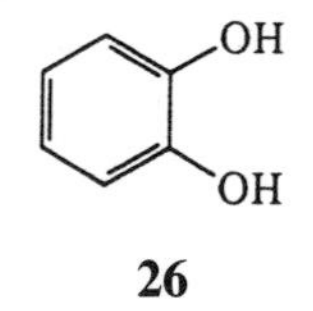

26

4-Chlorophenol gave a variety of photoproducts, including hydroquinone, benzoquinone, and several hydroxylated and/or chlorinated biphenyls (Boule et al., 1987). In addition, higher oligomers containing polyphenyls with up to 5 units were identified after prolonged irradiation.

Pentachlorophenol undergoes exchange of its Cl substituents by OH groups in water, presumably also by photonucleophilic aromatic substitution (Wong and Crosby, 1979; Figure 6.18). Dark reactions probably lead to the quinones and ring-opened products observed on prolonged irradiation.

Substantial rate enhancement for the photolysis of the pesticide methoxychlor

(16) was observed in natural waters; $t_{1/2}$ was 2–5 hr in the natural medium versus more than 300 hr in distilled water (Zepp et al., 1976). During the course of photolysis or metabolism of such chemicals, irreversible binding to humus is often observed (Bollag et al., 1982). Chaudhary et al. (1985) conducted experiments with solutions of methoxychlor in the presence of hydroquinone as a model compound for photoactive portions of humic material; their mechanism (Figure 6.19) indicated that an initial electron transfer from hydroquinone produced a dechlorinated methoxychlor radical and Cl^-, followed by reaction with solvent (reduction), rearrangement, or elimination. Coupling of the initially formed phenoxy radical with methoxychlor also occurred to give adducts containing phenyl ether and benzofuran groups.

Photolysis of 1-naphthol **(27)** in water and in organic solvents was shown to be highly solvent-dependent (Larson and Rounds, 1987). In water, the rate of disappearance increased with pH, being about a factor of 6 higher at pH 8 than at pH 5; the sunlight half-life at pH 8 was about 1 hr. This result is not consistent with the 1-naphtholate anion (pKa = 9.34) being the only reactive species. Products included a naphthoquinone derivative, lawsone **(28)**. The suggested mechanism of formation of lawsone is shown in Figure 6.20. In cyclohexane, the reaction was much faster (sunlight half-life was about 15 min) and the products were very different, being largely products derived from the solvent by apparent free-radical mechanisms.

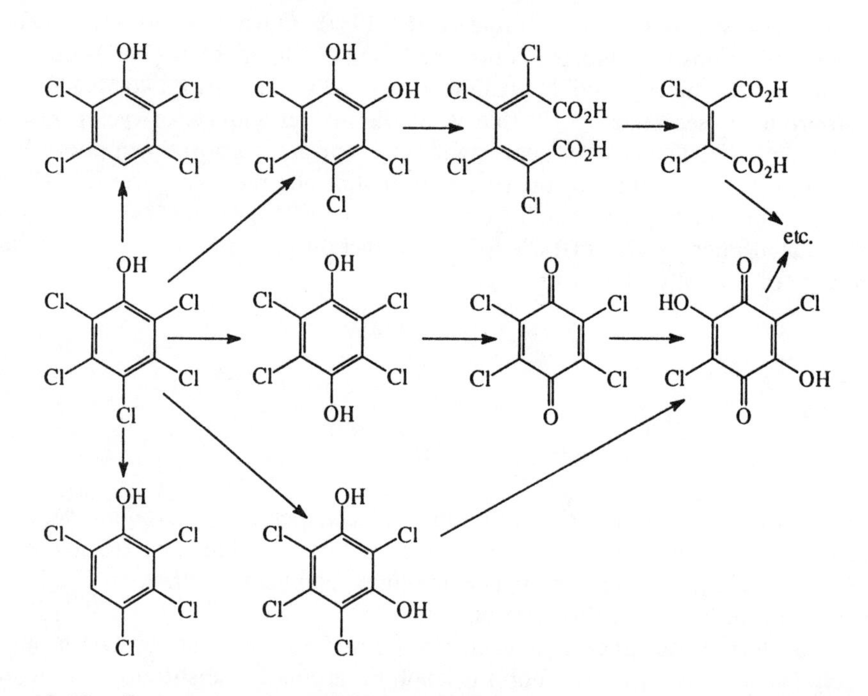

Figure 6.18. Proposed mechanism for the photodechlorination and ring opening reactions observed for pentachlorophenol. From Zafiriou et al. (1984). Reprinted by permission. Copyright 1984 American Chemical Society.

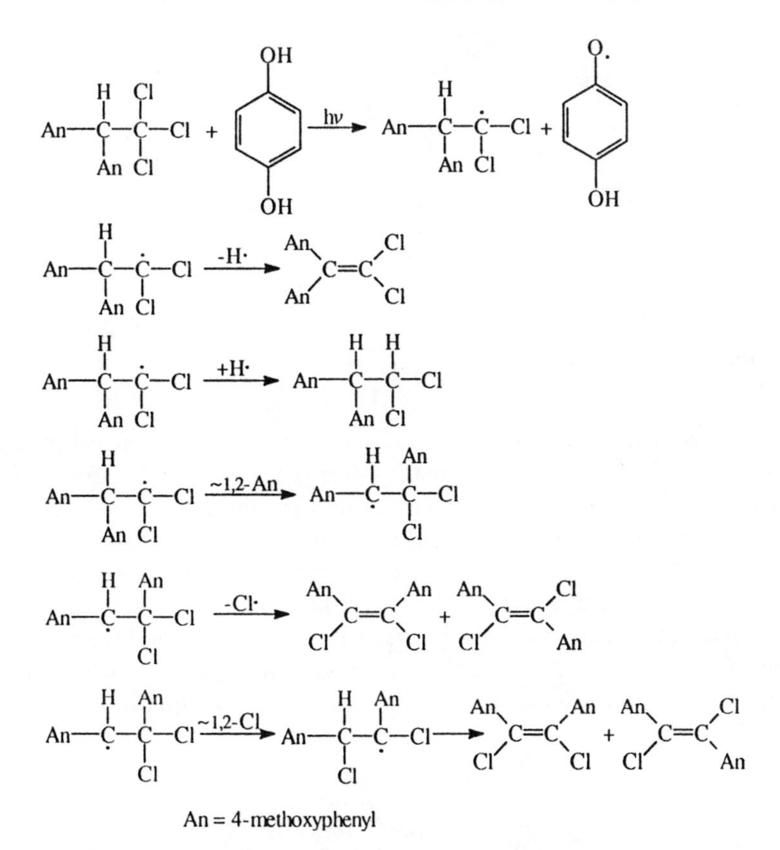

Figure 6.19. Proposed mechanism for the photodecomposition of methoxychlor (An = *p*-methoxyphenyl) in the presence of hydroquinone. From Chaudhary et al. (1985). Reprinted by permission of Pergamon Press Ltd.

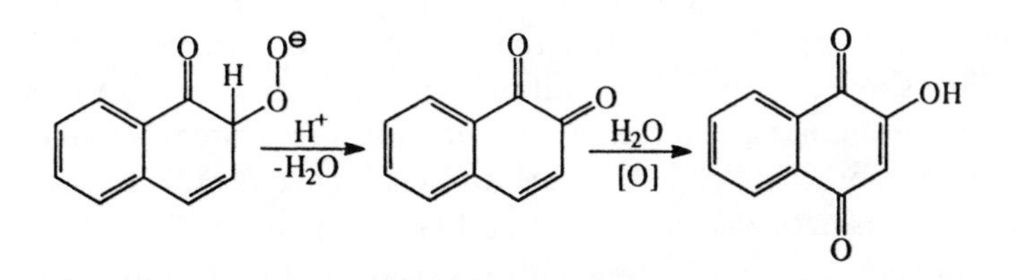

Figure 6.20. Proposed mechanism for the formation of lawsone in the photolysis of α-naphthol. From Larson and Rounds (1987). Reprinted by permission. Copyright 1987 American Chemical Society.

6. Anilines

Absorption of light energy by aniline derivatives can be accompanied by ejection of an electron and radical cation formation (Equation 6.18). (Anilines weakly absorb solar UV, with a long-wavelength maximum around 290 nm.) Chloroanilines, especially 4-chloroaniline and 2,4,5-trichloroaniline, were rapidly photolyzed in sunlight, with half-lives of a few hours, in distilled water or natural estuarine water (Hwang et al., 1987b). Photohydrolysis was the principal fate of 3,4-dichloroaniline in water, a reaction similar to those observed with chlorophenols (Miller et al., 1979; Miille and Crosby, 1983). Azobenzene, along with other, more polar compounds, was observed in photooxidation of aniline sensitized by aquatic humic material (Zepp et al., 1981).

$$C_6H_5\ddot{N}R_1R_2 \xrightarrow{h\nu} C_6H_5\dot{N}^{\oplus}R_1R_2 + e_{aq} \qquad (6.18)$$

$$\text{3,4-}Cl_2C_6H_3NH_2 \xrightarrow[H_2O]{h\nu} \text{4-Cl-3-OH-}C_6H_3NH_2 \qquad (6.19)$$

Chesta et al. (1986, 1988) have shown that the photolysis of N,N-dimethylaniline can be coupled with uptake of the electron (or with direct electron transfer from the excited aniline singlet or triplet) by chlorinated benzenes and fragmentation of the resulting radical anion to chloride ion. Hydrogen-substituted benzene derivatives appear to be the other products. Bunce and Gallagher (1982) have shown that a similar aniline-sensitized dechlorination occurs with polychlorinated biphenyls.

The promutagen 2-aminofluorene **(29)** was activated to an unknown mutagenic compound by exposure to sunlight; oxygen was required for activity. Mixtures of oxidized nitrogen species were suggested to be the active agents (White and Heflich, 1985).

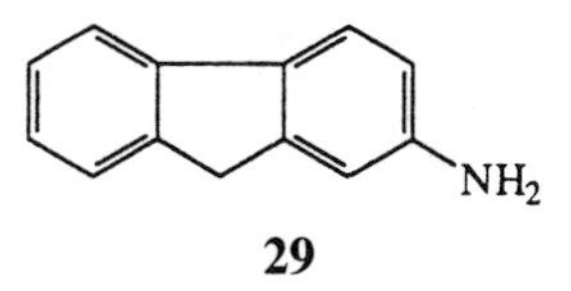

29

3,3′-Dichlorobenzidine (**30**), a known animal carcinogen, reacts rapidly in aqueous solution to form the dehalogenation products monochlorobenzidine and benzidine, in addition to unidentified, probably polymeric, water-insoluble materials (Banerjee et al., 1978). Photolysis did not take place in hexane or alcohols, and was faster at lower pH, suggesting an acid-catalyzed process. The disappearance quantum yields (in water at circumneutral pHs) for these compounds were very high; at 300 nm they were determined to be 0.43 for the dichlorobenzidine and 0.70 for monochlorobenzidine.

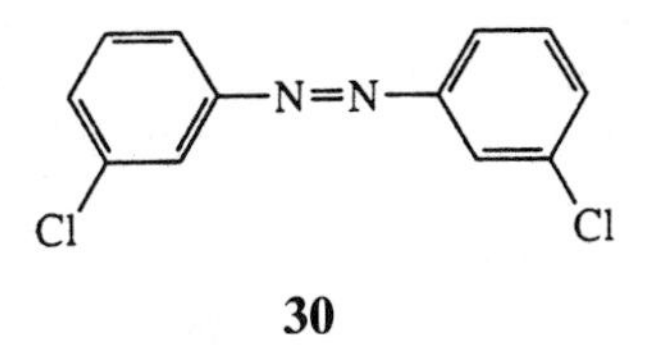

30

7. Nitro Compounds

The nitro group is a potent chromophore. Nitromethane, for example, absorbs strongly below about 410 nm. Nitroaromatic compounds, in particular, are usually intensely colored, and because of this strong absorption in the visible and adjacent UV region, should be susceptible to photolysis. Two types of excitation are possible since an intense $\pi \rightarrow \pi^*$ transition due to the aromatic ring tails from a maximum at around 260 nm into the solar UV region, where it merges with an $n \rightarrow \pi^*$ absorption that can be activated up to 350 nm or even higher. Although quantum yields are usually low (ca. 10^{-3}), some reactions have been documented. For example, parathion (**7**; Crosby, 1983), TNT (**31**; Mabey et al., 1983) and other mono- and dinitrotoluene isomers (Simmons and Zepp, 1986) underwent more or less rapid photodegradation in sunlight. Parathion was reported to afford *p*-nitrophenol as one product, possibly by direct cleavage of the parent compound to the phenoxy radical, followed by hydrogen atom abstraction. In distilled water, alkylated or methoxylated nitro compounds had photolysis half-lives of less than a day, whereas more electron-poor nitro compounds reacted slowly (half-lives were measured in weeks or months). In the case of *o*-methyl-substituted nitro aromatic compounds, dramatic rate enhancements (up to 100-fold) occurred in the presence of natural humic materials. The mechanisms of the observed rate effects are not certain.

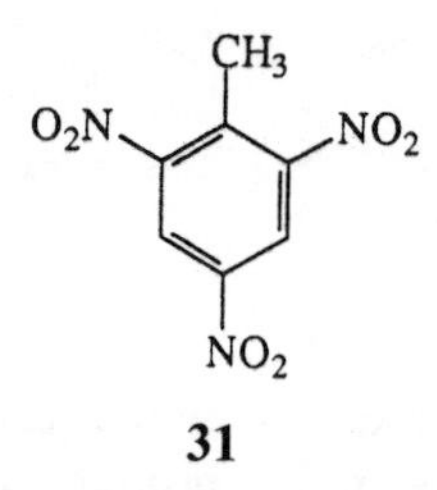

31

The photochemistry of aromatic nitro compounds has been reviewed by Dopp (1975) and by Morrison (1969). In brief, these compounds readily are photoreduced in the presence of suitable electron donors and also undergo photonucleophilic substitution. The long-wavelength absorption band at around 350 nm in nitro compounds appears to correspond to an n $\rightarrow$ π^* transition and accordingly the excited state should have some diradical character. Like the excited states of ketones, photoexcited nitro groups tend to abstract hydrogen atoms with the formation of an intermediate HO–$\dot{N}$=O free radical. The ultimate reduction products are usually the corresponding anilines but sometimes nitroso derivatives, hydroxylamines, or azo compounds (dimeric reduction products) can be isolated. Nitro groups can also add either intra- or intermolecularly to suitably activated double and triple bonds (for a review, see Morrison, 1969).

Nitro groups adjacent to alkyl or other substituents containing abstractable hydrogen atoms can undergo intramolecular reactions leading to diradicals that can cyclize or follow other reaction paths. An example is the photoreaction of 2-(2-nitrophenyl)-ethanol to an indoline derivative (Bakke, 1970).

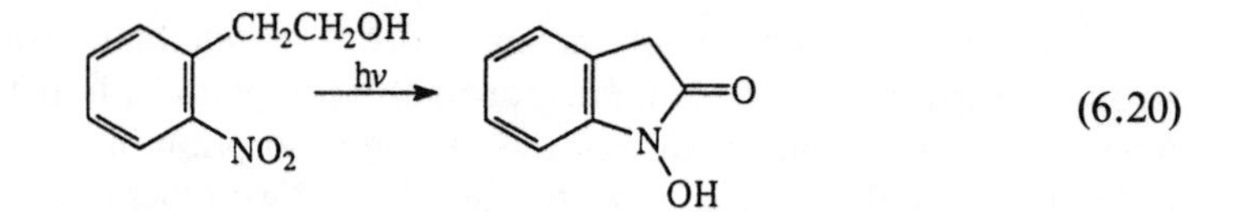

H. PHOTOCHEMISTRY IN WASTE TREATMENT

Attention to the possibilities of using photochemical techniques to degrade undesirable compounds in water has been intermittent, although interest has increased rapidly in recent years (Zepp, 1988). Techniques that have been studied in the laboratory or pilot plant include use of soluble sensitizers, metal ions or oxides, and "advanced oxidation processes" (see Section 5.C.2) involving some combination of light, ozone, and H_2O_2.

The use of sunlight, oxygen, and dissolved compounds capable of absorbing solar energy for the treatment of contaminated waters was first reported by Acher and Rosenthal (1977), who added methylene blue at $1.5\text{–}9 \times 10^{-5}$ *M* concentrations to samples of sewage, exposed the samples to sunlight, and followed the losses of chemical oxygen demand (COD) and detergents. Declines in COD of more than half (from 380 to 100 ppm) were reported, and detergents declined by more than 90% (12 ppm to <1 ppm). More recently (1985), an experimental plant using methylene blue and sunlight for wastewater disinfection began to operate in Livingston, Tennessee. The plant was built by Tennessee Technological University as part of a research effort funded by the U.S.-Israel Binational Agricultural Research and Development Fund. Preliminary findings on this project have been reported (Eisenberg et al., 1987).

A different approach to sensitized photodecomposition has been outlined by Larson et al. (1989, 1992a). Riboflavin, as described earlier (Section 6.E.2) is capable of forming complexes with some electron-rich donor molecules such as anilines, phenols, and polycyclic aromatic hydrocarbons and causing them to react in the presence of light. Although riboflavin is quite unstable ($t_{1/2}$ = a few minutes) in sunlight, under most conditions in water it is converted to lumichrome, which also has good sensitizing characteristics but less desirable light absorption properties (it no longer absorbs significantly in the visible region). A potentially more useful additive, riboflavin tetraacetate (**32**), has recently (Larson et al., 1992a) been shown to have similar light-absorbing properties and photocatalytic behavior to riboflavin, but to persist for much longer periods. The tetraacetate was used in tests with groundwaters and wastewaters contaminated with phenols and polycyclic hydrocarbons and shown to reduce the concentrations of many of the contaminants. Products of the reactions, however, are still not very well understood.

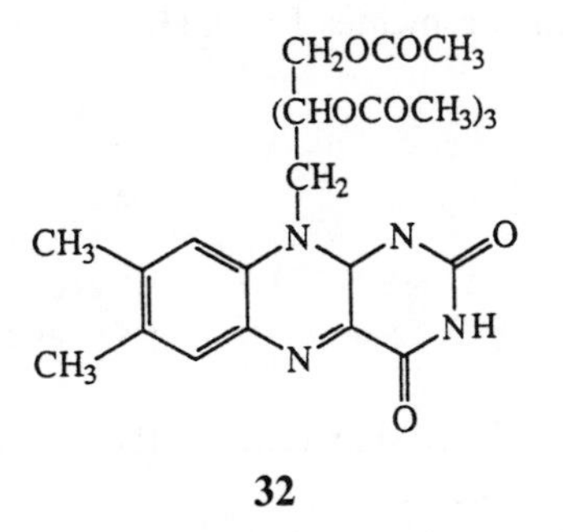

32

A joint project of the Solar Energy Research Institute and Sandia National Laboratories has devised a system in which sunlight is focused by parabolic or sun-tracking mirrors onto a reactor containing wastewater and suspended TiO_2. In some configurations, hydrogen peroxide is also added. Laboratory- and engineering-scale experiments have demonstrated the efficiency of the method (which probably depends on the formation of ·OH) for photodestruction of organic compounds ranging from salicylic acid to tetrachloroethylene (Tyner, 1990).

Larson et al. (1992b) discovered that ferric salts in weakly acidic solution were very significant promoters of photodecomposition of the triazine herbicides atrazine, ametryn, prometon, and prometryn (Figure 6.21). As described earlier (Equa-

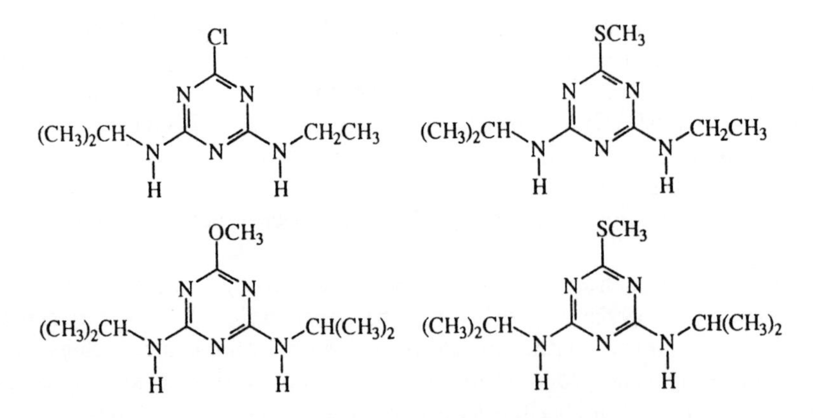

Figure 6.21. Triazine herbicides susceptible to photodecomposition in the presence of ferric iron salts. Reprinted with permission from Larson et al. (1992b). Copyright 1992 American Chemical Society.

tion 6.7), the ferric aquo complex $Fe(OH)^{2+}$ is capable of efficient HO· formation (Faust and Hoigné, 1990). Like other photoreactions (TiO_2 and "advanced oxidation processes," for example) that depend on hydroxyl radical formation, the ferric salt photolyses were greatly inhibited by the addition of buffers and organic compounds that could compete with the herbicides for ·OH.

REFERENCES

Acher, A. M. and I. Rosenthal. 1977. Dye-sensitized photooxidation—a new approach to the treatment of organic matter in sewage effluents. *Water Res.* 11, 557–562.

Anjaneyulu, A. S. and U. V. Mallavadhani. 1988. Photosensitised reactions of aniline. *Indian J. Chem.* 27B, 154–155.

Arnold, D. R. 1968. The photocycloaddition of carbonyl compounds to unsaturated systems: the synthesis of oxetanes. *Advan. Photochem.* 6, 301–423.

Asthana, A., R. A. Larson, K. A. Marley, and R. W. Tuveson. 1992. Mechanisms of citral phototoxicity. *Photochem. Photobiol.* 56, 211–222.

Bailey, G. W. and S. W. Karickhoff. 1973. Ultraviolet-visible spectroscopy in the characterization of clay mineral surfaces. *Anal. Lett.* 6, 43–49.

Baker, K. S. and R. C. Smith. 1982. Bio-optical classification and model of natural waters. *Limnol. Oceanogr.* 27, 500–509.

Bakke, J. 1970. The photocyclization of 2-(*o*-nitrophenyl)-ethanol to N-hydroxyindole. *Acta Chem. Scand.* 24, 2650–2651.

Balzani, V. and V. Carassiti. 1970. Photochemistry of coordination compounds. Academic Press, London.

Banerjee, S., H. C. Sikka, R. Gray, and C. M. Kelly. 1978. Photodegradation of 3,3-dichlorobenzidine. *Environ. Sci. Technol.* 12, 1425–1427.

Berenbaum, M. and R. A. Larson. 1988. Flux of singlet oxygen from leaves of phototoxic plants. *Experientia* 44, 1030 -1032.

Berends, W. and J. Posthuma. 1962. Energy transfer in aqueous solution. *J. Phys. Chem.* 66, 2547–2550.

Bertino, D. J. and R. G. Zepp. 1991. Effects of solar radiation on manganese dioxide reactions with selected organic compounds. *Environ. Sci. Technol.* 25, 1267–1273.

Bollag, J.-M., S.-Y. Liu, and R. D. Minard. 1982. Enzymatic oligomerization of vanillic acid. *Soil Biol. Biochem.* 14: 157–163.

Boule, P., C. Guyon, A. Tissot, and J. Lemaire. 1987. Specific phototransformation of xenobiotic compounds: chlorobenzenes and halophenols. In Zika, R. G. and Cooper, W. J., eds. Aquatic photochemistry. *Amer. Chem. Soc. Sympos. Ser.* 327, 10–26.

Bunce, N. J. and J. C. Gallagher. 1982. Photolysis of aryl chlorides with dienes and with aromatic amines. *J. Org. Chem.* 47, 1955–1958.

Burkhard, N. and J. A. Guth. 1976. Photodegradation of atrazine, atraton and ametryne in aqueous solution with acetone as a photosensitizer. *Pestic. Sci.* 7, 65–71.

Butz, R. G., C. C. Yu, and Y. H. Atallah. 1982. Photolysis of hexachlorocyclopentadiene in water. *Ecotoxicol. Environ. Safety* 6, 347–357.

Calkins, J. 1975. Measurements of the penetration of solar UV-B into various natural waters. *In* Impacts of climatic change on the biosphere. CIAP Monograph #5, U.S. Dept. of Transportation, Washington, DC. pp. 2-267.

Carey, J. H., J. Lawrence, and H. M. Tosine. 1976. Photodechlorination of PCBs in the presence of titanium dioxide in aqueous suspensions. *Bull. Environ. Contam. Toxicol.* 16, 697–701.

Castro, C. E. and N. O. Belser. 1981. Photohydrolysis of methyl bromide and chloropicrin. *J. Agric. Food Chem.* 29, 1005–1008.

Castro, C. E., S. Mayorga, and N. O. Belser. 1987. Photohydrolysis of 1,2-dibromo-3-chloropropane. *J. Agric. Food Chem.* 35, 865–870.

Chan, H. W.-S. 1977. Photo-sensitized oxidation of unsaturated fatty acid methyl esters. The identification of different pathways. *J. Amer. Oil Chem. Soc.* 54:100–104.

Chandrasekaran, K. and J. K. Thomas. 1985. Photophysical and photochemical properties of pyrene-doped TiO_2 particle suspensions in water. *J. Colloid Interf. Sci.* 106, 532–537.

Chaudhary, S. K., R. H. Mitchell, P. R. West, and M. J. Ashwood-Smith. 1985. Photodechlorination of methoxychlor induced by hydroquinone: rearrangement and conjugate formation. *Chemosphere* 14: 27–40.

Chesta C. A., J. J. Cosa, and C. M. Previtali. 1986. The N,N-dimethylaniline-photosensitized dechlorination of chlorobenzenes. *J. Photochem.* 32, 203–215.

Chesta, C. A., J. J. Cosa, and C. M. Previtali. 1988. Photoinduced electron transfer from N,N-dimethylaniline to chlorobenzene. The decomposition rate constant of the radical anion of chlorobenzene. *J. Photochem. Photobiol.* A45, 9–15.

Chien, J. C. W. 1965. On the possible initiation of photooxidation by charge-transfer excitation. *J. Phys. Chem.* 69, 4317–4325.

Chou, S.-F. J., R. A. Griffin, M.-I. M. Chou, and R. A. Larson. 1987. Photodegradation products of hexachlorocyclopentadiene (C-56) in aqueous solution. *Environ. Toxicol. Chem.*, 6, 371–376.

Choudry, G. G. and O. Hutzinger. 1984. Acetone-sensitized and nonsensitized photolyses of tetra-, penta-, and hexachlorobenzenes in acetonitrile-water mixtures: Photoisomerization and formation of several products including polychlorobiphenyls. *Environ. Sci. Technol.* 18, 235–241.

Conrad, R., W. Seiler, G. Bunse, and H. Giehl. 1982. Carbon monoxide in sea water (Atlantic Ocean). *J. Geophys. Res.* 87, 8839–8852.

Crosby, D. G. 1983. Atmospheric reactions of pesticides. In J. Miyamoto and P. C. Kearney, eds., *Pesticide chemistry: human welfare and the environment.* Pergamon Press, NY, pp. 327–332.

Crosby, D. G. and H. O. Tutass. 1966. Photodecomposition of 2,4-dichlorophenoxyacetic acid. *J. Agric. Food Chem.* 14, 596–599.

Crosby, D. G. and A. S. Wong. 1973. Photodecomposition of 2,4,5-trichlorophenoxyacetic acid (2,4,5-T) in water. *J. Agric. Food Chem.* 21, 1052–1054.

Cunningham, K. M., M. C. Goldberg, and E. R. Weiner. 1985. The aqueous photolysis of ethylene glycol adsorbed on goethite. *Photochem. Photobiol.* 41, 409–416.

Cunningham, K. M., M. C. Goldberg, and E. R. Weiner. 1988. Mechanisms for aqueous photolysis of adsorbed benzoate, oxalate, and succinate on iron oxyhydroxide (goethite) surfaces. *Environ. Sci. Technol.* 22, 1090–1097.

DeBoer, C. D. and R. H. Schlessinger. 1968. The multiplicity of the photochemically reactive state of 1,2-diphenylcyclobutene. *J. Amer. Chem. Soc.* 90, 803–804.

Dopp, D. 1975. Reactions of aromatic nitro compounds via excited triplet states. *Topics Curr. Chem.* 55, 49–85.

Draper, R. B. and D. G. Crosby. 1987. Catalyzed photodegradation of the herbicides molinate and thiobencarb. *In* Zika, R. G. and Cooper, W. J., eds. *Aquatic photochemistry, Amer. Chem. Soc. Sympos. Ser.* 327, 240–247.

Dulin, D., H. Drossman, and T. Mill. 1986. Products and quantum yields for photolysis of chloroaromatics in water. *Environ. Sci. Technol.* 20, 72–77.

Eisenberg, T. N., E. J. Middlebrook, and V. D. Adams. 1987. Sensitized photooxidation for wastewater disinfection and detoxification. *Water Sci. Technol.* 19, 1255–1258.

Epling, G. A., W. McVicar, and A. Kumar. 1987. Accelerated debromination of biphenyls by photolysis with sodium borohydride. *Chemosphere* 16, 1013–1020.

Epling, G. A., E. M. Florio, A. J. Bourque, X.-H. Qian, and J. D. Stuart. 1988. Borohydride, micellar, and exciplex-enhanced dechlorination of chlorobiphenyls. *Environ. Sci. Technol.* 22, 952–956.

Faust, B. C. and M. R. Hoffmann. 1986. Photoinduced reductive dissolution of α-Fe_2O_3 by bisulfite. *Environ. Sci. Technol.* 20, 943-948.

Faust, B. C. and J. Hoigné. 1990. Photolysis of Fe(III)-hydroxy complexes as sources of OH radicals in clouds, fog, and rain. *Atmos. Environ.* 23, 235-240.

Ferek, R. J. and M. O. Andrae. 1984. Photochemical production of carbonyl sulphide in marine surface waters. *Nature* 307, 148-150.

Finden, D. A. S., E. Tipping, G. H. M. Jaworski, and C. S. Reynolds. 1984. Light-induced reduction of natural iron(III) oxide and its relevance to phytoplankton. *Nature* 309, 783-784.

Foote, C. S., S. Wexler, E. J. Corey, and W. C. Taylor. 1964. Olefin oxidations with singlet oxygen. *J. Amer. Chem. Soc.* 86, 3879-3881.

Fujihara, M., Y. Satoh, and T. Osa. 1981. Heterogeneous photocatalytic oxidation of aromatic compounds on TiO_2. *Nature* 293, 206-208.

Giannotti, C., S. LeGreneur and O. Watts. 1983. Photo-oxidation of alkanes by metal oxide semiconductors. *Tetrahedron Lett.* 24, 5071-5072.

Gjessing, E. T. 1966. Enhancement of colour and chemical oxygen demand in a dammed-up lake. *Vattenhygien* 4, 181-183.

Gjessing, E. T. 1970. Reduction of aquatic humus in streams. *Vatten* 1, 14-23.

Haag, W. R. and J. Hoigné. 1985. Photosensitized oxidation in natural water via ·OH radicals. *Chemosphere* 14, 1659-1671.

Hansen, H. P. 1975. Photochemical degradation of petroleum hydrocarbon surface films. *Mar. Chem.* 3, 183-195.

Harbour, J. R., J. Tromp, and M. L. Hair. 1985. Photogeneration of H_2O_2 in aqueous TiO_2 dispersions. *Can. J. Chem.* 63, 204-208.

Hautala, R. 1978. Surfactant effect on pesticide photochemistry in water and soil. EPA Report EPA-600/3-78-060.

Hebert, V. R. and G. C. Miller. 1989. Depth dependence of direct and indirect photolysis on the soil surfaces. *J. Agric. Food Chem.* 37, 913-918.

Heelis, P. F. 1982. The photophysical and photochemical properties of flavins. *Chem. Soc. Rev.* 11, 15-39.

Hirsch, M. and O. Hutzinger. 1989. Naturally occurring proteins from pond water sensitize hexachlorobenzene photolysis. *Environ. Sci. Technol.* 23, 1306-1307.

Hwang, H.-M., R. E. Hodson, and R. F. Lee. 1986. Degradation of phenol and chlorophenols by sunlight and microbes in estuarine water. *Environ. Sci. Technol.* 20, 1002-1007.

Hwang, H.-M., R. E. Hodson, and R. F. Lee. 1987a. Photolysis of phenol and chlorophenols in estuarine water. *In* Zika, R. G. and Cooper, W. J., eds. *Aquatic photochemistry, Amer. Chem. Soc. Sympos. Ser.* 327, 27-43.

Hwang, H.-M., R. E. Hodson, and R. F. Lee. 1987b. Degradation of aniline and chloroanilines by sunlight and microbes in estuarine water. *Water Res.* 21, 309-316.

Ishimitsu, S., S. Fujimoto, and A. Ohara. 1985. The photochemical decomposition and hydroxylation of phenylalanine in the presence of riboflavin. *Chem. Pharm. Bull.* 33:1552-1556.

Joshi, P. C. 1985. Comparison of the DNA-damaging property of photosensitised riboflavin via singlet oxygen ($^1O_2$) and superoxide radical (O_2^-) mechanisms. *Toxicol. Lett.* 26:211–217.

Khalil, G.-E., and M. Kasha. 1978. Oxygen-interaction luminescence spectroscopy. *Photochem. Photobiol.* 28:435–443.

Kirk, J. T. O. 1977. Attenuation of light in natural waters. *Austr. J. Mar. Freshwat. Res.* 28, 497–508.

Knight, R. J. and R. N. Sylva. 1975. Spectrophotometric investigation of iron(III) hydrolysis in light and heavy water at 25°C. *J. Inorg. Nucl. Chem.* 37, 779–783.

Kormann, C., D. W. Bahnemann, and M. R. Hoffmann. 1988. Photocatalytic production of H_2O_2 and organic peroxides in aqueous suspensions of TiO_2, ZnO, and desert sand. *Environ. Sci. Technol.* 22, 798–806.

Kotzias, D., H. Parlar and F. Korte. 1982. Photoreaktivität organischer Chemikalien in wässrigen Systemen in Gegenwart von Nitraten und Nitriten. *Naturwissenschaften* 69, 444.

Kropp, P. J. 1984. Photobehavior of alkyl halides in solution: radical, carbocation, and carbene intermediates. *Acc. Chem. Res.* 17, 131–137.

Landymore, A. F. and N. J. Antia. 1978. White-light promoted degradation of leucopterin and related pteridines dissolved in seawater, with evidence for involvement of complexation from major divalent cations of seawater. *Mar. Chem.* 6, 309–325.

Langford, C. H., M. Wingham, and V. S. Sastri. 1973. Ligand photooxidation in copper(II) complexes of nitrilotriacetic acid. Implications for natural waters. *Environ. Sci. Technol.* 7, 820–822.

Larson, R. A. and L. L. Hunt. 1978. Photooxidation of a refined petroleum oil: inhibition by β-carotene and role of singlet oxygen. *Photochem. Photobiol.* 28, 553–555.

Larson, R. A. and S. A. Rounds. 1987. Photochemistry in aqueous surface layers: 1-naphthol. *In* Zika, R. G. and Cooper, W. J., eds. Aquatic photochemistry, *Amer. Chem. Soc. Sympos. Ser.* 327, 206–214.

Larson, R. A., L. L. Hunt, and D. W. Blankenship. 1977. Formation of toxic products from a #2 fuel oil by photooxidation. *Environ. Sci. Technol.* 11, 492–496.

Larson, R. A., T. L. Bott, L. L. Hunt, and K. Rogenmuser. 1979. Photooxidation products of a fuel oil and their antimicrobial activity. *Environ. Sci. Technol.* 13, 965–969.

Larson, R. A., D. D. Ellis, H.-L. Ju, and K. A. Marley. 1989. Flavin-sensitized photodecomposition of anilines and phenols. *Environ. Toxicol. Chem.* 8, 1165–1170.

Larson, R. A., P. L. Stackhouse, and T. O. Crowley. 1992a. Tetraacetylriboflavin, a potentially useful photosensitizing agent for water treatment. *Environ. Sci. Technol.* 26, 1792–1798.

Larson, R. A., M. B. Schlauch, and K. A. Marley. 1992b. Ferric ion promoted photodecomposition of triazines. *J. Agric. Food Chem.* 39, 2057–2062.

Leermakers, P. A. and H. T. Thomas. 1965. Electronic spectra and photochemistry of adsorbed organic molecules. I. Spectra of ketones on silica gel. *J. Amer. Chem. Soc.* 87, 1620–1622.

Leifer, A. 1988. The kinetics of environmental aquatic photochemistry: theory and practice. American Chemical Society, Washington, DC.

Lelieveld, J. and P. J. Crutzen. 1991. The role of clouds in tropospheric photochemistry. *J. Atmos. Chem.* 12, 229–267.

Lichtenthaler, R. G., W. R. Haag, and T. Mill. 1989. Photooxidation of probe compounds sensitized by crude oils in toluene and as an oil film on water. *Environ. Sci. Technol.* 23, 39–45.

Lockhart, H. B. and R. V. Blakely. 1975. Aerobic photodegradation of X (N) chelates of (ethylenedinitrilo)tetraacetic acid (EDTA): implications for natural waters. *Environ. Lett.* 9, 19–31.

Mabey, W. R., D. Tse, A. Baraze, and T. Mill. 1983. Photolysis of nitroaromatics in aquatic systems. I. 2,4,6-Trinitrotoluene. *Chemosphere* 12, 3–16.

Mansour, M. and H. Parlar. 1978. Gas chromatographic determination of several cyclodiene insecticides in the presence of polychlorinated biphenyls by photoisomerization reaction. *J. Agric. Food Chem.* 26, 483–485.

Masuda, Y. and M. Kuratsune. 1966. Photochemical oxidation of benzo[a]pyrene. *Air Water Pollut. Int. J.* 10, 805–811.

Matthews, R. W. 1983. Near-U. V.-light-induced competititve hydroxyl radical reactions in aqueous slurries of titanium dioxide. *J. Chem. Soc. Chem. Commun.* 177–179.

Matthews, R. W. 1986. Photo-oxidation of organic material in aqueous suspensions of titanium dioxide. *Water Res.* 20, 569–578.

Miille, M. J. and D. G. Crosby. 1983. Pentachlorophenol and 3,4-dichloroaniline as models for photochemical reactions in seawater. *Mar. Chem.* 14, 111–120.

Miles, C. J. and P. L. Brezonik. 1981. Oxygen consumption in humic-colored waters by a photochemical ferrous-ferric catalytic cycle. *Environ. Sci. Technol.* 15, 1089–1095.

Mill, T., W. R. Mabey, B. Y. Lan, and A. Baraze. 1981. Photolysis of polycyclic aromatic hydrocarbons in water. *Chemosphere* 10, 1281–1290.

Miller, G. C. and R. G. Zepp. 1979. Effects of suspended sediments on the photolysis rates of dissolved pollutants. *Water Res.* 13, 453–459.

Miller, G. C. and R. G. Zepp. 1983. Extrapolating photolysis rates from the laboratory to the environment. *Residue Rev.* 85, 89–110.

Miller, G. C., M. J. Miille, D. G. Crosby, S. Sontum, and R. G. Zepp. 1979. Photosolvolysis of 3,4-dichloroaniline in water. Evidence for an aryl cation intermediate. *Tetrahedron* 35, 1797–1800.

Moore, T. and R. M. Pagni. 1987. Unusual photochemistry of 4-chlorobiphenyl in water. *J. Org. Chem.* 52, 770–773.

Mopper, K. and R. G. Zika. 1987. Natural photosensitizers in sea water: riboflavin and its breakdown products. *In* Zika, R. G. and Cooper, W. J., eds. *Aquatic photochemistry, Amer. Chem. Soc. Sympos. Ser.* 327, 174–190.

Morita, M., M. Mukunoki, F. Okubo, and S. Tadakora. 1976. Lipid-oxidation catalyses by substances in water on lipid-water interface. *J. Amer. Oil Chem. Soc.* 53, 489-490.

Morrison, H.A. 1969. The photochemistry of the nitro and nitroso groups. In Feuer, H., ed., The chemistry of the nitro and nitroso groups. Part I. New York: Interscience.

Moser, J. and M. Grätzel. 1982. Photochemistry with colloidal semiconductors. Laser studies of halide oxidation in colloidal dispersions of TiO_2 and α-Fe_2O_3. *Helv. Chim. Acta* 65, 1436–1444.

Nicholls, C. H. and P. A. Leermakers. 1971. Photochemical and spectroscopic properties of organic molecules in adsorbed or other perturbing polar environments. *Adv. Photochem.* 8, 315–336.

Oelkrug, D., W. Flemming, R. Fülleman, R. Günther, W. Honnen, G. Krabliche, M. Schäfer, and S. Uhl. 1986. Photochemistry on surfaces. *Pure Appl. Chem.* 58, 1207–1218.

Ollis, D. F., E. Pelizzetti, and N. Serpone. 1991. Photocatalyzed destruction of water contaminants. *Environ. Sci. Technol.* 25, 1522–1529.

Omura, K. and T. Matsuura. 1971. Photolysis of halogenophenols in aqueous alkali and in aqueous cyanide. *Tetrahedron* 27, 3101–3109.

Onodera, K., H. Sakuragi, and K. Tokumaru. 1980. Effect of light wavelength of photooxygenation of hexamethylbenzene. *Tetrahedron Lett.* 21, 2831–2832.

Orvis, J., J. Weiss, and R. M. Pagni. 1991. Further studies on the photoisomerization and hydrolysis of chlorobiphenyls in water. Common ion effect in the photohydrolysis of 4-chlorobiphenyl. *J. Org. Chem.* 56, 1851–1857.

Pasternak, M. and A. Morduchowitz. 1983. Photochemical oxidation and dimerization of alkylbenzenes. Selective reactions of the alkyl side groups. *Tetrahedron Lett.* 24, 4275–4278.

Payne, J. R. and C. R. Phillips. 1985. Photochemistry of petroleum in water. *Environ. Sci. Technol.* 19, 569–579.

Peake, E., B. L. Baker, and G. W. Hodgson. 1972. The contribution of amino acids, hydrocarbons, and chlorins to the Beaufort Sea by the Mackenzie River system. *Geochim. Cosmochim. Acta* 36, 867–883.

Rabek, J. F. 1968. Photosensitized processes in polymer chemistry: a review. *Photochem. Photobiol.* 7, 5–57.

Rejto, M., S. Saltzman, A. J. Acher, and L. Muszkat. 1983. Identification of sensitized photoooxidation products of s-triazine herbicides in water. *J. Agric. Food Chem.* 31, 138–142.

Rizzuto, F., J. D. Spikes, and G. D. Coker. 1986. The lumiflavin-sensitized photooxidation of substituted phenylalanines and tyrosines. *Photobiochem. Photobiophys.* 10:149–162.

Rodgers, M. A. and P. T. Snowden. 1982. Lifetime of O_2 ($^1\Delta g$) in liquid water as determined by time-resolved infrared luminescence measurements. *J. Amer. Chem. Soc.* 104, 5541–5543.

Ross, R. D., and D. G. Crosby. 1975. The photooxidation of aldrin in water. *Chemosphere.* 5: 277–282.

Ross, R. D., and D. G. Crosby. 1985. Photooxidant activity in natural waters. *Environ. Toxicol. Chem.* 4, 773–778.

Ruzo, L. O., M. J. Zabik, and R. D. Schuetz. 1974. Photochemistry of bioactive compounds. Photochemical processes of polychlorinated biphenyls. *J. Amer. Chem. Soc.* 96, 3809–3813.

Sancier, K. M. and H. Wise. 1981. Photoassisted oxidation of organic material catalyzed by sand. *Atmos. Environ.* 15, 639–640.

Schmitzer, J., S. Gäb, M. Bahadir, and F. Korte. 1980. Photomineralisierung chlorierter Alkane, Alkene und Aromaten an Kieselgel. *Z. Naturforsch.* 35b, 182–186.

Schwack, W. 1988. Photoinduced additions of pesticides to biomolecules. 2. Model reactions of DDT and methoxychlor with methyl oleate. *J. Agric. Food Chem.* 36, 645–648.

Scott, A. I. 1964. Interpretation of the ultraviolet spectra of natural products. Pergamon, Oxford.

Scully, F. E. and J. Hoigné. 1987. Rate constants for reactions of singlet oxygen with phenols and other compounds in water. *Chemosphere* 16, 681–694.

Sigman, M. E., S. P. Zingg, R. M. Pagni, and J. H. Burns. 1991. Photochemistry of anthracene in water. *Tetrahedron Lett.* 41, 5737–5740.

Simmons, M. S. and R. G. Zepp. 1986. Influence of humic substances on photolysis of nitroaromatic compounds in aqueous systems. *Water Res.* 20, 899–904.

Smith, E. C. and D. E. Metzler. 1963. The photochemical degradation of riboflavin. *J. Amer. Chem. Soc.* 85, 3285–3288.

Spencer, W. F., J. D. Adams, T. D. Shoup, R. E. Hess, and R. D. Spear. 1980. Conversion of parathion to paraoxon on soil dusts and clay minerals as affected by ozone and ultraviolet light. *J. Agric. Food Chem.* 28, 366–371.

Srisankar, E. V. and L. K. Patterson. 1979. Reactions of ozone with fatty acid monolayers: a model system for disruption of lipid molecular assemblies by ozone. *Arch. Environ. Health* 34, 346–349.

Stearns, R. H. 1916. Decolorization of water by storage. *J. New Engl. Water Works Assoc.* 30, 20–27.

Sunda, W. G., S. A. Huntsman, and G. R. Harvey. 1983. Photoreduction of manganese oxides in seawater and its geochemical and biological implications. *Nature* 301, 234–236.

Suzuki, J., T. Sato, and S. Suzuki. 1985. Hydroxynitrobiphenyls produced by photochemical reaction of biphenyl in aqueous nitrate solution and their mutagenicities. *Chem. Pharm. Bull.* 33, 2507–2515.

Suzuki, J., T. Hayagino, and S. Suzuki. 1987. Formation of 1-nitropyrene by photolysis of pyrene in water containing nitrite ion. *Chemosphere* 16, 859–867.

Suzuki, J., T. Watanabe, K. Sato and S. Suzuki. 1988. Roles of oxygen in photochemical reaction of naphthols in aqueous nitrite solution and mutagen formation. *Chem. Pharm. Bull.* 36, 4567–4575.

Takahashi, N., N. Mikami, H. Yamada, and J. Miyamoto. 1985. Photodegradation of the pyrethroid insecticide fenpropathrin in water, on soil and on plant foliage. *Pestic. Sci.* 16, 119–131.

Takahashi, N., N. Ito, N. Mikami, T. Matsuda, and J. Miyamoto. 1988. Identification of reactive oxygen species generated by irradiation of aqueous humic acid solution. *J. Pestic. Sci.* 13, 429–435.

Trott, T., R. W. Henwood, and C. H. Langford. 1972. Sunlight photochemistry of ferric nitriloacetate complexes. *Environ. Sci. Technol.* 6, 367–368.

Turro, N. J. 1978. *Modern molecular photochemistry*. Benjamin-Cummings, Menlo Park, CA.

Tuveson, R. W., R. A. Larson, K. A. Marley, G.-R. Wang, and M. R. Berenbaum. 1989. Sanguinarine, a phototoxic H_2O_2-producing alkaloid. *Photochem. Photobiol.* 50, 733–738.

Tyner, C. E. 1990. Application of solar thermal technology to the destruction of hazardous wastes. *Solar Energy Mater.* 21, 113–129.

Wagner, I., H. Strehlow, and G. Busse. 1980. Flash photolysis of nitrate ions in aqueous solution. *Z. Phys. Chem.* 123, 1–33.

Waite, T. D. 1985. Photoredox chemistry of colloidal metal oxides. In J. A. Davis and K. F. Hayes, eds., *Geochemical processes at mineral surfaces*. ACS Sympos. Ser. #323, pp. 426–445.

Waite, T. D., I. C. Wrigley, and R. Szymczak. 1988. Photoassisted dissolution of a colloidal manganese oxide in the presence of fulvic acid. *Environ. Sci. Technol.* 22, 778–785.

Wang, W. H., R. Beyerle-Pfnur, and J. P. Lay. 1988. Photoreaction of salicylic acid in aquatic systems. *Chemosphere* 17, 1197–1204.

Wei, K., J. Mani, and J. N. Pitts Jr. 1967. The formation of polyenic dialdehydes in the photooxidation of pure liquid benzene. *J. Amer. Chem. Soc.* 89, 4225–4227.

Weiner, E. R. and M. C. Goldberg. 1985. Aquatic photochemistry: selected topics from current research. *Toxicol. Environ. Chem.* 9, 327–339.

Wheeler, J. 1972. Some effects of solar levels of ultra-violet radiation on lipids in artificial sea water. *J. Geophys. Res.* 77, 5302–5306.

White, G. L. and R. H. Heflich. 1985. Mutagenic activation of 2-aminofluorene by fluorescent light. *Teratogen. Carcinogen. Mutagen.* 5, 63–73.

Wong, A. S. and D. G. Crosby. 1979. Photodecomposition of pentachlorophenol in water. *J. Agric. Food Chem.* 29, 125–130.

Zadelis, D. and M. S. Simmons. 1983. Effects of particulates on the photodecomposition of polynuclear aromatic hydrocarbons in aquatic systems. In M. Cooke and A. J. Dennis, eds. *Polynuclear aromatic hydrocarbons: formation, metabolism, and measurement*. Battelle Press, Columbus, OH. pp. 1279–1291.

Zafiriou, O. C. 1983. Natural water photochemistry. In J. P. Riley and R. Chester, eds. *Chemical oceanography*, Vol. 8, pp. 339–379, Academic Press, London.

Zafiriou, O. C., J. Joussot-Dubien, R. G. Zep, and R. G. Zika. 1984. Photochemistry of natural waters. *Environ. Sci. Technol.* 18: 358A–371A.

Zepp, R. G. 1982. Experimental approaches to environmental photochemistry. In O. Hutzinger, ed. *Handbook of environmental chemistry*. Springer Verlag, Berlin. Vol. 2B, pp. 19–41.

Zepp, R. G. 1988. Factors affecting the photochemial treatment of hazardous waste. *Environ. Sci. Technol.* 22, 256–257.

Zepp, R. G. and D. M. Cline. 1977. Rates of direct photolysis in the aqueous environment. *Environ. Sci. Technol.* 11, 359–366.

Zepp, R. G. and P. F. Schlotzhauer. 1979. Photoreactivity of selected aromatic hydrocarbons in water. In P. W. Jones and P. Leber, eds. *Polynuclear aromatic hydrocarbons*. Ann Arbor Sci. Pub., Ann Arbor, MI. pp. 141–158.

Zepp, R. G. and P. F. Schlotzhauer. 1983. Influence of algae on photolysis rates of chemicals in water. *Environ. Sci. Technol.* 17, 462–468.

Zepp, R. G., N. L. Wolfe, J. A. Gordon, and R. C. Fincher. 1976. Light-induced transformations of methoxychlor in aquatic systems. *J. Agric. Food Chem.* 24, 727–733.

Zepp, R. G., N. L. Wolfe, G. L. Baughmann, and R. C. Hollis. 1977. Singlet oxygen in natural waters. *Nature* 267, 421–423.

Zepp, R. G., N. L. Wolfe, G. L. Baughman, P. F. Schlotzhauer, and J. N. Macallister. 1979. Dynamics of processes influencing the behavior of hexachlorocyclopentadiene in the aquatic environment. Abstr. Papers, 178th Mtg. Amer. Chem. Soc., ENVR-042.

Zepp, R. G., G. L. Baughman, and P. F. Schlotzhauer. 1981a. Comparison of photochemical behavior of various humic substances in water. 1. Sunlight induced reactions of aquatic pollutants photosensitized by humic substances. *Chemosphere* 10, 109–117.

Zepp, R. G., G. L. Baughman, and P. F. Schlotzhauer. 1981b. Comparison of photochemical behavior of various humic substances in water. 2. Photosensitized oxygenations. *Chemosphere* 10, 119–126.

Zepp, R. G., J. Hoigné, and H. Bader. 1987a. Nitrate-induced photooxidation of trace organic chemicals in water. *Environ. Sci. Technol.* 21, 443–450.

Zepp, R. G., Y. I. Skurlatov, and J. T. Pierce. 1987b. Algal-induced decay and formation of hydrogen peroxide in water: its possible role in oxidation of anilines by algae. *In* Zika, R. G. and Cooper, W. J., eds. Aquatic photochemistry, *Amer. Chem. Soc. Sympos. Ser.* 327, 215–224.

Zepp, R. G., A. M. Braun, J. Hoigné, and J. A. Leenheer. 1987c. Photoproduction of the hydrated electron from natural organic solutes in aquatic environments. *Environ. Sci. Technol.* 21: 485–490.

CHAPTER 7

MOLECULAR REACTIONS: THE DIELS-ALDER AND OTHER REACTIONS

A. SURFACE AND AQUEOUS CATALYSIS OF THE DIELS-ALDER REACTION

The Diels-Alder reaction, a classic of synthetic organic chemistry, is theoretically a 4 + 2 cycloaddition in which a diene and a "dienophile" with very different electron abundances combine to form a new six-membered ring. Equation 7.1 is a typical Diels-Alder synthesis. The reaction has great synthetic usefulness and has attracted many researchers interested in its practical applications and mechanistic aspects. It is not at all certain whether any Diels-Alder reaction occurs under environmental

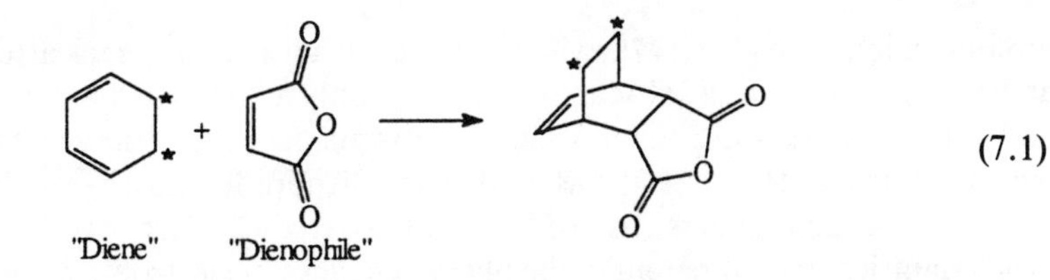

(7.1)

conditions, although it has been suggested as a mechanism for the formation of *dl*-limonene from polyisoprene in the pyrolysis of waste rubber tire material (Equation 7.2; Pakdel et al., 1991). Several recent studies which exploit water as a solvent or report dramatic rate accelerations by the use of natural clays suggest that this reaction perhaps should be investigated further.

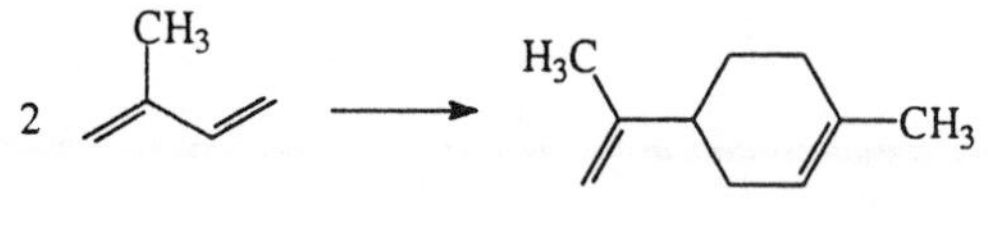

(7.2)

Although nearly all Diels-Alder reactions have been performed in the past by either heating the reactants together in the absence of a solvent, or using an inert and usually high-boiling solvent to initiate the reaction, more recent studies have shown that water may greatly increase reaction rates (Breslow et al., 1983; Grieco et al., 1983). In a typical example (Equation 7.3), water accelerates the reaction by a factor of about 60 over methanol and is more than 700 times faster than in a hydrocarbon solvent. Water has been suggested to promote the close approach of hydrophobic reagents to one another and thus to facilitate bond-forming processes due to high encounter rates.

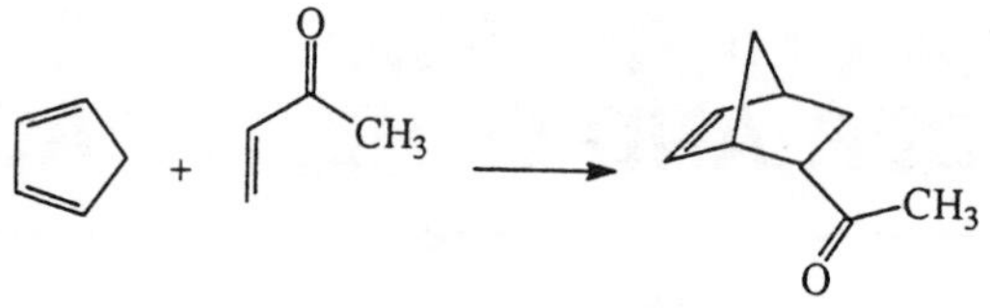

(7.3)

Many Diels-Alder reactions have been shown to be accelerated in the presence of Lewis acid species (Inukai and Kojima, 1971). Several workers have capitalized on the highly organized character of the clay surface, in conjunction with its ability to hold water and bind potentially catalytic Lewis acids, to successfully carry out Diels-Alder syntheses that have been difficult to accomplish in solution. For example, 1,3-cyclohexadiene dimerized at 0°C in dichloromethane containing Fe(III)-doped

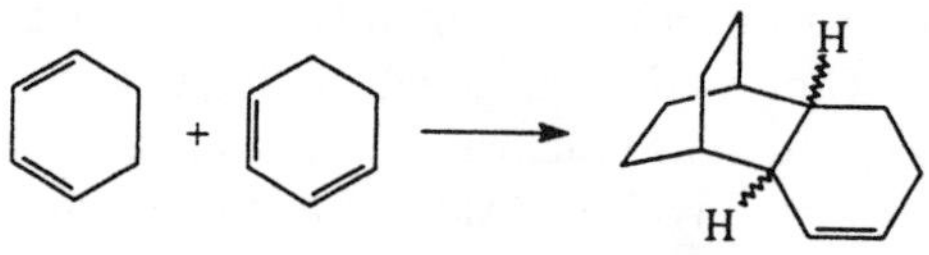

(7.4)

montmorillonite to give a 77% yield of the cycloadducts, whereas after it was heated at 200°C for 20 hr in the absence of the clay, only a 30% yield was achieved (Laszlo and Lucchetti, 1984). Similar yield improvements were observed for reactions in which acrolein ($CH_2{=}CHO$) was used as the dienophile (Laszlo and Moison, 1989). In the above experiments, use of a phenol as a co-catalyst was found to improve yields significantly. Apparently, the phenol engages in electron-transfer processes on the iron-doped surface, giving radical cations which are said to promote cycloaddition reactions. In keeping with the proposed Lewis acid mechanism for the clay-catalyzed Diels-Alder reaction, some studies have shown that the presence of water is inhibitory (Collet and Laszlo, 1991); dried kaolinite was an effective catalyst for the cyclopentadiene-methyl vinyl ketone condensation, but moist air diminished the rate and stereoselectivity of the reaction.

Zeolites are clays with rather large internal pore structures which have the property of concentrating nonpolar organic compounds within their cavities. Measurements of gaseous hydrocarbon equilibria have shown enhancements of several orders of magnitude within zeolite pores relative to the vapor phase. Cyclodimerization of butadiene to 4-vinylcyclohexene (Equation 7.5) at 250°C was catalyzed by large-pore zeolites in the sodium form (Dessau, 1986). Zeolites in the Cu(I) form also promoted Diels-Alder addition of furan and other dienes with electron-deficient dienophiles such as methyl vinyl ketone (Equation 7.6). Dichloromethane was the solvent in these reactions, which usually were carried out at 0°C or lower (Ipaktschi, 1986).

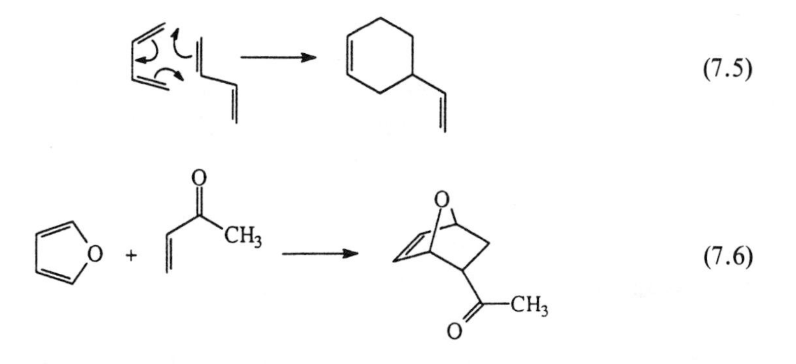

It is not certain whether these clays would have similar activity in the aqueous or solid phases.

B. SURFACE-CATALYZED REARRANGEMENTS

Clay surfaces have been shown to promote many other types of reactions such as oxidation, reduction, hydrolysis, elimination and polymerization (Voudrias and Reinhard, 1986). Some of these reactions have been discussed in earlier chapters. Another process probably influenced by clays is rearrangement, which would be favored by the presence of catalytic, positively charged sites on the mineral surface. These sites, by accepting one or two electrons from an adsorbed molecule, could promote radical- or cation-influenced rearrangements. Only a few examples of these reactions have been reported in the environmental literature, although geochemists have postulated that they would occur (along with microbial alteration reactions) during the diagenesis of sedimentary hydrocarbons and petroleum (Philp et al., 1976).

Sieskind and Albrecht (1985) showed that cholestene underwent a backbone rearrangement in the presence of montmorillonite at room temperature in cyclohexane to give an almost quantitative yield of the product (Equation 7.7). The same isomer had also been obtained, in 50% yield, in the absence of clay by refluxing cholestene in acetic acid in the presence of *p*-toluenesulfonic acid. Dry bentonite or kaolinite catalyzed the rearrangement of parathion (**1**) to O,S-diethyl-O-(*p*-nitrophenyl)-phosphate at room temperature (Mingelgrin and Salzman, 1979). Under some conditions, rearrangement was followed by hydrolysis of the rearranged compound.

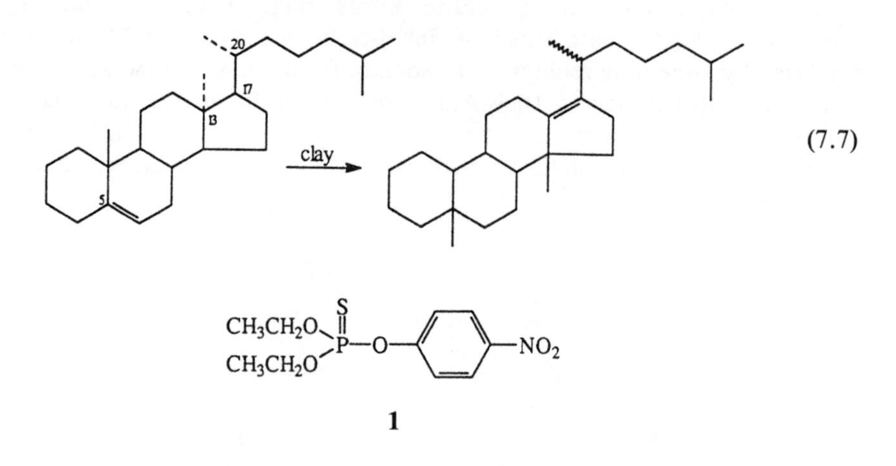

(7.7)

REFERENCES

Breslow, R., U. Maitra, and D. Rideout. 1983. Selective Diels-Alder reactions in aqueous solutions and suspensions. *Tetrahedron Lett.* 24: 1901–1904.

Collet, C. and P. Laszlo. 1991. Clay catalysis of the non-aqueous Diels-Alder reaction and the importance of humidity control. *Tetrahedron Lett.* 32: 2905–2908.

Dessau, R. M. 1986. Catalysis of Diels-Alder reactions by zeolites. *J. Chem. Soc. Chem. Commun.* 1167–1168.

Grieco, P. A., P. Garner, K. Yoshida, and J. C. Huffman. 1983. Aqueous intermolecular Diels-Alder chemistry: novel products derived from substituted benzoquinone-diene carboxylate adducts via tandem Michael reactions. *Tetrahedron Lett.* 24, 3807–3810.

Inukai, T. and T. Kojima. 1971. Aluminum chloride catalyzed diene condensation. *J. Org. Chem.* 36: 924–928.

Ipaktschi, J. 1986. Diels-Alder reaction in the presence of zeolite. *Z. Naturforsch.* 41b: 496–498.

Laszlo, P. and J. Lucchetti. 1984. Catalysis of the Diels-Alder reaction in the presence of clays. *Tetrahedron Lett.* 25: 1567–1570.

Laszlo, P. and H. Moison. 1989. Catalysis of Diels-Alder reactions with acrolein as dienophile by iron(III)-doped montmorillonite. *Chem. Lett.* 1031–1034.

Mingelgrin, U. and S. Salzman. 1979. Surface reactions of parathion on clays. *Clays Clay Min.* 27, 72–78.

Pakdel, H., C. Roy, H. Aubin, G. Jean, and S. Coulombe. 1991. Formation of *dl*-limonene in used tire vacuum pyrolysis oils. *Environ. Sci. Technol.* 25: 1646–1649.

Philp, R. P., J. R. Maxwell, and G. Eglinton. 1976. Environmental geochemistry of aquatic sediments. *Sci. Progr. (Oxf.)* 63: 521–545.

Sieskind, O. and P. Albrecht. 1985. Efficient synthesis of rearranged cholest-13(17)-enes catalysed by montmorillonite clay. *Tetrahedron Lett.* 26: 2135–2136.

Voudrias, E. A. and M. Reinhard. 1986. Abiotic organic reactions at mineral surfaces: a review. In J. A. Davis and K. F. Hayes, eds., Geochemical processes at mineral surfaces. *Amer. Chem. Soc. Sympos. Ser.* 323: 462–486.

Index

The basic principle of this index is that hydrocarbons or compounds that have only one functional group are indexed together under the generic name of the compound class. Therefore, chloroform and methyl iodide will both be found under "Aliphatic halogen compounds." Also, compounds such as chloroform that are mentioned more than twice in the text are cross-indexed. Pesticides and compounds with multiple functional groups are indexed under their common names. Occasional exceptions, such as polychlorinated dibenzofurans and dioxins, are also cross-indexed, so that their actual location should not be too difficult to find.